A Changing Environment for Human Security

Environmental change presents a new context and new opportunities for transformational change. This timely book will inspire new ways of understanding the relationship between environmental change and human security. *A Changing Environment for Human Security* both supports and informs a call for new, transformative approaches to research, policy and action. The chapters in this book include critical analyses, case studies and reflections on contemporary environmental and social challenges, with a strong emphasis on those related to climate change. Human thoughts and actions have contributed to an environment of insecurity, manifested as multiple interacting threats that now represent a serious challenge to humanity. Yet humans also have the capacity to collectively transform the economic, political, social and cultural systems and structures that perpetuate human insecurities.

Fresh perspectives on global environmental change from an interdisciplinary group of international experts will inspire readers – whether students, researchers, policy makers, or practitioners – to think differently about environmental issues and sustainability. The contributions show that in a changing environment, human security is not only a possibility, but a choice.

Linda Sygna is a Project Leader at the Department of Sociology and Human Geography, University of Oslo, Norway.

Karen O'Brien is a Professor of Human Geography at the Department of Sociology and Human Geography, University of Oslo, Norway.

Johanna Wolf is an Associate Faculty Member at Royal Roads University, Canada and Visiting Fellow at the Tyndall Centre for Climate Change Research, UK.

A Changing Environment for Human Security

Transformative approaches to research, policy and action

Edited by Linda Sygna, Karen O'Brien and Johanna Wolf

Routledge
Taylor & Francis Group
LONDON AND NEW YORK

earthscan
from Routledge

First published 2013
by Routledge
2 Park Square, Milton Park, Abingdon, Oxon OX14 4RN

Simultaneously published in the USA and Canada
by Routledge
711 Third Avenue, New York, NY 10017

Routledge is an imprint of the Taylor & Francis Group, an informa business

British Library Cataloguing in Publication Data
A catalogue record for this book is available from the British Library

Library of Congress Cataloging-in-Publication Data
A changing environment for human security : transformative approaches to research, policy and action / edited by Karen O'Brien, Johanna Wolf and Linda Sygna.
pages cm
1. Human beings – Effect of climate on. 2. Human security.
3. Human rights. 4. Climatic changes – Social aspects.
I. O'Brien, Karen L.
GF71.C42 2013
363.738′74–dc23 2012051048

ISBN13: 978-1-84971-301-6 (hbk)
ISBN13: 978-1-84971-302-3 (pbk)
ISBN13: 978-0-203-10988-5 (ebk)

Typeset in Baskerville
by HWA Text and Data Management, London

This book is dedicated to
Stephen Schneider
(1945–2010)

Professor Stephen Schneider gave the keynote speech during the midsummer dinner on June 23, 2009 at the Oslo conference on 'Human Security in an Era of Global Change'. His acknowledgement that we are not just scientists but human beings who would like to see the world a better place has inspired much of our research on environmental change and human security.

Photograph by Frank Paul Silye

Through insightful analyses of the systems, structures, paradigms, values, hopes and beliefs that underlie the ways in which we define and shape human security in a world of global change, the authors lead us to recognize our potential to deliberately drive processes of transformation towards equitable, sustainable societies. Herein lies the truly creative, essential response of the social sciences to some of the most urgent global environmental challenges and questions of our time.

Heide Hackmann, Executive Director,
International Social Science Council, France

Dramatic changes require radical moves towards a sustainable and peaceful future with a human face, gender equity and environmental security. This important volume combines innovative thinking by leading environmental experts on the linkages between environmental change and human security, focusing on transformative approaches to research, policy and action. This book addresses issues of global responsibility, uncertainty, bottom up approaches to resilience, adaptation, community and gender empowerment.

Ursula Oswald Spring, National University of Mexico,
Regional Multidisciplinary Research Centre, Mexico

Diverse social processes combine with global environmental change to affect human security. Inherent perceptions, beliefs, and attitudes often constrain our ability to proactively manage such changes. This well-organized, highly informative and engaging book is a balanced and sobering expose of capabilities needed to effect transformative changes essential for sustained human well-being in a fast changing environment; indeed a must-read for all thoughtful people.

Hassan Virji, Executive Director, START, USA

Contents

Figures

Tables

Contributors

Helen Adams is an Associate Research Fellow at the University of Exeter where she researches the role of ecosystem services in poverty alleviation in delta environments. Helen's PhD focused on migration decision-making under environmental change in Peru. She previously worked on adaptation to climate change with the OECD and the UNFCCC.

W. Neil Adger is Professor in Geography at the University of Exeter. He researches social dimensions of vulnerability, adaptation and resilience in the face of environmental change. He was an Associate of the Global Environmental Change and Human Security Programme and convenes the chapter on human security for the Fifth Assessment Report of the IPCC.

Carolina E. Adler is a geographer and environmental scientist, working as a Research Fellow at the Swiss Federal Institute of Technology in Zurich. Her PhD on climate change adaptation was awarded the 2010 Harold D. Lasswell Prize for best dissertation in policy sciences, and her research focuses on problem-oriented global environmental change, science and policy interface, and inter-transdisciplinary collaborations.

Paulina Aldunce is Assistant Professor at the University of Chile and Deputy-PI of the Human Dimensions at the Center for Climate and Resilience Research. She is a social scientist and her research interests are social and institutional dimensions of Disaster Risk Management and Adaptation to Climate Change. She has been involved as Lead Author and Review Editor for the IPCC

Jon Barnett is a Professor in the Department of Resource Management and Geography at Melbourne University. He is a political geographer whose research investigates the impacts of and responses to climate change on social systems, with a focus on risks to human insecurity, hunger, violent conflict, and water stress.

Eduardo Behrentz is an Associate Professor and a former director of the Environmental Engineering Research Center and of the Centre of Studies on Urban and Regional Sustainability, Colombia. He is an author of multiple peer-reviewed articles.

Katrina Brown is an interdisciplinary environmental social scientist, and specializes in environmental change, vulnerability and resilience. She is co-editor of the journal *Global Environmental Change* and a member of the Resilience Alliance. She is Professor of Social Sciences at the Environment and Sustainability Institute, University of Exeter.

Halvard Buhaug is a Research Professor at the Peace Research Institute Oslo (PRIO) and Director of PRIO's Department on Conditions of Violence and Peace. His work has been published in *International Organization*, *International Security*, *Journal of Conflict Resolution*, *Political Geography*, and *PNAS*.

Mercy Cordova Borbor is an oceanographer by training. Mercy is studying the interactions among air pollution, heat waves, and vulnerability urban areas. She has also explored the social and ecological factors determining dengue transmission in Ecuador.

Simon Dalby is CIGI Chair in the Political Economy of Climate Change at the Balsillie School of International Affairs, and author of *Creating the Second Cold War* (Pinter and Guilford, 1990), *Environmental Security* (University of Minnesota Press, 2002) and *Security and Environmental Change* (Polity, 2009).

Halvor Dannevig's research focuses on various issues around the relationship between society and climate. He has done research on adaptation to climate change in natural resource industries in Northern Norway, adaptation policies at multiple levels of governance and on planning and natural hazards.

Laura Dawidowski is a physicist and is a Professor at the Institute for Research in Environmental Engineering of the National San Martín University. She is also a Professor at the Faculty of Engineering of the Buenos Aires National University and a researcher in the Chemical Department of the National Atomic Energy Commission of Argentina.

Andrew J. Dugmore is a geographer and Professor at the University of Edinburgh. His principal research interests include human ecodynamics and the development and application of tephrochronology (a dating technique based on the identification and correlation of layers of volcanic ash) to questions of environmental reconstruction, human–environment interactions, adaptations and change.

Siri Eriksen is an Associate Professor at the Norwegian University of Life Sciences. Her research interests focus on social vulnerability, in particular how people manage climatic variability and change in eastern and southern Africa as well as in Norway, and the politics involved in adaptation processes.

Jason Flanagan is an Assistant Professor of International Studies at the University of Canberra. His research cuts across the fields of security, development, and political and diplomatic history. He is the author of *Imagining the Enemy: Presidential War Rhetoric from Woodrow Wilson to George W. Bush* (Regina 2009) and the forthcoming *American Foreign Relations since Independence* with Joseph M. Siracusa and Richard Dean Burns.

Oscar A. Gómez is a Japan Society for the Promotion of Science Postdoctoral Fellow at Doshisha University's Graduate School of Global Studies.His chapter was prepared while at the Graduate School of Environmental Studies, Tohoku University. He is also Deputy Secretary General of the Japan Association for Human Security Studies. His work deals with the theory and practice of human security, with emphasis on the environment, health and disasters. His fieldwork includes Japan, Colombia and the Philippines.

Marisa Goulden is a Lecturer in Climate Change at the Tyndall Centre for Climate Change Research and School of International Development at the University of East Anglia. Her research interests are in climate change adaptation and natural resource management, resilience of rural livelihoods, and cooperation and conflict in transboundary river basins.

Roberto P. Guimarães is a Brazilian political scientist. He is Coordinator of MBA Programs on Environment and Sustainability of FGV and Visiting Professor on the Doctoral Program on Environment and Society of UNICAMP, both in Brazil, and Principal Researcher of Socio-Ecological Dimensions of Desigualdades.net in Berlin. He served as Chair and Vice-Chair of the Scientific Committee of the IHDP. He was a staff member of the UN (1983-2007) and was its Chief of Social and Policy Analysis, having also served as Technical Coordinator of all UN International Conferences on the Environment (Rio-92, Rio+5, Johannesburg-2002).

John Handmer leads RMIT's Centre for Risk and Community Safety and its Human Security Program. He is also Convener of the National Climate Change Adaptation Research Network for Emergency Management, and Principal Scientific Advisor for the Bushfire CRC. He holds adjunct professorial positions at the Australian National University and the Flood Hazard Research Centre in London.

Betsy Hartmann is Director of the Population and Development Program and Professor of Development Studies at Hampshire College in Amherst, MA, USA. Her work explores the interactions between population, migration, environment and security concerns. Her books include *Reproductive Rights and Wrongs*, *A Quiet Violence* (co-authored with James Boyce), and the co-edited anthology *Making Threats: Biofears and Environmental Anxieties.*

Bronwyn Hayward is a New Zealand author and Senior Lecturer in Political Science at the University of Canterbury, New Zealand; a Visiting Fellow at the Sustainable Lifestyles Research Group, University of Surrey, UK and a Trustee of the Foundation for Democracy and Sustainable Development, London. Her work focuses on the challenges for children, youth and democracy in a changing world.

Johan Hedrén is an Associate Professor at Water and Environmental Studies at Linköping University. His main research areas are political organization, ideologies and discourses on the environment and sustainable development, utopian thought on the same issues and the relationship between politics and science.

Thomas Heyd teaches Philosophy, is Adjunct Associate Professor of Environmental Studies at the University of Victoria, Canada, and is member of the Centre for Ecosocial Studies, University of La Laguna, Spain. His research focuses on cultural dimensions of environmental change, environmental ethics and aesthetics. He has organised/chaired 7 sessions at international meetings (2008–2010) and an interdisciplinary workshop, 'Culture and Climate Change' (2010). He edited a special issue of the Human Ecology Review (2010) on this topic.

Henk B.M. Hilderink is Senior Policy Researcher at PBL Netherlands Environmental Assessment Agency where he leads various projects on global sustainability. His research focus is on global demography and population health, especially in the context of sustainable development.

Svein Jarle Horn is currently a Professor of Bioprocess Technology at the Norwegian University of Life Sciences. His research focuses on biomass processing for products such as chemicals, feed and fuels. His interests also include climate change, peak oil, the emerging biobased economy and applications of the Integral Model developed by Ken Wilber.

Grete K. Hovelsrud's research interests include studies of vulnerability and adaptation to environmental and societal changes (coupled with social-ecological systems); adaptive capacity across societal scales; barriers and constraints of adaptation in the northern regions, and aspects of societal transformation in the face of global change.

Philip Ireland has worked across the NGO sector as a policy and advocacy advisor on development. He has recently completed a PhD from Macquarie University, Australia, where he researched adaptation to climate change and development at UN negotiations in Nepal and in Bangladesh. He is a published author across the academic literature and mainstream media.

Peter Keegan is an international development practitioner and has worked in the NGO sector for the last 12 years. He is currently responsible for programs across Africa and South-East Asia. Peter has degrees from the Australian National University and the University of Sydney, and is completing a postgraduate research degree.

Christian Keller is a Professor at the University of Oslo and works in Nordic archaeology and history, mainly focusing on the Vikings and the Middle Ages, the human eco-dynamics of the pre-industrial world and human responses to climate change. In addition, he has carried out underwater archaeology in Cyprus, Turkey and Thailand.

Marcel T.J. Kok is a Senior Researcher and Project Leader on global governance at PBL Netherlands Environmental Assessment Agency. His main focus is on sustainable development, global governance and vulnerability analysis. He was a coordinating lead author for the chapter on vulnerability in the UNEP 'Global Environmental Outlook IV'.

Vikram Kolmannskog holds degrees in Law and the Humanities, including a specialized LLM in Human Rights Law. He has researched, lectured and advised extensively on the rights of 'climate refugees. He is an independent scholar and adviser as well as a psychotherapist.

Christian Kuhlicke has a background in Human Geography. His research focuses on how vulnerabilities and risk are constructed in governance and management processes as well as in practices of everyday life. He is Chairperson of the Working Group 'Natural Hazards/Natural Risks' of the German Association of Geography (DGfG).

Victoria Lawson is a Professor and former Chair of Geography at the University of Washington. She is co-founder of the Relational Poverty Network project. A past president of the Association of American Geographers, she is also the author of *Making Development Geography* (2007) and served as editor for *Progress in Human Geography*. She works on the social and political effects of global economic restructuring across the Americas.

Hayley Leck is a post-doctoral researcher at the Grantham Research Institute on Climate Change and the Environment, London School of Economics and Political Science. She is a geographer and her research focuses on the relationship between global environmental change and urbanism, with a particular interest in social and institutional dimensions of municipal and community-based adaptation in diverse contexts.

Robin M. Leichenko is an Associate Professor in the Department of Geography at Rutgers University. She is co-author with Karen O'Brien of the book *Environmental Change and Globalization: Double Exposures* (New York: Oxford University Press, 2008), which received the 2009 Meridian Book Award for Outstanding Scholarly Contribution in Geography from the Association of American Geographers.

Alejandro León is an Associate Professor at the University of Chile, and Director of the Department of Environmental Sciences and Renewable Natural Resources. He teaches Environmental Economics, and his research has focused on institutional and economic arrangements regarding water issues and the desertification process.

Asunción Lera St. Clair, a philosopher and sociologist, is Research Director at the Center for International Climate and Environmental Research-Oslo (CICERO). Her research interests are focused on critical poverty studies, climate change, development ethics, human rights, global justice, epistemology, processes of knowledge and multilateral organizations.

Virginia Levín has worked as a Research Fellow at the University of Chile and currently she is employed in the private sector. Her research has focused on disaster management and environmental planning

Paul L. Lucas is a Policy Researcher and Integrated Assessment Modeler at PBL Netherlands Environmental Assessment Agency. His work focuses on international climate policy and sustainable development issues, and more specifically on the link between environment and human development. He was involved as a lead author for UNEP's fifth Global Environmental Outlook.

Matthias K.B. Lüdeke has been Senior Scientist at the Potsdam Institute for Climate Impact Research and Lecturer at Potsdam University since 1995. Since 2008 he has been a Project Leader in the research domain 'Climate Impacts and Vulnerability', focusing on climate change, urban development and global patterns of vulnerability.

Amanda H. Lynch is Professor of Geological Sciences and Director of the Environmental Change Initiative at Brown University. Her interests lie in the application of climate and meteorological research to concrete problems of policy relevance. Her approaches include regional and global climate models of the contemporary and past climates, weather prediction models, statistical models, and quantitative and qualitative analysis.

Christian Koch Madsen is an archaeologist and a senior doctoral student focusing on the Norse Eastern Settlement of Greenland. Christian has made a central contribution to the Vatnahverfi Project at the National Museum of Denmark (which began in 2005) and has been involved in both extensive survey work and archaeological excavation.

David Manuel-Navarrete is an Assistant Professor at the School of Sustainability, Arizona State University, where he studies the socio-ecological boundaries produced by tourism development and climate adaptation regimes. He has carried out extensive research in Latin America on socio-ecological inequality, climate governance and sustainable development through appointments at the Free University of Berlin, King's College London, and the UN Economic Commission for Latin America and the Caribbean.

Peter J. Marcotullio is an Associate Professor of Geography at Hunter College and Co-Deputy Director of the City University of New York Institute for Sustainable Cities. His research focuses on the relationships between urban growth, urbanization, globalization, urban planning and the environment.

Thomas H. McGovern is a zoöarchaeologist and researcher with a major focus on Viking-Age European expansion into a very diverse set of North Atlantic island ecosystems and the subsequent dynamics of human impact, climate change, and inter-cultural contacts. Since 1992 he has served as Coordinator for the NABO Research Cooperative.

Desmond McNeill is Professor, and a former Director at SUM (Centre for Development and Environment), University of Oslo. He heads the research area on Governance for Sustainable Development. He has worked in many countries in Africa, Asia and Latin America and written extensively on aid and global governance.

Susanne C. Moser is Director of Susanne Moser Research and Consulting and a Social Science Research Fellow at Stanford's Woods Institute for the Environment.

She is an internationally recognized expert on adaptation, communication, and science-policy interactions.

Lars Otto Naess is a Research Fellow in the Climate Change Team at the Institute of Development Studies (IDS). His research interests include social and institutional dimensions of adaptation to climate change, policy processes on climate change and agriculture at national and sub-national levels, the role of local knowledge for adaptation to climate change, and adaptation planning in the context of international development.

Karen O'Brien is Professor at the Department of Sociology and Human Geography, University of Oslo. She is interested in how transdisciplinary and integral approaches to global change research can contribute to a better understanding of how societies respond to change, and in particular how deliberate, ethical, equitable and sustainable transformations can be catalysed in response to climate change.

Elspeth Oppermann is a Research Fellow at the Northern Institute, Charles Darwin University, Australia. Building on her doctoral exploration of the problematization of climate change adaptation in the UK, her post-doctoral research is on transformation and the role of social networks and tacit knowledge in the North Australia region.

Gudrun Østby is Senior Researcher at the Peace Research Institute Oslo (PRIO). Her work has appeared in *International Studies Quarterly*, *Journal of Conflict Resolution*, *Journal of Development Studies*, and *Journal of Peace Research*.

Elena Ostrovskaya has extensive experience in the field of environmental monitoring and protection, natural resources management and environmental policy research. She has participated in several international projects focusing on river basin, wetland and coastal management. She has also been involved in capacity development projects as a trainer and a group moderator.

Paola Ramallo is an environmental engineer with an MSc in Environmental Planning and Management. She has seven years of experience working with public and private institutions in the fields of water resources management, sustainable development, environmental impact assessment and environmental education in developing countries. She has mainly worked as an assessor for project development as well as project developer in the areas of IWRM and climate change in the Andean region.

Patricia Romero-Lankao is an interdisciplinary sociologist by training working at NCAR since 2006. Her research concentrates on three questions: how cities affect the climate system, and the water and carbon cycles; how cities are affected by those changes and what factors shape urban political and policy responses to mitigate and adapt.

Fiona J.Y. Rotberg is an Assistant Professor at Uppsala University, Sweden, and directs a project on the humanitarian challenges of climate change. Her research interests include human security, accountable governance and land use. Fiona holds a PhD in Natural Resources and Law from the Fletcher School of Law and Diplomacy, USA.

David Simon is Professor of Development Geography at Royal Holloway, University of London and a member of the Academy of Social Sciences. He serves on the Scientific Steering Committee of the IHDP's Urbanization and Global Environmental Change project, and has published widely on development, environment and urban issues.

Konrad Smiarowski is an archaeologist and an advanced doctoral student at the City University of New York. His graduate studies on Norse archaeo-faunal remains involve survey and excavation in Greenland where he has most recently been involved in excavations of the Bishop's residence in Gardar (Igaliku).

William D. Solecki is a Professor of Geography at Hunter College and Director of the City University of New York Institute for Sustainable Cities. He is also Co-chair of the New York City panel on Climate Change, and a member of the Scientific Steering Committee of the Urbanization and Global Environmental Change Project (UGEC) of the International Human Dimensions Program on Global Environmental Change (IHDP).

Sofie Storbjörk is a Senior Lecturer at Linköping University, Sweden. Her research focus on drivers and barriers for climate adaptation, with an emphasis on the institutional and organizational dimension of adaptive capacity, as well as the role of spatial planning in facilitating transformations. Further research includes the theory and practice of environmental policy-integration at national, regional and local levels.

Richard Streeter is a geographer who recently completed his PhD at the University of Edinburgh developing tephrochronology in order to understand the landscape implications of plague outbreaks in fifteenth-century Iceland. He is currently investigating early warning signals of threshold crossing events in complex socio-ecological systems.

Linda Sygna is an economist by training and works as a Project Leader at the Department of Sociology and Human Geography, University of Oslo. Her work focuses on the social and human dimensions of climate change, and especially how individuals and society perceive and respond to a changing climate.

Michael Thompson studied anthropology at University College London and the University of Oxford while also following a career as a Himalayan mountaineer. Early research on how something secondhand becomes an antique (*Rubbish Theory*, OUP, 1979) diverted him into teaching at the Slade School of Fine Art, London and at Portsmouth University's School of Architecture, and from there to the International Institute for Applied Systems Analysis, a global change think-tank in Austria.

Henrik Urdal is a Senior Researcher at the Peace Research Institute Oslo (PRIO) and Editor of the *Journal of Peace Research*. His work has appeared in

International Interactions, *International Studies Quarterly*, *Journal of Conflict Resolution*, *Journal of Development Studies*, and *Political Geography*.

Katharine Vincent is a Director of Kulima Integrated Development Solutions (Pty) Ltd, and a visiting Associate Professor at the School of Architecture and Planning at the University of the Witwatersrand, Johannesburg. She is interested in vulnerability and adaptation to climate change in southern Africa.

Jennifer West is a Research Fellow at the Center for International Climate and Environmental Research (CICERO) in Oslo, Norway, with interests and experience in climate change vulnerability and adaptation policy and practice in a range of natural resource contexts and countries.

Ben Wisner has worked since 1966 on livelihoods, governance, environment and risk. While teaching in Africa, the US, the UK and Switzerland, he conducted fieldwork world wide. Now retired, he advises, advocates and rants. Writings include *At Risk*, *Routledge Handbook of Hazards and Disaster Risk Reduction* and *Toward a New Map of Africa*.

Johanna Wolf is an Associate Faculty Member at Royal Roads University, Victoria, Canada. Her research interests concern the human dimensions of environmental and climate change. She has examined responses to climate change and related policies in the UK and Canada and her work has been published in international journals including *Global Environmental Change*, *Environment and Planning A*, *Environmental Politics and Climatic Change*.

Preface and acknowledgements

One does not have to go many years back to remember a time when environmental change was mainly viewed and studied separately from the human and social dimensions. Even today, some still consider environmental changes in isolation from the wider context in which they take place. This separation presents a misleading picture of not only what drives environmental change, but also how such changes affect the larger context in which humans and other species live. Viewing environmental change from an integrated systems perspective provides much-needed insights on the drivers of change, the distribution of vulnerability, and the long-term consequences that emerge from the complex and non-linear interactions between environment and society.

Motivated by the need for a better understanding of the relationship between environmental change and human security, in 1997 a group of researchers submitted a scoping report to the International Human Dimensions Programme (IHDP) proposing a new core research project. A Scientific Planning Committee was established, and a Science Plan was developed under the leadership of Steve Lonergan. The Global Environmental Change and Human Security (GECHS) project came into being in 1999, with its first International Project Office (IPO) located at the University of Victoria, Canada. As a core project of the IHDP, GECHS was led by a Scientific Steering Committee made up of researchers from a variety of disciplines. In 2002, when GECHS came under the scientific leadership of Mike Brklacich, the IPO moved to Carleton University in Ottawa, Canada. Karen O'Brien served as the Scientific Chair from 2005–2010, and during that time the IPO was situated at the University of Oslo in Norway.

The main goal for the GECHS project was to conduct research on relationships between environmental change and human security. The project recognized environmental change as an equity issue that it is linked to development pathways, and it interrogated some of the dominant generalizations about relationships between environmental change and conflict, migration and security. Through research, policy briefs, workshops, conferences, and other activities, human security has gradually become a core theme within the global environmental change arena. First and foremost, there has been considerable progress in the way that global environmental change is conceptualized. It is now widely recognized that individuals respond to multiple interacting stressors that threaten their social, environmental and human rights. The socioeconomic and political context has become more central to understandings of the causes and consequences of biophysical changes, whether in relation to vulnerability and resilience, or conflict and cooperation. This understanding is based on a large body of empirical research demonstrating how human security is affected by environmental change. Lastly, there have been important advances in linking research with policy and practitioner activities, with the aim of identifying ways of enhancing human capacities to respond to environmental change and create positive social change.

The GECHS project drew to an end in 2009/2010, culminating with the Synthesis Conference on "Human Security in an Era of Global Change" in June 2009. This conference, which took place in Oslo, Norway, brought together about 150 researchers, policy makers and practitioners from 30 countries to synthesize a decade of research on the societal consequences of and responses to global environmental change and implications for human security.

This book is an outcome of the Synthesis Conference, and most of the chapters included in this book were first presented at the conference in June 2009. There is no doubt that many of the ideas and material originally presented at the conference have matured considerably in the interim period. It has been particularly inspiring to follow the many young researchers who presented at the conference and have contributed to this book.

The message that emerged from the GECHS project has over time become increasingly clear: "It is time to stop talking about human security, and start creating it." This means not only questioning current models of developments, but also focusing on the political, economic, social, technological, institutional, cultural and personal transformations that together can create a more sustainable and secure future.

There are some key individuals and institutions that were vital to the GECHS project over the years. We owe many thanks first to the IHDP Secretariat, and especially to Maarit Thiem, Douglas Williamson, and Falk Schmidt for their support over the years. The GECHS project also recognizes the important role of committed researchers around the world, particularly those that served on the Scientific Steering Committee. These include, in alphabetical order: Jon Barnett, Hans-Georg Bohle, Mike Brklacich, Kanchan Chopra, Chris Cocklin, Ken Conca, Nils Petter Gleditsch, Edgar Gutierrez-Espeleta, Elizabeth Hartmann, Okechukwu Ibeanu, Patricia Kameri-Mbote, Fred Langeweg, Steve Lonergan, Alexander Lopez, Richard Matthew, Lyla Mehta, Elena Nikitina, Kwasi Nsiah-Gyabaah, Joni Seager, Indra de Soysa, Marvin Sooros, Chou Meng Tarr and Jinxia Wang. GECHS Associates have also been an important capacity for the project, and these included: Neil Adger, Hans Guenter Brauch, Stewart Cohen, Geoff Dabelko, Saswati Dutta, Hallie Eakin, Siri Eriksen, Gunhild Hoogensen, Richard Klein, Robin Leichenko, Peter Liotta, Andrei Marin, Susanne Moser, Karen O'Brien, Mark Pelling, Julie Silva, Úrsula Oswald Spring, Asuncion Lera St. Clair and Maureen Woodrow. Finally, Executive Officers and research assistants Ann Zurbrigg, Maureen Woodrow, Lynn Rosentrater, Øystein Kristiansen and Kirsten Ulsrud played significant roles in developing the GECHS project.

There are many institutions and people that made it possible for GECHS to find a home at the University of Oslo in Norway. The Research Council of Norway funded the GECHS IPO from 2005–2010. We would especially like to thank Kirsten Broch Mathisen and Inger-Ann Ulstein for prioritizing global change research in Norway, and for inspiring collaborations over the years. We would also like to thank Norad for co-funding the IPO in the initial years, and for supporting several events over the years. The Ministry of Foreign Affairs has been critical to several of the larger events and initiatives, including serving as the main funder of the GECHS Synthesis Conference.

Now to those to whom we are indebted in this book project. First, we are grateful to all of the contributors to this book, a group of inspiring and committed individuals who brought their impressive capacities to this project. We would also like to thank the many reviewers who provided valuable feedback to the authors. We would also like to thank Amanda McCavour for allowing us to use her art work for the front cover of this book. It gave us energy and inspiration to bring out the need for both agency and collaboration in creating human security. We are very appreciative of the team at Routledge – Earthscan, and especially Editorial Assistant Helen Bell and Associate Editor Louisa Earls for their clear guidance and positive and quick responses. Finally, we would like to thank our families for supporting us through this very long book project. Thank you Annika, Espen, Jens Erik and Kristian, thank you Linnea and Rick, and thank you Stella, Maria, Lea and Håvard.

Linda Sygna
GECHS Executive Officer (2007–2010)
Oslo, December 2012

1 A changing environment for human security

Karen O'Brien, Linda Sygna and Johanna Wolf

Introduction

Whether in relation to climate, food, water, health, energy, or livelihoods, dramatic changes are underway in the world today, with significant implications for human security. There are very few places that are untouched by a convergence of global crises, many of which are directly or indirectly associated with environmental change. While climate change receives much of the attention these days, biodiversity loss, air and water pollution, ocean acidification, degradation and salinization of soils, deforestation, and threats to ecosystem services together create one systemic problem that is often referred to as 'global environmental change'. This is a deceptively benign label, for it hides the asymmetrical responsibilities for such changes, as well as the unequal outcomes that result from them. Although global environmental changes are often perceived as 'future' problems, the human consequences are highly visible today, through risky living conditions, lost livelihoods, displaced populations, and unnecessarily high levels of morbidity and mortality. While the current situation is disturbing, future scenarios are alarming.

The challenges facing humanity are towering over our meagre attempts to respond. In fact, almost all strategies to address environmental change have fallen short, failing to change current trajectories, whether in relation to the number of endangered species, increases in global average temperatures, pH levels of oceans, or other challenges (MEA, 2005; IPCC, 2007; UNEP, 2012). Even the successful reduction of ozone depleting substances in the atmosphere is challenged by changes in climate, which are expected to have an increasing influence on stratospheric ozone in coming decades (WMO, 2010). As a result, human security – both in the short and long term – is increasingly threatened by environmental change. We are approaching 'the end of the world as we know it', and existing approaches seem to be incongruous to the challenges.

The environment is changing, and this calls for not only different ways of thinking about human security, but also new and more daring ways of approaching it. Human security has been defined in many ways, including as 'a state that is achieved when and where individuals and communities have the options necessary to end, mitigate or adapt to threats to their human, environmental and social rights; have the capacity and freedom to exercise these options; and actively participate in pursuing these options' (GECHS, 1999). While there has been a considerable amount of research on the many interacting threats associated with environmental change, and on their consequences for individuals and communities, it is now time to focus on the latter part of this definition, and engage with the capacities to pursue options that address the root causes of multiple, interrelated threats.

The objective of this book is to present insights from research that inspires new ways of understanding the relationship between environmental change and human security. The chapters include critical analyses, case studies, and reflections on contemporary environmental and social challenges, with a strong emphasis on those related to climate change. Taken together, the contributions both support and inform a call for new, transformative approaches to research, policy and action. The focus on transformative approaches recognizes a compelling need for change. The success of such approaches, however, depends on understanding and addressing the root causes of human insecurity. This involves engaging directly with the systems, structures and development paradigms that perpetuate insecurities, but also with the deeper human dimensions of global environmental change. These deeper human dimensions relate to the subjective attitudes, emotions, identities and beliefs that underlie how humans consciously or unconsciously influence and respond to change. Transforming how we perceive, experience, respond to and create change is a powerful leverage point for promoting human security.

We have divided the book into three parts, with the first part presenting a 'reality check' about the relationship between environmental change and human security. This section shows that while environmental change presents unprecedented threats to human security, this understanding needs to be balanced by a more nuanced consideration of the social and human factors. This sets the stage for the second part of the book, which discusses the 'breakthrough conditions' that underlie transformative approaches to human security in a changing environment. Included here are changes in perspectives and paradigms, a sense of individual and collective empowerment, and the integration and emergence of new ways of approaching environmental change and human insecurity. The third part of the book summarizes the story by considering transformations in the past and future, and discusses the best 'next steps' for creating human security in a changing environment.

In this first chapter, we present a brief review of the concept of human security, including how it has influenced research on global environmental change. The human security literature has strongly emphasized the importance of context, drawing attention to the social, economic, political, institutional, technological and cultural factors that influence both environmental change and its consequences (e.g., addressing questions of who, where, why, when and how). However, we also consider the subjective dimensions of human security, which influence how humans engage with this broader context (Gasper, 2010). These include values, beliefs, worldviews, and behaviours that are grounded in both ethical and rights perspectives on human development (Gasper and St. Clair, 2010; O'Brien *et al.*, 2010).

Human security: a concept 'coming of age'

The idea of human security became prominent in the 1990s, and has grown in importance over the years, not only in debates about environmental change, but also in relation to topics such as globalization, migration, poverty, and development (Tadjbakhsh and Chenoy, 2007; Goucha and Crowley, 2008; Gasper, 2010). With its normative appeal to equity, justice, and ethics, the concept of human security is 'coming of age' in a period of history where boundaries of all sorts are shifting (Beck, 1999; Held *et al.*, 1999). Personal, social, cultural, political, and planetary boundaries are being pushed, resisted, reinforced and questioned, and in some cases overstepped, redrawn or even dissolved. It is a shaky time, and there is a growing recognition that the changes underway will have profound implications for humanity. The concept of human security brings together the many threats, fears and uncertainties of the present with a strong recognition of agency over the future. It draws attention not only to the insecurities that are prevalent today and perpetuated through deeply embedded systems, structures, behaviours and beliefs, but also to the capacities of individuals and communities to change them.

Over the past two decades the concept of human security has experienced some growing pains, and these are widely reflected in the literature. It has faced sharp critiques within 'securitization' debates, for human security is but one small word away from national security, state security, and military security (Paris, 2001; Tadjbakhsh and Chenoy, 2007; Barnett, 2009). The juxtaposition of 'human' and 'security' can certainly create dissonant images, but this may be a strength rather than a weakness. For example, one such image of human security might show individuals and communities living happily together with no unmet needs or fears, yet with this sense of security maintained by strict borders and controls, as depicted in the novel and film 'The Hunger Games'. Human security is, in fact, highly political, for the security of some individuals, groups, nations or species often comes at the expense of the security of others (Barnett, 2009). Another dissonant picture might involve individuals and communities so focused on alleviating their fears and wants that in the process of 'being human', they actually lose sight of their connections to the biosphere and each other – the connections that secure their existence (Folke *et al.*, 2009). This, unfortunately, is not merely fiction; it is a stunning representation of our current situation. In drawing attention to discomforting contradictions and inconsistencies, the concept of human security compels those who are engaging with threats to individuals and communities in a global context to question boundaries and explore the assumptions that go with them.

Human security has also at times been viewed suspiciously by those focusing on human well-being, human rights, and human development, for it risks corrupting what is already well attended to in other discourses (Gasper, 2010). Indeed, the core of human security is arguably captured in Sen and Nussbaum's capability approach, and reflected in the discourse on development ethics (Sen, 1999; Gasper, 2004; Gasper and St. Clair, 2010; Nussbaum, 2011). It is about the needs, rights and values of individuals and communities that are facing multiple stressors (Barnett *et al.*, 2010). What human security adds, however, is an explicit recognition of the spatial, temporal and social dynamics between threats and responses. In a changing global environment, human security is about *both* the well-being of individuals and communities *and* the collective security of humanity (O'Brien and Leichenko, 2006). The connections between human security at the micro and the macro scales shows that many of the 'secure' boundaries that have been established by humans may indeed be illusions.

A changing context for human security

Research on global environmental change and human security has been presented in numerous books, edited volumes, journal articles, and policy papers (Barnett, 2001; Dodds and Pippard, 2012; Dalby, 2009; Brauch *et al.*, 2009; Matthew *et al.*, 2010; O'Brien *et al.*, 2010). Taken together, this literature represents a considerable body of knowledge about why some groups are more vulnerable to environmental change than others, and how capacities to respond are influenced by multiple threats and stressors. Research on environmental change and human security has emphasized that environmental problems and 'solutions' must be considered within their social, economic, political, technological, cultural and philosophical context, for it is this very context that creates both human security and insecurity. The cultural and philosophical contexts are important here, for they often justify and sanctify existing systems and structures, and maintain perceptions and ideas of security that are increasingly challenged by environmental change.

Security can be defined as the state of being or feeling secure (i.e., safe and unthreatened, free from fear and anxiety). By definition, security has both objective and subjective dimensions related to 'what is' and 'what is felt'. The objective aspects of human security can be easily measured and monitored; hence they often form the basis for indices and comparative rankings (Gasparatos *et al.*, 2009; Singh *et al.*, 2009). Access to food, water and shelter, literacy rates, life expectancy, and other such indicators provide a first entry point as to who is secure. However, security has significant subjective dimensions as well, and although they are more difficult to measure, they have a profound influence on actions in the world. The freedom to express oneself or one's opinions, to move freely, or to differ from cultural norms and expectations can be a result of both experiences and perceptions, but they nonetheless form an intrinsic part of human security. Likewise, experiences and relationships that are individually and collectively valued are part of human security. Understanding these deeper aspects of human security calls attention to the emotional, perceptual, and existential groundings of development ethics (Gasper and Truong, 2005).

Economic and political systems are closely linked to value systems and worldviews, and changes in values have implications for human security (Inglehart, 1997; O'Brien and Wolf, 2010). Traditional, modern, and post-modern worldviews and value systems coexist in the world today (Inglehardt and Welzel, 2005). The values prioritized by modern value systems have yielded unprecedented benefits and extraordinary achievements, but there have been some significant costs involved with models of economic development and notions of progress that 1) consider some human lives to be of lesser value; 2) treat the environment as an externality; and 3) sever the relationship between the individual and collective good. Although there are certainly traces of pre-modern attitudes (for example, linked to fatalism and social hierarchies), the shortcomings of this approach to human progress and development have resulted in a highly uneven gap between 'the good life' accessible to a relative few, and the stark insecurities experienced by most people on earth today (Pogge, 2008). Although the social and economic consequences of this 'equality' gap have been shown to be unfair, unethical and dangerous (UN, 2005; Wilkinson and Pickett, 2009), research on environmental change suggests that this particular dichotomy may, in fact, be unsustainable.

Yet sustained it is, with the subjective context of modernity continuing to view environmental change as a separate issue from economic, social and human development. Human security research has certainly moved discussions beyond the overly simplistic 'technical' approaches exemplified by the IPAT equation (where Impacts are related to Population, Affluence and Technology), to consider the role of social, economic and political relations, including how access, entitlements and power influence processes, responses and outcomes (Leichenko and O'Brien, 2008; Matthew *et al.*, 2010). This body of research recognizes that environmental problems are closely linked to human insecurities, and that the root causes of these insecurities must be addressed when confronting environmental challenges, for environmental issues *are* political issues, social issues, economic issues, and development issues (Adger, 2010; O'Brien *et al.*, 2010). This is important, for although there are indeed many practical actions or measures that can be taken to reduce exposure, sensitivity and vulnerability to climate change (e.g., building defences against sea-level rise) or to protect species and genetic diversity (e.g., establishing conservation areas or facilitating the translocation of species), taken to an extreme this technical and managerial approach may reinforce the values and interests that underlie environmental, social and development problems.

While increasing human security can be considered an enduring social goal that originates in efforts to abolish slavery, poverty, war, and gross inequalities, whether it is achievable depends to a large extent also upon how people make sense of the world, i.e., their worldviews. Some argue that we live in a world where winners and losers are natural, inevitable and evolutionary facts of life; others view most negative outcomes as far from inevitable, but rather socially and politically generated and sustained by powerful interests (O'Brien and Leichenko, 2003). These contrasting viewpoints may seem difficult to reconcile, and indeed they have often contributed to both polarization and paralysis. However, a third viewpoint may be worthy of consideration: that global environmental change is currently challenging our basic premises about human security, including understandings of nature-human interactions, winners and losers, relationships among individuals and collectives, and the capacity of humans to influence outcomes.

Conceptualizations of human security can no longer ignore the fact that people and places are connected in a globalized world, and that environmental change processes will affect not only the vulnerable and insecure, but also the collective security of humanity, both in the present and future (O'Brien and Leichenko, 2008). This may be difficult to comprehend, particularly for those whose identities are dependent upon fixed boundaries between 'us' and 'the other', or upon maintaining a clear separation between 'humans' and 'nature'. Nonetheless, by introducing the notion of planetary boundaries (see Rockström *et al.*, 2009), research on social-ecological systems is challenging many of the personal and cultural boundaries that hinder real progress on human security. It may be time to tell a new story.

In an edited volume with 38 chapters about global environmental change and human security, it is challenging to extract only one story. Yet this is precisely the role of a synthesis process, which involves combining components to form a connected whole. In synthesizing the contents of the individual contributions to this book, each of which has its own story to tell, we hope to trigger some insights and discoveries as to what can be done to create human security in a changing environment. The larger story we have to tell can be summed up as follows: Human thoughts and actions have contributed to an environment of insecurity, manifested as multiple interacting threats that represent a collective challenge to humanity. Yet humans also have the capacity to collectively transform systems and structures that are based on dangerously outdated paradigms. There are, in fact, many openings and opportunities for deliberate, transformative actions, but these call first and foremost for breaking through some entrenched thoughts and attitudes about human-environment relationships, boundaries between 'us' and 'the other', and the relationship between personal and collective responsibility. Human security is not an option, but a choice.

In the story that we present in this book, humans move from being the antagonists that are destroying the global environment, to being protagonists who are capable of working together to 'change the change'. However, this type of transformation will not happen magically or through wishful thinking. It is also unlikely to happen within the paradigms that dominate current understandings of sustainable development. It will take conscious change, which includes the courage to continually question the assumptions and beliefs that justify both inaction and action (O'Brien, 2013a, 2013b). Responding from a different action logic, it becomes possible to identify actionable and effective leverage points for systems transformations toward sustainability (Torbert *et al.*, 2004; Meadows, 2008).

A reality check

Part I of the book represents a reality check. Understanding the consequences of global environmental change for human security means 'getting real' about the significance from a large-scale, long-term perspective. At a practical level, realism is defined as the attitude or practice of accepting a situation as it is and dealing with it accordingly. Realism involves coming to grips with and accepting the current state of the global environment, rather than denying it, ignoring it, or trivializing it. Acknowledging and accepting the state of things as they actually exist, rather than an idealized version of them, can be cognitively and emotionally challenging. As Moser (2012) argues, the bravest thing to do at this time is to get real, which includes facing the truth and letting it sink in. This means accepting that the changes that we are facing are not fiction, but rather facts that should inform present-day decisions and actions. While there have been many post-modern discussions about the social construction of reality, as well as metaphysical and existential debates about 'whether reality exists,' there is nonetheless a reality that is actually experienced and shared by billions of people and species around the world – a reality that is being observed, measured and monitored, and which indicates that the rate, magnitude and scale of environmental change is unprecedented in human history and has impacts on human security (Steffen *et al.*, 2004).

The seven chapters in this section describe the reality that humanity is currently facing, and consider the challenges this reality poses for human security. They present a sobering view of the current 'state of the world', and expose some of the dangers of continuing along current pathways. They also challenge dominant interpretations of causality, questioning some of the mainstream views and drawing attention to the dangers of deterministic thinking, as well as the need for more careful policies that address more nuanced understandings of human security. Whether the realities are linked to climate change, urbanization, population displacement, hazards, or access to food, clean drinking water, basic sanitation and modern energy sources, these chapters provide a strong rationale for developing new and more daring approaches to human security.

Human actions are contributing to transformations toward a very different world, and our actions need to reflect this new understanding. In *Chapter 2* Simon Dalby discusses human security in the Anthropocene, a geologic era that acknowledges the significant impacts of humans on the global environment. Drawing on findings from earth systems science, he stresses that the transformations that are coming are unprecedented in human history, and that the past is not a good guide to what is coming. In fact, how we respond to change will be decisive for human security in the coming decades. He calls for new thinking about security, where people and their changing ecological contexts are at the heart of the analysis, and where human security is closely linked to what we choose to make, build and use. The chapter makes it clear that we are living in a new reality, and that our actions are influencing that reality.

Climate change and security is the focus of *Chapter 3* by Jon Barnett. Starting with the likelihood that efforts to slow the rate of greenhouse gas emissions will fail to prevent significant warming by the end of the current century, he reviews the state of knowledge about climate change, human security, migration and violent conflict. While there are many interpretations of the climate change and security nexus, Barnett shows that research results are inconclusive. What is more certain, however, is that current policies and practices can actually enhance security if they focus on human security and particularly on people's needs, rights and values. The chapter shows that migration and violent conflict in a changing climate may be a concern, but prioritizing human security may be a practical and effective means of addressing these concerns.

Elaborating on some of the issues raised by Barnett in Chapter 3, Betsy Hartmann argues in *Chapter 4* that the linkages that are increasingly made between climate change, conflict and natural disasters not only lack empirical evidence, but they also contribute to a further militarization of development and humanitarian assistance. For example, the evolution of security narratives on climate change and conflict have justified US military strategies in Africa, at the same time ignoring how the local, regional, national and international political economy influences both coping and adaptation. The challenges of climate change, she argues, call not for the blurring of military/civilian boundaries, but for a more clear distinction between national security and human security.

Urbanization has significant implications for human security. In *Chapter 5*, Halvard Buhaug and his co-authors point out that urban growth and migration to urban areas are both drivers and by-products of economic development, and they assess the implications of recent and likely future trends in urban population growth. The authors show that although linkages have been made between population pressure and political violence, the connection is weak and inconclusive. Likewise, the potential for urban social disorder resulting from population increases is also unsubstantiated by research. Instead of focusing on these fears, the authors argue for pursuing broader developmental policies that will increase the economic contribution of urban populations, such as education, employment, basic health care, and safe water and sanitation services. The chapter shows that urbanization may not be a threat, but rather an opportunity for increasing human security.

Health risks are often presumed to exist in specific vulnerable groups linked to specific socio-economic and age profiles. Patricia Romero-Lankao and her co-authors suggest in *Chapter 6* a more complex reality, presenting research results that illustrate how both social and environmental factors shape risk and vulnerability. The authors draw attention to the dangers of focusing on one paradigm or approach to risk, and suggest a more integrated framework. Equally important, their research shows that increasing air pollution may interact with climate change and pose risks to rich and poor alike. In other words, with 50 per cent of the world's population currently living in urban areas, the emerging reality may be a shared reality.

Reality can provide some tough lessons on human security. In *Chapter 7*, Oscar Gomez argues that much can be learned about human security by directly engaging with the problem of uncertainty. Using the examples of two disasters in Japan, he considers how incomplete knowledge and social interactions influenced efforts to regain security and establish justice after the outbreak of Minimata disease in 1956, and in the aftermath of the Fukushima disaster in 2011. Gomez emphasizes the importance of processes, including the roles of democracy and a free press in providing information in identifying threats and prioritizing actions. This suggests that human security in a changing environment characterized by great uncertainty can be best engaged with as a process, rather than as merely as a desired outcome.

Improving human security has no single set of options. In *Chapter 8*, Paul Lucas and co-authors use child and infant mortality as proxy indicators of human security to examine long-term prospects for food, clean drinking water, basic sanitation and modern energy sources at a global scale. They then identify several archetypical patterns of vulnerability, focusing on small-holder farmers in dryland areas. The variety in vulnerability patterns that emerge from their analysis suggest that multiple options exist for improving human security in dryland areas.

Breakthrough conditions

The reality check presented in Part I of this book provides a sobering view of environmental challenges, but it also shows that many of the generalized consequences for human security are in fact contested by empirical research. A focus on human insecurity as a predetermined outcome of environmental change must be balanced by understandings that these outcomes are influenced by social, economic, political and technological systems, and by the individual and collective behaviours, mindsets and worldviews that support and perpetuate them. Systems are, in other words, linked to individual and collective 'action logics', which Torbert and associates (2004) define as strategies used by individuals and organizations to analyse the world and react to it. How does one transform action logics? This is a developmental process that occurs at the personal level, but it nonetheless influences organizations and cultures (Torbert *et al.*, 2004; Kegan and Lahey, 2009). Transformation of worldviews may involve radical or subtle shifts in perspectives, but they often call for first breaking through the entrenched habits, loyalties, identities, assumptions and beliefs that hold worldviews firmly in place (Heifetz *et al.*, 2009; Schlitz *et al.*, 2010).

We consider 'breakthrough conditions' to be part of the subjective backdrop that must be significantly present for transformative approaches to research, policy and practice to emerge. Breakthrough conditions facilitate a deeper set of responses to global environmental change and human security that challenge conditioned ways of thinking, feeling and acting, whether about vulnerability, adaptation, resilience, innovation, or sustainability. Breakthrough conditions are about transforming mindsets or action logics to move beyond rethinking, reassessing, revising, reforming, or reworking current approaches. These often represent little more than business-as-usual, for they typically occur within the same mindsets as earlier approaches, without challenging assumptions and beliefs about the way that the world works (O'Brien, 2012).

While there are many potential breakthrough conditions that can contribute to transformative approaches to human security, we focus in this book on the importance of 1) seeing things from multiple *perspectives*, which includes linking issues and scales; 2) questioning dominant *paradigms* and framings of problems; 3) individual and collective *empowerment*; and 4) *integration* of knowledge and action. The breakthrough conditions identified here enable transformation of action logics through their careful engagement with the underlying assumptions that support dominant approaches and responses.

Perspectives

The first breakthrough condition that we discuss involves seeing problems from different perspectives. This includes taking a systems perspective, and making connections across a diversity of issues and scales (Flood and Romm, 1996; Meadows, 2008). However, it also involves recognizing that there are many other perspectives, including different types of knowledge and different types of validity claims (Berkes, 2008). While this may sound simple and straightforward, is not always evident in policies and actions taken in response to complex global challenges. International solutions such as REDD+ may have negative impacts on local livelihoods, creating resistance and resentment for what might otherwise be considered a common goal of reducing the impacts of climate change (Sunderland, 2011; Beymer-Farris and Bassett, 2012). Although research itself can often be reductionist, approaches based on political ecology have long highlighted the linkages among scales and issues (Forsyth, 2003). Increasingly, calls for post-normal science and transdisciplinary research recognize the importance of acknowledging and including multiple perspectives in policy framings and solutions (Mezirow, 2000; Ravetz, 2006).

Transformability and transformative capacity are important characteristics of social ecological systems. In *Chapter 9*, Katrina Brown highlights how social ecological systems thinking can inform and enrich understandings and analyses of human security. Resilience is a concept that is gaining popularity within scientific and policy debates on environmental change, for it is fundamentally about how systems can deal with disturbance and surprise. This systems view, she argues, is important to human security, for it introduces a dynamic perspective that facilitates more process-oriented analyses across scales. As Brown points out, change is an integral part of how systems function, and disturbance has the potential to create opportunities for doing new things. The chapter reminds us that rather than coping with change and disturbance, it is possible to transform systems.

In a world facing multiple crises, recognizing the connections among diverse issues is critical. In *Chapter 10*, Robin Leichenko and her co-authors use the double exposure framework to explore pathways of interaction between environmental change and globalization. Focusing on California's Central Valley, they describe connections and potential feedbacks between drought, the collapse of the housing market, and the fiscal crisis in California, which together had a measurable impact on human security. They also discuss how interactions between these processes play out across space and over time. The chapter stresses that while both processes contribute to growing risks and uncertainties, they also create new opportunities for transformative change. This, however, requires policy responses that connect multiple processes with local outcomes.

Recognizing the importance of cross-scale interactions, in *Chapter 11* Katharine Vincent and co-authors look at national level policies versus local level realities in Africa, and consider whether they can be reconciled to promote sustainable adaptation. Based on findings from Uganda, Tanzania and South Africa that examine national policies for climate change adaptation and local practices, the authors point out that observed local level adaptation strategies are not recognized in national policies. This discord emerges from a centralized selection of adaptation priorities that favour technical interventions that cast local communities as recipients rather than as agents of adaptation, and from governments wishing to define desirable and undesirable adaptations so that valuable local strategies may be hindered. They argue that policy frameworks need to be flexible to ensure that they do not inhibit local climate-resilient livelihoods, and that the final choice over which adaptation actions are most effective should be up to those who are affected.

Connections between multiple processes and scales are also the focus in *Chapter 12*, where Grete Hovelsrud and her co-authors emphasize the important role that local knowledge plays in helping to frame appropriate research questions. Communities themselves are often aware of the social factors that shape their flexibility to respond; in some cases it is the fishing regulations that need to be adapted, rather than the fishermen themselves. Connecting both local and scientific knowledge systems leads to a broader and more holistic understanding of the complexities related to vulnerability and adaptation. This, they argue, is also of importance to community empowerment, which is a critical component of adaptive capacity. The chapter shows that local knowledge and community empowerment are prerequisites for successful and legitimate actions and interventions.

Sometimes it is easier to promote a new perspective than to actually implement it. In *Chapter 13*, Sofie Storbjörk and Johan Hedrén analyse the potential for integrated coastal zone management and consider some of the difficulties in its implementation in a small coastal community in Sweden. ICZM is an integrated and strategic approach aimed at involving stakeholders on issues related to social, cultural, economic and environmental sustainability. It challenges the reactive approach to coastal management, which focuses on technical fixes. In their case study, the authors uncovered a collision of values and priorities, with divergent interests and stakes leading to divergent policy agendas. They argue that a lack of coherent policies at regional and national levels gives power to individual interpretations that can influence practical management decisions. The chapter shows that different perspectives influence policies and actions, and that consensual policy framings may be limited when sustainability is, in fact, inherently political and contested.

Rights-based perspectives have been underemphasized in policy responses to environmental change. In *Chapter 14*, Vikram Kolmannskog first acknowledges that states, which have primary responsibility for people on their territory, are sometimes unwilling or unable to protect them. Drawing upon a synthesis of international laws to protect displaced persons, he considers how concepts such as non-discrimination, dignity and participation were considered in the cases of Hurricane Katrina and the Asian tsunami. He recognizes the risk that states and other non-local actors may use climate change and disasters as a pretext for relocating groups to other areas, and points out that involuntary relocation and resettlement rarely leads to improvements in the quality of life. The chapter suggests a need for greater attention to contribution of human rights violations to vulnerability, and greater recognition of the responsibility of the international community to protect human dignity and develop new social identities based on interdependence and shared goals.

In *Chapter 15*, David Simon and Hayley Leck consider the intersection between global environmental change and human security in urban areas. The impacts of environmental change on human security in the near and long-term future are likely to be felt most severely in urban areas of the developing world. The authors argue that this intersection of human security and urban change and development deserves significantly more attention in research, policy and practice in ways that can account for the specific urban challenges. One possible approach to achieve this is to focus on conceptions of, communication of, and responses to risk by different actors in an urban context, and to engage with local authorities.

Human security perspectives can have drawbacks, particularly if they lead to securitization at the expense of people at risk. This is evident in *Chapter 16*, where Jason Flanagan and John Handmer focus on the securitization of wildfire risk in Australia. The 'war on wildfire' clearly has winners, creates heroes, helps enlarge budgets, and creates a clear mission as well as commercial opportunities. Yet the threat-defence logic inherent in securitization cannot deliver if the real objective of human security is the protection of people and key assets. A focus on individual safety could cast people as agents in the solution rather than as part of the problem. They suggest that wildfires need to be recognized as a significant and indeed growing danger, yet that the approach to address them needs to be defined by the concepts of resilience, risk sharing and a focus on the safety of the people at risk.

Perspectives on growth and development are at the heart of human security. In *Chapter 17*, Roberto Guimarães argues that the most important challenge facing humanity is in fact not the quantity but the quality of growth, tied up intricately with current styles of development that are socially unsustainable. Reforms to increase access of the poor to health, education, and employment are needed in order to shape social cohesion. The emphasis, he suggests, must be on restructuring such access and opportunities as this reduces the intergenerational transmission of poverty and inequality. Structural distortions of the economy and inequalities of social and political power need to be addressed if sustainability is to be achieved.

Paradigms

Challenging dominant discourses, framings and paradigms involves questioning 'what is', including the foundational assumptions and beliefs that hold particular paradigms in place. While most people are 'subject' to their particular viewpoint, i.e., unaware that they are operating from within a paradigm, or immune to the idea that their own perspective may be partial or limited, others can readily see the limitations and identify the dangers of blindly pursuing the 'action logics' of that paradigm. Unfortunately, those who criticize other paradigms are often equally blind to their own assumptions and beliefs, leading to further criticisms by others (Torbert *et al.*, 2004). Contesting paradigms, then, is a process of continuously challenging assumptions and beliefs, eventually recognizing that all paradigms are nothing more than constructions (Meadows, 2008).

This has practical implications for how individuals and groups respond to complex, non-linear problems. The many proposals and ideas for responding to environmental change are often competing, but some gain more traction in public debates than others. In terms of climate change, there is no agreement about what mix of social, economic, regulatory and individual responses would be effective in averting the current trajectory towards 4, 5 or 6°C global average temperature change (Anderson and Bows, 2011). The different types of responses are often underpinned by different ways of conceiving the problems that draw on distinct sets of knowledge and emerge from different worldviews and ontologies (Hulme, 2009). Many dominant discourses of how best to respond to environmental change sit comfortably within existing paradigms. Yet given what is at stake, the diversity of responses and their competing and contradictory nature clearly warrant – even necessitate – that their premises and underlying assumptions be examined carefully. This is particularly significant in the case of climate change adaptation, which has enormous implications for human security, where both the effects of adaptation policies and the effects of a changing climate can compound existing challenges and produce maladaptive outcomes, including for development (Barnett and O'Neill, 2010).

The importance of challenging existing paradigms is perhaps nowhere greater than in models of development. In *Chapter 18*, Asunción Lera St. Clair and Victoria Lawson present critical arguments that challenge dominant approaches to understanding and addressing poverty. They argue that the dominant poverty knowledge is ill-suited to the challenges of global environmental change, particularly as it assumes dependence upon continued economic growth and consumption. This knowledge neglects the consequences for the environment, and ignores relations between poverty and prosperity. They criticize this knowledge for overemphasizing the economic causes of poverty, for shifting the burden of responsibility for poverty to individuals, and for blinding us to the ethical and justice dimensions of poverty by treating it as a technical matter that is moved from social relations. Dominant models of development perpetuate this knowledge, and limit the possibilities for creating real solutions.

The solutions to global challenges often indeed reflect dominant discourses and paradigms. In *Chapter 19*, Desmond McNeill reflects on 'REDD+ and the Global Discourse on Climate Change and Poverty'. He argues that global problems such as climate change call for new thinking about relationships among the nations of the world. Many solutions to climate change emissions reductions, such as Reducing Emissions from Deforestation and Forest Degradation in Developing Countries (REDD+), will transform established relationships, for REDD+ money represents payment for services rather than aid. It is directed at poor countries, but not necessarily at poor people, and in fact it might counteract efforts to reduce poverty, as it does not take into account the rights and land tenure of poor and marginalized communities. The chapter suggests that it is not just the 'North/South' dichotomy that is being challenged, but the powerful discourse on environment and development.

Dominant approaches to climate change adaptation are leading to adaptation finance being absorbed into business-as-usual approaches to development, argue Philip Ireland and Peter Keegan in *Chapter 20*. The authors challenge mainstream thinking on climate change adaptation and note that adaptation is significantly influencing the development sector through large and growing volumes of financing, but without taking into consideration the unique challenges of climate change. The practice of co-opting and integrating new development ideas within existing practices is not new, but it risks framing adaptation as a matter of aid and altruism rather than justice or international responsibility. Ireland and Keegan suggest instead a postdevelopmental approach to adaptation that values diverse practices that are already occurring at the local level, which build on the needs, priorities and capacities of local communities. Rather than promoting pre-existing and ideologically driven development aid efforts in the name of adaptation, this chapter seeks to identify viable alternatives and an openness to new possibilities for adaptation.

Discourses powerfully enshrine paradigms prevalent in environmental change. In *Chapter 21*, Elspeth Oppermann presents a discourse analysis of adaptation as problematized in the United Kingdom's Climate Impacts Programme (UKCIP). Recognizing that threats such as climate change are discursively produced through language, she considers the implications for both adaptation and human security. Her analysis of an early technical report by UKCIP considers how a particular understanding can implicitly or explicitly exclude alternative conceptualizations. The UKCIP's techno-scientific, risk-based problematization excludes critical engagement with the social and political dimensions of the problem, and fails to consider the role of identity, norms, values, and response capacity. Although alternative notions that raise the concerns of human security are evident in later UKCIP documents, they have not changed the overall rational for adaptation. Oppermann reminds us not to underestimate the importance of discourses. Without an awareness of the limitations of established discourses, our capacity to critically conceptualize and respond to climate change may also be quite limited.

Institutions are constantly produced and reproduced based on a shared definition of social reality. In *Chapter 22*, Christian Kuhlicke challenges the dominant conceptualization of vulnerability, arguing that it all too often results in simplified and reductionist reasoning based on stereotypical assumptions. Drawing on an analysis of the 2002 floods in Central and Eastern Europe that caught most villages, cities, authorities and citizens by surprise, he focuses on the distribution of knowledge between experts and citizens, including the underlying beliefs and assumptions that contribute to feelings of trust and dependency. He shows that the effective control of the river paradoxically prepared the ground for the experience of a radical surprise. The chapter suggests that there are dangers in leaving knowledge to the experts, and that the 'promise of security' needs to be questioned. Any attempt to develop transformational strategies needs to directly engage with institutionalized patterns that shape both interpretations and practices.

New paradigms call for new agendas, and nowhere is this more evident than in current approaches to disaster risk and vulnerability reduction. In *Chapter 23*, Ben Wisner discusses the role of capacities, knowledge and skills, and participation in disaster risk reduction and vulnerability reduction. Drawing on a case study in Northern Tanzania, he presents empirical evidence that illustrates the dangers of the generic and undifferentiated narratives on risk and vulnerability. He argues that systems thinking is important in this context, and considers how understandings of demographic, economic, socio-cultural, and political systems can inform transformational responses to risk and vulnerability specific to places. The chapter calls for new agendas for research, policy and action, noting that 'the devil is in the details'.

Empowerment

Human security draws attention not only to what is at stake in relation to environmental problems, but also to those who hold the stakes and can actually move them. A stakeholder is a person with an interest or concern in something, and when it comes to the global environment, arguably everyone holds a stake. However, some interests and values are prioritized, promoted, or protected over others through existing systems, structures, institutions, policies and practices. These interests and values are seldom directly discussed, yet they play a significant role in determining who and what is considered worthy of securing. This generates a power gradient extending from those whose values and interests are secured to those who are excluded from decisions and whose values and interests are often not even acknowledged or represented. While there are different types of power, including hard vs. soft power and resource power vs. relational power (Nye, 2011), responding successfully to complex global challenges calls for *both* individual and collective empowerment *and* a recognition of the connections between individual and collective human security.

Global environmental change may exacerbate existing vulnerability, and it may also introduce new threats to communities. Change often results in conflicts of values, interests, and ideas, but how these conflicts are managed is essential to human security. Here the concepts of power and empowerment become important. In *Chapter 24*, Susanne Moser presents valuable insights on individual and community empowerment for human security. Empowerment, she argues, is about attaining the power to make choices, but it calls for more than just making different choices – it calls for a change in outcomes. The definition of human security does not necessarily imply an emancipatory change process, for it does not always take into account the common good. Human security for some can lead to insecurity for others, which tends to lead to further disempowerment, particularly in oppressive social systems that dehumanize people. To contribute to human security, empowerment of an individual or group should not diminish the power, humanity, dignity and integrity of another. Support from those in true solidarity may be vital to the empowerment of individuals and communities, with solidarity based on a recognition of social and ecological interdependence.

Empowerment clearly can be enabled when social reforms allow for the meaningful inclusion of those previously marginalized. In *Chapter 25*, Paola Ramallo and Elena Ostrovskaya examine the role that conflicts over water have played in the empowerment of indigenous communities in Bolivia. Drawing on lessons from a water project near Sucre (the official capital), the authors show how recent national social reforms have contributed to reshaping relations among stakeholders in the water sector through a redistribution of power. The legal inclusion of indigenous villagers in water resource management processes has raised their consciousness and brought about long sought-after social change. These changes nonetheless present new challenges, and not all actors are prepared to embrace or acknowledge the rights of indigenous people. Their chapter aptly reminds us that social and political change may call for changes in mindsets across the wider society, and that great challenges remain in order to close the gaps between promises and practices related to human security.

Achieving empowerment is not straightforward and requires engaging critically with theories of social power. In *Chapter 26*, David Manuel-Navarrete considers human-environmental integration and social power in global environmental change research. He argues that the time may be ripe to reformulate global environmental change as a human-environmental phenomenon, instead of as a geophysical, chemical and bio-ecological event with a 'human dimension'. He highlights the need to combine systems analyses of global environmental change with critical theories of social power, recognizing that simply advocating for the empowerment of 'the vulnerable' does not necessarily lead to fundamental changes in power relations. The chapter discusses promising opportunities for the social sciences in global environmental change research. In fact, he concludes that responses to global environmental change may have more to do with meaning-making than policy or decision-making.

Empowerment is intricately tied up with the engagement of citizens and their recognition of responsibility. In *Chapter 27*, Thomas Heyd explores the notion of personal responsibility, and considers it in the context of what appears to be growing climate change complacency and defeatism. He argues that active involvement by concerned citizens is necessary for putting pressure on institutions to respond to climate change, and critical to new civil society initiatives. Responsibility can be thought of in terms of justice, solidarity, and harm prevention, which together translate into a commitment to take responsibility for human security. This is nonetheless challenging, particularly as many people are prone to 'all or nothing' or 'wait and see' attitudes that ignore what can actually be done right now to reduce the risks of climate change. Responsibility to act on climate change, Heyd argues, translates to a responsibility to help transform the cultural frameworks underlying these defeatist attitudes, towards one that recognizes individual and collective responsibility for human security.

Yet empowerment of the citizenry can in and of itself be no panacea. In fact, the focus on individuals' behaviour may be in part counterproductive. In *Chapter 28*, Bronwyn Hayward critically examines the social handprint of sustainable citizenship. The handprint metaphor is increasingly used to highlight individual agency, personal responsibility, and a moral imperative to address environmental and social injustice. However, she points out that such images can also reinforce unexamined notions of self-help citizenship, leaving wider systemic patterns and processes untouched. Such an image ignores how individual actions are constrained or mediated by their context, including social institutions, norms, habits, and economic structures. An emphasis on personal responsibility can overstate the value of agency as action, and obscure the potential for collaborative efforts deemed necessary for collective human security. Mutual dependency and human connectedness, Hayward suggests, may be better expressed by the metaphor of 'holding hands' in solidarity to create desirable change.

Social capital can be a powerful enabler of action, but it also has its limits. In *Chapter 29*, Paulina Aldunce and co-authors examine the role of social capital in disaster risk reduction, and consider its role in adapting to climate change. Drawing on a case study from Chile, they consider social capital as both an enabler and barrier to effective responses to disasters. A sense of community and place attachment, for example, can increase the effectiveness of responses to disasters, but may inhibit relocation to less risky areas. Citizen participation and the presence of community leaders has also been shown to be an enabling condition for adaptation, but this should not be at the expense of having supporting institutions at the local and national level. The chapter shows that social capital is critical to vulnerability reduction, but it cannot replace efforts to address poverty, social justice, and environmental integrity.

Individual actors within social networks can act as effective leaders and be instrumental in shaping capacity in communities to respond to risks. In *Chapter 30*, Fiona Rotberg considers how innovative information or knowledge can be introduced into communities to cope with or adapt to climate variability and change. Drawing on social network theory and social network analysis, she explores the role of actors in social networks, and analyses how key individuals in a Bangladeshi village contribute to coping and adaptive capacity. She finds that informal leadership was important, and that honesty, acceptance by others and trust were valued leadership qualities. She also found that young spontaneous leaders played an important role during extreme events, and suggests that their capacities can contribute to future development and adaptation policies, as they can be sources of local innovation and change. The chapter shows that individuals can play an important role in building collective security.

Integration of knowledge and action

There have been many calls for 'new ways of thinking' about environmental problems, as well as appeals for more reflexivity, greater awareness, expanded consciousness and universal values (Rifkin, 2010; Kasser, 2011). While these subjective dimensions are arguably very important aspects of human security, they are of little value until they are translated into actions that influence outcomes. The integration of knowledge and action is thus key to the creation of human security. This is a fluid process that calls for learning, innovation, adaptive management, and leadership, which together can represent a transformative process (O'Brien *et al.*, 2012). A mere fusion of fixed knowledge to action is likely to result in new problems, and potentially in increased vulnerability for others or for future generations (Eriksen *et al.*, 2011). Indeed, Meadows (2008: 145) notes that 'although people deeply involved in a system often know intuitively where to find leverage points, more often than not they push the change in the *wrong direction*'. Integrating knowledge and action often calls for different frameworks and approaches, and even for different paradigms.

There is a direct link between the values that drive responses to environmental change and how responses are enacted in practice. In *Chapter 31*, Siri Eriksen discusses climate change responses and transformational change, with a focus on the concept of sustainable adaptation. She argues that critical thinking about development processes is not reflected in most discussions of adaptation, thus adaptation is seldom considered as part of larger social change processes. Sustainable adaptation emphasizes social equity as a critical part of adaptation, and focuses on the social processes that contribute to vulnerability. Sustainable adaptation, she contends, embraces a normative approach to change, which includes addressing the context for vulnerability, acknowledging differing values and interests and how they are negotiated through social and political relations. Sustainable adaptation is a process characterized by an ability to shape the social and environmental conditions of the future.

Actions to address urban sustainability in a changing environment can no longer be based on fragmented knowledge. In *Chapter 32*, Peter Marcotullio and William Solecki focus on approaches to the sustainability of cities, which have traditionally been based on a fragmented literature that prioritizes different drivers of change, understands feedbacks and internal processes differently, and uses different criteria for sustainability. This contributes to competing visions of the sustainable city based on different ideologies and models. Although attempts have been made to reconcile the diverse models, the authors call for the development of a new science of urbanization based on a systems perspective that recognizes interactions across scales and the role of emergent phenomena. This new science can help not only to identify and understand urbanization processes, but also identify alternative pathways for sustainability.

New frameworks are needed for integrating objective and subjective dimensions of responses to environmental change. In *Chapter 33*, Svein Jarle Horn presents a holistic framework for analysis, planning and actions to move society away from its dependence on fossil fuel energy. Arguing that fossil fuels are finite resources, he considers the implications of peak oil and climate change for long-term planning and investment decisions. However, he stresses that these changes will not be enough: the transition away from fossil fuels calls for cultural change, psychological change, and behavioural change as well. Drawing on integral theory, he argues that both peak oil and climate change can be considered exterior symptoms of an interior reality that reflects collective choices linked to cultural stories and worldviews, including stories about progress. Importantly, he points out that changes in worldviews take time, and promoting change instead requires flexibility in communication, and new approaches to solving complex problems. Critical to these approaches is attention to subjective aspects such as feelings, values, motivations and worldviews.

In the practice of adaptation to climate change, diverse goals are underpinned by distinct interests, driven by particular values. In *Chapter 34*, Carolina Adler and Amanda Lynch draw on the policy sciences framework as an example of a problem-oriented approach to move adaptation forward in a context of irreducible uncertainties. The authors examine a case study of the tourism sector in Australian Alpine Shire to draw out how actors' distinct interests shape adaptation. Local stakeholders found extreme events to hamper the potential for developing the tourism sector in the area. However, these experiences were situated in a much broader context of non-climatic factors. This highlights the broad contextual vulnerability of the community. An analysis of the myths associated with the issues of concern raised by participants can help to distil concerns that serve narrow interests from those that serve common goals.

Myths and misconceptions are pervasive, and can influence responses to environmental change, including approaches to adaptation. In *Chapter 35*, Helen Adams and W. Neil Adger challenge conventional thinking about the relationship between adaptation and migration, arguing that the focus on migration in climate change policy debates is currently misplaced. They argue that migration is a critical adaptation strategy that should be seen as beneficial, rather than as a threat or challenge to state and regional security. Throughout history, relocation and population mobility have been important responses to environmental change. It can be seen as a 'last resort' action, or it can be seen as a mechanism used to cope with risk in the face of uncertainty. They argue that the debate on migration needs to consider the benefits of population migration on social resilience and human security.

Integrating diverse perspectives into policy and action is a goal of cultural theory. In *Chapter 36*, Michael Thompson presents clumsy solutions to environmental change that approach the environment from the perspective of 'elegance out, clumsiness in'. The focus here is on holistic approaches that consider system complexity. According to cultural theory, hierarchical actors place their trust in expert knowledge, believe in stewardship, and view nature as forgiving within certain limits. Discovering where those limits are is central to a hierarchical view. Individualists see nature as entirely robust and place their trust in market institutions because they view man as inherently competitive and self-seeking. In this view, the focus is on ensuring nothing interferes with those institutions that enable individualism. Egalitarian actors see nature as precarious and consider man as essentially benign until corrupted by coercive and unjust institutions. Fatalistic actors take a passive view of humans and see nature as erratic and fickle. Drawing on examples from Nepal, Thompson shows that clumsy solutions to complex problems do not represent a compromise or consensus, but rather a constructive engagement with different ways of organizing.

Transformations: past and future

The final section of the book presents a powerful story of past and future changes – one story focuses on what could have been, as much as on what actually happened centuries ago in Greenland; the other story focuses on the potential for transformations to create alternative futures. Transformations can be defined as physical and/or qualitative changes in form, structure, or meaning-making (Folke *et al.*, 2010). As pointed out earlier, transformations are coming whether we want them or not, and the point now is to choose the type of outcomes that we prefer and deliberately create transformations, which is likely to involve a combination of personal, cultural, organizational and systems transformations (O'Brien, 2012). While it is easy to argue for the need for transformation, how do they actually come about? This is not an easy question to answer, for some transformations happen gradually (e.g., through changes in educational opportunities for girls), while others emerge rapidly (e.g., through the introduction of a new technologies such as mobile telephones). Generally, however, one can say that transformative changes involve challenging existing assumptions and beliefs (e.g., girls should not go to school; telephones are luxury items for the wealthy), and acting upon the new beliefs, with an awareness of the values that are prioritized and the implications of actions for others and future generations.

Lessons about the future are rarely more compelling than when they emerge from 'completed experiments' of the past that can be tracked over century scales. In *Chapter 37*, Andrew Dugmore and co-authors discuss the utility of a long-term perspective when assessing human security. They examine how the seemingly well-adapted society of the Greenland Norse failed in responding to complex problems. They argue that too much specialization in harvesting marine resources, rather than a mix of marine and land-based resources, in trade and diet combined with volcanically induced climate cooling challenged the limited flexibility of the society. The extremely dispersed Greenlandic settlements and extensive networks could also not adjust to the increased societal complexity of the fifteenth century that emerged from economic change in the aftermath of the Black Death in Europe and from Thule peoples arriving in Greenland. This case therefore illustrates 'elegant failure' as Greenlandic settlements reached a critical threshold in social coherence during a time of multiple interacting changes. Rather than considering this as an inevitable outcome, they consider 'what might have been' by examining more successful outcomes for human security in the Faroe Islands and Iceland. The chapter emphasizes the dangers of focusing on single points of view and simple solutions to complex problems that may seem effective in the short run, but which can also produce unexpected outcomes that rapidly threaten human security.

In *Chapter 38*, we conclude the book by discussing the future, not as a distant and faraway ideal, but as a manifestation of the present. From this perspective, human security can be seen as a choice that is tied to our values, to the decisions that we make, and to the actions that we take. We consider what it means to transform knowledge, and stress the importance of transdisciplinary research. We consider what it takes to transform policy, including a desirable vision for the future. Finally, we reflect on what is involved with transforming action. We emphasize that this is not about changing values, but rather about promoting global citizenship and new types of leadership. We end by recognizing that human security is a deliberate choice that reflects not only understandings of the past and present, but also an emergent capacity to transform situations that are currently portrayed by many as 'hopeless and inevitable' towards one that enables humans to thrive.

Conclusion

This book brings together research that came out of a 2009 international synthesis conference ending a 10-year project on Global Environmental Change and Human Security (GECHS). As a core project of the International Human Dimensions Programme (IHDP), GECHS asked some critical questions about what environmental change means for human security. The project provided many valuable insights, but the ultimate answers lie in the actions that are taken today, tomorrow, and in the coming years.

As the implications of environmental change become more evident from both scientific findings and experiences accumulating from around the world, the research frontiers are shifting. Transformative approaches to research, policy and action are needed now more than ever, and research on transformation itself is critical. Do we know how to make deliberate transformations happen? Do we have a sufficient knowledge base to inform strategies and actions for equitable, ethical and sustainable transformation at the rate and scale that is called for? Can we increase human security in the face of rapid environmental change, increasing uncertainty, and the potential for tipping points and surprises? The insights and perspectives presented in this book can be considered a starting point for addressing these questions, but transformation itself is a process that calls for changes within science itself. A changing environment for human security calls for networks of networks that connect research, policy and action at all levels. It is about collective action, and connective action to create a secure future. If there is anything that reading this book should provide, it is the understanding that this is actually possible.

References

Adger, W.N. (2010) 'Climate change, human well-being and insecurity', *New Political Economy*, 15: 275–292.

Anderson, K. and Bows, A. (2011) 'Beyond "dangerous" climate change: emission scenarios for a new world', *Philosophical Transactions of the Royal Society A: Mathematical, Physical and Engineering Sciences*, 369 (1934): 20–44.

Barnett, J. (2001) *The Meaning of Environmental Security: Ecological Politics and Policy in the New Security Era*, London: Zed Books.

Barnett, J. (2009) 'The prize of peace (is eternal vigilance): a cautionary editorial essay on climate geopolitics', *Climatic Change*, 96 (1–2): 1–6.

Barnett, J. and Adger, W.N. (2007) 'Climate change, human security and violent conflict', *Political Geography*, 26: 639–655.

Barnett, J. and O'Neill, S. (2010) 'Maladaptation', *Global Environmental Change*, 20 (2): 211–213.

Barnett, J., Matthew, R.A. and O'Brien, K. (2008) 'Globalization and environmental challenges, reconceptualizing security in the 21st Century', in H.G. Brauch, Ú. Oswald Spring, C. Mesjasz, J. Grin, P. Dunay, N. C. Behera, B. Chourou, P. Kameri-Mbote and P. H. Liotta (eds) *Globalization and Environmental Challenges*, Hexagon Series on Human and Environmental Security and Peace, Heidelberg, Berlin: Springer: 355–361.

Barnett, J., Matthew, R. and O'Brien, K. (2010) 'Introduction: global environmental change and human security', Chapter 1, in R.A. Matthews, J. Barnett, B. McDonald and K.L. O'Brien (eds) *Global Environmental Change and Human Security*, Cambridge, MA: MIT Press.

Beck, U. 1999. *World Risk Society*, Cambridge: Polity Press.

Berkes, F. (2008) *Sacred Ecology*, London: Routledge.

Beymer-Farris, B. and Bassett, T. (2012) 'The REDD menace: resurgent protectionism in Tanzania's mangrove forests', *Global Environmental Change*, 22: 331–341.

Brauch, H.G., Oswald Spring, Ú., Grin, J., Mesjasz, C., Kameri-Mbote, P., Chadha Behera, N., Chourou, B. and Krummenacher, H. (eds) (2009) *Facing Global Environmental Change: Environmental, Human, Energy, Food, Health and Water Security Concepts*, Berlin: Springer.

Buhaug, H., Gleditsch, N. and Theisen, O. (2008) *Implications of climate change for armed conflict*, Paper presented at the World Bank Workshop on the Social Dimensions of Climate Change, The World Bank, Washington, 5–6 March 2008.

Dabelko, G.D. (2008) 'An uncommon peace: environment, development, and the global security agenda', *Environment*, 50 (3): 32–45.

Dabelko, G D. (2009) 'Planning for climate change: the security community's precautionary principle', *Climatic Change*, 96 (1–2): 13–21.

Dalby, S. (2009) *Security and Environmental Change*, Cambridge: Polity.

Dodds, F. and Pippard, T. (2012) *Human and Environmental Security: An Agenda for Change*, London: Earthscan.

Eriksen, S. Aldunce, P., Bahinipati, C., Martins, R., Molefe, J. Nhemachena, C., O'Brien, K., Olorunfemi, F., Sygna, L., Park, J., and Ulsrud, K. (2011) 'When not every response to climate change is a good one: Identifying principles for sustainable adaptation', *Climate and Development*, 3: 7–20.

Flood, R.L. and Romm, N.R.A. (1996) *Diversity Management: Triple Loop Learning*, Chichester: John Wiley and Sons.

Folke, C., Chapin III, F.S. and Olsson, P. (2009) 'Transformations in ecosystem stewardship', in F.S. Chapin III, G.P. Kofinas and C. Folke (eds) *Principles of Ecosystem Stewardship: Resilience-Based Natural Resource Management in a Changing World*, Springer Verlag, New York, USA: 103–125.

Folke, C., Carpenter, S.R., Walker, B., Scheffer, M., Chapin T. and Rockström, J. (2010) 'Resilience thinking: integrating resilience, adaptability and transformability', *Ecology & Society*, 15 (4): 20.

Forsyth, T. (2003) *Critical Political Ecology: The Politics of Environmental Science*, London: Routledge.

Gasparatos, A., El-Haram, M. and Horner, M. (2009) 'The argument against a reductionist approach for measuring sustainable development performance and the need for methodological pluralism', *Accounting Fourm*, 33: 245–256.

Gasper, D. (2004) *The Ethics of Development*, Edinburgh: Edinburgh University Press.

Gasper, D. (2010) 'The idea of human security', in K. O'Brien, A.L. St. Clair and B. Kristoffersen (eds) *Climate Change, Ethics, and Human Security*, Cambridge: Cambridge University Press: 23–46.

Gasper, D. and Truong, T.-D. (2005) 'Deepening development ethics: from economism to human development to human security', *European Journal of Development Research*, 17 (3): 372–384.

Gasper, D. and St. Clair, A. (2010) *Development Ethics*, Aldalore: Ashgate Press.

GECHS (1999) Global Environmental Change and Human Security (GECHS) Science Plan, IHDP Report No. 11, Bonn, Germany: IHDP.

Goucha, M. and Crowley, J. (2008) *Rethinking Human Security*, Volume 2 of International Social Science Journal Monograph Series, Oxford: Wiley-Blackwell.

Hamilton, C. (2010) *Requiem for a Species*, London: Earthscan.

Heifetz, R., Grashow, A. and Linsky, M. (2009) *The Practice of Adaptive Leadership: Tools and Tactics for Changing Your Organization and the World*, Boston, MA: Harvard Business Press.

Held, D., McGrew, A., Goldblatt, D. and Perraton, J. (1999) *Global Transformations: Politics, Economics and Culture*, Cambridge: Polity Press.

Hulme, M. (2009) *Why We Disagree About Climate Change – Understanding Controversy, Inaction and Opportunity*, Cambridge, Cambridge University Press.

Inglehart, R. (1997) *Modernization and Postmodernization: Cultural, Economic, and Political Change in Forty-Three Societies*, Princeton, NJ: Princeton University Press.

Inglehart, R. and Welzel, C. (2005) *Modernization, Cultural Change, and Democracy: The Human Development Sequence*, Cambridge: Cambridge University Press.

IPCC (2007) Climate Change 2007: Synthesis Report, Contribution of Working Groups I, II and III to the Fourth Assessment Report of the Intergovernmental Panel on Climate Change, Core Writing Team, Pachauri, R.K and Reisinger, A. (eds.), IPCC, Geneva, Switzerland, 104 pp.

Kasser, T. (2011) 'Cultural values and the well-being of future generations: a cross-national study', *Journal of Cross-Cultural Psychology*, 42: 206–215.

Kegan R. and Lahey, L.L. (2009) *Immunity to Change: How to Overcome It and Unlock the Potential in Yourself and Your Organization*, Cambridge, MA: Harvard Business Press.

Leichenko, R.M. and O'Brien, K.L. (2008) *Global Environmental Change and Globalization: Double Exposures*, New York: Oxford University Press.

Matthew, R.A., Barnett, J., McDonald, B. and O'Brien, K.L. (eds) (2010) *Global Environmental Change and Human Security*, Cambridge, MA: MIT Press.

Meadows, D.H. (2008) in D. Wright (ed.) *Thinking in Systems: A Primer*, London: Earthscan.

Mezirow, J. (2000) *Learning as Transformation: Critical Perspectives on a Theory in Progress*, San Francisco, CA: Jossey-Bass Publishers.

MEA (2005) *Ecosystems and Human Well-being: Synthesis, Millennium Ecosystem Assessment*, Washington, DC: Island Press.

Moser, S.C. (2012) 'Getting real about it: navigating the psychological and social demands of a world in distress', in Rigling Gallagher, Deborah, Richard N.L. Andrews, and Norman L. Christensen (eds) Sage Handbook on Environmental Leadership, SAGE Reference Series on Leadership, Sage: 432-440

NRC (1992) *Global Environmental Change: Understanding the Human Dimensions*, National Research Council, Washington, DC: National Academy Press.

Nussbaum, M. (2011) *Creating Capabilities: The Human Development Approach*, Cambridge, MA: Belknap Press of Harvard University Press.

Nye, Jr., J.S. (2011) *The Future of Power*, New York: BBS Public Affairs.

O'Brien, K. (2012) 'Global environmental change (2): from adaptation to deliberate transformation', *Progress in Human Geography*, 36 (5): 667–676.

O'Brien, K. (2013a) 'Global environmental change (3): closing the gap between knowledge and action', *Progress in Human Geography*, first published online on December 20, 2012. http://phg.sagepub.com/content/early/2012/12/20/0309132512469589.abstract.

O'Brien, K. (2013b) 'The courage to change: adaptation from the inside-out', in S. Moser and M. Boykoff (eds) *Successful Adaptation: Linking Science and Practice in Managing Climate Change Impacts*, London: Routledge.

O'Brien, K. and Leichenko, R. (2003) 'Winners and losers in the context of global change', *Annals of the Association of American Geographers*, 93 (1): 99–113.

O'Brien, K. and Leichenko. R. (2006) Climate change, equity and human security, *Die Erde*, 137 (3): 223–240.

O'Brien, K. and Leichenko, R. (2008) *Human Security, Vulnerability and Sustainable Adaptation*, Background report to the UNDP Human Development Report 2007/2008: Fighting climate change: Human solidarity in a divided world.

O'Brien, K. and Wolf, J. (2010) 'A values-based approach to vulnerability and adaptation to climate change', *Wiley Interdisciplinary Reviews: Climate Change*, 1: 232–242.

O'Brien, K., Wolf, J. and Sygna, L. (2009) 'Global environmental change and human security in a decade that matters', IHDP UPDATE, 2, 2009.

O' Brien, K.L., St. Clair, A. L. and Kristoffersen, B. (eds) (2010) *Climate Change, Ethics, and Human Security*, Cambridge: Cambridge University Press.

O'Brien, K., Pelling, M., Patwardhan, A., Hallegatte, S., Maskrey, A., Oki, T., Oswald-Spring, U., Wilbanks, T. and Yanda, P.Z. (2012) 'Toward a sustainable and resilient future', in C.B. Field, V. Barros, T.F. Stocker, D. Qin, D.J. Dokken, K.L. Ebi, M.D. Mastrandrea, K.J. Mach, G.-K. Plattner, S.K. Allen, M. Tignor, and P.M. Midgley (eds) *Managing the Risks of Extreme Events and Disasters to Advance Climate Change Adaptation*, a Special Report of Working Groups I and II of the Intergovernmental Panel on Climate Change (IPCC), Cambridge University Press, Cambridge, UK, and New York: 437–486.

Paris, R. (2001) 'Human security: paradigm shift or hot air', *International Security*, 26 (2): 87–102.

Pogge, T. (2008) *World Poverty and Human Rights: Cosmopolitan Responsibilities and Reforms*, 2nd ed., Cambridge: Polity Press.

Ravetz, J. (2006) 'Post-normal science and the complexity of transitions towards sustainability', *Ecological Complexity*, 3 (4): 275–284.

Rifkin, J. (2010) *The Empathic Civilization*, Cambridge: Polity.

Rockström, J., Steffen, W., Noone, K., Persson, Å., Chapin, III, F.S., Lambin, E., Lenton, T.M., Scheffer, M., Folke, C. Schellnhuber, H., Nykvist, B., De Wit, C.A., Hughes, T., van der Leeuw, S., Rodhe, H., Sörlin, S., Snyder, P.K., Costanza, R., Svedin, U., Falkenmark, M., Karlberg, L., Corell, R.W., Fabry, V.J., Hansen, J., Walker, B., Liverman, D., Richardson, K., Crutzen, P. and Foley, J. (2009) 'Planetary boundaries: exploring the safe operating space for humanity', *Ecology and Society*, 14 (2): 32.

Schlitz, M.M., Vieten, C. and Miller, E.M. (2010) 'Worldview transformation and the development of social consciousness', *Journal of Consciousness Studies*, 17 (7–8):18–36.

Sen, A. (1999) *Development as Freedom*, Oxford: Oxford University Press.

Singh, R.K., Murty, H.R., Gupta, S.K. and Dikshit, A.K. (2009) 'An overview of sustainability assessment methodologies', *Economic Indicators*, 9 (2): 189–212.

Steffen, W., Sanderson, A., Tyson, P.D., Jäger, J., Matson, P.A. Moore III, M., Oldfield, F., Richardson, K., Schellnhuber, H.J., Turner II, B.L. and Wasson, R.J. (2004) *Global Change and the Earth System: A Planet Under Pressure*, Berlin: Springer.

Sunderland, T. (2011) '*Win-win' is too Simplistic a Description for REDD+ – and Possibly Wrong*, Indonesia: CIFOR.

Tadjbakhsh, S. and Chenoy, A. (2007) *Human Security: Concepts and Implications*, London: Routledge.

Torbert, B. and Associates (2004) *Action Inquiry: The Secret of Timely and Transforming Leadership*, San Francisco: Berrett-Koehler Publishers, Inc.

UN (2005) *The Inequality Predicament: Report on the World Social Situation 2005*, United Nations, New York: United Nations Department of Economic and Social Affairs.

UNEP (2012) *Global Environmental Outlook GEO5: Environment for Development*, United Nations Environment Programme, Nairobi: UNEP.

Wilkinson, R. and Pickett, K. (2009) *The Spirit Level: Why More Equal Societies Almost Always Do Better*, London: Allen Lane.

WMO (2010) *Scientific Assessment of Ozone Depletion: 2010*, Global Ozone Research and Monitoring Project – Report No. 52, World Meteorological Organization, Geneva, Switzerland.

Part I
Reality check

2 Human security in the Anthropocene

The implications of earth system analysis

Simon Dalby

The anthropocene and security

The accelerating transformation of the globe by human action has meant that earth scientists now frequently refer to the present period of geological history as 'the Anthropocene'. Although this is not yet the official term for the present geological era, there is a widespread sense among those scientists who study the earth and its lengthy history, that we do live in times where the transformation of natural processes set in motion by humanity is 'on a scale comparable with some of the major events of the ancient past' (Zalasiewicz *et al.*, 2010: 2228). The construction of major cities and their infrastructures is new in terms of the sedimentary patterns on the planet; concrete and asphalt are effectively new rocks in the earth system. The chemical and biological transformations in the atmosphere, the rise in temperatures especially in the Arctic, the rapid acceleration of species extinction, and the acidification of the oceans that will soon cause major changes to marine ecosystems are all new too. 'Geologically, this is a remarkable episode in the history of this planet' (Zalasiewicz *et al.*, 2010: 2231).

While the terminology used by geologists and other earth scientists is rarely a matter of much concern to social scientists and policy makers studying the human condition, the fact that earth scientists, who in turn are usually relatively unconcerned about the human present, are seriously discussing a new historical designation for the planet, suggests clearly that something momentous is afoot (Steffen *et al.*, 2004). We are, albeit mostly inadvertently, changing the context for humanity both directly by building an artificial infrastructure of roads, railways, ports, pipelines, dams, power-stations, computer networks and numerous other artefacts that make our increasingly urban life possible, and indirectly by setting ecological perturbations in motion, precisely because that infrastructure and the waste products of the vehicles and fuels used to run it are changing how the biosphere operates.

These changes are on a large enough scale, and are beginning to have such far-reaching consequences for life on earth that most earth scientists have come to recognize that we are indeed now living in a new geological era. The rest of human history will unfold in these new circumstances, whether we like it or not. This is the context within which we now have to face global environmental change

(Scheffran *et al.*, 2012). The past is not a good guide to what is coming; things are in motion that will change the context for our lives, shifting storm tracks, precipitation patterns and hence agricultural possibilities among many other things. Species and humans are already on the move responding to environmental changes; they will continue to do so, but in conditions that we can help to shape in coming decades. Analogies with previous civilizations don't help much given the extraordinary capabilities of current technology and the global scope of our economy.

We can no longer take for granted something called the environment 'out there' as the backdrop to the human drama (Wapner, 2010). Culture is no longer separate from nature, but the cultural choices we make, by deciding what we build and what we buy, are shaping nature in profound ways. Ecosystems and species in turn will attempt to adapt to these changing circumstances, and will frequently do so by migrating. How we respond to these changes will be at the heart of what security means in the coming decades – including human security; how we think about linking ecological change and what we feel is important enough to need securing will soon shape future human prospects profoundly. The Anthropocene really is a new era, and we need to work that key insight into all our analyses and policy prescriptions.

Traditionally, security has often been about trying to keep things the same so as to preserve the political and economic status quo. Trying to prevent change may be, in our new circumstances, the wrong thing to do, especially if it is a matter of trying to prevent species and people migrating. The contrast between Indian and Bangladeshi perspectives on security in the face of people fleeing storms in the Bay of Bengal, 'distress migration' as it is sometimes called, suggests what is at stake now in rethinking security. While building a fence round Bangladesh to control migration might look like an excellent way to deal with a threat to national security from refugees in New Delhi, precisely that fence is seen as an obstacle to reaching safety on the part of those forced to flee their homes as storm surges inundate the Bengal coast (Ahmed, 2009).

Security thinking needs to be substantially updated to be relevant to the new era of the Anthropocene, although this new agenda requires much more theoretical work than can be discussed in this short chapter (see Brauch *et al.*, 2011). What this chapter does argue is that putting people and their changing ecological contexts at the heart of the analysis, rather than modern territorial states and assumptions of nature as merely a backdrop to human activity, is a necessary step to formulate a human security agenda adequate to cope with the changes we are starting to live through. At least it is necessary if anything more than the short-term prerogatives of the rich and powerful are to be considered as a matter of security.

Global environmental change

The Anthropocene is usually said to have begun with the industrial revolution, when the steam engine in particular started both the widespread use of coal as a fuel for industry, and subsequently when these engines were put in motion, both as railway locomotives and ship propulsion units, to move huge quantities of materials around the world (Steffen *et al.*, 2011). In place of forests, grasslands, and many other ecosystems we have built cities, farms, highways, and numerous industrial and agricultural systems. Rivers have been dammed, valleys flooded, water diverted and whole new artificial hydrologies created by using pipes and pumps. The steam engine powered by coal combustion set this all in motion, and the carbon dioxide level in the atmosphere started to rise.

In the process of doing all this we have taken our fate into our own hands, even if we have been slow to realize that this is indeed what we have done. Petroleum-powered economic activity has effectively set in motion a second phase of the Anthropocene where the transformations are more dramatic and which now present us with the climate change crisis (Steffen *et al.*, 2011). These new circumstances require that we seriously think about what kind of a biosphere we are making for future generations. Rising sea levels are inevitable in coming decades as glaciers and icecaps melt (Church and White, 2011), but more severe weather events, droughts, floods and hurricanes are also hazards we face in the immediate future (IPCC, 2012). If we continue the dramatic changes to natural systems that we have set in motion we may end up facing much larger, abrupt disruptions of the conditions necessary for human security. While the rich may be better equipped to adapt in the face of these disruptions the poor and marginal will become even more endangered unless policies actively devoted to their safety are put in place (Brauch *et al.*, 2008).

The point is not that humanity has recently started changing the biosphere; we've been doing that for thousands of years (Ruddiman, 2005). Earth system science, the study of the planet as a single system, has come to the realization that life has been much more important as a driving force in shaping the earth's surface than was until fairly recently widely realized (Steffen *et al.*, 2011). While traditional views of geology suggested that the large forces that moved the continents, and responded to the radiation from the sun, were fundamental, and life was merely a superficial addition to this given planetary context, recent research has made it increasingly clear that the atmosphere in particular is partly a result of life processes. The oxygen that we all breathe is a product of life, and the carbon dioxide that is of concern in the discussion of climate change is likewise a key part of the living forces of the planet. Suddenly, in geological terms, there is a new dominant life form loose in the biosphere and it is changing numerous facets of life rapidly. What matters now is that the scale on which we are now doing this and the pace of contemporary transformation is new; so too are the likely disruptions as climate change accelerates in coming decades.

The discussion of the Anthropocene focuses attention on the largest scale phenomena, and the behaviour of the earth climate system in particular, but numerous other parts of the biosphere are being transformed by human action simultaneously. As rising levels of carbon dioxide and methane show very clearly, the whole earth system is now being changed. The planet is now undergoing one of the most dramatic extinction events in its geologic history due to human activity, and the dramatic decline in the world's biodiversity started well before climate change had a major impact (Pereira *et al.*, 2012). Few ecosystems in the oceans are untouched by the voracious fishing practices of the industrial harvesting machines that we now use on an increasingly acidified ocean (Anticamara *et al.*, 2011). Deforestation not only reduces the potential for absorption of carbon dioxide, it simultaneously eradicates habitat for many species; this too has had a dramatic impact on changing the biosphere and is a central issue in contemporary discussions of conservation (Phelps *et al.*, 2012).

A quick glance at the fifth edition of the United Nations publication, *Global Environmental Outlook* (GEO5) makes this point clear (UNEP, 2012). Numerous earlier UN documents repeatedly invoked themes of environment or development, of environmental constraints as a limit on development, or tried to suggest either some notion of balance between the two. In GEO5 and elsewhere these formulations are gradually being superseded by the realization that humanity is an active change agent in the environment. This insight has shifted the discussion of development so that having a functional environment is now understood as a necessary precondition for development; this precondition is what has to be secured if civilization in a form that we recognize is to continue into the future. In case the significance of their findings isn't obvious, the GEO5 authors added a pointed subtitle: 'Environment for the Future We Want'.

Securing what future?

Not only are we living in a new geological era, we have also become an urban species, and in many places a coastal dwelling species, while setting these transformations of the biosphere in motion. We are increasingly living in artificial structures and are dependent on long commodity chains, as well as the infrastructure that allow food, fuel, clothing and other essentials to get to where we live. Anthropocene vulnerabilities are about both the increasing severity of hazards, and the fact that we are vulnerable because these hazards disrupt the infrastructure that makes urban life possible. In these circumstances human security is now about simultaneously slowing the rate of change by reducing carbon and methane emissions into the atmosphere and planning infrastructure that is much more robust in the face of storms, droughts, heat waves and other hazards, as well as securing lives and livelihoods in a changing climate.

Predicting precisely what the future holds in store for us is impossible, but some of the high profile disasters of recent years suggest the importance of thinking through this double logic of human security (Dalby, 2009). For example, Hurricane Katrina and the subsequent inundation of much of New Orleans in 2005 may be a precursor of things to come, and the consequences of Hurricane Sandy in the New York metropolitan region underscore this point. Unless careful planning and attention to infrastructure is a priority, and unless humanity seriously reduces the total amount of greenhouse gas emissions, such disasters may become the norm, rather than exceptions. Katrina was a disaster, not because of the immediate impact of the hurricane, but because the combination of land-use changes and inadequate maintenance of the flood prevention infrastructure, made the city vulnerable to a storm water surge. All of which was greatly exacerbated by the lack of effective evacuation routines, and the inability to deal with people who needed help once the flooding began. Human security clearly should now be about building cities that are not so vulnerable to extreme events, and simultaneously preparing to help people who, despite these precautions, still end up in harm's way.

Many of the climate models that have been worked on for the last couple of decades suggest that a number of large problems will emerge in the coming century, and do so more quickly and more severely if nothing is done to reduce greenhouse gas emissions soon. Among the large-scale changes of concern that caught the attention of the German Advisory Council on Global Change in their 2008 report on 'Climate Change as a Security Risk' are the melting of glaciers and the polar icecaps, possible disruptions of the monsoon climate, and the drying out of the Amazon rainforest. Each of these major disruptions of the earth system will have effects on agricultural production in particular; droughts and storms affect crops directly and the price of food on the world markets indirectly, a matter of major concern, especially to poor people in many places whose security is first and foremost a matter of gaining sustenance.

The nightmare scenarios, where failures to deal with greenhouse gas emissions lead to accelerated climate changes and abrupt reorganization of oceanic and atmospheric systems, loom in the more distant future. If the planet responds to substantially heightened greenhouse gas levels by switching to a new hotter and much drier arrangement, as James Lovelock (2009), who developed the Gaia hypothesis, has warned is a real possibility in the coming century, then all bets for human security are off. In such drastically altered circumstances humanity might be violently reduced from billions to at best millions of survivors trying to make crops grow in Greenland and Antarctica after the ice has gone. This worst-case scenario suggests that agriculture would only be possible at the poles in the new climate regime. Clearly such a runaway mode of climate change would be a situation where all the carefully worked out ideas for adapting to climate change (Adger *et al.*, 2009) would be useless. Preventing such runaway climate change is obviously the first priority for any long-term consideration of the future of human security.

Towards a new human security framework

In its initial formulation, human security was understood as a universal concern, affecting all humans, and an interconnected one in the contemporary world where famines and social disintegration have consequences that travel the globe (UNDP, 1994). Prevention was key to the early formulation of human security, both in the medical sense of disease avoidance rather than treatment, but also in the more general sense of planning and putting measures in place to avert potential dangers. Human security also emphasized that dangers to people were often unintentional and had origins in distant places and in actions that had unforeseen consequences. This all obviously applies to climate change in particular, and to Anthropocene disruptions more generally. While the 1994 formulation suggested that human security was a narrower formulation than development, in that it focused on people's abilities to safely enjoy the possibilities of development, now the changing ecological context of our times emphasizes the point that the dominant mode of economic development is itself a central concern in plans to build for the future (Scheffran *et al.*, 2012).

If we are literally making the future of the biosphere by what we choose to produce, which is the key practical insight encapsulated in the term 'Anthropocene', then human security is now a matter of what we choose to make, build and use. It is about investment decisions, only most obviously in terms of energy choices and transportation modes. Clearly the petroleum-powered civilization we have built over the last century has to be redesigned to run on renewable energy sources, such as solar and wind energy, and rebuilt to accommodate the growing number of people born in the last few generations. The long-term shape of the Anthropocene isn't yet decided; but focusing human energies on making ecologically sensible modes of living in the next few decades will probably determine both whether reasonable levels of human security are possible in the long run, and whether the Anthropocene will have a climate loosely similar to the conditions that have supported human civilization.

Getting this clearly in focus in both scholarly analysis and policy prescription is important if appropriate questions are to be researched and institutions empowered to deal with coming difficulties (Matthew *et al.*, 2010). The focus on security in the Anthropocene is especially important because the contrasts with traditional notions of national security are frequently stark, not only in the case of India and Bangladesh but in many other cases too (Brauch *et al.*, 2011). The shift from a perspective that takes rival territorial states as the key to global order to one that looks to low carbon modes of economy providing basic needs for all the world's people is dramatic. But such are the requirements facing policy makers and scholars in rethinking security agendas in light of the scientific insights of earth system science. We have finally begun to understand that we live in the Anthropocene, a world that is increasingly quite literally of our own making, and whose future depends on what we make it.

References

Adger, W.N., Lorenzoni, I. and O'Brien, K. (eds) (2009) *Adapting to Climate Change: Thresholds, Values, Governance*, Cambridge: Cambridge University Press.

Ahmed, I. (2009) 'Environmental refugees and environmental distress migration as a security challenge for India and Bangladesh' in H.G. Brauch, U. Oswald Spring, J. Grin, C. Mesjasz, P. Kameri-Mbote, N. Chadha Behera, B. Chourou, and H. Krummenacher (eds) *Facing Global Environmental Change: Environmental, Human, Energy, Food, Health and Water Security Concepts*, Berlin, Heidelberg, New York: Springer-Verlag: 295–308.

Anticamara, J.A., Watson, R., Gelchu, A. and Pauly, D. (2011) 'Global fishing effort (1950–2010): trends, gaps, and implications', *Fisheries Research*, 107: 131–136.

Brauch, H.G., Grin, J., Mesjasz, C., Dunay, P., Chadha Behera, N., Chourou, B., Oswald Spring, U., Liotta, P.H. and Kameri-Mbote, P. (eds) (2008) *Globalisation and Environmental Challenges: Reconceptualising Security in the 21st Century*, Berlin, New York: Springer-Verlag.

Brauch, H. G., Dalby, S. and Oswald Spring, U. (2011) 'Political geoecology for the Anthropocene' in H.G. Brauch, U. Oswald Spring, P. Kameri-Mbote, C. Mesjasz, J. Grin, B. Chourou, P. Dunay and J. Birkmann (eds) *Coping with Global Environmental Change, Disasters and Security Threats, Challenges, Vulnerabilities and Risks*, Berlin, Heidelberg, New York: Springer-Verlag.

Church, J.A. and White, N.J. (2011) 'Sea level rise from the late 19th to the early 21st century', *Surveys in Geophysics*, 32: 585–602.

Dalby, S. (2009) *Security and Environmental Change*, Cambridge: Polity.

German Advisory Council on Global Change (2008) *Climate Change as a Security Risk*, London: Earthscan.

IPCC (2012) *Managing the Risks of Extreme Events and Disasters to Advance Climate Change Adaptation*, A Special Report of Working Groups I and II of the Intergovernmental Panel on Climate Change, C.B. Field,, V. Barros, T.F. Stocker, D. Qin, D.J. Dokken, K.L. Ebi, M.D. Mastrandrea, K.J. Mach, G.-K. Plattner, S.K. Allen, M. Tignor, and P.M. Midgley (eds), Cambridge: Cambridge University Press.

Lovelock, J. (2009) *The Vanishing Face of Gaia: A Final Warning*, London: Penguin.

Matthew, R., Barnett, J., McDonald, B. and O'Brien, K. (eds) (2010) *Global Environmental Change and Human Security*, Cambridge, MA: MIT Press.

Pereira, H.M., Navarro, L.M. and Martins, I.S. (2012) 'Global biodiversity change: the bad, the good and the unknown', *Annual Review of Environment and Resources*, 37: 16.1–16.26.

Phelps, J., Webb, E.L and Adams, W.M. (2012) 'Biodiversity co-benefits of policies to reduce forest-carbon emissions', *Nature Climate Change*, 2: 497–503.

Ruddiman, W.F. (2005) *Plows, Plagues, and Petroleum: How Humans Took Control of Climate*, Princeton, NJ: Princeton University Press.

Scheffran, J., Broszka, M., Brauch, H.G., Link, P.M. and Schilling, J. (eds) (2012) *Climate Change, Human Security and Violent Conflict: Challenges for Societal Stability*, Berlin: Springer Verlag.

Steffen, W., Sanderson, A., Tyson, P.D., Jäger, J., Matson, P.A., Moore III, B., Oldfield, F., Richardson, K., Schellnhuber, H.J., Turner, B.L. and Wasson, R.J. (2004) *Global Change and the Earth System: A Planet under Pressure*, Berlin, New York: Springer.

Steffen, W., Persson, A., Deutsch, L., Zalasiewicz, J., Williams, M., Richardson, K., Carole Crumley, C., Crutzen, P., Folke, C., Gordon, L., Molina, M., Ramanathan, V., Rockström, J., Scheffer, M., Schellnhuber, H.J. and Svedin, U. (2011) 'The anthropocene: from global change to planetary stewardship', *Ambio*, 40: 739–761.

UNDP (1994) *Human Development Report 1994*, United Nations Development Programme, New York: Oxford University Press.

UNEP (2012) *Global Environmental Outlook GEO5: Environment for Development*, United Nations Environment Programme, Nairobi: UNEP.

Wapner, P. (2010) *Living Through the End of Nature: The Future of American Environmentalism*, Cambridge, MA: MIT Press.

Zalasiewicz, I., Williams, M., Steffen, W. and Crutzen, P. (2010) 'The new world of the Anthropocene', *Environmental Science and Technology*, 44 (7): 2228–2231.

3 Climate change and security

Jon Barnett

Introduction

It seems increasingly likely that efforts to slow the rate of greenhouse gas emissions will fail to prevent mean global warming in excess of 4°C above pre-industrial levels by the end of this century (Anderson and Bows, 2008). When warming reaches 2°C above pre-industrial levels, the security of supply of freshwater from many of the world's major river basins will diminish, many coral reefs will bleach on an annual basis, crop productivity in lower latitudes will decline, the frequency and severity of coastal flooding will increase, extreme hydroclimatic events will increase in intensity, and the risk of infectious diseases will increase (IPCC, 2007). These changes in environmental conditions pose significant challenges to social systems.

Because the magnitude of change is unprecedented in the history of human civilization, climate change has been and is increasingly being identified as a risk to global, national, and human security. Intergovernmental organizations have identified climate change as a security issue, with the UN Secretary General arguing that climate change "can impact on security, namely by: increasing human vulnerability; retarding economic and social development; triggering responses that may increase risks of conflict, such as migration and resource competition; causing statelessness; and straining mechanisms of international cooperation" (UNGA, 2009: 8). Governments, including those in Europe (Beckett, 2007; Carius *et al.*, 2008) and the United States (Blair, 2009), are also concerned about the security implications of climate change, and leaders of Pacific Island Countries have stated that "for Pacific Island states, climate change is the great challenge of our time. It threatens not only our livelihoods and living standards, but the very viability of some of our communities" (PIF 2009: 12). In addition, think tanks and consulting firms, non-governmental organizations, and numerous scholars have also picked up on the topic (for example: Christian Aid, 2006; Dupont and Pearman, 2006; CNA, 2007; CSIS, 2007; Gleditsch *et al.*, 2007; Nordås and Gleditsch, 2007; Smith and Vivekananda, 2007; Busby, 2008; Scheffran, 2008).

Those familiar with earlier debates about environmental security identify parallels with recent concerns about climate change and security (Floyd, 2008; Dabelko, 2009). This is not surprising given that research on environmental security has noted the potential for climate change to cause "security" problems (or at least to be framed as a cause of security problems). As with environmental security, debates about climate change and security are characterized by concerns about the dangers of military appropriation of environmental issues, empirical uncertainties, conceptual ambiguities, and complex boundary politics (Barnett, 2009; Hartmann, this volume). Also similar to environmental security debates are the considerable differences in the way that the climate change and security nexus is interpreted. These stem from different interpretations of the concept of security; some see it as a broad multisectoral problem that concerns all levels of society, while others see it more narrowly as a matter of defending states against strategic (usually military) threats. Accordingly, in the literature on climate change and security some focus on relatively narrow strategic issues (Dupont, 2008), others focus on a broadly defined national security (Busby, 2007), while others focus on human security (Smith and Vivekananda, 2007).

Despite these differences, almost all discussions of climate change and security are concerned about the effects of climate change on human security, on migration, and on conflict. Indeed, these three issues define the core content of research and policy on climate change and security, and in this respect there appears to be greater agreement among those interested in climate change and security than there was (or is) among those concerned with the broader subject of environmental security. There is also growing attention to the interactions between these three processes. There is general agreement that there are causal relationships between climate change, human security, migration, and violent conflict. It is possible that if climate change stimulates negative change in one or more of these issues, then a process of mutually reinforcing increases in human insecurity, migration and conflict may ensue (Barnett and Adger, 2007). However, the evidence for most of the causal connections between these issues is sparse and unconvincing, and considerably more research is required before confident statements about causality can be made.

In this chapter, I briefly review the state of knowledge about climate change and each of the following, human security, migration, and violent conflict, arguing that while there is reason for concern, research results are inconclusive. I argue that while there is significant knowledge about the linkages between climate change and human insecurity, generalized conclusions still cannot be drawn. What can be concluded, however, is that policies and practices in response to climate change can significantly enhance security by focusing on needs, rights and values, taking into account the wellbeing of people; i.e., by prioritizing human security.

Climate change and human security

The concept of human security has been used in critical, normative, and applied ways. Its critical function is to highlight important human dimensions of well-being that national security practices typically fail to address, and indeed often undermine (e.g. Shaw, 1993). Its normative function is to articulate an alternative and more peaceful approach to security (e.g. Booth, 1991). Its applied function is to highlight dimensions of well being and processes that give rise to it that are typically not well recognized in development or security policy (e.g. UNDP, 1994).

This diversity of uses can also be found in the literature on climate change and security. O'Brien (2006) is perhaps the strongest advocate of the normative approach, using the concept to highlight the risks that climate change poses to important material and symbolic dimensions of well being. This powerful argument is substantiated through empirical research (e.g. Leichenko and O'Brien, 2008), and is bolstered by the work of the Global Environmental Change and Human Security project (see Matthew *et al.*, 2009). It has driven the growing realization that climate change is a social problem with environmental characteristics. A recognition of the implications of climate change for human security has resulted in a chapter on human security in the fifth assessment report of the Intergovernmental Panel on Climate Change (IPCC), due to be published in 2014.

Normative arguments such as those of O'Brien are also critical arguments. In this case a human security position is critical in three respects. First, it is used to critique the effects of national security practices on emissions of greenhouse gases and vulnerability to climate change (see Barnett 2003, 2006). Second, it is used to critically analyze the ways in which climate change research itself, which is excessively driven by models of environmental processes, tends to exclude recognition of people and the way climate change puts at risk their needs, rights, and values (O'Brien *et al.*, 2010). Included here too is a critique of the assumption included in many climate impacts models that people will behave in predictable and rational ways, which is not justified by evidence of behavioral responses to environmental changes (Morgan and Dowlatabadi, 1996; Grothmann and Patt, 2005).

The third critical application of human security is to highlight the limits to current debates about the mitigation of greenhouse gases, which, it is argued, is excessively concerned with the distribution of the benefits and costs of decision making to economies, rather than to people's needs, rights and values. It also argues that the decision-making institutions that have been used thus far do not give sufficient voice to local people (Adger *et al.*, 2009). This gives rise to a critique of the logics and processes for deciding on and implementing adaptation, which, it is argued, are disproportionately focused on engineering and planning responses that seek to reduce the exposure of things to risk (such as moving people, and building things like sea-walls), which are themselves likely to impact on human security (Adger and Barnett, 2009; Barnett and O'Neill, 2010). A human security approach to adaptation would instead favor no-regrets strategies that seek to reduce sensitivity (such as changing design codes or better soil management) or increase adaptive capacity (such as improving access to health care, or the provision of microfinance).

A cross-cutting theme of these human-security based critiques of climate change research and policy is the recognition that not all people are equally at risk from climate change, nor are people equally responsible for its causes. People have different levels of exposure, sensitivity, and adaptive capacity, so some are far more at risk from climate change than others. It is argued therefore that the measure of the dangerous climate change that the UNFCCC is supposed to avoid should be the impacts on the most vulnerable people (Adger *et al.*, 2006). In terms of responsibility, Baer (2006) shows that the wealthiest people in developed countries produce 155 times more greenhouse gases than the poorest people in developing countries.

Drawing on the experience of applied research and policy with respect to human security as understood in the field of international development, human security offers a guide to research on the human dimensions of climate change. So, for example, studies of famine and food security (Sen, 1981; Bohle *et al.*, 1994), social vulnerability to disasters (Blaikie and Brookfield, 1987; Mortimore, 1989; Adger, 1999), and sustainable livelihoods (Scoones, 1998; Bebbington, 1999) have all heavily informed understanding of the ways in which climate change poses risks to human security. These and other studies show that people and groups that are already insecure are the most at risk of harm from environmental change. The poor tend to be highly exposed to environmental risks, sensitive to damage from them, and have few assets with which to recover (Adger, 1999). Excessive dependence on natural resources is also associated with increased risk of damage or loss due to environmental change (Peluso *et al.*, 1994). There is evidence to suggest that women tend to suffer more, and longer, during times of environmental change because they have lower incomes, and because of the gendered division of labor (Enarson and Morrow, 1998). Economically or politically marginalized groups, which tend to have low incomes and minimal access to social services, also seem to be more sensitive and less able to adapt to environmental changes (Narayan *et al.*, 2000), as are people with insecure access to land (Liverman, 1990). Drawing on these insights, a strong finding of research on climate change and human security shows that poverty and hunger are most likely to arise when multiple stressors – such as trade liberalization or violent conflict – interact with climate change (Eakin, 2005; Leary *et al.*, 2006; Lind and Eriksen, 2006; Ziervogel *et al.*, 2006; Leichenko and O'Brien, 2008).

A focus on climate change as a human security issue suggests the need to screen climate change policies for their potential adverse social impacts. Policies on climate change mitigation and adaptation that seem likely to undermine people's enjoyment of basic needs, human rights, and core values should be considered to be maladaptive, and avoided or amended. A human security approach suggests the need for adaptation policies that seek to sustain and enhance access to and enjoyment of needs, rights and core values in the face of climate risks, and these are likely to have numerous co-benefits and synergize well with approaches that promote both basic needs (such as water and sanitation) and human rights (such as the International Covenants on Economic, Social and Cultural Rights, and on Civil and Political rights).

Climate change and migration

There are clearly potential linkages between climate change and both human security and *insecurity* created through the process of migration. Increasing migration seems to be a likely outcome of climate change, as people may choose to move to minimize the impacts of climate change on their needs, rights and values. Whether or not these moves are an impact of environmental change (in the sense that it is a response to that migrants might rather have avoided) or an adaptation (in the sense that it is not necessarily an unwelcome response to avoid or adjust to an even more undesirable outcome), and a problem or a benefit to the places the migrants move to, depends very much on the degree to which adaptation policies accept and plan for migration as an adaptation strategy (Barnett and Webber, 2009).

Nevertheless, evidence about the influence of environmental change on migration is limited (Piguet, 2010). There is some consensus that sudden or slow onset changes in environmental conditions can be indirect factors in decisions to move, but that social processes that create poverty and marginality are more important factors than environmental changes per se (Hugo, 1996; Kireab, 1997; Lonergan, 1998; Black, 2001; Castles, 2002). It is also recognized that decisions to move in response to environmental change (real or perceived) are rarely entirely "forced" or "voluntary" (Hugo, 1996), and that the consequences of migration for migrants and the sustainability of the places they move between are highly dependent on the social and ecological contexts in each place (Black and Sessay, 1998; Jacobsen, 1997; Locke *et al.*, 2000; Kliot, 2004; Unruh, 2004).

Still, there are estimates and projections of numbers of future migrants impelled by climate change, and these vary between 200 million and 1 billion people by 2050 (Myers, 2002; Christian Aid, 2007). These estimates are based on broad-scale assessments of exposure to risk, and not systematic evidence about the sensitivity of migration to environmental change, or the degree to which adaptation may minimize climate impacts (Brown, 2008; Kniveton *et al.*, 2008; Barnett and Webber, 2009). These and other epistemological difficulties mean that attempts to quantify the flows of migrants impelled by climate change are unlikely to produce meaningful results. Nevertheless, there are many good reasons for concern about the effects of climate change on migration (Meze-Hausken, 2000; McLeman and Smit, 2006; Tacoli, 2007; Perch-Nielson *et al.*, 2008; Piguet, 2008; Warner *et al.*, 2008; ADB, 2009; Mortreux and Barnett, 2009).

Barnett and Webber (2009) suggest that in the coming decades, climate change is most likely to exacerbate existing migration patterns more than it will create entirely new flows. The barriers to migration are critical influences on these patterns. This means that where climate change exacerbates migration, it is likely to be predominantly internal migration away from rural areas within developing countries. It may be that a larger proportion of international migrants will not be the rural poor, as their ability to move to developed countries will be restricted by the financial costs of movement.

Migration is pursued by hundreds of millions of people around the world because it frequently improves the lives of migrants, their families, and the communities they move both from and to (De Haan, 1999; Skeldon, 2002; Ellis, 2003; De Haas, 2005). Migration is therefore a strategy that can help people, including the families and the communities that they move between, to adapt to climate change in ways that sustain human security (Agrawal, 2008; Barnett and Webber, 2009). There is a slowly emerging consensus in the international community that while migration in response to climate change poses risks to social stability, carefully considered policy measures can minimize these risks, and may indeed be harnessed to promote adaptation in places most at risk.

Climate change and violent conflict

The links between climate change, human insecurity, and violent conflict are tenuous. Much of what is known has come from research on the linkages between violent conflict and environmental change more broadly, about which the following findings are generally considered to be robust: environmental change is rarely even a proximate factor in conflict among states (Baechler, 1999; Homer-Dixon 1999); scarcity of resources (including water) is not a major driver of violent conflict (Liverman, 1994; Wolf, 1999; Theisen, 2008); economies dependent on lucrative natural resources are more at risk than others of violent conflict (Collier, 2000; de Soysa, 2000); absolute and relative poverty coupled with poor leadership increase the risk of violent conflict (David, 1997; Fearon and Laitin, 2003; Goodhand, 2003); a recent history of violent conflict, or violent conflict in neighboring states, increases the risk of further conflict (Collier, 2000; Buhaug and Gleditsch, 2008); and weak states and those undergoing economic and political transitions are relatively more prone than others to internal violent conflict (Esty *et al.*, 1999; Kahl, 2006). Yet this is an emerging field of study and more research is required to increase confidence in these findings (Levy, 1995; Conca, 2002).

More recent qualitative case studies conducted by geographers and anthropologists offer more nuanced insights into environmental problems and violence (Bobrow-Swain, 2001; Peluso and Harwell, 2001; Walton and Barnett, 2008). They stress the importance of actual or perceived unequal outcomes of environmental changes in stimulating conflict. This emphasis on perceptions contrasts with earlier studies that imply that material changes translate directly into social actions.

Drawing on insights from these earlier studies, research on climate change and violent conflict is producing tentative findings. Some of the likely outcomes of climate change, such as dwindling resource stocks, livelihood decline, decreasing state revenues, and increasing inequality, may create opportunities for elites to mobilize people to fight, and this is more likely when regimes are weakened by decreasing revenues from resource based rents or taxes (Kahl, 2006; Barnett and Adger, 2007). If climate change causes migration, this too may be a factor in violent conflict in certain circumstances (Gleditsch *et al.*, 2007; Reuveny, 2007). Three studies find associations between decreases in rainfall and violent conflict (Miguel *et al.*, 2004; Hendrix and Glaser, 2007; Meier *et al.*, 2007). Rapid onset natural disasters also increase the risk of violent conflict (Nel and Ringharts, 2008). These findings are not robust, but they indicate that climate change may increase the risk of violent conflict under particular circumstances (Buhaug *et al.*, 2008; Salehyan, 2008).

Conclusions

The state of knowledge about climate change and security can be summarized in four statements. First, there is a considerable if incomplete body of evidence to suggest that climate change poses risks to human security. Second, there is a modest body of evidence about climate change and migration, from which it may be inferred that changes in climate may indirectly increase migration. Third, there is inconclusive and contested evidence about the linkages between climate change and violent conflict, from which it may be inferred that climate change may increase some of the factors that increase the risk of violent conflict. Fourth, there is very little evidence about the mutual interactions between human security, migration, and violent conflict. Taken together, there is an abundance of concern, yet much that remains unknown about climate change and security. Hence, the report of the United Nations Secretary General on climate change and security states that: "the nature and full degree of the security implications of climate change are still largely untested" (UNGA, 2009: 4). There is therefore a need for more research and synthesis of findings. In the interim, statements about climate insecurity require careful qualification, and cautious policy responses that consider the needs, rights and values of people should be developed.

References

ADB (2009) *Climate Change and Migration in Asia and the Pacific*, Mandaluyong City: Asian Development Bank.

Adger, W.N. (1999) 'Social vulnerability to climate change and extremes in coastal Vietnam', *World Development*, 27 (2): 249–269.

Adger, W.N. and Barnett, J. (2009) 'Four reasons for concern about adaptation to climate change', *Environment and Planning A*, 41 (12): 2800–2805.

Adger, W.N. Paavola, J. Mace, M. and Huq, S. (eds) (2006) *Fairness in Adaptation to Climate Change*, Cambridge, MA: MIT Press.

Adger, W.N., Barnett, J. and Ellemor, H. (2009) 'Unique and valued places at risk', in S. Schneider, A. Rosencranz, and M. Mastrandrea (eds) *Climate Change Science and Policy*, Washington, DC: Island Press.

Agrawal, A. (2008) 'The role of local institutions in adaptation to climate change', paper prepared for the Social Dimensions of Climate Change, Social Development Department, Washington DC: The World Bank.

Anderson, K. and Bows, A. (2008) 'Reframing the climate change challenge in light of post-2000 emission trends', *Philosophical Transactions of the Royal Society A: Mathematical, Physical and Engineering Sciences*, 366: 863–3882.

Baechler, G. (1999) *Violence Through Environmental Discrimination: Causes, Rwanda Arena And Conflict Model*, Dordrecht: Kluwer Academic Publishers.

Baer, P. (2006) 'Adaptation: who pays whom?' in N. Adger, J. Paavola, M. Mace and S. Huq (eds) *Fairness in Adaptation to Climate Change*, Cambridge, MA: MIT Press.

Barnett, J. (2003) 'Security and climate change', *Global Environmental Change*, 13 (1): 7–17.

Barnett, J. (2006) 'Climate change, insecurity and justice', in W.N. Adger, J. Paavola, M. Mace and S. Huq (eds) *Fairness in Adaptation to Climate Change*, Cambridge, MA: MIT Press.

Barnett, J. (2009) 'The prize of peace (is eternal vigilance): a cautionary essay on climate geopolitics', *Climatic Change*, 96 (1–2): 1–6.

Barnett, J. and Adger, N. (2007) 'Climate change, human security and violent conflict', *Political Geography*, 26 (6): 639–655.

Barnett, J. and O'Neill, S. (2010) 'Maladaptation', *Global Environmental Change*, 20: 211–213.

Barnett, J. and Webber, M. (2009) *Accommodating Migration to Promote Adaptation to Climate Change*, A policy brief prepared for the Secretariat of the Swedish Commission on Climate Change and Development and the World Bank World Development Report 2010 team, World Bank, Washington, and SCCCD, Stockholm.

Bebbington, A. (1999) 'Capitals and capabilities: a framework for analyzing peasant viability, rural livelihoods and poverty', *World Development*, 27 (12): 2021–2044.

Beckett, M. (2007) *The case for climate security*, lecture by the Foreign Secretary, the Rt. Hon. Margaret Beckett MP, to the Royal United Services Institute, 10 May 2007. Online. Available HTTP: <http://www.rusi.org/events/past/ (accessed 10 November 2012).

Black, R. (2001) *Environmental Refugees: Myth or Reality?*, New Issues in Refugee Research, Working Paper No.34, University of Sussex, UNHCR Evaluation and Policy Analysis Unit, Geneva.

Black, R. and Sessay, M. (1998) 'Refugees and environmental change in West Africa: the role of institutions', *Journal of International Development*, 10: 699–713.

Blaikie, P. and Brookfield, H. (1987) *Land Degradation and Society*, New York: Methuen.

Blair, D.C. (2009) *Annual Threat Assessment of the Intelligence Community for the Senate Select Committee on Intelligence*, 12 February 2009. Online. Available HTTP: http://intelligence.senate.gov/090212/blair.pdf (accessed 10 November 2012).

Bobrow-Strain, A. (2001) 'Between a ranch and a hard place: violence, scarcity, and meaning in Chiapas, Mexico', in N. Peluso and M. Watts (eds) *Violent Environments*, Ithaca, NY: Cornell University Press.

Bohle, H-G., Downing, T. and Watts, M. (1994) 'Climate change and social vulnerability: Toward a sociology and geography of food insecurity', *Global Environmental Change*, 4 (1): 37–48.

Booth, K. (1991) 'Security and emancipation', *Review of International Studies*, 17 (4): 313–326.

Brown, O. (2008) *Migration and Climate Change*, IOM Migration Research Series, No. 31, Geneva: International Organization for Migration.

Buhaug, H. and Gleditsch, K. (2008) 'Contagion or confusion? Why conflicts cluster in space', *International Studies Quarterly*, 52 (2): 215–233.

Buhaug, H., Gleditsch, N. and Theisen, O. (2008) *Implications of Climate Change for Armed Conflict*, Washington, DC: World Bank.

Busby, J. (2007) *Climate Change and National Security: An Agenda for Action*, New York: Council on Foreign Relations.

Busby, J. (2008) 'Who cares about the weather? climate change and U.S. National Security', *Security Studies*, 17 (3): 468–504.

Carius, A., Tanzler, D. and Maas, A. (2008) *Climate Change and Security: Challenges for German Development Cooperation*, Eschborn: GTZ.

Castles, S. (2002) *Environmental Change and Forced Migration: Making Sense of the Debate*, New Issues in Refugee Research, Working Paper No.70, University of Oxford Refugee Studies Centre, UNHCR Evaluation and Policy Analysis Unit, Geneva.

Christian Aid (2006) *The Climate of Poverty: Facts, Fears and Hopes*, London: Christian Aid.

Christian Aid (2007) *Human Tide: The Real Migration Crisis: A Christian Aid Report*, London: Christian Aid.

CNA (2007) *National Security and the Threat of Climate Change*, Alexandria, VA: CNA Corporation.

Collier, P. (2000) *Economic Causes of Civil Conflict and their Implications for Policy*, Washington, DC: The World Bank.

Conca, K. (2002) 'The case for environmental peacemaking', in K. Conca and G. Dabelko (eds), *Environmental Peacemaking*, Baltimore, MD: John Hopkins University Press.

CSIS (2007) *The Age of Consequences: The Foreign Policy and National Security Implications of Global Climate Change*, Washington, DC: Centre for Strategic and International Studies,

Dabelko, G. (2009) 'Planning for climate change: the security community's precautionary principle', *Climatic Change*, 96 (1–2): 13–21.

David, S. (1997) 'Review article: internal war: causes and cures', *World Politics*, 49 (4): 552–576.

De Haan, A. (1999) 'Livelihoods and poverty: the role of migration – a critical review of the migration literature', *Journal of Development Studies*, 36 (2): 1–31.

De Haas, H. (2005) 'International migration, remittance and development: myths and facts', *Third World Quarterly*, 26 (8): 1269–1284.

de Soysa, I. (2000) 'The resource curse: are civil wars driven by rapacity or paucity?', in M. Berdal and D. Malone (eds) *Greed and Grievance: Economic Agendas and Civil Wars*, Boulder, CO: Lynne Rienner.

Dupont, A. (2008) 'The strategic implications of climate change', *Survival*, 50 (3): 29–54.

Dupont, A. and Pearman G. (2006) *Heating up the Planet: Climate Change and Security*, Lowy Institute Paper 12, Double Bay: The Lowy Institute.

Eakin, H. (2005) 'Institutional change, climate risk and rural vulnerability: cases from central Mexico', *World Development*, 33 (11): 1923–1938.

Ellis, F. (2003) 'A livelihoods approach to migration and poverty reduction', paper commissioned by the Department for International Development, London.

Enarson, E. and Morrow, B. (1998) *The Gendered Terrain of Disaster*, New York: Praeger.

Esty, D., Goldstone, J., Gurr, T., Harff, B., Levy, M., Dabelko, G., Surko, P. and Unger, A. (1999) *State Failure Task Force Report: Phase II Findings*, Environmental Change and Security Project Report No. 5: 49–72.

Fearon, J. and Laitin, D. (2003) 'Ethnicity, insurgency and civil war', *American Political Science Review*, 97 (1): 75–90.

Floyd, R. (2008) 'The environmental security debate and its significance for climate change', *The International Spectator*, 43 (3): 51–65.

Ford, J. (2009) 'Dangerous climate change and the importance of adaptation for the Arctic's Inuit population', *Environmental Research Lett*ers, 4: 1–9.

Gleditsch, N., Nordas, R. and Salehyan, I. (2007) *Climate Change and Conflict: The Migration Link*, International Peace Academy Coping with Crisis Working Paper, International peace Academy, New York.

Goodhand, J. (2003) 'Enduring disorder and persistent poverty: a review of linkages between war and chronic poverty', *World Development*, 31: 629–646.

Grothmann, T. and Patt, A. (2005) 'Adaptive capacity and human cognition: the process of individual adaptation to climate change' *Global Environmental Change-Human and Policy Dimensions*, 15 (3): 199–213.

Hendrix, C. and Glaser, S. (2007) 'Trends and triggers: climate, climate change and civil conflict in Sub-Saharan Africa', *Political Geography*, 26 (6): 695–715.

Homer-Dixon, T. (1999) *Environmental Scarcity and Violence*, Princeton, NJ: Princeton University Press.

Hugo, G. (1996) 'Environmental concerns and international migration', *International Migration Review*, 30 (1): 105–131.

IPCC (2007) *Climate Change 2007, Synthesis Report*, Contribution of Working Groups I, II and III to the Fourth Assessment Report of the Intergovernmental Panel on Climate Change, IPCC, Geneva.

Jacobsen, K. (1997) 'Refugees' environmental impact: the effect of patterns of settlement', *Journal of Refugee Studies*, 10, 1: 19–36.

Kahl, C. (2006) *States, Scarcity, and Civil Strife in the Developing World*, Princeton, NJ: Princeton University Press,

Kireab, G. (1997) 'Environmental causes and impact of refugee movements: a critique of the current debate', *Disasters*, 21 (1): 20–38.

Kliot, N. (2004) 'Environmentally induced population movements: their complex sources and consequences. A critical review', in J. Unruh, M. Krol and N. Kliot (eds) *Environmental Change and its Implications for Population Migration*, Dordrecht: Kluwer Academic Publishers.

Kniveton, D., Schmidt-Verkerk, K., Smith, C. and Black, R. (2008) *Climate Change and Migration: Improving Methodologies to Estimate Flows*, IOM Migration Research Series, No.33, Sussex University, Brighton: International Organization for Migration.

Leary, N., Adejuwon, J., Bailey, W., Barros, V., Caffera, M., Chinvanno, S., Conde, C., De Comarmond, A., De Sherbinin, A., Downing, T., Eakin, H., Nyong, A., Opondo, M., Osman, B., Payet, R., Pulhin, F., Pulhin, J., Ratnasiri, J., Sanjak, E., von Maltitz, G., Wehbe, M., Yin, Y. and Ziervogel, G. (2006) *For Whom the Bell Tolls: Vulnerability in a Changing Climate*, A Synthesis from the AIACC project, AIACC Working Paper No. 21, Florida: International START Secretariat, Florida.

Leichenko, R. and O'Brien, K. (2008) *Environmental Change and Globalisation: Double Exposures*, Oxford: Oxford University Press.

Levy, M. (1995) 'Is the environment a national security issue?', *International Security*, 20 (2): 35–62.

Lind, J. and Eriksen, S. (2006) 'The impacts of conflict on household coping strategies: Evidence from Turkana and Kitui districts in Kenya', *Die Erde*, 137 (3): 249–270.

Liverman, D. (1990) 'Drought impacts in Mexico: climate, agriculture, technology, and land tenure in Sonora and Puebla', *Association of American Geographers*, 80 (1): 49–72.

Liverman, D. (1994) 'Environment and security in Mexico', in B. Bagley and S. Aguayo Quezada (eds), *Mexico: In Search Of Security*, New Brunswick, NJ: Transaction Publishers: 230–264.

Locke, C., Adger, W.N. and Kelly, P. (2000) 'Changing places: migration's social and environmental consequences', *Environment*, 42 (7): 24–35.

Lonergan, S. (1998) 'The role of environmental degradation in population displacement', *Environmental Change and Security Project Report*, 4: 5–15.

Matthew, R., McDonald, B., Barnett, J. and O'Brien, K. (eds) (2009) *Global Environmental Change and Human Security*, Cambridge, MA: MIT Press.

McLeman, R. and Smit, B. (2006) 'Migration as an adaptation to climate change', *Climatic Change*, 76: 31–53.

Meier, P., Bond, D. and Bond, J. (2007) 'Environmental influence on pastoral conflict in the Horn of Africa', *Political Geography*, 26 (6): 716–735.

Meze-Hausken, E. (2000) 'Migration caused by climate change: how vulnerable are people in dryland areas? A case-study in Northern Ethiopia', *Mitigation and Adaptation Strategies for Global Change*, 5: 379–406.

Miguel, E., Satyanath, S. and Sergenti, E. (2004) 'Economic shocks and civil conflict: an instrumental variables approach', *Journal of Political Economy*, 112 (4): 725–753.

Morgan, M. and Dowlatabadi, H. (1996) 'Learning from integrated assessment of climate change', *Climatic Change*, 34 (3–4): 337–368.

Mortimore, M. (1989) *Adapting to Drought: Farmers, Famines, and Desertification in West Africa*, Cambridge: Cambridge University Press.

Mortreux, C. and Barnett, J. (2009) 'Climate change, migration and adaptation in Funafuti, Tuvalu', *Global Environmental Change*, 19 (1): 105–112.

Myers, N. (2002) 'Environmental refugees: a growing phenomenon of the 21st century', *Philosophical Transactions of the Royal Society*, 357 (1420): 609–613.

Narayan, D., Patel, R., Schafft, K., Rademacher, A. and Koch-Schulte, S. (2000) *Voices of The Poor: Can Anyone Hear Us?*, Oxford: Oxford University Press.

Nel, E. and Ringharts, M. (2008) 'Natural disasters and the risk of violent civil conflict', *International Studies Quarterly*, 52 (1): 159–185.

Nordås, R. and Gleditsch, N. (2007) 'Climate conflict: common sense or nonsense?', *Political Geography*, 26 (6): 627–638.

O'Brien, K. (2006) 'Are we missing the point? Global environmental change as an issue of human security', *Global Environmental Change*, 16 (1): 1–3.

O'Brien, K., St.Clair, A. and Kristoffersen, B. (eds) (2010) *Climate Change, Ethics and Human Security*, Cambridge: Cambridge University Press.

Peluso, N. and Harwell, E. (2001) 'Territory, custom, and the cultural politics of ethnic war in West Kalimantan, Indonesia', in N. Peluso and M. Watts (eds), *Violent Environments*, Ithaca, NY: Cornell University Press: 83–116.

Peluso, N., Humphrey, C. and Fortmann, L. (1994) 'The rock, the beach and the tidal pool: people and poverty in natural resource dependent areas', *Society and Natural Resources*, 7 (1): 23–38.

Perch-Nielson, S., Bättig, M. and Imboden, D. (2008) 'Exploring the link between climate change and migration', *Climatic Change*, 91 (3–4): 375–393.

PIF (Pacific Islands Forum) (2009) *Forum Communique 2009*, Fortieth Pacific Islands Forum, Cairns, Australia, August 5–6. Online. Available HTTP: http://www.forumsec.org/pages.cfm/newsroom/press-statements/2009/final-communique-of-40th-pacific-islands-forum-cairns.html (accessed 10 November 2012).

Piguet, E. (2008) *Climate Change and Forced Migration*, New Issues in Refugee Research, Research Paper No. 153, Policy Development and Evaluation Service, Geneva: United National High Commissioner for Refugees.

Piguet, E. (2010) 'Linking climate change, environmental degradation, and migration: a methodological overview', *WIRES Climate Change*, 1 (4): 517–524.

Reuveny, R. (2007) 'Climate change induced migration and violent conflict', *Political Geography*, 26 (6): 656–673.

Salehyan, I. (2008) 'From climate change to conflict? No consensus yet', *Journal of Peace Research*, 45 (3): 315–326.

Scheffran, J. (2008) 'Climate change and security', *Bulletin of the Atomic Scientists*, 64 (2): 19–25.

Scoones, I. (1998) *Sustainable Rural Livelihoods: A Framework for Analysis*, Working Paper 72, Brighton: Institute for Development Studies.

Sen, A. (1981) *Poverty and Famines: An Essay on Entitlement and Deprivation*. Oxford: Clarendon Press.

Shaw, M. (1993) 'There is no such thing as society: beyond individualism and statism in international security studies', *Review of International Studies*, 19 (2): 159–175.

Skeldon, R. (2002) 'Migration and poverty', *Asia Pacific Population Journal*, 17 (4): 67–82.

Smith, D. and Vivekananda, J. (2007) *A Climate of Conflict: The Links Between Climate Change, Peace and War*, London: International Alert.

Tacoli, C. (2007) *Migration and Adaptation to Climate Change*, Sustainable Development Opinion, International Institute for Environment and Development.

Theisen, O. (2008) 'Blood and soil? Resource scarcity and internal armed conflict revisited', *Journal of Peace Research*, 45 (6): 801–818.

UNDP (1994) *Human Development Report 1994*, United Nations Development Program, Oxford and New York: Oxford University Press.

UNGA (2009) *Climate Change and Its Possible Security Implications*, Report of the Secretary-General, United Nations General Assembly, 11 September 2009, A/64/350. Online. Available HTTP: http://www.unhcr.org/refworld/docid/4ad5e6380.html (accessed 10 November 2012).

Unruh, J. (2004) 'Migration induced legal pluralism in land tenure. Implications for environmental change', in J. Unruh, M. Krol and N. Kliot (eds) *Environmental Change and its Implications for Population Migration*, Dordecht, Boston, London: Kluwer Academic Publishers.

Walton, G. and Barnett, J. (2008) 'The ambiguities of environmental degradation and violence: the case of the Tolukuma Gold Mine', *Society And Natural Resources*, 21 (1): 1–16.

Warner, K., Afifi, T., Dun, O., Stal, M. and Schmidl, S. (2008) *Human Security, Climate Change and Environmentally Induced Migration*, Institute for Environment and Human Security, United Nations University.

Wolf, A. (1999) '"Water wars" and water reality: conflict and cooperation along international waterways', in S. Lonergan (ed.), *Environmental Change, Adaptation, and Security*, Dordrecht: Kluwer Academic Publishers: 251–265.

Ziervogel, G., Bharwani, S. and Downing, T. (2006) 'Adapting to climate variability: pumpkins, people and policy', *Natural Resources Forum*, 30 (4): 294–305.

4 Lines in the shifting sand

The strategic politics of climate change, human security and national defense

Betsy Hartmann

In September 2009 the U.S. Central Intelligence Agency (CIA) launched its new Center on Climate Change and National Security. According to the agency's press release, the center will study the national security impacts of desertification, rising sea levels, population shifts and conflict over natural resources, and will be "aggressive in outreach to academics and think tanks working on the issue" (CIA, 2009). The center's establishment reflects mounting concern in U.S. national security circles about the threats climate change poses to American interests and international stability (Broder, 2009).

These threats were elaborated further in the Defense Department's 2010 Quadrennial Defense Review (QDR). This is the first QDR to tackle climate change. While it cites various practical ways climate change could impact national security, such as its potential effects on defense installations and capabilities, the first reason given is its potential to act as an "accelerant of instability or conflict" (DOD, 2010: 85). In addition to contributing to poverty, environmental degradation, and the weakening of fragile governments, climate change will exacerbate resource scarcity, the spread of disease and mass migration (ibid.). Moreover, "extreme weather events may lead to increased demands for defense support to civil authorities for humanitarian assistance or disaster response both within the United States and overseas" (DOD, 2010: 85).

In this chapter, I argue that the linkages U.S. defense interests are making between climate change, conflict, and natural disasters threaten not only to distort climate policy, but to further militarize development and humanitarian assistance.[1] Since the early 1990s, the humanitarian mission has become increasingly politicized and militarized. Especially in conflict situations, humanitarian aid agencies are deeply worried about the evolving doctrine of military-civilian operations promoted by the U.S. and NATO. Doctors Without Borders, for example, has taken a strong stand against military involvement in humanitarian aid in Afghanistan (Hofman, 2011). Under this doctrine, military forces are seeking a larger and larger role in the delivery of aid, undermining the independence, neutrality and safety of humanitarian aid workers (Barry and Jefferys, 2002). They are also pushing for more involvement in humanitarian assistance during natural disasters (Hofmann and Hudson, 2009).

As a result, it is vitally important for those working at the intersection of global environmental change, development and human security to be attentive to the role of defense and intelligence interests in the climate field and the perils of securitizing climate change (Floyd, 2008). The boundaries between human security, environmental security and national security have often been fuzzy, and ought to be recognized explicitly as such.

Section one of this chapter explores the evolution of emerging security narratives about climate change and conflict focusing in particular on those that emphasize the risk of violent conflict among poor and displaced people. Section two looks at current U.S. defense strategies and how climate change could factor, rhetorically and practically, into future military and development interventions, especially in Africa.

Operation enduring narrative

Although long in the making, climate conflict narratives burst decisively onto the world stage in 2007. First the Atlantic Monthly (Faris, 2007), then the UN Environment Programme (UNEP, 2007), then even UN Secretary General Ban Ki Moon (2007) attributed violence in Darfur to a combination of demographic pressures, resource scarcities and climate change. Along with the Darfur stories came other dire predictions about the threat of so-called "climate refugees" spilling across the globe and wreaking havoc, and according to a Christian Aid report called "Human Tide," creating "a world of many more Darfurs" (Christian Aid, 2007a; 2007b). Well-known security pundits like Jeffrey Sachs (2007) and Thomas Homer-Dixon (2007) jumped on the theme with apocalyptic opinion pieces in popular media.

In April of 2007 the British government brought the issue of climate change before the UN Security Council for the first time with then foreign secretary Margaret Beckett identifying climate change as one of the main causes of conflict in Darfur (Harvey, 2007). In the U.S., the defense think tank, Center for Naval Analysis (CNA), gathered a team of 11 retired U.S. generals and admirals to produce a report titled "National Security and the Threat of Climate Change," which argued that global warming could help trigger widespread political instability in poor regions and large refugee movements to the U.S. and Europe (CNA, 2007). Toward the end of the year, the Norwegian Nobel committee (2007) warned of the threat of climate-induced violent conflict and war when it awarded the Nobel Peace Prize to Al Gore, Jr. and the Intergovernmental Panel on Climate Change (IPCC).

The 2007 climate conflict blitz could be read in purely pragmatic terms: Journalists and pundits wanted to make headlines. NGOs and researchers were looking for new winds to shake the money trees. The Nobel committee found a rationale to link peace and climate change. Climate scientists hoped security arguments would heighten the need to reduce carbon emissions. The UN decided to divert attention from its own peacekeeping failures in Darfur by shifting the focus to a shortage of rainfall rather than a long drought of political will. National security interests were looking for new threats to provide a sense of historical purpose and to justify increasing defense expenditures.

No doubt linking climate change to national security is a way to get the issue more attention at the highest levels of government. "The glum reality is that governments tend to take security threats more seriously than any other kind," writes British journalist Jonathan Freedland (2007).

> So this makes political sense: cast global warming as an environmental or science issue, and it will be given a budget to match. Cast it as a problem for the big boys, on a par with nuclear proliferation or international terror, and then it should get a big-boy budget and attention.
>
> (Freedland, 2007)

However, while all these factors may have played a role, the pragmatic explanation is not sufficient. Missing from the picture is the fact that many people believe climate conflict narratives because they are based on old, widely accepted assumptions about poverty, scarcity, overpopulation and migration that in many Western academic and policy circles are virtual articles of faith.

Climate conflict narratives draw on neo-Malthusian environmental security discourses of the 1980s and 1990s, in particular the concept of environmental conflict developed by Canadian political scientist Thomas Homer-Dixon (Hartmann, 2009). He maintains that:

> Population growth and unequal access to good land force huge numbers of people onto marginal lands. There, they cause environmental damage and become chronically poor. Eventually, they may be the source of persistent upheaval, or they may migrate yet again, helping to stimulate ethnic conflicts or urban unrest elsewhere.
>
> (Homer-Dixon, 1999: 155)

According to Homer-Dixon, this conflict can potentially fragment states or make them more authoritarian, destabilizing the international order. At the root of Homer-Dixon's model of environmental conflict, and others closely related to it, is an unquestioned acceptance of old colonial and neo-colonial stereotypes of destructive, over-breeding peasants and herders. Meanwhile, there is very little attention paid to the resource degradation caused by powerful economic and political interests, such as extractive industries (Hartmann, 2001). The rich only make cameo appearances in a tragic play where poor people get a few roles as victims, but are cast mainly as villains.

Add climate change to the mix, and you get the next iteration of environmental conflict. UNEP's 2007 report on Sudan, for example, draws on Homer-Dixon's model to make claims that overpopulation of both people and livestock, coupled with environmental stresses such as water shortages related to climate change, is at the root of conflict in Northern Darfur (UNEP, 2007). Thus, it doesn't take much to make the leap, or rather small step, from environmental to climate conflict. The same is true in the case of the step from environmental to climate refugees.

Also with a problematic neo-Malthusian history, the term 'environmental refugee' gained widespread currency in the international policy arena in the 1980s and 1990s (Saunders, 2000). In 1995 a report by Norman Myers for the Climate Institute in Washington, DC made the claim that globally there were

25 million environmental refugees, "persons who can no longer gain a secure livelihood in their traditional homelands because of environmental factors of unusual scope," including water shortages, desertification, natural disasters, and climate change (Myers, 1995: 18–19). Although the 25 million figure was arrived at more by conjecture than scientific method, and the term "environmental refugee" is analytically flawed, the figure became "fact" and was widely cited in policy documents (Black, 1998; Nordas and Gleditsch, 2007).

Then Myers later claimed there will be 200 million climate migrants by 2050, a figure which is similarly making the rounds in policy documents, even though Myers himself acknowledges that the estimate is based on "heroic extrapolations" (cited in Brown, 2008: 8). As Oli Brown notes, "The simple fact is that nobody really knows with any degree of precision what climate change will mean for human population distribution" (Brown, 2008: 8).

This is not to deny that climate change is likely to cause displacement, but the extent of that displacement will not only depend on how much the temperature rises and affects sea-levels, rainfall patterns and the severity of storms, but on the existence and effectiveness of adaptation measures that help individuals and communities cope with environmental stresses. Whether or not such measures are in place in turn depends on political economies at the local, regional, national and international levels that are often conveniently left out of the discussion of so-called "climate refugees." And as one report points out, larger climate-related humanitarian emergencies may be in places "where people cannot afford to move, rather than the places to which they do move" (O'Brien *et al.*, 2008: 24). There is a growing consensus that climate-related migration is more likely to take place within national borders than across them (UNHCR, 2008).

In the end, migration is too complex a process to label simply as environmental or climate-induced (Morrissey, 2008: 28; Farbotko, 2010; Tacoli, 2011; White, 2011). It is also too complex to link *ipso facto* to violent conflict. Yet this is the overwhelming message of the climate conflict literature: poor and displaced people are dangerous. They violently fight over scarce resources and threaten security in the Global North.

One cannot underestimate the tremendous power of this narrative, despite the rich body of research that challenges it. For example, a World Bank study by Norwegian researchers found that current alarms about climate conflict are not based on substantive evidence (Buhaug *et al.*, 2008). In Africa, violent conflict is actually connected more closely to competition over resource abundance (rich oil and mineral reserves, valuable timber, diamonds, etc.) than resource scarcity (Fairhead, 2001). Portraying Darfur as a climate conflict ignores basic elements of Sudanese political economy. These include gross inequalities in wealth and power between the elite in the capital and the rural population; government agricultural policies that favor large mechanized farms and irrigation schemes over rain-fed, small-farmer agriculture; forced migration, such as the 1990s removal of Nuba farmers from their lands into so-called "peace villages"; and what Alex de Waal calls "militarized tribalism" (de Waal, 2007, see also Manger, 2005).

This long history of oppression gets lost in the climate conflict version of events. Precisely for that reason, researcher Harry Verhoeven writes, the Sudanese government "loves the 'climate war' rhetoric" because it obscures the regime's role in causing the conflict through policies of exclusion, patronage and violence (Verhoeven, 2011: 695) Verhoeven sees a reverse logic at work between climate change and conflict. State-sponsored violence, whether in the form of direct military action or more indirect attacks on people's land and livelihoods, makes local communities less able to adapt to climate-related stresses such as long-term declines in rainfall.

Narratives about the risk of climate conflict and climate refugees could pave the way for the appropriation of the climate change and human security agenda by military interests. The warning signs are already here.

All terrain security

While 2007 may have represented the high water mark of climate conflict alarm, the waves were already washing in several years before. In 2003 a Pentagon-sponsored scenario of the impacts of abrupt climate change painted a neo-Malthusian nightmare of poor, starving populations overshooting the reduced carrying capacity of their lands, engaging in violent conflict over scarce resources, and storming en masse towards U.S. and European borders (Schwartz and Randall, 2003).

A next round of scenarios on the foreign policy and national security implications of climate change was undertaken in 2006–2007 by the Center for a New American Security (CNAS) and the Center for Strategic and International Studies (CSIS) in Washington, DC. Those involved were supposedly a diverse group of experts from a wide range of disciplines, yet from a reading of the list there appear to be very few, if any, international development or environment scholars representing perspectives from the Global South. Instead on the foreign policy side, the list reads like a who is who of former Clinton-era officials and advisors waiting in the wings for a Democratic presidential victory (Campbell, 2008: 1–3). Indeed, the *Wall Street Journal* described CNAS in November 2008 as a "top farm team" for the incoming Obama administration's national security apparatus (Ackerman, 2008). Michele Flournoy, co-founder of CNAS, was named Undersecretary of Defense for Policy in January 2009. In this capacity, she spoke about how climate change was going to accelerate state failure, mass migration, the spread of disease and possibly insurgency (CSIS, 2009).

The CNAS-CSIS project predicts that as the mercury rises, so will the violence of the poor, especially in Africa. With a rise of 4.7 °F (2.6°C) by 2040 (Campbell and Parthemore, 2008: 19), governments with resources will have to engage in triage and decide which of the poor are worth saving. The author of this scenario envisions that war, disease and draconian population control measures might restore an environmentally sustainable relationship between people and nature (Fuerth, 2008). The overall message of the project is that climate change may be the biggest security challenge the U.S. faces, and that it presents "surprisingly similar" challenges as terrorism, a powerful reason why "groups whose interests center on either the environment or on national security have cause to come together and act in tandem" before the world turns into a harrowing, Hobbesian dystopia (Campbell and Parthemore, 2008: 19–20).

This beating of the climate conflict drums has to be viewed in the context of larger orchestrations in U.S. national security policy. While development assistance and humanitarian aid have long been strategically deployed as an element of defense policy (Keen, 2008), in recent years the military has encroached much further into civilian territory. Observers are beginning to speak of an "aid-military complex" – in 2005 the share of official U.S. development assistance dispersed by the Pentagon was 22 percent, up from 6 percent three years before (Easterly, 2008). The State Department's role in both diplomacy and development has been severely weakened as a consequence, and disaster response is increasingly becoming the purview of the Pentagon at home and abroad (Berrigan, 2008).

These trends reflect a strategic shift in defense thinking toward a focus on stability operations and the taming of "ungoverned spaces." In 2005 the Department of Defense (DOD) issued a directive stating that "stability operations" shall be given equal priority to combat operations:

> The immediate goal often is to provide the local populace with security, restore essential services, and meet humanitarian needs. The long-term goal is to help develop indigenous capacity for securing essential services, a viable market economy, rule of law, democratic institutions, and a robust civil society.
> (DOD, 2005: 2)

To this end, the U.S. military should mainly work through "indigenous, foreign, or U.S. civilian officials" or "military-civilian teams" which shall be open to representatives of International Organizations, NGOs, and the private sector (DOD, 2005: 3). The Army's 2008 manual on stability operations cites climate change as a driver of conflict (Department of the Army, 2008).

The concept of "ungoverned spaces" derives from a 2007 study of "ungoverned territories" done by the Rand Corporation for the U.S. Air Force, which identified these areas as failed or failing states, poorly controlled borders, or locations within "otherwise viable states where the central government's authority does not extend" (Rabasa and Peters, 2007: 1). As critics point out, many of these spaces are actually governed, but not by groups favorable to U.S. interests (Clunan and Trikinus, 2008). Ungoverned spaces are perceived as a threat because they can serve as recruiting and organizing grounds for terrorists, criminal networks and other illicit activities. Discursively and strategically, the concept of ungoverned spaces provides a point of convergence for anti-terrorism efforts, stability operations and development assistance.

Since 2007, Africa has been the primary focus of climate conflict discourse. Coincidence or not, this development has coincided with the establishment of the new U.S. military command for Africa, AFRICOM. The reasons for the creation of AFRICOM are multi-faceted and include the protection of U.S. access to

African oil and other strategic resources, the War on Terror, and countering increasing Chinese influence in the region (Volman, 2008). Popular resistance within Africa has meant that the U.S. has not been able to locate AFRICOM's headquarters on the continent; it is currently stationed in Stuttgart, Germany. By its very institutional structure, AFRICOM represents the blurring of military/civilian boundaries. Among its staff AFRICOM includes senior USAID officials to "help us plan our own military tasks supportive of USAID efforts" (USAID, 2009: 14). In general, AFRICOM seeks to integrate U.S. military objectives more firmly with economic and political ones.

Constructing climate conflict as a particularly African security threat meshes well with these objectives. CNA's 2007 report on the threat of climate change specifically linked potential insecurity caused by climate change to the proposed mission of AFRICOM (CNA, 2007). While it is highly unlikely that the U.S. would send in the troops or base strategic development assistance solely on a perceived risk of climate conflict, the promotion of that risk helps to make such interventions more palatable, especially in liberal foreign policy circles. Indeed, a report by the Center for American Progress, another think tank close to the Obama administration, calls for protecting America through "sustainable security" (Brigety and Dewan, 2009: 1). It seeks to tie U.S. development assistance to strategic defense and intelligence objectives. "Climate-induced resource conflicts" are cited as a potential "significant source of political instability and violence" (Brigety and Dewan, 2009: 14). CNAS, meanwhile, is promoting a similar concept of "natural security" (Burke, 2009).

What is emerging is a rearticulation of Clinton/Gore style liberal stewardship, a vision of a win-win world where the U.S. not only makes the world safe through democracy and free markets and from terrorism, but protects the environment and wins the hearts and minds of potentially dangerous poor people too. This time around, however, the ideological and institutional convergence between military and civilian objectives is more complete, and the ungoverned spaces of Africa are the new, and perhaps last, frontier.

Conclusion: drawing the line

On the simplest level, the deployment of climate conflict and associated discourses is a serious distraction from the urgent need for the international community to come up with just and effective measures to increase the resilience of the most vulnerable people in the face of climate change. It does those people no favor to portray them as potentially dangerous and violent threats to Western borders and national security. And as Rita Floyd (2008) points out, the focus on climate conflict could inhibit cooperation between countries on climate policy.

On a deeper level, it is time to draw a sharp line between national security and human security. We do not live in a win-win world. The real battle ahead is not for the hearts and minds of people in enemy territory, but for the meaning of development itself as defense interests encroach further and further into what should be civilian territory.

Note

1 There are diverse reasons for the Pentagon's concerns about climate change (see Dabelko, 2009), but this chapter focuses primarily on the purported risk of climate conflict.

References

Ackerman, S. (2008) 'Obama's Pentagon-in-waiting', *The Washington Independent*, 10 November. Online. Available HTTP: http://washingtonindependent.com/17710/obama (accessed 24 April 2009).

Barry, J. with Jefferys, A. (2002) *A bridge too far: Aid agencies and the military in humanitarian response*, Humanitarian Practice Network, Paper 37, London: Overseas Development Institute.

Beebe, S. (2007) Guest contributor Shannon Beebe on AFRICOM and environmental security, 19 July 2007, *The new security beat.* Online. Available HTTP: http://newsecuritybeat.blogspot.com/2007/07/guest-contributor-shannon-beebe-on.html (accessed 15 September 2008).

Berrigan, F. (2008) 'Entrenched, embedded and here to stay: the Pentagon's expansion will be Bush's lasting legacy', Tomdispatch.com, 27 May. Online. Available HTTP: at http://www.tomdispatch.com/post/174936 (accessed on 27 September 2009).

Black, R. (1998) *Refugees, environment and development*, New York: Longman.

Brigety, E. and Dewan, S. (2009) *A national strategy for global development: Protecting America and our world through sustainable security*, Washington, DC: Center for American Progress.

Broder, J.M. (2009) 'Climate change seen as a threat to U.S. security', *New York Times*. 9 August. Online. Available HTTP: <http://www.nytimes.com/2009/08/09/science/earth/09climate.html?_r=1&pagewanted=2 (accessed 13 November 2009).

Brown, O. (2008) 'The numbers game', *Forced Migration Review*, 31: 8–9.

Brown, O., Hammill, A. and McLeman, R. (2007) 'Climate change as the new security threat: Implications for Africa', *International Affairs*, 83 (6): 1141–1154.

Buhaug, B., Gleditsch, N.P. and Theisen, O.M. (2008) *Implications of climate change for armed conflict*, Paper presented to the World Bank Workshop on Social Dimensions of Climate Change, 25 February 2008, World Bank, Washington, DC.

Burke, S. (2009) 'Security implications of climate scenario 3: catastrophic climate change over the next one hundred years', in K.M. Campbell (ed.) *Climatic Cataclysm*, Washington, DC: Brookings Institution Press.

Campbell, K.M. (ed.). (2008) *Climatic Cataclysm: The foreign policy and national security implications of climate change*, Washington, DC.: Brookings Institution Press.

Campbell, K.M. and Parthemore, C. (2008) 'National security and climate change in perspective', in K.M. Campbell (ed.), *Climatic cataclysm*, Washington, DC: Brookings Institution Press.

Christian Aid (2007a) *Human tide: The real migration crisis*, London, May. Online. Available HTTP: http://www.christian-aid.org.uk (accessed 31 May 2007).

Christian Aid (2007b) *World facing worst migration crisis*, Press release, 14 May. Online. Available HTTP: http://www.christianaid.org.uk/news/media/pressrel/070514p.htm (accessed 31 May 2007).

CIA (2009) *CIA opens center on climate change and national security*, Press release, 25 September. Online. Available HTTP: <https://www.cia.gov/news-information/press-releases-statements/center-on-climate-change-and-national-security.html (accessed 13 November 2009).

Clunan, A. and Trinkunas, H. (2008) *Ungoverned spaces? Alternatives to state authority in an era of softened sovereignty*, Paper delivered at the International Studies Association 48th Annual Meeting, San Francisco, CA, 26–30 March, 2008.

CNA Corporation (2007) *National security and the threat of climate change*, Alexandria, VA: CNA Corporation. Online. Available HTTP: http://securityandclimate.cna.org/report/ (accessed 7 May 2007).

CNAS (2009b) Natural resources + national security = natural security. Online. Available HTTP: http://www.cnas.org/print/26 (accessed on 18 May 2009).

CSIS (2009) *Rebalancing the force: Major issues for QDR 2010*, 29 April, 2009. Washington, DC: Federal News Service Transcript.

Dabelko, G.D. (2009) 'Planning for climate change: The security community's precautionary principle', *Climatic Change*, 96(1): 13

Department of the Army (2008) *Stability operations*, Field manual No. 3-07, Washington, DC: Headquarters, Department of the Army.

de Waal, A. (2007) 'Sudan: the turbulent state', in A. de Waal (ed.) *War in Darfur and the search for peace*, Cambridge, MA: Justice Africa and the Harvard University Global Equity Initiative: 1–38.

DOD (2005) *Military support for stability, security, transition and reconstruction*, Directive 3000.05, 28 November. Online. Available HTTP: at http://www.dtic.mil/whs/directives/corres/pdf/300005p.pdf (accessed 6 October 2007).

DOD (2010) *Quadrennial defense review report*, Washington, DC, 128 pp. Online. Available HTTP: http://www.defense.gov/QDR/QDR%20as%20of%2029JAN10%201600.pdf (accessed 10 February 2010).

Dun, O. and Gemenne, F. (2008) 'Defining environmental migration', *Forced Migration Review*, 31: 10–11.

Easterly, W. (2008) 'Foreign aid goes military', *New York Review of Books*, 55 (19). Online. Available HTTP://www.nybooks.com/articles/archives/2008/dec/04/foreign-aid-goes-military/?pagination=false (accessed 26 November 2012).

Fairhead, J. (2001) 'International dimensions of conflict over natural and environmental resources', in N.L. Peluso and M. Watts (eds) *Violent Environments*, Ithaca, NY: Cornell University Press.

Farbotko, C. (2010) 'Wishful sinking: Disappearing islands, climate refugees, and cosmopolitan experimentation', *Asia Pacific Viewpoint*, 51 (1): 53–54.

Faris, S. (2007) 'The real roots of Darfur', *Atlantic Monthly*, April.

Floyd, R. (2008) 'The environmental security debate and its significance for climate change', *The International Spectator*, 43 (3): 51–65.

Freedland, J. (2007) 'There is no doubt that global warming is a security threat to us all', *The Guardian*, 4 April. Online. Available HTTP: http://www.guardian.co.uk/comment/story/0,,2059564,00.html (accessed 7 October 2007).

Fuerth, L. (2008) 'Severe climate change over the next thirty years: security implications of climate scenario 2', in K.M. Campbell (ed.) (2008) *Climatic cataclysm: The foreign policy and national security implications of climate change*, Washington, DC: Brookings Institution Press: 133–154.

Gregory, W. (2011) *Climate change and migration: Security and borders in a warming world*, Oxford: Oxford University Press.

Hartmann, B. (2001) 'Will the circle be unbroken? A critique of the project on environment, population and security', in N.L. Peluso and M. Watts (eds) *Violent environments*, Ithaca, NY: Cornell University Press.

Hartmann, B. (2009) 'From climate refugees to climate conflict: Who is taking the heat for global warming?' in M. Salih (ed.) *Climate change and sustainable development: New challenges for poverty reduction*, Northampton, MA: Edward Elgar Publishers.

Harvey, F. (2007) 'UN climate panel detailed potential for global conflict', *Financial Times*, 13 October.

Hofman, M. (2011) Dangerous aid in Afghanistan. 14 January. Online. Available HTTP: http://www.doctorswithoutborders.org/publications/article.cfm?id=4970&cat=op-eds-articles (accessed 26 November 2012).

Hofmann, C. and Hudson, L. (2009) 'Military responses to natural disasters: Last resort of inevitable trend?' *Humanitarian Exchange*, 44: 29–31.

Homer-Dixon, T. (1999) *Environment, Scarcity and Violence*, Princeton, NJ: Princeton University Press.

Homer-Dixon, T. (2007) 'Terror in the weather forecast', *New York Times*, 24 April.

Keen, D. (2008) *Complex Emergencies*, Cambridge, MA: Polity Press.

Manger, L. (2005) 'Understanding resource management in the Western Sudan: a critical look at the new institutional economics', in Q. Gausset, M.A.Whyte and T. Birch-Thomsen (eds) *Beyond Territory and Scarcity: Exploring Conflicts Over Natural Resource Management*, Stockholm: Nordiska Afrikainstitutet: 135–148.

Marks, E. (2009) Why USAFRICOM? *Joint Forces Quarterly*, 52, First Quarter 2009: 148–151.

Moon, B.K. (2007) 'A climate culprit in Darfur', *Washington Post*: 16, p. A15.

Morrissey, J. (2008) 'Rural-urban migration in Ethiopia', *Forced Migration Review*, 31: 28–29.

Myers, N. (1995) *Environmental exodus: An emergent crisis in the global arena*, Washington, DC: The Climate Institute.

Nordas, R. and Gleditsch, N.P. (2007) 'Climate change and conflict', *Political Geography*, 26 (6): 627–638.

Norwegian Nobel Committee (2007) *Press release*. Online. Available HTTP: http://nobelprize.org/nobel_prizes/peace/laureates/2007/press.html (accessed 26 November 2012).

O'Brien, K., Sygna, L., Leichenko, R., Adger, W.N., Barnett, J., Mitchell, T., Schipper, L., Tanner, T., Vogel, C. and Mortreux, C. (2008) *Disaster risk reduction, climate change adaptation and human security*, a commissioned report for the Norwegian Ministry of Foreign Affairs, Oslo: University of Oslo.

Rabasa, A. and Peters, J.E. (2007) 'Understanding lack of governance', in A. Rabasa, S. Boraz, P. Chalk, K. Cragin, T.W. Karasik, J.D.P. Moroney, K. O'Brien and J.E. Peters (eds) *Ungoverned territories: Understanding and reducing terrorism risks*, Santa Monica, CA: RAND Corporation: 1–6.

Rabasa, A., Boraz, S., Chalk, P., Cragin, K., Karasik, T.W., Moroney, J.D.P., O'Brien, K.A., and Peters, J.E. (2007) *Ungoverned territories: Understanding and reducing terrorism risks*, Report prepared for the United States Air Force. Santa Monica, CA: RAND Corporation.

Sachs, J.D. (2007) *Climate change refugees*, ScientificAmerican.com, 20 May. Online. Available HTTP: http://www.sciam.com (accessed 9 July 2007).

Saunders, P.L. (2000) 'Environmental refugees: The origins of a construct', in P. Stott and S. Sullivan (eds) *Political Ecology: Science, Myth and Power*, London: Arnold.

Schwartz, P. and Randall, D. (2003) *An abrupt climate change scenario and its implications for United States National Security*, Washington, DC: Environmental Media Services. Online. Available HTTP: http://www.gbn.com/articles/pdfs/Abrupt%20Climate%20Change%20February%202004.pdf (accessed 26 November 2012).

Tacoli, C. (2011) *Not only climate change: Mobility, vulnerability and socio-economic transformations in environmentally fragile areas of Bolivia, Senegal and Tanzania*, IIED Human Settlements Working Paper Series, Rural-Urban Interactions and Livelihood Stategies, No. 28, London: IIED, Online. Available HTTP: http://pubs.iied.org/10590IIED.html (accessed 26 November 2012).

UNEP (2007) *Sudan post-conflict environmental assessment*, United Nations Environmental Programme, June. Online. Available HTTP: http://www.unep.org/Sudan/ (accessed 7 July 2007).

UNHCR (2008) 'Climate change, natural disasters and human displacement: a UNHCR perspective', Online. Available HTTP: http://www.unhcr.org/refworld/pdfid/492bb6b92.pdf (accessed 26 November 2012).

USAID (2009) 'AFRICOM general hails USAID-military links', *Frontlines*, December–January 2009: 1, 14.

Verhoeven, H. (2011) 'Climate change, conflict and development in Sudan: Global neo-Malthusian narratives and local power struggles', *Development and Change*, 42 (3): 679–707.

Volman, D. (2008) *Why is the pentagon marching into Africa?* African Security Research Project, 27 June, 2008.

White, G. (2011) *Climate Change and Migration: Security and Borders in a Warming World*, Oxford: Oxford University Press.

5 Sustainable urbanization and human security

Halvard Buhaug, Henrik Urdal and Gudrun Østby

Introduction

For the first time in history, the majority of the world population now lives in cities. By 2050, two in every three persons will live in urban areas, and virtually all population growth during this period, around 3 billion people, will be absorbed by cities, primarily in developing countries (UN, 2008). Considering the impending consequences of global warming, such as accelerating sea level rise and more extreme weather patterns, we might see an even stronger urbanization trend. Adverse climate is a potential additional contributor to rural-urban migration everywhere, and in particular in poorer countries where livelihoods, economies and political structures may be inept at local adaptation (e.g. Barnett and Agder, 2007; Tacoli, 2009; Raleigh and Jordan, 2010). Various forms of environmental change can also directly limit adaptive capacities and wellbeing of urban dwellers.

City growth and migration to urban areas are both drivers and byproducts of economic development. Yet, a growth that is too strong can be a challenge. Rapid urbanization puts significant demands on a society's ability to provide public services like adequate housing, electricity, water supply, health care, education, and jobs. Widespread shanty towns around major cities in the developing world epitomize the challenges of accommodating a fast expanding population. According to a recent survey, many governments of developing countries now explicitly discourage urbanization; 78 percent of African and 71 percent of Asian countries have implemented policies to reduce migrant flows to large cities (UN, 2008: 12).

Rural-to-urban migration is often seen as a consequence of high and increasing population pressure in the countryside, with associated scarcities of renewable resources like cropland, forests, and freshwater (Homer-Dixon, 1999; Barrios *et al.*, 2006). Various forms of environmental degradation, including desertification, prolonged droughts, and soil salinization, are other factors that might deteriorate agricultural livelihoods and push people to the cities. On the other hand, rapid urbanization may cause many of the same environmental problems that plague densely populated and poorly managed countryside: water scarcity and contamination, land shortage, and insufficient sanitation (Brennan, 1999). And while opportunities for employment are usually better in urban areas, the labor market will struggle to absorb fast-growing populations. Higher income inequalities and differences in privileges among city dwellers is another likely source of urban frustration and despair. While urbanization in itself does not necessarily constitute a threat to peace and stability, urban pressures within the context of economic stagnation and poor governance could increase the risks of violence and political turmoil (Gizewski and Homer-Dixon, 1995).

A commonly voiced goal in the arena of international sustainable development is to provide the coming generations with a secure future, formally expressed and popularized in the *World Commission on Environment and Development* (WCED, 1987). So far, however, there has been little consideration of what responses will be needed in urban areas affected by environmental change and rapid population growth. Indeed, the debate on social consequences of climate change is disproportionately focusing on rural communities and concerns. In an effort to fill this gap, this chapter focuses on the ways through which aspects of human security can be developed into a central framework for organizing the debate over the sustainability of metropolitan areas.

The rest of the chapter proceeds as follows. It first describes recent and likely future trends in urban population growth with a particular focus on developing countries. It then discusses whether and how environmental change in conjunction with rapid population growth may affect urban security. Findings from emerging research on determinants of urban social disorder suggest that rapid urbanization may be less of a security threat than often assumed. The chapter then briefly discusses what responses are needed at the city-level to cope with the joint challenges of growing populations and climate change, and concludes and identifies some priorities for a future research agenda on sustainable cities.

Trends and projections in global urban population

The world's population crossed the 7 billion mark in 2011 and the number is rising by about 75 million per year. Current projections indicate that there will be about 9.3 billion people on earth by 2050. However, in absolute terms the growth rate of the global population peaked in the late 1980s (at about 88 million per year) and the relative growth rate (currently at 1.1 percent per year) has halved since it peaked in the early 1960s (US Census Bureau, 2010). The "population explosion" originally set the stage for numerous Malthusian doomsday claims (e.g. Ehrlich, 1968) and continues to dominate the environmental security discourse today (e.g. Redding, 2007; see Urdal, 2005 for a review). Yet, there is another less acknowledged major demographic trend that could have very significant security implications: urbanization. According to UN statistics, the global share of the urban population increased more than fourfold during the twentieth century, and while population growth rates are gradually stagnating, urbanization continues at high speed. In coming decades, the largest urban population changes are expected to occur in developing countries in Asia and Africa, particularly in coastal areas (Goldstone, 2010; Hunt and Watkiss, 2011).

Urban population growth is the joint product of three phenomena: natural increase, rural-to-urban migration, and reclassification of rural land (Cohen,

2006; UNFPA, 2007). Although the relative weights of these contributing factors vary significantly between different urban areas, the most important source of urban population growth in developing countries is natural population increase – i.e. a high birth-to-death ratio (UNFPA, 2007: 5, 13). As fertility rates are expected to drop with social and economic development (Mason, 1997), rural-to-urban migration may become the most powerful driver of urbanization in the future. Urbanization is tightly linked to development; some of the most rapidly urbanizing countries over the last fifty years, such as Botswana and the United Arab Emirates, have experienced exceptionally high economic growth rates. Indeed, no country in the industrial age has managed sustained economic growth without urbanization. As the economies in the developing world are expanding, so are the urban populations. While people may flee a countryside dominated by personal insecurity, poverty, and environmental degradation, urban centers tend to offer a higher level of services such as health care and education. Greater urbanization could mean a more efficient use of energy, and the combined demographic trends of global population ageing and urbanization have been suggested to lead to lower global emissions (O'Neill *et al.*, 2010).

With global warming and accompanying challenges, such as sea level rise, more extreme weather, and general deterioration of agricultural productivity in parts of the world, the global rate of migration and urbanization might increase even further (McGranahan *et al.*, 2007). Picking up on this possibility, a report by Christian Aid (2007: 1) warns of a "human tide" and considers migration "the most urgent threat facing poor people in developing countries." Other studies are much more cautious in their wording (e.g. Bardsley and Hugo, 2010; Raleigh and Jordan, 2010). The multi-causal nature of human migration implies that any attempt to quantify the extent of future "environment-induced" urbanization will be fraught with uncertainty. Regardless of the importance of environmental push factors relative to economic and other motives, however, the crucial reality remains: nearly all population growth within the next few decades will be absorbed by cities.

Population pressure and political violence

Proper management of urban growth and changing population structures is key to preserving human security. Rapid population growth can seriously constrain local governments' ability to provide basic services, including employment, housing, electricity, water, and enforcement of law and order and development of social capital, thus greatly affecting the quality of life of the citizens. According to Goldstone (2002), it is exactly when overurbanization combines with underdevelopment – where the job market and the economy cannot keep up with urban population growth – that violence and instability may arise. Similarly, the mixing of ethnicities and shifting demographic composition of urban centers are cited as central destabilizing factors in urban environments. Climate-induced urbanization and temporary rural-to-urban distress migration are likely to add to these challenges (Raleigh and Jordan, 2010). People fleeing the disaster-struck –

or environmentally unsustainable – countryside may have limited resources and social networks to rely on if the social safety-net offered is inadequate (Gizewski and Homer-Dixon, 1995; Sygna, 2005). Although the nature and consequences of such population dynamics are poorly understood, recent research indicates that environmental conditions are more closely associated with local than long-distance mobility that often takes the form of (temporary) rural-to-urban migration. The environment is very rarely the most important reason for migration (UNDP, 2009), and where environmental conditions matter, migration is more often conducted as a proactive diversification strategy to cope with anticipated long-term declines in livelihood than a result of immediate stress from a natural disaster (Bardsley and Hugo, 2010; Massey *et al.*, 2010; Tacoli, 2011).

Environmental change is relevant to population-security dynamics beyond merely acting as a push factor for migration. For example, concentration of people and economic activities in cities increase exposure of lives and assets to extreme weather and, in the longer run, to sea level rise. According to the Fourth Assessment Report of the UN's Intergovernmental Panel on Climate Change (IPCC), a 40 cm rise in sea level will result in flooding of more than 100 million people every year under business as usual scenarios (Nicholls *et al.*, 2007). The potential for societal instability would appear significant as the scale of urban poverty in low-income countries is growing rapidly (Tacoli, 2011).

The broad literature on social movements and collective action offers a number of insights into how increasing urban population pressure might transform into political violence, and the notion of population pressure is central in the environmental security literature (Homer-Dixon and Blitt, 1998; Homer-Dixon, 1999). Possible theoretical approaches range from almost deterministic assumptions about ethnic hatreds and associated security dilemmas (e.g. Horowitz, 1985; Posen, 1993; Kaufmann, 1996) via modernization-based arguments of radicalization of aggrieved, unemployed youths (e.g. Huntington, 1996; Goldstone, 2002) to classic theories of structural inequalities and relative deprivation (e.g. Gurr, 1970). Common to all of these contributions is their attention to the distribution of opportunities and privileges among the urban population.

Often framed in a rural context, rapid population growth may lead to a reduced per-capita access to subsistence resources, such as water, pasture, and food, as resource reproduction is unable to keep up with the growing demand. High population pressure may also lead to a decline in the overall supply of some resources, for example, due to pollution, deforestation, overgrazing, and unrestrained fishing. Further, resource scarcity may spark or escalate inter-group competition (Østby *et al.*, 2011). Under unfavorable economic and political conditions, such competition may take the form of violent conflict. Poor countries are argued to be particularly susceptible to violent resource conflicts as they have limited capacity to adapt to changing environments and often lack institutional arrangements for peaceful conflict resolution (e.g. Smith and Vivekananda, 2007).

While some individuals and communities manage to adapt or enlarge their access to diminishing resources, others will exit and settle in more promising environments, including in urban areas. In this way, urbanization may act as a safety valve, relieving the countryside of the impending population pressure. On the other hand, large-scale migration to the cities might merely translate the problem of overpopulation into an urban setting. There is no shortage of studies that see rapid urbanization as a potentially destabilizing force (e.g. Homer-Dixon and Blitt, 1998; Homer-Dixon, 1999; Kahl, 2006). Potential risk factors include economic marginalization of migrants, poor integration and adjustment to city life, political radicalization, and greater opportunities for collective action (Gizewski and Homer-Dixon, 1995).

The empirical evidence for a connection between urbanization, environmental change, and violent conflict so far is weak and inconclusive. In an analysis of urban ethnic violence in Kenya in the 1990s, Kahl (2006: 132) finds that rapid rural-to-urban migration pushed the urban infrastructure and social services beyond their limits, and that this was the prime cause of the conflict. Huntington (1996: 113), providing a more sweeping account of the urbanization-violence nexus, argues that the newly uprooted masses of rural-to-urban migrants throughout the Islamic World in the 1970s and 1980s were attracted to radical Islamic movements as these provided slum dwellers with social services and offered a "dignified identity." In contrast, quantitative cross-national studies find little evidence that high urban growth rates or poor environmental conditions increase the baseline risk of violent conflict. If anything, urban population growth appears to be associated with less conflict (see e.g. Collier and Hoeffler, 2004; Urdal, 2005). However, one should bear in mind that these literatures often consider different forms of conflict and vary in their assessment of the environmental context so that seemingly opposing findings need not be contradictive.

Urban social disorder: trends and new findings

According to the Urban Social Disorder dataset from the Peace Research Institute Oslo (PRIO), there has been a slight increase in the average number of urban disorder events over the last fifty years (Urdal and Hoelscher, 2012). This weak trend is evident for lethal as well as non-lethal events, albeit with considerable inter-annual variation (Figure 5.1). Although increasing populations may have contributed to the weak upward trend, better news coverage and more systematic reporting in recent times are probably also important explanatory factors. The rate of the contemporaneous population growth in these cities is many times higher, however, implying a considerable drop in per-capita rate of disorder events throughout this period.

So far, statistical analysis of the urban social disorder data has revealed little evidence of a systematic connection between increasing urban populations and increasing risk or frequency of social disorder (Urdal and Hoelscher, 2012; Buhaug and Urdal, 2013). If anything, it seems that population growth is negatively associated with unrest. Notably, the most rapidly expanding cities in Asia and Sub-Saharan Africa generally have very few disorder events. This trend does not in principle dismiss the possibility that higher pressure on urban environments and public services increase social unrest. It could be that the demography-security

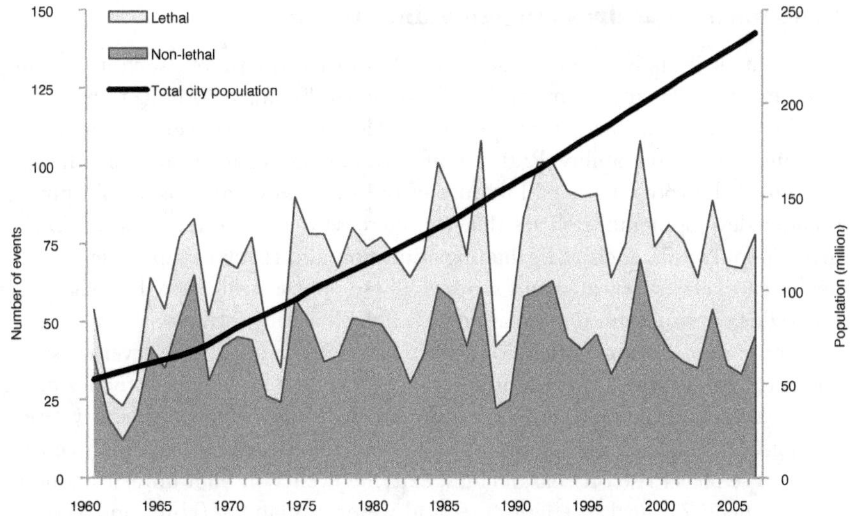

Figure 5.1 Frequency of urban social disorder and city population by year, 1960–2006. The figure illustrates the frequency of lethal and non-lethal social disorder events and the total city population in the Asian and Sub-Saharan African sample over time. Source: Urdal and Hoelscher (2012)

relationship is strictly non-linear whereby a certain critical mass is required for mobilization and population growth beyond this threshold matters relatively less. Of course, we cannot exclude the possibility of a different threshold effect, whereby cities are able to absorb the burgeoning population only up to a given point (which may not yet have been reached in most cases), after which environmental degradation, underperforming public goods delivery, and inadequate labor market lead to societal disintegration and widespread political violence. Current studies also do not attempt to assess the counterfactual – what are the most likely security scenarios under different urbanization patterns? Moreover, data scarcity limits contemporary empirical analysis to using population figures as proxies for a host of related city-specific conditions (unemployment, poverty, quality of housing and infrastructure, social inequalities, etc.) that by no means are perfectly correlated with high/increasing urban population pressure. Direct measures of city-level environmental conditions and infrastructure are largely absent.

Given that rapid urban population growth appears unrelated to the frequency and severity of social disorder, it is in no way substantiated that future adverse environmental change – to the extent it may amplify the strong rural-urban population movement in developing countries – will be a major driver of political violence in cities. Rather, we believe urban political violence will continue to emerge in response to general societal features and events, such as economic shocks, low secondary educational attainment, absence of democratic institutions, and ongoing intrastate conflict (Urdal and Hoelscher, 2012; Buhaug and Urdal, 2013).

Responses to address urban vulnerability

Government initiatives to discourage rural-to-urban migration, such as evicting squatters and denying them services, have generally failed to halt urbanization (UNFPA, 2007). Instead, such policies could hinder improvements in urban planning and sustainability. Rather than introducing negative sanctions that fail to address the root causes of high urban pressures, governments should pursue broader developmental policies that will increase the economic contribution of urban populations, such as facilitating education and regular employment, basic health care, and safe water and sewage. These policies will, over the long term, also reduce urban natural population growth by lowering fertility.

Even if high urban population growth may not increase the level of social conflict directly, physical concentration of people and economic activities make cities particularly vulnerable to environmental change (Hallegatte and Corfee-Morlot, 2011). The most important effects of climate change on urban centers relate to sea level rise on coastal cities; extreme events on built infrastructure; as well as effects on health; energy use, and water availability (Hunt and Watkiss, 2011). Urban centers in low-elevation coastal zones in the developing world are of particular concern as they tend to be very densely populated and expanding rapidly (McGranahan *et al.*, 2007). In terms of asset exposure, however, major coastal cities in the developed world, such as the USA, the Netherlands, and Japan dominate the list.

The cost of rebuilding infrastructure and buildings, as well as less tangible social and cultural ties to the land, means that relocation of exposed coastal cities is unlikely. Instead, risk-reducing adaptation strategies need to be developed. Required initiatives may include strengthening of barrages, flood embankments and natural barriers, improved early warning systems and plans for evacuation, climate-sensitive building regulations and land use planning, and risk-sharing through insurance and reinsurance (Hanson *et al.*, 2011). A crucial adaptation challenge for local actors is to understand the nature of future climate change risks in their region and to identify the main causes of urban vulnerability. To date such adaptation has been scarce. A few cities in high-income countries appear to be at the forefront of adaptation planning and implementation, such as Chicago, London, Miami, New York, Paris, and Toronto (Hallegatte and Corfee-Morlot, 2011).

A better understanding of societal vulnerabilities to climate change impacts can assist local authorities to communicate with local, other sub-national, and national decision-makers in order to mobilize political will, assess adaptation options and design cost-effective and timely responses. This is a complex challenge since many cities are likely to be hit by shifting climate patterns as well as potentially increased intensity and frequency of extreme events, such as flooding and droughts. However, in many cases measures may be of "no regret," i.e. they would yield benefits if implemented today, even in the absence of climate change. This is especially true in developing countries, in which basic urban infrastructure is often absent, leaving room for actions that both increase immediate well-being and reduce vulnerability to future climate change (Hallegatte and Corfee-Morlot, 2011).

Conclusion

Global urbanization will continue at high speed in the foreseeable future and urban growth will be especially strong in the developing world. The total urban population in the world is expected to nearly double over the next 40 years. While urban populations in principle are more easily reached by government policies and population concentration often implies economies of scale, rapid urban growth in combination with direct impacts of environmental change and poor urban management may produce considerable strains on the capacity to provide services like housing, water, sewage, power, transportation, health care, and education. Clustering of people and assets may also increase vulnerability to extreme weather events and sea level rise. A key challenge is maintaining high economic growth to accommodate the burgeoning urban youth into the labor market without forsaking the environment. At present, strategic planning of sustainable cities of tomorrow is largely absent, and climate adaptation of urban centers is still primarily a feature of developed states.

Thus far, there is little evidence to suggest that rapid urbanization is a significant contributing factor to political violence in cities. Less is known about whether and how local environmental conditions and degradation affect urban security. In fact, comparative research on determinants of urban social unrest is still in its infancy so it is not hard to find areas of improvement for future research. We conclude that there are three major challenges for existing quantitative scholarship.

First, most previous quantitative research is limited to investigating direct relationships or simple interaction effects, thereby enforcing rather crude causal assumptions. The general ignorance of conditional effects – such as underlying social cleavages, quality of governance, and the environmental context – has been a ready criticism of quantitative analyses among case-oriented researchers (e.g. Homer-Dixon, 1999). Future studies should specify the conditions under which urbanization and high urban population growth is more likely to cause human insecurity.

Second, quantitative cross-national research has been hampered by poor and overly coarse data, and empirical analyses are often conducted at a too high level of aggregation. Whereas the theoretical arguments typically concern local-level relationships between increasing population pressure and local violence, statistical models typically apply a country-level approach where national urbanization statistics are regressed on country-level conflict data. Better city-specific data on socio-economic, demographic, and environmental conditions may be obtained through more extensive use of comprehensive surveys, such as the Demographic and Health Surveys (DHS) program. A major priority for future research should be to re-examine drivers of urban violence using indicators measured precisely at the city level.

Third, previous quantitative research on urban population growth and violence is almost exclusively limited to studying conventional armed conflicts (e.g. Collier and Hoeffler, 2004; Urdal, 2005). In contrast, there is an increasing awareness that urban violence, often constitutes a mixture of criminal and political motives, and where the violence may take the form of homicides, organized crime, sexual violence, and terrorist acts (e.g. Jütersonke *et al.*, 2007). The new Urban Social Disorder data presented above (Urdal and Hoelscher, 2012) go some way toward this goal but further systematic data collection is needed. Future research on urban insecurity should expand the scope of conflict and look at wider patterns of political and social unrest below conventional armed conflict.

Finally, adapting cities in the south, and indeed in the north, to climate change may contribute to a more sustainable development of urban centers. In the south, developing urban infrastructure like housing, water and sanitation, transportation and education is likely to have long-term positive impacts that could even eventually contribute to curb social disorder and violence. Overall, the adaptive policies most likely to succeed are those that respond to existing local needs, contribute to other development goals, and can be locally driven (McGranahan *et al.*, 2007). Vulnerable urban settlements in low-income countries clearly deserve international support in measures to adapt to environmental change. Likewise, the shift in focus from discouraging to sustainably governing urban population growth is not the sole responsibility of developing country governments. International organizations and developed country donor agencies are crucial in supporting broad strategies for urban development.

References

Bardsley, D. K. and Hugo, G. J. (2010) 'Migration and climate change: Examining thresholds of change to guide effective adaptation decision-making', *Population and Environment*, 32 (2–3): 238–262.

Barnett, J. and Agder, W.N. (2007) 'Climate change, human security and violent conflict', *Political Geography*, 26 (6): 639–655.

Barrios, S., Bertinelli, L. and Strobl, E. A. (2006) 'Climatic change and rural-urban migration: The case of Sub-Saharan Africa', *Journal of Urban Economics*, 60 (3): 357–371.

Brennan, E. (1999) 'Population, urbanization, environment, and security: A summary of the issues', *Comparative Urban Studies Occasional Papers Series*, 22.

Buhaug, H. and Urdal, H. (2013) 'An urbanization bomb? Population growth and social disorder in cities', *Global Environmental Change*, 23 (1): 1–10.

Christian Aid (2007) *Human Tide: The Real Migration Crisis*, United Kingdom: Christian Aid. Online. Available HTTP: www.christianaid.org.uk/Images/human-tide.pdf (accessed 7 November 2012).

Cohen, B. (2006) 'Urbanization in developing countries: Current trends, future projections, and key challenges for sustainability', *Technology in Society*, 28 (1–2): 63–80.

Collier, P. and Hoeffler, A. (2004) 'Greed and grievance in civil war', *Oxford Economic Papers*, 56 (4): 563–595.

Ehrlich, P. R. (1968) *The Population Bomb*, Ballantine, New York: Books.

Gizewski, P. and Homer-Dixon, T. F. (1995) 'Urban growth and violence: Will the future resemble the past?', Occasional Paper, Project on Environment, Population and Security, American Association for the Advancement of Science, Washington, D.C., and the University of Toronto.

Goldstone, J. A. (2002) 'Population and security: How demographic change can lead to violent conflict', *Journal of International Affairs*, 56 (1): 3–21.

Goldstone, J. A. (2010) 'The new population bomb: Four megatrends that will change the world', *Foreign Affairs*, 89 (1): 31–43.

Gurr, T. R. (1970) *Why Men Rebel*, Princeton, NJ: Princeton University Press.

Hallegatte, S. and Corfee-Morlot, J. (2011) 'Understanding climate change impacts, vulnerability and adaptation at city scale: An introduction', *Climatic Change*, 104 (1): 1–12.

Hanson, S., Nicholls, R., Ranger, R., Hallegatte, S., Corfee-Morlot, J., Herweijer, C. and Chateau, J. (2011) 'A global ranking of port cities with high exposure to climate extremes', *Climatic Change*, 104 (1): 89–111.

Homer-Dixon, T. F. (1999) *Environment, Scarcity, and Violence*, Princeton, NJ: Princeton University Press.

Homer-Dixon, T. F. and Blitt, J. (eds) (1998) *Ecoviolence: Links among Environment, Population and Security*, Lanham, MD: Rowman & Littlefield.

Horowitz, D. L. (1985) *Ethnic Groups in Conflict*, Berkeley, CA: University of California Press.

Hunt, A. and Watkiss, P. (2011) 'Climate change impacts and adaptation in cities: A review of the literature', *Climatic Change*, 104 (1): 13–49.

Huntington, S. P. (1996) *The Clash of Civilizations and the Remaking of World Order*, New York: Simon & Schuster.

Jütersonke, O., Krause, K. and Muggah, R. (eds) (2007) *Small Arms Survey 2007: Guns and the City*, Cambridge: Cambridge University Press.

Kahl, C. H. (2006) *States, Scarcity, and Civil Strife in the Developing World*, Princeton, NJ: Princeton University Press.

Kaufmann, C. (1996) 'Possible and impossible solutions to ethnic civil wars', *International Security*, 20 (spring): 136–175.

Mason, K. O. (1997) 'Explaining fertility transitions', *Demography* 34 (4): 443–454.

Massey, D. S. (1996) 'The age of extremes: Concentrated affluence and poverty in the twenty-first century', *Demography*, 33 (4): 395–412.

Massey, D. S., Axinn, W. G. and Ghimire, D. J. (2010) 'Environmental change and out-migration: Evidence from Nepal', *Population and Environment*, 32 (2–3): 109–136.

McGranahan, G., Balk, D. and Anderson, B. (2007) 'The rising tide: Assessing the risks of climate change and human settlements in low-elevation coastal zones', *Environment and Urbanization*, 19 (1): 17–37.

Montgomery M. R., Stren, R., Cohen, B. and Reed, H. E. (eds) (2003) *Cities Transformed: Demographic Change and its Implications in the Developing World*, Washington, DC: National Academies Press.

Nicholls, R.J., Wong, P.P., Burkett, V.R., Codignotto, J.O., Hay, J.E., McLean, R.F., Ragoonaden, S. and Woodroffe, C.D. (2007) 'Coastal systems and low-lying areas', in M.L. Parry, O.F. Canziani, J.P. Palutikof, P.J. van der Linden and C.E. Hanson (eds) *Climate Change 2007: Impacts, Adaptation and Vulnerability. Contribution of Working Group II to the Fourth Assessment Report of the Intergovernmental Panel on Climate Change*, Cambridge: Cambridge University Press.

O'Neill, B. C., Dalton, M., Fuchs, R., Jiang, L., Pachauri, S. and Zigova, K. (2010) 'Global demographic trends and future carbon emissions', *Proceedings of the National Academy of Sciences of the United States of America*, 107 (41): 17521–17526.

Østby, G., Urdal, H., Tadjoeddin, Z., Murshed, S. M. and Strand, H. (2011) 'Population pressure, horizontal inequality and political violence: A disaggregated study of Indonesian provinces, 1990–2003', *Journal of Development Studies*, 47 (3): 377–398.

Posen, B. (1993) 'The security dilemma and ethnic conflict', *Survival*, 35 (spring): 27–47.

Raleigh, C. and Jordan, L. (2010) 'Climate change and migration: Emerging patterns in the developing world', in R. Mearns and A. Norton (eds) *Social Dimensions of Climate Change*, Washington, DC: The World Bank.

Redding, T. M. (2007) 'The population challenge: Key to global survival', *21st Century Papers*, no. 2, Washington, DC: The Population Institute.

Smith, D. and Vivekananda, J. (2007) *A Climate of Conflict: The Links between Climate Change, Peace and War*, London: International Alert.

Stewart, F. (2000) 'Crisis prevention: Tackling horizontal inequalities', *Oxford Development Studies*, 28 (3): 245–262.

Sygna, L. (2005) 'Climate vulnerability in Cuba: The role of social networks', *CICERO Working Paper*, no2005:01, Center for International Climate and Environmental Research, Oslo (CICERO).

Tacoli, C. (2009) 'Crisis or adaptation? Migration and climate change in a context of high mobility', *Environment and Urbanization*, 21 (2): 513–525.

Tacoli, C. (2011) 'Not only climate change: Mobility, vulnerability and socio-economic transformations in environmentally fragile areas of Bolivia, Senegal and Tanzania', *Human Settlements Working Paper Series*, no. 28, London: International Institute for Environment and Development (IIED).

UN (2008) *World Urbanization Prospects: The 2007 Revision: Highlights*, New York: United Nations Population Division.

UNDP (2009) Human Development Report 2009. *Overcoming Barriers: Human Mobility and Development*, United Nations Development Programme.

UNFPA (2007) *State of the World Population 2007: Unleashing the Potential of Urban Growth*, New York: United Nations Population Fund.

Urdal, H. (2005) 'People vs. Malthus: Population pressure, environmental degradation, and armed conflict', *Journal of Peace Research*, 42 (4): 417–434.

Urdal, H. and Hoelscher, K. (2012) 'Explaining urban social disorder and violence: An empirical study of event data from Asian and Sub-Saharan African cities', *International Interactions*, 38 (4): 512–528.

US Census Bureau (2010) International Data Base. Online. Available HTTP: http://www.census.gov/ipc/www/idb/ (accessed 7 November 2012).

WCED (1987) *Our Common Future, World Commission on Environment and Development*, Oxford: Oxford University Press.

6 Urban risk and vulnerability

Insights and lessons from Latin American cities

Patricia Romero-Lankao, Mercy Cordova Borbor, Eduardo Behrentz and Laura Dawidowski

Introduction

Climate change has increasingly become recognized as one of the greatest challenges to global efforts to meet sustainability and human security goals. However, climate change is a deeply local issue as well. Given that more than 50 percent of the world's population is located in urban areas, cities bear both the opportunities and challenges for the integration of climate considerations into efforts towards meeting sustainability goals (Wheeler and Beatley, 2008). While previous scholarship has explored how hazards such as air pollution and climate variability affect the risks of negative health impacts on urban populations (e.g., Bell *et al.*, 2008), research on urban risk has yet to develop a full understanding of the complexity of factors shaping the vulnerability of cities and urban populations and creating differences in health impacts within and between cities even when similar stressors are applied. This is especially true of Latin America where, compared to other parts of the world, urban vulnerability and risk have received little attention from scholars (Romero-Lankao *et al.*, 2012b).

This chapter addresses this gap by exploring the complexity of factors and urbanization processes shaping urban vulnerability and risk in three Latin American cities. Drawing on a review of globally relevant literature that examines the linkages between urbanization and climate change, the chapter provides an analytical review of the climate relevance of urbanization. Notwithstanding the increasing number of studies on vulnerability, there is comparatively little research focusing specifically on the conceptualization of urban vulnerability and risk. Within existing scholarship, the interdisciplinary differences in definition and scope result in divergent narratives on the determinants and dimensions of urban vulnerability, and thus in diverging policy implications. In our project, "Adaptation to the Health Impacts of Air Pollution and Climate Extremes in Latin American Cities" (ADAPTE), we developed and tested our own framework, and explored the complex nature of urban risk and vulnerability to air pollution and climate change in the Latin American cities of Bogotá, Buenos Aires, Mexico City and Santiago. In this chapter, we will describe three of the study cities and present some research highlights from ADAPTE on whether health risks within cities are differentiated between neighborhoods by socioeconomic factors alone or whether, environmental factors must also be considered.

Relevance of urban change

More than half of the world's population now lives in urban areas. Therefore, any assessment of what factors may increase or reduce urban vulnerability and risk needs to begin from an understanding of four processes of and linkages between urbanization and climate change. First, the scale of urban population growth is unprecedented, with a quadrupling of urban populations between 1950 and 2007 and cities becoming ever larger and more crowded. For example, the number of cities with populations greater than 1 million increased from 75 in 1950 to 447 in 2011, while the average size of the world's largest cities increased from 2 to 7.6 million (Romero-Lankao and Gnatz, 2011: 3). This unprecedented growth is critical to considerations of urban vulnerability because in the absence of adequate governance structures, the concentration of populations, economic activities and built environments can turn urban centers into hotspots of vulnerability and risk.

Second, the urban transition in the twentieth century in Europe, North and Latin America, where nations changed from predominantly rural to increasingly urbanized, is now being replicated also in low- and middle-income countries in Asia and Africa. While these countries currently lack many of the institutions, resources and – in some cases – commitment to enhancing adaptive capacity and human security, more than 90 percent of the world's urban population growth is occurring there, and they are currently home to nearly three-quarters of the world's urban populations. These socioeconomic and institutional conditions, coupled with an increased intensity and frequency of adverse hydro-meteorological events brought by climate change, might further challenge these urban centers' already low capacities to address development goals.

Third, smaller urban centers are registering the fastest rates of growth. In fact, it is in urban centers with fewer than 500 000 people, currently home to just over 50 percent of the world's urban population, where, between 2009 and 2025, 45 percent of the world's total increase in urban population is expected (Romero-Lankao and Gnatz, 2011: 2). Smaller urban centers, in low- and middle-income countries, are often particularly institutionally weak, and unable to promote effective mitigation and adaptation actions (Romero-Lankao and Gnatz, 2011). As the intensity and frequency of adverse weather events increases with climate change, devastating impacts may occur precisely where the capacity to adapt is weaker (IPCC, 2007). Conversely, because most of the development in these areas has yet to occur, there will be opportunities to reduce their vulnerability to climate hazards and other stresses while building necessary infrastructures, for example, by including climate proof features and effective response systems, or by the creation of the educational and political systems that support adaptive capacity.

In many low- and middle-income countries extremely high densities are associated with locally detrimental environmental health conditions while, elsewhere, urban form is becoming increasingly expansive with a high degree of sprawl (Seto *et al.*, 2010). Urban form and density therefore play pivotal, yet not fully understood roles in the arena of human security. For instance, extremely low densities of many suburban areas (particularly in North America and Australia) are associated with high levels of household energy consumption and emissions. On the other hand, many health risks – including location in flood-prone areas, exposure to the urban heat-island effects or high levels of indoor air pollution– can be exacerbated by the high urban densities experienced in many low-income cities. At the same time, these high urban densities can create opportunities for simultaneously improving health and quality of life through policies related to effective public transport systems, urban planning, and affordable housing (Romero-Lankao and Gnatz, 2011).

Fourth, urbanization has taken place in tandem with many other apparently insurmountable development problems (e.g., high levels of poverty, indigence and informality). Although around 30 percent of the world's urban population lives in slums, the percentage ranges from 6 percent in developed nations to 78 percent in least developed countries (UN-Habitat, 2003: xxv). Because urban planning tends to support land development for formal settlements, informal settlements tend to be excluded and their needs for secure tenure, infrastructures and services neglected. Thus, the protection from floods and other hazards afforded by these infrastructures is often lacking in informal settlements (Birkmann *et al.*, 2010).

Affluence and poverty, the two faces of the urban development coin, create a clear set of winners and losers, with different levels of responsibility for environmental degradation and differential capacities to respond to human-security threats. These differences ultimately crystalize in an inverse relationship clearly separating those who are most responsible for global environmental and climate change and those who will suffer its most profound consequences (Romero-Lankao and Gnatz, 2011). One of the prime mechanisms of this inverse relationship occurs as uneven development and inadequate infrastructure and governance structures (e.g., insufficient sources of revenue for local authorities) constrain the ability of urban populations and authorities to adapt to existing and future climate change and to other environmental and societal stresses (Holgate, 2007).

Understanding urban vulnerability to climate change

Climate variability and change have a variety of implications for urban dwellers, economic sectors and on the network-infrastructures of urban centers, such as energy, waste water and transport systems (Gasper *et al.*, 2011). However, a focus solely on the impacts, hazards and affected sectors does not allow for an understanding of the entire array of dimensions and determinants of urban vulnerability and risk. While risk refers to possible harmful consequences, vulnerability relates to the degree to which urban populations are susceptible to, or able to adapt to the adverse effects of one or more hazards. In a very vulnerable city (or neighborhood) more people would die during a summer heat wave than in a less vulnerable city (or neighborhood); the buildings and infrastructures of more vulnerable communities would be more damaged by floods than in a less vulnerable community. The task is, hence, to explore what makes a given city, neighborhood, population, or household more or less vulnerable, and to gain an understanding of these differences that is meaningful for policy and decision making at relevant levels.

As is the case with vulnerability in general, research on urban vulnerability is fragmented by disciplinary differences in definitions and methods. Furthermore, it is dominated by the urban vulnerability as impact paradigm, which suggests that more integrated studies should be undertaken (Romero-Lankao *et al.*, 2012a). The numerous lineages of vulnerability research these methodologies and definitions represent (e.g., disaster risk community, political ecology, climate change community) have been summarized previously (Birkmann, 2006; Kienberger, 2010; Romero-Lankao *et al.*, 2012a; Cardona *et al.*, 2012). Therefore, here we will outline three approaches that have informed the framework developed for the ADAPTE project: urban vulnerability as impact, inherent vulnerability, and risk-society theory (Romero-Lankao and Qin, 2011).

The urban vulnerability as impact (or natural hazard) tradition conceives urban vulnerability as an outcome based on a population's exposure, hazards and individual socioeconomic characteristics (Bell *et al.*, 2008; Makri and Stilianakis, 2008). A large proportion of these studies have focused on cities in North America and Western Europe, whereas only limited research has been conducted in Latin America (O'Neill *et al.*, 2005; Bell *et al.*, 2008). This approach has shown the importance of hazards' attributes in defining the quantitative relationship between hazards and outcomes (often defined by mortality). For example, while a very small change in ambient pollution levels can have substantial impacts on large populations (Makri and Stilianakis, 2008: 2), the temperature mortality relationship has a V or J shape, with mortality generally increasing both above and below some temperature threshold (Michelozzi *et al.*, 2006; McMichael *et al.*, 2008; Muggeo and Hajat, 2009).

Urban vulnerability as impact research has also shown that socioeconomic characteristics such as age, poverty and gender can modify or confound the severity of the health effects of hazards such as extreme temperatures and air pollution (Basu, 2009; Reid *et al.*, 2008). For instance, the elderly, the very young and people with pre-existing medical conditions have been shown to be more sensitive to environmental hazards (Dear *et al.*, 2005; Pope and Dockery, 2006). Lower levels of education are also associated with higher levels of mortality from extreme temperatures and air pollution (Smoyer, *et al.*, 2000; Medina-Ramón and Schwartz, 2007). However, the effects of such socioeconomic features as income, poverty and ethnicity are mixed (O'Neill *et al.*, 2005; Stafoggia *et al.*, 2006; Ishigami *et al.*, 2008; D'Ippoliti *et al.*, 2010).

The inherent urban vulnerability lineage illuminates the structural drivers creating differences in vulnerability and adaptive capacity among and within urban populations. Yet, its scholars tend to differ in their emphasis on entry points for analysis and intervention. For example, a livelihoods' approach focuses on the assets and options individuals and households draw from to cope with hazards (Pelling, 2003; Moser and Satterthwaite, 2010). Other scholars point to the fundamental role the state plays in shaping adaptive capacity, for instance, by promoting economic growth and poverty reduction; providing or warranting access to infrastructures and services, and fostering land and housing, and public emergency-response systems (Romero-Lankao and Qin, 2011).

Finally, among the many approaches to risk (Thywissen, 2006; Kienberger, 2010), Beck's (1986) "risk society theory" deserves attention because it states that in the current era of capitalism, science and technology become the central mechanism to increase the production of goods, and thus to reduce material needs. Yet, at the same time, they become sources of "bads," the negative byproducts of industrialization, such as climate change and air pollution. These "bads," in turn, create risks that operate over large geographic scales, cut across sectors, and constrain or overwhelm the possibilities, tools and resources wealthy populations have previously had to escape from risk or compensate for risks, thus increasing their vulnerability. Beck refers to this dynamic as the "boomerang effect" and concludes that while "hunger is hierarchical, smog is democratic".[1] The risk society approach to urban vulnerability is therefore distinctive in that it suggests differences in urban vulnerability that are not only due to socioeconomic and institutional characteristics of the city and its population but also physical characteristics of the hazard itself.

In the ADAPTE project, the risk of adverse effects results from the combination of multiple social and environmental processes and dimensions that need to be addressed using more integrated approaches (Romero-Lankao *et al.*, 2013). It depends on the nature of the heat waves, floods, fires and other *hazards* to which urban populations are *exposed* or subjected. It also depends on the dynamics of these populations' sensitivity, capacity to respond and actual responses (O'Brien *et al.*, 2007; Gallopín, 2006; Romero-Lankao *et al.*, 2012a). Each of the dimensions has different components and can be approached at different levels. For instance, increases in cities' average temperatures result not only from global and long-term anthropogenic GHG driven climate change, but also from local changes in air quality and temperature induced by urbanization. *Sensitivity* or the degree to which subsets of urban populations are susceptible to hazards depends on patterns of susceptibility based on both individual characteristics (e.g., age, gender, medical conditions) and neighborhood- or city-level social conditions responsible for differential hazards' effects (e.g., equity in access to resources). The *capacity* and actual *adaptations* to avoid or lessen the negative consequences of hazards are determined by multilevel factors ranging from individual/household access to assets and options (e.g., income, good quality housing, air conditioning or heating appliances) to the extent and quality of urban infrastructures and services; and the quality and inclusiveness of governance structures, planning instruments and community organizations that provide or manage safety nets and other short- and longer-term responses (Figure 6.1).

We combined a set of quantitative and qualitative methods and data that draw from the described approaches to determine whether and under what conditions the people in these cities are vulnerable to temperature and air pollution hazards. The results presented here focus on the municipal level. Daily temperature data from the meteorological stations of each city provide information on maximum, mean, and minimum temperature. Air pollution data on three primary criteria pollutants, particulate matter (PM_{10}), nitrogen dioxide (NO_2), and ozone were

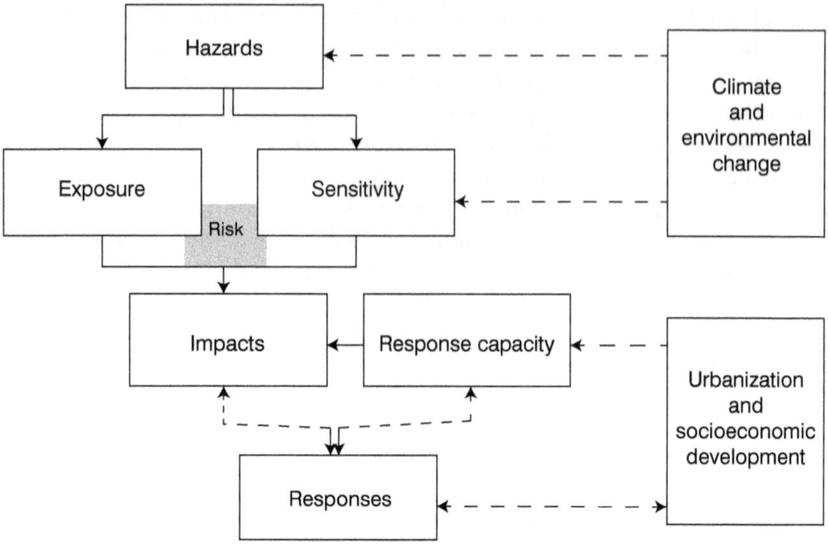

Figure 6.1 A conceptual framework of urban vulnerability and risk. The dynamics and drivers of urban vulnerability determine the impacts of environmental hazards. Source: Adapted from Romero-Lankao and Qin (2011)

collected from each city's environmental agency. Mortality data on respiratory mortality and cardiovascular mortality were obtained from the public health agencies of each city. We used socio-demographic data such as education, poverty, income, age structure and housing condition from the study cities' census offices to construct municipality-level measures of vulnerability based on the livelihoods approach. We developed a multidimensional vulnerability index (MVI) using a multi-criteria model of socioeconomic vulnerability that is based on four different types of livelihood assets: social, human, physical, and financial capitals (Baud *et al.*, 2008; Romero-Lankao *et al.*, 2013).

We used various statistical tools to identify the characteristics of and interactions among hazards, exposure, sensitivity, adaptive capacity, and health impacts. Exploratory Time Series Analysis helped us identify the temporal patterns of air pollution and temperature. A Generalized Linear Model (GLM) with Poisson log-linear distribution was used to explore the relative risks (RR) of mortality from exposure to hazards. We also tested the statistical correlations of exposure to major air pollutants and human mortality with socio-economic vulnerability to examine whether municipalities with different vulnerability levels can be differentiated with respect to health risks.

The cities

The climate of the cities ranges from Mediterranean (Santiago) to subtropical highland (Bogotá and Mexico City). While none of these cities experiences extreme variations in temperature, there are more seasonal variations in Mexico City and Santiago than in Bogotá (Table 6.1). Health risks due to atmospheric emissions are a concern in all three cities, particularly because atmospheric and meteorological conditions can be conducive to air pollutant retention and ozone formation (Molina and Molina, 2002). These hazards might be further intensified with climate change.

Bogotá, Mexico City and Santiago are the primary centers of their national economies, with each generating 25, 34 and 43 percent of its national GDP, respectively; each also concentrates populations, economic activities, energy and atmospheric emissions; and each is especially affected by hazards climate change is expected to aggravate, such as air pollution and changes in average temperatures. However, the effects of hazards and stresses on cities are not always negative. Socioeconomic, political and cultural factors help give urban populations, infrastructures and economic activities the ability to bounce back, recover from, and even take advantage of climatic and non-climatic stresses.

Differentiated urban development shapes adaptive capacities among urban populations. With all their dynamism and high levels of integration in the global economy, these cities are still faced with high levels of poverty, income inequality and informality. In aggregate, Table 6.1 demonstrates the strong presence of poverty, income inequality and informality in the cities, but, for a fuller picture of how these factors play out in each, we look here at municipality level differences. Although the local patterns of population and economic activities have changed in recent decades (e.g., with lower rates of demographic growth), spatial segregation within these cities remains a key characteristic. Core areas have registered slower growth and in some cases decay; high income, gated communities have grown in suburban and peri-urban areas; and low income, often informal, settlements occupy the periphery. Uneven development and inadequate infrastructure and governance structures constrain the ability of urban populations and authorities to adapt to existing and future hazards and stresses.

The cities have deficits in such key determinants of adaptive capacity as health (with high infant mortality rates in Mexico and Bogotá), education (with socially segregated school systems in all cities) and in housing with inadequate housing stock, informal settlements (with the exception of Santiago) and problems of housing affordability in all cities. Sometimes, decaying central areas and peri-urban areas are being inhabited by marginalized populations with inadequate services, a portfolio of precarious livelihood mechanisms, and the existence of a risk-management tradition that is long enough to address these underlying determinants of vulnerability. Last but not least, the levels of crime and violence are high in Mexico and Bogotá, which might undermine the influence of social capital (i.e., individual levels of social trust and participation in networks) on adaptive capacity.

ADAPTE: research findings

This section presents findings on some of the features and linkages between the components of urban vulnerability in Mexico City, Santiago and Bogotá. It explores the risks of mortality from exposure to hazards as well as whether

Table 6.1 Selected environmental, socioeconomic and demographic indicators. Source: ADAPTE; Jordan *et al.* (2010)

	Bogotá	*Buenos Aires*	*Mexico City*	*Santiago*
Latitude	4°32 N	34° 35 S	9° 26 N	33° 28 S
Average annual temperature	14.0	17.96	15.0	14.0
Average and max. summer temp (°C)	13.7 & 19.1	22.9 & 27.7	17.8 & 24.5	19.8 & 26.8
Average and min. winter temp (°C)	13.6 & 8.2	11.9 & 8.3	16.1 & 9.9	12.8 & 8.0
Annual PM_{10}	68.75	N.D.	51.60	70.18
Annual O_3	11.57	N.D.	32.4	29.84
Population	6,776,009	12,828,669	17,946,313	5,392,804
GDP per capita in US$ (ppp)	12,692	19,098	17,552	16,826
Elderly (64 and over)	6%	15%	3.9%	6.4%
Persons without high school education	39.50%	22.8%	45.7%	35%
GINI coefficient	0.61	0.44	0.56	0.55
Infant mortality rates (per thousand)	13.5	7.9	17.8	7.6
Informal employment (% of total workforce)	n.d.	44	45.7	34
Slum population (% of inhabitants)	16.8	26.2	14.4	5
Homicides per 100 thousand	18.7	6.9	17.6	2.1

socioeconomic factors between municipalities differentiate these risks within cities. Air quality data confirmed that air pollution is one of the most important sources of health risks to populations in urban Latin America (Bell *et al.*, 2008). After organizing the air quality data, and comparing them with the World Health Organization recommendations for air quality, we found that Santiago has the highest concentration of PM_{10} while Mexico City presents the highest levels of ozone on average (Table 6.1). PM_{10} exceeded WHO standards by about 90 percent, and levels of nitrogen oxide exceeded WHO standards by up to 73 percent. Temporal trends for PM_{10} showed that the number of times that Mexico City's records surpassed international standards was lower than that of Santiago and Bogotá (Romero-Lankao *et al.*, 2013).

We used temporal series and a GLM to explore the risks of mortality from exposure to hazards and found that these relationships are complex and dependent on several factors such as temperature and unfavorable atmospheric conditions, high pressure or anti-cyclonic systems, which create higher pollution due to thermal inversions at night and during the first hours of the morning. Furthermore, while we found that higher outdoor temperatures led to lower mortality for both respiratory and cardiovascular causes (Figure 6.2), these positive effects could be offset by higher levels of particulate matter with an aerodynamic diameter of up to 10 μm (PM_{10}), which are associated with higher mortality (Figure 6.3).

Our quantification of urban populations' likelihood of mortality from exposure to hazards showed that higher temperatures during the warm season caused an increased risk of cardiovascular and respiratory mortality in Mexico City but only increased the risk of cardiovascular mortality in Bogotá (Romero Lankao *et al.*, 2012b). Although some of our results were not statistically significant, overall they

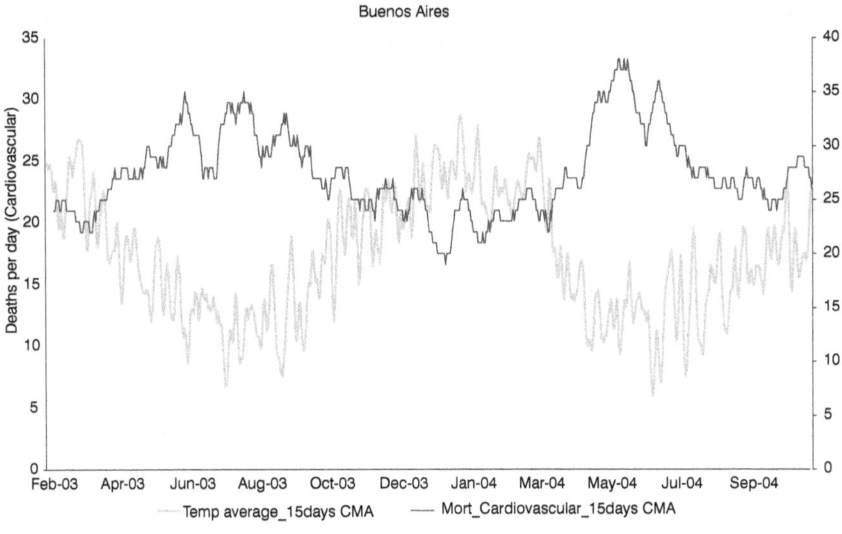

Figure 6.2 Buenos Aires: negative correlation between temperature and mortality

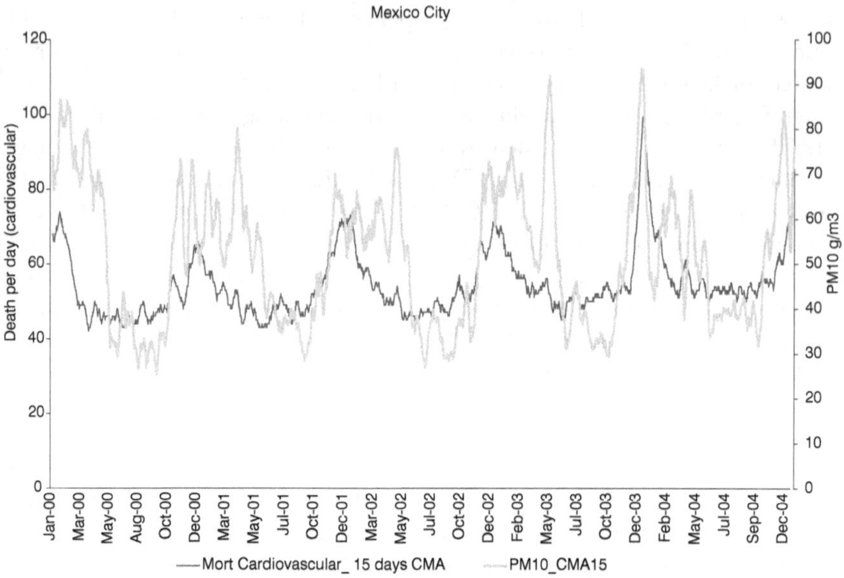

Figure 6.3 Mexico City: positive correlation between PM_{10} and mortality

showed a positive correlation between mortality and air pollution, associating air pollution in a city with more respiratory and cardiovascular related deaths. However, the exact patterns of the association differ by city and weather conditions. For instance, while adverse impacts of PM_{10} are especially evident during the cold season in all three cities, only in Mexico City were there positive associations between the levels of PM_{10} and mortality during the warm season (Romero-Lankao *et al.*, 2012b).

Scholarship tends to agree that the age, gender, and pre-existing health conditions may affect *sensitivity* to hazards. However, our measures of the relative risk of dying from exposure to PM_{10} show mixed and even unexpected results. For example, women and the elderly seem to be more sensitive than the total population only in some cases (e.g., respiratory mortality during the cold season in Mexico City), while they are equally or less sensitive in others (e.g., cardiovascular mortality in Bogotá and Mexico City). One possible explanation for this discrepancy is that neither women nor the elderly have a universal physiological susceptibility to air pollution and temperature, but rather that social conditions (e.g., occupation and equity in access to resources and social networks) are responsible for differential effects (Romero-Lankao *et al.*, 2012a).

We explored whether the health risks related to air pollution and temperature are equally distributed or spatially and socio-economically differentiated within cities. We found differences in socioeconomic vulnerabilities, as measured by the MVI, within each city. We found larger disparities in socioeconomic vulnerability in Bogotá (with MVI ranging from 0.06 to 0.72), than in the other two cities. The MVI index ranges from 0.37 to 0.69 and 0.33 to 0.62 between the least and the most vulnerable municipalities respectively in Mexico City and Santiago.

This suggests that wealthy populations may have the socioeconomic and political means, such as education, good quality houses and health services that allow them to escape from, or at least mitigate, some environmental risks.

From an environmental justice perspective, we would expect that spatial differences in health risks from environmental hazards relate to the socioeconomic characteristics of a population. However, within our studied cities, following the findings of other scholars studying different areas (Szasz and Meuser, 1997; Marshall, 2008), we have found this not always to be the case. In fact, only one district of Bogotá showed a positive correlation between the level of NO_2 and social vulnerability, and exposure to PM_{10} was not correlated with vulnerability in any of the three cities (Romero-Lankao *et al.*, 2013). Furthermore, some of both the most and the least vulnerable districts in the three cities are at similar relative risk of cardiovascular and respiratory mortality from exposure to PM_{10}. It follows that the spatial differences in socioeconomic vulnerabilities within cities do not necessarily correspond with the spatial distribution of cardiovascular and respiratory mortality rates and at times these results were quite unexpected. For instance, some of the least vulnerable districts in the three cities had the highest mortality rates, while several of the most deprived communities had the lowest. We suggest that this finding was driven, at least in part, by the nature of the hazards and scales we were examining given that, at the city level, atmospheric pollutants tend to move about freely, without regard to socioeconomic and neighborhood boundary.

Concluding remarks

This chapter offered some insights on a series of challenges and opportunities posed by the complexity of societal and environmental factors shaping urban vulnerability and risk. It pointed to some aspects of urban change, such as the unprecedented scale and rate of growth of urban populations in recent decades that may be of relevance for actions seeking to reduce urban vulnerability and enhance human security. For example, the fastest rates of urbanization are currently taking place in low- and middle-income countries, with the bulk of this growth taking place in smaller urban areas, and these processes offer both challenges and opportunities. Smaller urban areas, particularly in developing countries, are often institutionally weak, and unable to promote effective adaptation and development policies. At the same time, the flourishing development of these centers may offer an opportunity if they can be redirected in ways that can enhance their sustainability in general and their ability to effectively respond to hazards and stresses, for instance through the development of effective disaster management systems and climate proof urban infrastructures that can simultaneously satisfy the needs of rich and poor alike. The recent change in urban form and density may also be a source of challenges and opportunities. High densities can exacerbate many factors determining climate risks (e.g., heat-island effect, air pollution). These same factors, however, can become sources of opportunities for simultaneously improving health and cutting GHG emissions through policies related to transport systems and more effective urban planning.

The nuanced nature of urban vulnerability and risk poses challenges that are further highlighted by the characteristics of existing scholarship. For example, the dominance of one paradigm and the discrepancies in scope, methods and policy implications constrain researchers' ability to develop a more integrated understanding of the multiple dimensions and determinants involved. To fill the gap, the ADAPTE project included the design of an integrated framework to explore some of the key features and linkages among the dimensions and determinants of urban vulnerability and risk. Our findings suggest that researchers need to consider the attributes of the hazards to which urban populations are exposed. In our cities, for instance, the risks associated with high concentrations of PM_{10}, ozone and other criteria pollutants depended, first, on whether atmospheric and meteorological conditions were conducive to air pollutant retention, ozone formation and more intense heat islands. Each of our assessed hazards had different health impacts; while lower temperatures relate to higher cardiovascular and respiratory mortality (pointing to a negative mortality/temperature relationship), higher levels of PM_{10} relate to higher mortality (pointing to a positive mortality/ PM_{10} relationship).

Our findings on the risk of mortality among sensitive groups do not support existing literature that emphasizes a universal physiological susceptibility to air pollution and temperature within vulnerable populations such as the elderly or women. On the contrary, we found divergent trends in the data on how age and gender related to mortality risk. We suggest, hence, more empirical work needs to be done on the influence of occupation, education, equity in access to resources, cultural roles, social networks and other factors that may explain differences in risk mortality among sensitive groups.

Even considering these additional determinants of mortality risk, however, our study found that threats such as air pollution reach threshold levels at which the importance of socioeconomic factors falls away. Although wealthy populations in the three cities have access to education, good quality housing and health services to mitigate some environmental risks, overall, health risks from air pollution and temperature in the study cities do not necessarily depend on socio-economic differentiations. We suggest therefore, that health risks from atmospheric conditions and pollutants act without boundaries or social distinctions and show characteristics of a boomerang effect at least at the relatively small spatial scales of urban areas. This finding suggests that in a plausible future increased levels may interact with more intense urban heat islands, heat waves and other climate change effects, and pose risks to rich and poor alike.

In the final analysis, this may turn out to offer the greatest reason for hope, as anything perceived as a universal threat is more likely to gain the attention needed to drive effective action to reduce the urbanization drivers of emissions, vulnerability and risk. To be successful, policies will need to be developed to turn the hazards resulting from human pressures on the environment into sources of opportunities and innovations aimed at building more resilient and sustainable cities.

Note

1 "Not ist hierarchisch, Smog ist demokratisch" (Beck, 1986: 48).

References

Basu, R. (2009) 'High ambient temperature and mortality: A review of epidemiologic studies from 2001 to 2008', *Environmental Health*, 8 (1): 40.

Baud, I., Sridharan, N., and Pfeffer, K. (2008) 'Mapping urban poverty for local governance in an Indian mega-city: The case of Delhi', *Urban Studies*, 45: 1385–1412.

Beck, U. (1986) *Risikogesellschaft: Auf dem Weg in eine andere Moderne*, Frankfurt: Suhrkamp Verlag.

Bell, M.L., O'Neill, M.S., Ranjit, N., Borja-Aburto, V.H., Cifuentes, L.A., and Gouveia, N.C. (2008) 'Vulnerability to heat-related mortality in Latin America: A case-crossover study in São Paulo, Brazil, Santiago, Chile and Mexico City, Mexico', *International Journal of Epidemiology*, 37: 796–804.

Birkmann, J. (2006) 'Measuring vulnerability to promote disaster-resilient societies: Conceptual frameworks and definitions', in J. Birkmann (ed.) *Measuring Vulnerability to Natural Hazards*, New Delhi, India: TERI Press: 9–54.

Birkmann, J., Garschagen, M., Kraas, F. and Quang, N. (2010) 'Adaptive urban governance: New challenges for the second generation of urban adaptation strategies to climate change', *Sustainability Science*, 5 (2): 185–206.

Cardona, O.D., van Aalst, M.K., Birkmann, J., Fordham, M., McGregor, G., Perez, R., Pulwarty, R.S., Schipper, E.L.F. and Sinh, B.T. (2012) 'Determinants of risk: Exposure and vulnerability', in C.B. Field, V. Barros, T.F. Stocker, D. Qin, D.J. Dokken, K.L. Ebi, M.D. Mastrandrea, K.J. Mach, G.-K. Plattner, S.K. Allen, M. Tignor, and P.M. Midgley (eds) *Managing the Risks of Extreme Events and Disasters to Advance Climate Change Adaptation*, A Special Report of Working Groups I and II of the Intergovernmental Panel on Climate Change (IPCC). Cambridge, and New York: Cambridge University Press: 65–108.

Curriero, F.C., Heiner, K.S., Samet, J.M., Zeger, S.L., Strug, L. and Patz, J.A. (2002) 'Temperature and mortality in 11 Cities of the Eastern United States', *American Journal of Epidemiology*, 155: 80–87.

Dear, K., Ranmuthugala, G., Kjellström, T., Skinner, C. and Hanigan, I. (2005) 'Effects of temperature and ozone on daily mortality during the August 2003 heat wave in France', *Archives of Environmental & Occupational Health*, 60: 205–212.

D'Ippoliti, D., P. Michelozzi, C. Marino, F. de'Donato, B. Menne, K. Katsouyanni, U. Kirchmayer, A. Analitis, M. Medina-Ramón, A. Paldy, R. Atkinson, S. Kovats, L. Bisanti, A. Schneider, A. Lefranc, C. Iñiguez and C. Perucci. (2010). 'The impact of heat waves on mortality in 9 European cities: results from the EuroHEAT project', *Environmental Health* 9, DOI:10.1186/1476-069X-9-37.

Dodman, D. (2009) 'Blaming cities for climate change? An analysis of urban greenhouse gas emissions inventories', *Environment and Urbanization,* 21(1): 185–202.

Fouillet, A., Rey, G., Jougla, E., Frayssinet, P., Bessemoulin, P. and Hemon, D. (2007) 'A predictive model relating daily fluctuations in summer temperatures and mortality rates', *BMC Public Health*, 7: 114.

Füssel, H. (2007) 'Vulnerability: a generally applicable conceptual framework for climate change research', *Global Environmental Change*, 17: 155–167.

Füssel, H. M. (2009) *Review and Quantitative Analysis of Indices of Climate Change Exposure, Adaptive Capacity, Sensitivity, and Impacts*, Background Note developed for World Development Report 2010: Development and Climate Change, Potsdam: Potsdam Institute for Climate Impact Research.

Gallopín, G. C. (2006) 'Linkages between vulnerability, resilience, and adaptive capacity', *Global Environmental Change*, 16 (3): 293–303.

Gasper, R., Blohm, A. and Ruth, M. (2011) 'Social and economic impacts of climate change on the urban environment', *Current Opinion in Environmental Sustainability*, 3 (3): 150–157.

Holgate, C. (2007) 'Factors and actors in climate change mitigation: A tale of two South African cities', *Local Environment: The International Journal of Justice and Sustainability*, 12 (5): 471–484.

IPCC (2007) *Climate Change 2007: Impacts, Adaptation and Vulnerability*: Contributions of the working group II to the fourth assessment report of the Intergovernmental Panel on Climate Change. Cambridge: Cambridge University Press.

Ishigami, A., Hajat, S., Kovats, R.S., Bisanti, L., Rognoni, M., Russo, A. and Paldy, A. (2008) 'An ecological time-series study of heat-related mortality in three European cities', *Environmental Health: A Global Access Science Source*: 7: 50.

Jerrett, M., Burnett, R.T., Pope, C.A., Ito, K., Thurston, G., Krewski, D., Shi, Y., Calle and E. Thun, M. (2009) 'Long-term ozone exposure and mortality', *New England Journal of Medicine*, 360: 1085–1095.

Jordan, R., Rehner, J. and Samaniego, J. (2010) *Regional Panorama Latin America: Megacities and Sustainability*, Santiago de Chile: Economic Commission for Latin America and the Caribbean, United Nations.

Kienberger, S. (2010) 'Spatial vulneraiblity assessment: Methodology for the community and district level applied to floods in Buzim Mozambique', PhD Thesis, University of Salzburg .

Magrin, G., Gay García, C., Cruz Choque, D., Giménez, J.C., Moreno, A.R., Nagy, G.J., Nobre, C. and Villamizar, A. (2007) 'Latin America', in M.L. Parry, O.F. Canziani, J.P. Palutikof, P.J. van der Linden and C.E. Hanson (eds) *Climate Change 2007: Impacts, Adaptation and Vulnerability. Contribution of Working Group II to the Fourth Assessment Report of the Intergovernmental Panel on Climate Change*, Cambridge: Cambridge University Press: 581–615.

Makri, A., and Stilianakis, N. I. (2008) 'Vulnerability to air pollution health effects', *International Journal of Hygiene and Environmental Health*, 211: 326–336.

Marshall, J.D. (2008) 'Environmental inequality: Air pollution exposures in California's South Coast Air Basin', *Atmospheric Environment*, 42: 5499–5503.

McMichael, A.J.,P. Wilkinson, R.S. Kovats, S. Pattenden, S. Hajat, B. Armstrong, N. Vajanapoom, E.M. Niciu, H. Mahomed and C. Kingkeow. (2008) 'International study of temperature, heat and urban mortality: the 'ISOTHURM'project', *International Journal of Epidemiology*, 37 (5): 1121–1131.

Medina-Ramón, M. and Schwartz, J. (2007) 'Temperature, temperature extremes, and mortality: a study of acclimatisation and effect modification in 50 US cities', *Occupational and Environmental Medicine*, 64 (12): 827–833.

Medina-Ramón, M., Zanobetti, A., Cavanagh, D.P. and Schwartz, J. (2006) 'Extreme temperatures and mortality: Assessing effect modification by personal characteristics and specific cause of death in a multi-city case-only analysis', *Environmental Health Perspectives*, 114: 1331.

Michelozzi, P., De Sario, M., Accetta, G., de'Donato, F., Kirchmayer, U., D'Ovidio, M. and Perucci, C.A. (2006) 'Temperature and summer mortality: Geographical and temporal variations in four Italian cities', *Journal of Epidemiology and Community Health*, 60 (5): 417–423.

Molina, M. and Molina, L. (eds) (2002) *Air Quality in the Mexico Megacity: An Integrated Assessment,* Netherlands: Kluwer Academic publishers.

Moser, C. and Satterthwaite, D. (2010) 'Toward pro-poor adaptation to climate change in the urban centers of low- and middle-income countries', in R.M.A. Norton (ed.) *Social Dimensions of Climate Change: Equity and Vulnerability in a Warming World,* Washington, DC: The International Bank for Reconstruction and Development/The World Bank.

Muggeo, V. M. and Hajat, S. (2009) 'Modelling the non-linear multiple-lag effects of ambient temperature on mortality in Santiago and Palermo: A constrained segmented distributed lag approach', *Occupational and Environmental Medicine*, 66 (9): 584–591.

O'Brien, K., St. Clair, A. and Kristoffersen, B. (eds) (2010) *Climate Change, Ethics and Human Security*, Cambridge: Cambridge University Press.

O'Brien, K., Eriksen, S., Nygaard, L.P., and Schjolden, A. (2007) 'Why different interpretations of vulnerability matter in climate change discourses', *Climate Policy*, 7: 73–88.

O'Neill, M., Zanobetti, A. and Schwartz, J. (2005) 'Disparities by race in heat-related mortality in four US cities: The role of air conditioning prevalence', *Journal of Urban Health*, 82: 191–197.

Pelling, M. (2003) The vulnerability of cities: Natural disasters and social resilience, London: Earthscan.

Pope, C.A. and Dockery, D.W. (2006) 'Health effects of fine particulate air pollution: lines that connect', *Journal of the Air & Waste Management Association*, 56: 709–742.

Reid, C.E., O'Neill, M.S., Brines, S.J., Gronlund, C., Diez-Roux, A.V., Brown, D.G. and Schwartz, J. (2008) 'Mapping community determinants of heat vulnerability', *Epidemiology*, 19 (6): S229.

Romero-Lankao, P. and Gnatz, P.D. (2011) 'Urbanization and the challenge of climate change', in UN-Habitat (eds) *2011 Human Settlements Report on Cities and Climate Change, London: Earthscan.*

Romero-Lankao, P. and Qin, H. (2011) 'Conceptualizing urban vulnerability to global climate and environmental change', *Current Opinion in Environmental Sustainability*, 3: 142–149.

Romero-Lankao, P., Tribbia, J.L. and Nychka, D. (2009) 'Testing theories to explore the drivers of cities' atmospheric emissions', *Ambio* 38: 236–244.

Romero-Lankao, P., Qin, H. and Dickinson, K., (2012a) 'Urban vulnerability to temperature-related hazards: A meta-analysis and meta-knowledge approach', *Global Environmental Change*, 22 (3): 670–683.

Romero-Lankao, P., Qin, H., Hughes, S., Borbor-Cordova M., and Haeffner, M. (2012b) *Urban Vulnerability and Adaptation to the Health Impacts of Air Pollution and Climate Extremes in Latin American Cites, Urban Areas and Global Climate Change Research in Urban Sociology*, 12, UK: Emerald Group Publishing: 247–274.

Romero-Lankao, P., Qin, H. and Borbor-Cordova, M. (2013) 'Exploration of health risks related to air pollution and temperature in three Latin American cities', *Social Science & Medicine*, 83: 110–118.

Schipper, L. (2009) 'Meeting at the crossroads? Exploring the linkages between climate change adaptation and disaster risk reduction', *Climate and Development*, 1:16–30.

Seto, K.C., Sanchez, R. and Fragkias, M. (2010) 'The new geography of contemporary urbanization and the environment', *Annual Review of Environment and Resources*, 35: 167–194.

Smoyer, K.E., Kalkstein, L.S., Greene, J.S., and Ye, H. (2000) 'The impacts of weather and pollution on human mortality in Birmingham, Alabama and Philadelphia, Pennsylvania', *International Journal of Climatology*, 20: 881–897.

Stafoggia, M., Forastiere, F., Agostini, D., Biggeri, A., Bisanti, L., Cadum, E., Caranci, N., de'Donato, F., De Lisio, S., De Maria, M., Michelozzi, P., Miglio, R., Pandolfi, P., Picciotto, S., Rognoni, M., Russo, A., Scarnato, C., Perucci, C.A., (2006) Vulnerability to heat-related mortality: A multicity, population-based, case-crossover analysis', *Epidemiology*, 17: 315–323.

Szasz, A. and Meuser, M. (1997) 'Environmental inequalities: Literature review and proposals for new directions in research and theory', *Current Sociology*, 45: 99–120.

Thywissen, K. (2006) *Components of Risk: A comparative Glossary*, SOURCE No. 2. Bonn: United Nations University Institute for Environment and Human Security.

UN/ISDR (2009) *Terminology on Disaster Risk Reduction*, Geneva: United Nations International Strategy for Disaster Reduction.

UN-Habitat (2003) *The Challenge of Slums: Global Report on Human Settlements*, London: Earthscan Publications.

United Nations Human Settlements Programme (2009) *Planning Sustainable Cities: Global Report on Human Settlements 2009*, London, Sterling, VA: Earthscan.

Wheeler, S. and T. Beatley (2008) *Sustainable Urban Development Reader*, London: Routledge.

World Bank (2009) *World Development Report 2010: Development and Climate Change*, Washington, DC: World Bank.

7 Uncertainty as insecurity

Lessons on preparing for contingencies in Japan

Oscar A. Gómez

If there is a type of disaster for which the Japanese are prepared, it is earthquakes. In particular the Tohoku (northeast) region is shaken every thirty to forty years by a tremor of around 7.5 in magnitude, and since the last one was in 1978, most of its dwellers were, to a certain extent, expecting it. They did not know exactly when and where it was going to happen, but there was no doubt about its eventual occurrence. What was not known was that seemingly every thousand years a full 9.0 earthquake could trigger a massive tsunami well beyond contingency plans, and that nuclear power plants might not be ready to endure such stress.

There are plenty of ways to analyze catastrophes like this, e.g. as an engineering challenge, as an epidemiological case, as an economic shock, etc.. However one elemental view often overlooked is to study them as a case of humans facing uncertainty and the unknown. Such a perspective is important because realizing human security in a changing environment depends on the timely identification of threats so that harm can be avoided or minimized, and the uncertainty of contingencies plays against this objective. In this chapter, I use two paradigmatic emergencies that have affected modern Japan, Minamata disease in the second half of the twentieth century and the Fukushima nuclear crisis after the large earthquake in March 2011, to illustrate the connections between uncertainty and human security. This chapter argues that the human security literature needs to engage more broadly and systematically with the task of understanding the *processes* behind social recognition of threats, moving from plain trial and error approach to one that focuses on resilience.

The uncertainty of the threat

Uncertainty refers to the degree of incompleteness in our knowledge about, in this case, a threat. In fact, it could be said that some issues are threats precisely because of this ignorance. Rosa (cited by Renn, 2008) distinguishes two sources of this indeterminacy in terms of aleatory and epistemic uncertainty: the former refers to the inherent variability of events, similar to the fact that we never know the result of the next time we roll the dice; and the latter to a basic lack of knowledge. Thus, humans constantly strive to outwit their epistemic ignorance, yet randomness underlines the limits of this task. Renn adds that uncertainty is worsened by problems of *complexity*, an argument that constantly appears when talking about threats and their "non-linearity" or "tipping points," and of *ambiguity*, which relates to how different people interpret certain inputs – e.g. a small tremor can be an isolated phenomenon or the precursor of a stronger one, as was famously the case in Italy in 2009. These are not mutually exclusive problems, so each threat usually presents complexity, randomness, etc. in its own mixture.

In order to deal with the dynamics of uncertainty, it is important to understand its basic characteristics. Following Kuhlicke (2010), who studies vulnerability to natural disasters, there are unexpected events that fall beyond experience, anticipation or recognition; unknown unknowns or *nescience*, the maximum degree of uncertainty. It is not possible *a priori* to deal with these threats, but once people are exposed to them, the effects can become *non-knowledge*, something not properly understood but which people can try to assess in order to counteract. Success turns into knowledge, yet it is also possible that some of those symptoms are noted but later disregarded, what Kuhlicke (2010: 31–32) calls *negative knowledge*. As a consequence, the security of individuals and communities depends on the capacity to transform nescience into non-knowledge, and then into knowledge in order to react. It is important to note that this flow is not unidirectional from unknown to known. As de Vries (2007) observes, some threats regain their capacity to surprise because people forget about the risks they represent, as seems to be the case after tsunamis in Japan, when frequency can even jump generations.

These descriptions of uncertainty highlight the deficiencies in information when identifying and prioritizing threats. Thus, an alternative to studying uncertainty is to focus on the agents in charge of processing the information. As Foucault (2007) remarked on his own explorations around security, the example of the market is useful to identify the crux of the problem. The market is the place where the truth of the price is produced, where the different factors affecting goods valuation become a single number. Yet in understanding the market, the importance is not whether the price is true or false, but the regime of *veridiction* that enables us to assert what is true. Similarly, a deeper understanding of human security addressing contingency could be drawn from identifying where and how the truth of the threat is realized. While it is possible to come up with generic lists of these actors and institutions, we can spot more specific places to start this aim through human security literature.

Uncertainty and the study of human security

The first generation of human security studies – i.e. from 1994 to circa 2006 – did not engage with the problem of uncertainty. The unknown was, of course, part of the original proposition of human security, implicit in the original principle of interdependency (UNDP, 1994). The idea that the consequences of insecurity "travel the globe" is partly based on our ignorance about where and who exactly is going to be affected. Nevertheless, the initial proposition of human security was already troubled with the daunting task of gaining recognition, a task for which abstract discussions do no good. So the alternative was to present the new concept using a list of threats/securities. From there on, the first generation of human security studies can be identified by their frantic argument of whether it comprised only threats related to violence or about both fear and want. The epitome of this discussion can be found in the mixed messages coming from the United Nations Intellectual History Project Series, of which the human security volume produced by MacFarlane and Khong (2006) was totally devoted to support the fear view, while one of the chairs of the project, Richard Jolly, favored the broader view (Jolly and Basu-Ray, 2007).

Since then, researchers have gradually moved away from this discussion. Through the first decade of the twenty-first century very dissimilar threats have made it to the top of the security agenda, irrespective of the abovementioned intellectual impasse. Be it a pandemic disease, an unusually powerful hurricane, a terrorist attack, or an oil spill, emerging threats have shaken actors around the world and motivated priority of action within and beyond politics. Therefore, human security researchers have had to reconsider their question. One alternative is present in 2009 Thailand's human security report (UNDP, 2010), which devotes half of its effort to identify threats that are not currently critical in the country but can become a priority in the near future. Another notable example is the work of Mary Kaldor (2007) which concentrates on rethinking the traditional security apparatus in a way that is flexible enough to react to multifarious crises – whatever those crises are.

Work promoted by the UN University Institute for Environment and Human Security, as well as the work of Leichenko and O'Brien (2008; Leichenko and O'Brien 2007) are notable examples of places to start a formal analysis on the problem posed by uncertainty. After its creation, the UNU-EHS devoted its initial research to the connection between its two pillar concepts using the literature on vulnerability and risk. The study of both concepts takes into consideration the incompleteness of knowledge and the importance of social interaction when making decisions. Leichenko and O'Brien (2008) add to this a framework in terms of "double exposures", which analyzes the interaction of multiple variables in the emergence of catastrophic threats to humanity.

While the work of Leichenko and O'Brien (2008) is more concerned with the interaction of globalization and global environmental change, they provide a precious insight into the question of dealing with uncertainty by pointing out Amartya Sen's differentiation of opportunity and *process* aspects of freedom (1999, 2002). Recently, Sen himself (2009) has made explicit the connection of process and human security, equating it to his famous finding about famine prevention. That is, when Sen claims that no famines have taken place in a *functioning democracy with active opposition parties* and a *free press*, he is pointing out the importance of the processes behind threat recognition, even over variables that seem intuitively more related to a famine – e.g. droughts and price hikes. But are these two institutions, democracy and a free press, enough for other threats? Can we use emerging threats to broaden the spectrum of essential security processes? Observe, for instance, that for Kaldor (2007) it is not democracy but a legitimate political authority that it is more important in the short term to actualize human security.

Satisfactory answers to such questions are beyond the reach of this chapter, yet in the remaining pages I point out some attractive overtures for future studies.

Before that, one methodological comment: Sen makes clear that there are two components of the process aspect of freedom: personal and systemic (1999). The former refers to processes inside an individual's sphere and the latter to the general institutions of society. These two are interrelated because some personal processes may reflect social institutions, while persons may also have the opportunity to influence the system, yet the study of each of these requires different methodologies. In the following pages I adopt a systemic approach, drawing on a meta-analysis of the lessons from contingency cases in Japan. Such an approach has to be complemented with a personal viewpoint (see Kuhlicke (2010)), which stresses the agency of people confronting threats, and thus balances the patronizing nuance of protective security with considerations of autonomy and empowerment.

Pollution strikes: Minamata disease

As the first "developmental state," Japan presented some of the most acute environmental problems related to industrialization during the last century. Minamata disease is one of the four big pollution diseases of Japan, which hit the archipelago starting in the 1950s, leading to the creation of the environment as a field of study and a branch of government, a movement that gradually disseminated worldwide. Minamata disease is the result of methyl-mercury poisoning through the consumption of contaminated fish, affecting principally the central nervous system. The pollutant produces chronic and acute effects, as well as teratogenic sequels – namely, cerebral palsy – greatly enlarging the magnitude of the harm. The cause of the Minamata outbreak was the effluent of a local acetaldehyde plant, indiscriminately poured into the city's bay, where local fishermen got their catch. Initially, cats started dying, and then several residents, mostly fishermen, were admitted to local hospitals, presenting poor motor coordination, sensory disturbances in the extremities, and the loss of speech and hearing capacity. As the intoxication worsened, those symptoms intensified, with the patient's condition deteriorating until death.

The government acknowledged the outbreak in May 1956, and this led to the formation of research teams and mediating commissions, as well as mobilization by victims. But the protest was appeased in 1959 through paltry payments to patients and fishermen, while the cause of the ailment remained unaddressed, precisely because of the uncertainty surrounding it (JICA, 2005). However, after a second outbreak in Niigata in 1965, manipulation of the research in the initial phase was revealed and social movements regained their vigor. A lawsuit was filed in 1967 by Niigata patients, and a ruling was won in 1971; Minamata patients filed their case in 1969 and the verdict was rendered in 1973. In Niigata, 690 persons were certified and compensated, and 2,265 in Minamata. The story is far from over, since the process of patient certification presents multiple complications, but the elements so far presented are enough for the analysis here.

One of the most detailed accounts of the events surrounding Minamata disease is framed in terms of democracy: George's "Minamata: Pollution and the Struggle for Democracy in Postwar Japan" (2001) makes explicit how difficult it could be for a vulnerable population to gain recognition while the priority of the state is elsewhere. The work of George confirms Sen's conclusion over the importance of democracy, and shows how it was inhibited in Minamata. Locally, the company that owned the acetaldehyde plant was too big to fail – in 1954 it accounted for 45.5 percent of the city's total tax income (George, 2001: 35) – so patients could not get recognition, while nationally, opposition parties did not make use of the disease issue to generate action. In this sense, George notes the failure of Japanese newspapers to raise general attention, which either ignored the issue or took a "neutral" stance, benefiting the company. Under such conditions, the meaning of democracy overlaps with Kaldor's (2007) legitimate political authority, which allows patients to fight for recognition, a struggle that becomes the turning point for threat recognition, even if it is *ad hoc* as George concludes. In the end, the movement produced deep changes in the Japanese system – i.e. the official concern for *the environment* – and elsewhere.

Nonetheless, the other institution highlighted by Sen (1999), the free press, needs to be reframed broadly to reflect the experience in Minamata. There the bigger picture of the struggle was hindered and resolved by different means of communicating the truth of the threat. On the one hand, central to the fifteen-year delay in recognizing the gravity of the situation was the uncertainty of the disease's cause. The scientific establishment was co-opted to favor confusion, with Tokyo professors and ministries advancing hypotheses about the cause of the disease unrelated to the company, thus preventing intervention in the plant, while local experts were deemed unscientific by academic elites (George, 2001: 45–70). Expert knowledge generated *in situ* was a moral support for the struggle, arguably a must for modern technologically-influenced threats, but it was not enough. On the other hand, Minamata disease is renowned for its photographs (Smith and Smith, 1975), among other artistic records, which helped raise the issue among the national and international public through pamphlets and other means. In short, the *process* of modern threat identification and prioritization goes beyond the free press, encompassing the society's complete knowledge apparatus, of which the arts play an important role.

There are at least two other processes shaping the reaction to uncertainty that deserve to be mentioned briefly. First, it is outstanding that patients wait for more than ten years before considering legal action. Traditionally in Japan, the courts have been seen as an external institution, so conflicts tend to be resolved through informal mechanisms. The case of Minamata disease suggests that threat identification does not necessarily depend on whether or not there is a functioning rule of law, but rather on the features of the actual conflict resolution mechanisms, on whether they encourage the system to react after the stress generated by new threats. If a company can pay its way out of a disease, as with the paltry payments made to the first Minamata patients, it is most likely that no new legislation or other action would be generated to address similar threats in the future.

Second, each society implicitly consents to some sacrifices in practice, for example deaths, or at least each reacts differently depending on the circumstances. Some of the patients from Minamata, mainly fishermen, resigned themselves to their disease, contemplating the idea that they were the necessary price to pay for the country's economic success (Upham, 1976); this attitude was reinforced by the Japanese "iron triangle" in power – i.e. politicians, bureaucrats and companies – which made it very clear that GDP growth was above anything else – as stated in the first version of environmental legislation (JICA, 2005: 150–151). Literature on risk appraisal recognizes the role of values when identifying threats (Renn, 2008: 3–4), but a more specific inquiry on death and justified loss can make the analysis more accurate.

Uncertainty in Fukushima

The uncertain effects of a new pollution are threatening the Japanese archipelago once more, but this time the social interactions have been quite different. Radiation from the Fukushima power plant has reminded us that all threats are double-edged: there is the threat and the ensuing panic. It was the same for Minamata patients who did not gain support from their neighbors in part because the unknown disease made pariahs of them. Yet, after the Fukushima crisis the broader knowledge apparatus has played the expected crucial role. Because of doubts about the government's transparency, universities and other domestic and international organizations promptly made their own radiation measurements public. One week after the earthquake, Tohoku University already had daily measurements of radiation on its home page, helping reassure the population in Miyagi, eighty kilometers away from the plant. Mass media were full of experts from different backgrounds offering different judgments about the situation, helping to recover a minimum of trust; meanwhile companies gradually made meters available for common citizens who wanted to confirm their exposure themselves. It is not surprising then to find that the most salient cases of panic were mainly in China and California. Also worth mentioning is that this time broad, long-term epidemiological follow-ups are ongoing in order to properly evaluate the effects of the exposure. A lack of such long-term epidemiological study has deeply undermined reconciliation efforts in Minamata.

Another important highlight has been the emergence of anti-nuclear social movements in Japan. While politicians find it difficult to take drastic measures given their closeness to the industry and reconstruction work, once more citizens on the streets are likely to affect the future policies of the country. These groups constantly react to news about the impacts of the disaster in Fukushima, and may guarantee that no symptom goes unnoticed. Unsurprisingly, it took la ong time before affected populations considered using courtrooms, the victims patiently waiting for the top-down calculations to take place – though concerns over the victim's certification process were quickly voiced (Asahi Shinbun, 2011), a part of the recovery process that also critically affected reconciliation in Minamata (Harada 2004).

Finally, this time the emergency seems less prone to inflict direct casualties. Though it is too early to know the fate of the workers heavily exposed during the peak of the emergency, there is no shame on the victims' side, and signs of stigma have been heavily criticized by the mass media. Alas, the same cannot be said about food products from the affected areas: fears keep affecting consumption but not every product can be tested; farmers and fishermen have to make a living; not to mention that low prices and the invisibility of the threat can allow unscrupulous practices. Society's resilience has to prove its strength once more against uncertainty.

Tackling uncertainty: the process of threat recognition

The cases of Minamata and Fukushima offer precious insights about how to engage human security in the study of uncertainty and the emergence of new threats. Both cases illustrate how the unknown can hit societies and the different *processes* that converge while trying to regain security and find justice. It is framed in terms of *processes* because the novelty of the situations challenges existing divisions of labor inside society and among different knowledge branches. For instance, before Minamata the environment was not something the government and citizens had to deal with explicitly; Fukushima affected what we regard as energy security. Human security studies have to go beyond traditional threats to security, and considering the state its sole provider, to make the most of these two learning experiences.

An appealing starting point to tackle uncertainty is expanding the *processes* identified by Sen to prevent famine around the world: democracy and the free press. Democracy has been the major focus of attention after Minamata and Fukushima, more related to the possibility of protest than to the role of opposition parties,[1] but it is hardly the whole story of human security under pervasive uncertainty. The other process, the free press, has to be understood in a wider perspective of how society constructs knowledge, including the prominent role played by the arts and humanities. Informal conflict resolution specific to each society, as well as attitudinal change towards community development vis-à-vis individual sacrifice, are other candidates for core *processes* behind social threat recognition. Human security researchers should take note as this may contribute to enlarging our understanding of threat identification to tackle the unknown.

Note

1 Fukushima and social movements have received a great share of the international attention after 2011 disaster. One extensive compilation of materials can be found at http://www.japanfocus.org/japans-3.11-earthquake-tsunami-atomic-meltdown (accessed 6 October 2012).

References

Asahi Shinbun (2011) 'TEPCO should cut nuclear compensation red tape', editorial 22 September. Online. Available HTTP: http://ajw.asahi.com/article/views/editorial/AJ2011092211613 (accessed 6 October 2012).

de Vries, D.H. (2007) 'Being temporal and vulnerability to natural disasters', in K. Warner (ed.) *Perspectives on Social Vulnerability*, Bonn: UNU-EHS.

Foucault, M. (2007) *Nacimiento de la biopolítica* [The birth of biopolitics], Buenos Aires: Fondo de Cultura Económica.

George, T. (2001) *Minamata: Pollution and the Struggle for Democracy in Postwar Japan*, Cambridge: Harvard University Press.

Harada, M. (2004) *Minamata Disease*, Minamata: Minamata Disease Patients Alliance.

JICA (2005) *Japan's Experiences in Public Health and Medical Systems – Towards Improving Public Health and Medical Systems in Developing Countries*, Tokyo: JICA.

Jolly, R. and Basu-Ray, D. (2007) 'Human security – national perspectives and global agendas', *Journal of International Development*, 19 (4): 457–472.

Kaldor, M. (2007) *Human Security. Reflections on Globalization and Intervention*, Cambridge: Polity.

Kuhlicke, C. (2010) 'The dynamics of vulnerability: some preliminary thoughts about the occurrence of "radical surprises" and a case study on the 2002 flood (Germany)', *Natural Hazards*, 55 (3): 671–688.

Leichenko, R.M. and O'Brien, K.L. (2008) *Double Exposure: Global Environmental Change in an Era of Globalization*, New York: Oxford University Press.

MacFarlane, S.N. and Khong, Y.F. (2006) *Human Security and the UN*, Bloomington, IN: Indiana University Press.

O'Brien, K.L. and Leichenko, R.M. (2007) *Human Security, Vulnerability and Sustainable adaptation*, Background paper for the UNDP Human Development Report 2007, UNDP: New York.

Renn, O. (2008) *Risk Governance*, London: Earthscan.

Sen, A. (1999) *Development as Freedom*, New York: Anchor Books.

Sen, A. (2002) *Rationality and Freedom*, Cambridge, MA: Belknap Harvard.

Sen, A. (2009) *The Idea of Justice*, Cambridge, MA: Belknap Harvard.

Smith, W.E. and Smith, A. (1975) *Minamata*, New York: Holt, Rinehart and Winston.

UNDP (1994) *Human Development Report: New Dimensions of Human Security*, New York: United Nations Development Program.

UNDP (2010) *Human Security, Today and Tomorrow – Thailand Human Development Report 2009*, Bangkok: United Nations Development Program.

Upham, F. (1976) 'Litigation and moral consciousness in Japan: an interpretative analysis of four Japanese pollution suits', *Law and Society Review*, 10 (4): 579–619.

8 Human security at risk

Development impacts of global environmental change in drylands

Paul L. Lucas, Marcel T.J. Kok, Henk B.M. Hilderink and Matthias K.B. Lüdeke

Introduction

Where human development is the process of enlarging people's choices by expanding human capabilities and functioning (Alkire, 2010; UNDP, 2010), human security is a variable condition under which people and communities have the capacity to manage stresses according to their needs, rights and values (Barnett *et al.*, 2010). In this sense, the concept of human security complements the concept of human development by focusing on basic human needs and addressing the concern for steady fulfillment of these needs. Reframing environmental change as an issue of human security requires analyzing environmental change within the wider context of the co-occurring economic, social, institutional, political, cultural, and technological changes taking place in the world today (O'Brien, 2006).

Human security is influenced by slowly evolving and interrelated changes in population, in economic development and in the environment. Unmitigated global environmental change, including climate change, land degradation and water scarcity, increase pressure on the well-being of especially the poor (MA, 2005; IPCC, 2007; UNEP, 2007; World Bank, 2008). As some of these environmental changes will only become apparent in the longer term, they increasingly may backlash on progress already made. Therefore, to ensure that development is achieved in a sustainable manner, these changes need to be taken into account in designing development policies (Bourguignon *et al.*, 2008).

This chapter assesses the role of the environment as an important precondition for human security. Two distinct but complementary analysis and related methodologies are used, with child and infant mortality as indicators of human security. To examine long-term prospects for human security, we focus on future developments in access to food, clean drinking water, basic sanitation and modern energy sources, and their implications for child mortality. This analysis is done on a largely aggregated regional scale, using the Global Integrated Sustainability Model (GISMO) (Hilderink *et al.*, 2008). As this global analysis only broadly provides insight into the underlying health risks, we complement the analysis by assessing the mechanisms that contribute to increasing vulnerability. This analysis is done on a more disaggregated subnational scale, using the methodology of archetypical patterns of vulnerability (UNEP, 2007; Kok and Jager, 2009). This methodology is based on the notion that there are similar vulnerability situations around the world that also share similar features in terms of mechanisms that drive them. We focus on the vulnerability of small-holder farmers in dryland areas, as these are amongst the most critical, from a development and environmental perspective, with high infant mortality rates and a fragile environment (Levy *et al.*, 2005).

Methodology

Global Integrated Sustainability Model (GISMO)

The Global Integrated Sustainability Model (GISMO1.0) operationalizes the concept of sustainable development by interlinking global environmental change and human development (Hilderink *et al.*, 2008). It addresses quality of life that results from changes in human, economic and environmental domains, using the systems approach to describe these complex dynamic systems and their interrelations. We applied the model in combination with the IMAGE model (Bouwman *et al.*, 2006), to address a broad range of global environmental change issues related to human development.

Quality of life is modeled focusing on poverty, education, human health and their underlying dynamics, linked through a cohort-component population model. Here, we focus on the role of food (malnutrition), water (access to safe drinking water and basic sanitation) and energy (use of solid fuels for cooking and heating) with respect to human health (child mortality).

The occurrence of malnutrition is determined by food availability, women's education and status relative to that of men, and access to save drinking water (Smith and Haddad, 2000). Access to safe drinking water and basic sanitation are determined by per-capita income, levels of urbanization and population densities, distinguishing between no access, access to an improved source and household connection (Cairncross and Valdmanis, 2006). For solid fuel use, a simple function using per-capita income as an explanatory variable is used (Mehta *et al.*, 2006). Child mortality is determined according to the causal chain, from health-risk factors towards disease and eventually death (WHO, 2002), based on socio-economic (e.g., poverty, education) and environmental health risks (e.g., child malnutrition, unsustainable access to safe drinking water, basic sanitation and solid fuel use).

Archetypical patterns of pulnerability

The analysis of the prospects for child mortality provides us with insights into trends in world regions under different scenarios. This, however, does not tell us much about the dynamics within specific human-environment systems. Vulnerability analysis is a valuable tool for a better understanding of these dynamics and for identifying potential impacts on human security caused by socio-economic and environmental changes. Such an analysis contributes to a better understanding of local conditions as it examines exposures and sensitivities of people to these changes, and their ability to cope with or adapt to the potential impacts (Turner *et al.*, 2003).

Recognizing the need and potential for looking at similarities between related situations around the globe, a method was developed to identify and analyze "archetypical patterns of vulnerability" (Jäger and Kok, 2007: 317), which generalizes the outcomes of local vulnerability studies (here we primarily use meta-studies) and builds on global assessment insights and tools. The method looks for common vulnerability creating mechanisms in a multitude of cases that, in that sense, are archetypical. It flags the most vulnerable groups, and provides regional perspectives and connections between regions within a global context. A pattern of vulnerability is defined as: "a specific representative pattern of the interactions between environmental change and human well-being" (Jäger and Kok, 2007: 317). The method not only looks at environmental changes but also includes the wider socio-economic context in which these changes take place, addressing the integrated human-environment system from a global perspective.

Taking the most important common properties and dynamics from case studies and meta-studies, the core mechanisms that cause specific vulnerabilities are derived. Next, these core mechanisms are quantified using indicators from global Integrated Assessment Models, here GISMO and the IMAGE model. In the final step, different manifestations of the patterns (so called vulnerability profiles) and their geographic locations are identified through cluster analysis (Janssen *et al.*, 2012). See Kok *et al.* (2010) for a further elaboration of the methodology and its application on different human-environment systems.

Prospects in child mortality

Child mortality is an important indicator of human security. Sustainable access to food, safe drinking water and basic sanitation (water) and modern energy sources (energy) may decrease child mortality significantly (PBL, 2009). Of the estimated 8.8 million annual child deaths in 2008, the vast majority were related to preventable and treatable infectious diseases and conditions that are related to their environment, including the use of solid fuels (fuel wood, dung, coal and charcoal) for cooking and heating (pneumonia) and unsafe drinking water and the lack of basic sanitation (diarrhea) (Black *et al.*, 2010) with malnutrition (underweight) being an important underlying factor responsible for about 35 percent of child deaths (Black *et al.*, 2008). Here, we assess future development in access to food, water and energy and the related impacts on child mortality. Reducing child mortality is also one of the eight Millennium Development Goals (MDGs): reduce by two-thirds, between 1990 and 2015, the under-five mortality rate.

In general, future projections are based on a range of related developments, all encompassing major uncertainties. To assess future developments in child mortality and progress towards achieving the related MDG target, we used three scenario variants, mainly differing in economic growth and food availability. Variant I is based on FAO projections (FAO, 2006) that uses data on economic growth from the World Bank (2005) that shows medium economic growth compared to the other variants. Variant II is based on the baseline scenario of the OECD Environmental Outlook (MNP/OECD, 2008), which shows the highest economic growth. Variant III is based on the reference scenario of the International Assessment of Agricultural Science and Technology for Development (IAASTD, 2009), which shows the lowest economic growth. In all three variants, the population is calibrated on the UN medium projection (UNDESA, 2007).

For the three variants child malnutrition (food), access to safe drinking water and basic sanitation (water), and access to improved energy sources (energy) are projected to increase significantly, with main improvements in East Asia and the Pacific. However, in absolute terms, improvements are much less, particularly in Sub-Saharan Africa and South Asia, and especially with respect to energy, as the relative improvements in access levels are largely compensated by large population growth rates. As a result of the improvements overall child mortality is projected to decrease significantly (as is shown in Figure 8.1, for Variant I). Nevertheless, in Sub-Saharan Africa and South Asia a lack of these basic services is still responsible for many child deaths. In Variant I, by 2030, about 65 in every

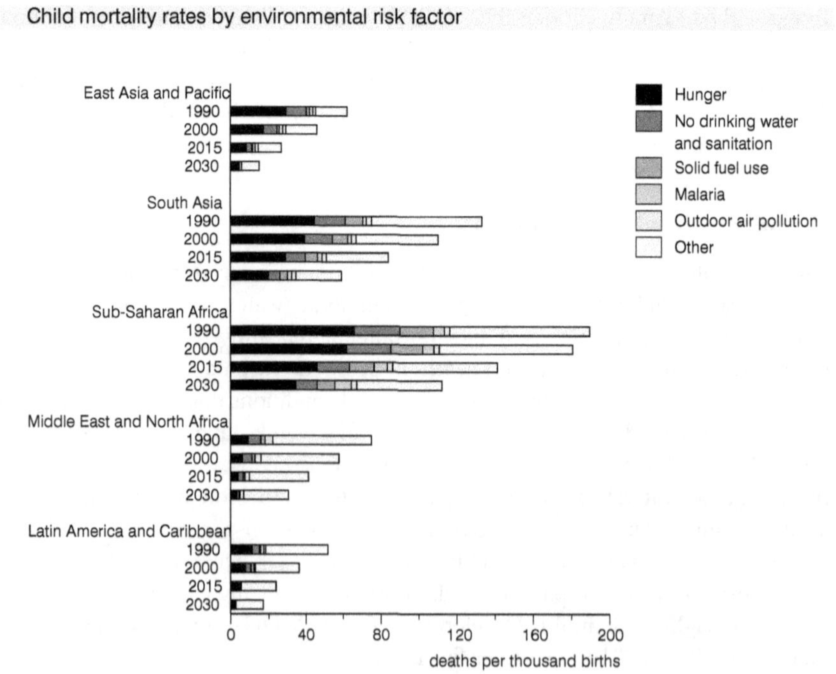

Figure 8.1 Projected child mortality due to inadequate access to food, safe drinking water and basic sanitation, and modern energy sources, for variant I. Source: PBL (2009)

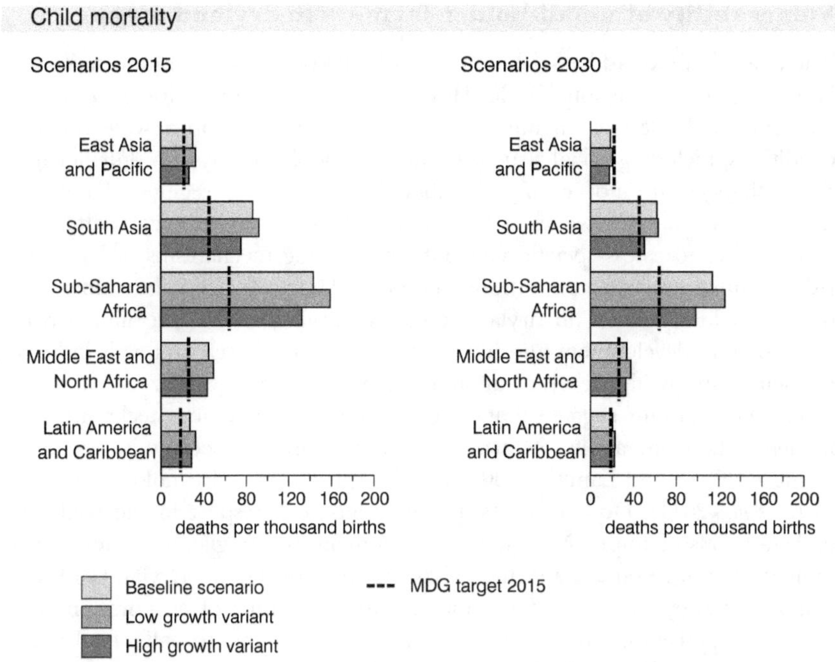

Figure 8.2 Projected child mortality levels for the three variants, for 2015 and 2030. Source: PBL (2009)

1,000 children are still expected to die before the age of five, with more than 45 percent of these deaths related to inadequate access to food, water and energy.

Figure 8.2 presents child mortality for 2015 and 2030, in all three variants. It clearly shows that, although substantial improvements are projected, the improvements in access levels are far from enough to reach the MDG target. Furthermore, only a few regions are projected to reach this target by 2015. High food prices in Variant III with low economic growth have a huge impact on overall child mortality, especially in Sub-Saharan Africa, while the much stronger economic growth in Variant II remains insufficient for reducing child mortality by two-thirds, by 2030.

The analysis of child mortality prospects is exemplary of the entry points where global environmental change may have an impact on human security. Global environmental change may potentially affect future access to food, water and energy, thereby undermining human security. The following section explores the dynamics of environmental change and human security, in more detail, by focusing on small-holder farmers in dryland areas.

Vulnerability of small-holder farmers in dryland areas

The analysis above addresses human security by projecting child mortality on a largely aggregated regional scale. However, the impact of socio-economic and environmental changes on human security is dependent on a wide range of conditions, including social structures and the biophysical system. Furthermore, these changes are not evenly distributed across world regions. To unpack the regional analysis, we used the methodology of archetypical patterns of vulnerability to assess specific vulnerability creating mechanisms, i.e. potential risks to human security, on a more local scale. Here, we assess the vulnerability of smallholder farmers in dryland areas. In dryland areas, the link between environment, development and human security is highly relevant as is indicated by their relatively high levels of infant mortality (Levy *et al.*, 2005).

Current literature suggests that there are typical and common mechanisms that influence the vulnerability of farmers in dryland areas, especially in developing countries (Geist and Lambin, 2004; Jäger and Kok, 2007; Reynolds *et al.*, 2007; Sietz *et al.*, 2011). Dryland areas are hot spots with respect to the challenges and trade-offs of improving human development in a fragile environment with limited natural resources and high risks of overexploitation. Dryland areas are characterized by low levels of precipitation and high rates of evaporation. They occupy 41 percent of the earth's land area and are home to half of all people living in poverty (Dobie, 2001). Most dryland countries have a large proportion of their labor force working in the agricultural sector, with the poor highly dependent on environmental services. Land degradation and climate change endanger agricultural production and environmental sustainability.

The pattern of vulnerability is characterized by increasing pressures on natural resources from a growing population, limited and insecure access to water and fertile soils, and soil degradation resulting from overuse. Combined with the breakdown of traditional coping mechanisms and barriers to alternative livelihoods (Geist and Lambin, 2004; Safriel *et al.*, 2005; UNEP, 2007) this drives a negative feedback on agricultural production and income generation (e.g. Safriel and Adeel, 2008). Poor infrastructure impedes market access (e.g. Shiferaw *et al.*, 2008), which together with the unavailability of capital influences the improvement of agricultural productivity (Twomolow *et al.*, 1999; Thomas, 2008). All these factors lead to situations in which rural households potentially become enmeshed in poverty traps.

A set of seven global indicators has been identified, as proxies for the basic vulnerability creating mechanisms and their impacts on human security. For the sake of availability, sub-national data on infant mortality were used instead of those on child mortality (CIESIN, 2005). Cluster analysis of these indicators has distinguished eight vulnerability profiles, six of which relate to dryland areas in developing countries (Figure 8.3).[1] These vulnerability profiles define different vulnerability patterns, i.e. combinations of indicator values that shape specific vulnerabilities in different dryland areas. Below we discuss the different vulnerability patterns and their spatial distribution (Figure 8.4), supplemented with relevant case studies taken from the work of Sietz *et al.* (2011).

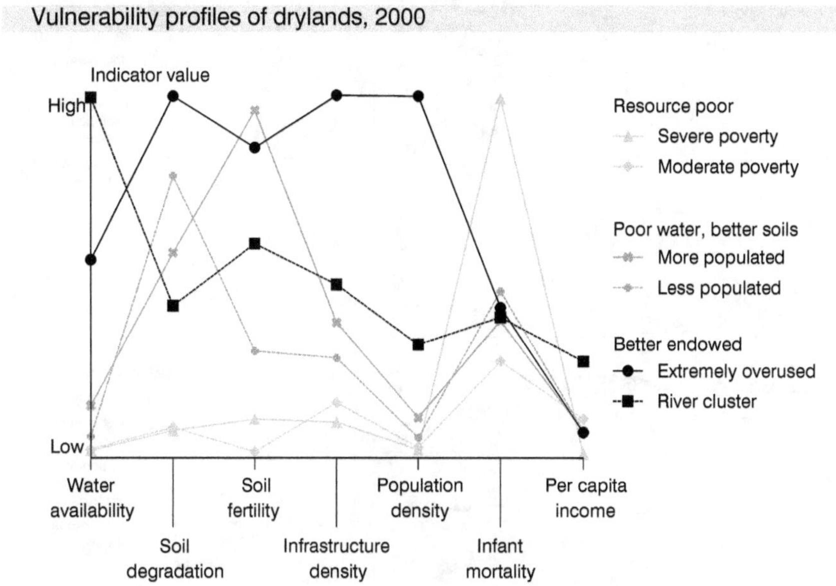

Figure 8.3 Vulnerability profiles of small-holder farmers in dryland areas. Source: Kok *et al.* (2010)

The two *resource-poor* vulnerability patterns identify the most resource-constrained and isolated areas of the world. The harsh dryland conditions explain the still relatively low levels of soil degradation, as agricultural and grazing practices are not favored. The very limited renewable water resources pose the risk of groundwater overuse. The two profiles differ mainly in level of human well-being (income and infant mortality). Severe poverty occurs in areas dominated by pastoral land use. More moderate poverty occurs in zones between pastoral and sporadic, sparse forms of agriculture on the desert fringes and in areas where national economies allow for improved living conditions, for example, because of fossil-fuel exploitation. A good illustration of the severe poverty pattern is the case study of Le Sage and Majid (2002) in Somalia. The poorest people there are not able to benefit from occasionally better rainfall due to the depleted asset base and war-related constraints on access to productive resources. Even though better situated people may produce more crops, debt repayment and recurrent droughts continue to exhaust their livelihood assets.

The two *poor water, better soils* vulnerability patterns are less resource-constrained, but are confronted with much higher soil degradation. They differ mainly in agricultural conditions and population densities. More favorable agricultural conditions are combined with higher population densities, but also with higher soil degradation, while poverty levels are similar. These patterns are found in parallel areas, neighboring deserts, with the less populated areas closer to the desert, corresponding with a land-use gradient from pastoral to agro-pastoral. The more populated pattern is illustrated by Kassahun *et al.* (2008) in East Ethiopia, where rangeland degradation has increased in severity and magnitude since the 1970s.

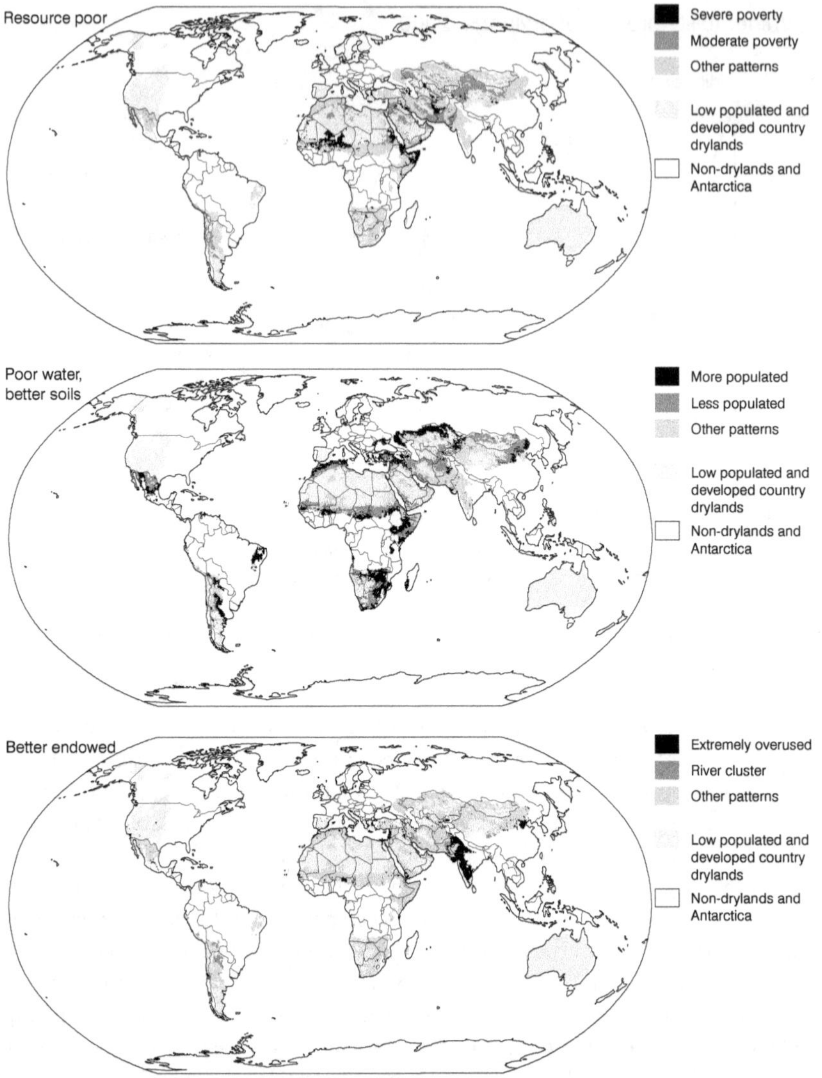

Figure 8.4 Distribution of vulnerability patterns of small-holder farmers in dryland areas in developing countries. Source: Kok *et al.* (2010)

Overgrazing and overexploitation of woody plants accelerate the pace of soil degradation, while also water bodies were affected by agricultural activities. This ongoing overuse of natural resources induces declining agricultural yields, food insecurity and increased poverty, and generates conflicts over grazing areas and water resources.

The *better endowed* vulnerability patterns are least resource-constrained. However, differences are more distinct than in the previous two groups of patterns. Although, the resource situation is best in the *extremely overused* vulnerability pattern, the human well-being is comparable to the *poor water, better soils* patterns. Furthermore, soils have been extremely degraded by a very dense population, also putting future generations under increased pressure. In contrast, the *river* vulnerability pattern is best endowed with water resources, while soil degradation is moderate in comparison. However, the pattern combines relatively high income levels with relatively high infant mortality rates, suggesting a very uneven distribution of income opportunities, probably due to differences in access to irrigation and grassland. The former pattern dominates the arid areas of India, but is also found in north-eastern China and on the African Mediterranean coast. The latter pattern is found around the lower reaches of the Indus, Euphrates, Tigris and Volga rivers, and in other irrigation areas, such as around the Aral Sea. A good illustration of the *extremely overused* pattern is the case study of Ram *et al.* (1999) describing the situation in Khabra Kalan (Rajasthan). The study shows increasing population, shrinking land holdings and shortfall of food on small farms which results in the deterioration of the land productivity. Mustafa and Qazi (2007) verify the conclusions drawn from the *river* vulnerability pattern with a case study from Balochistan (part of the Indus basin). They show how the transition from a sustainable, traditional irrigation system ("karez") to groundwater pumping leads to increasing social disparities and degradation of environmental resources.

Increasing vulnerability due to global environmental change

Unsustainable use and increasing competition over natural resources (including land, water and fossil energy sources), as well as global environmental changes (including climate change and biodiversity loss) increase the pressure on already stressed livelihoods and will increase vulnerabilities – especially in dryland areas where the resource situation is generally constrained. Countries facing the most serious environmental degradation are also making the slowest progress toward achieving the MDGs (MA, 2005). Furthermore, deteriorating environmental conditions can lead to fewer opportunities for generating income, and exhaust survival strategies (Oxfam, 2006). Finally, resource scarcities increase the pressure on available resources, putting upward pressure on prices of especially food and energy (PBL, 2011).

Climate change eventually reduces crop yields and increases the likelihood of short-term crop failure and long-term production declines (IPCC, 2007). The most vulnerable are poor and food-insecure countries at lower latitudes, especially in seasonally dry and tropical regions that largely depend on rain-fed farming (IPCC, 2007). In drylands, this is especially the case for the *resource poor* pattern for Sub-Saharan Africa and the *extremely overused* pattern for South Asia, but people in the *poor water, better soils* pattern also should take these future changes into account.

Increased land use and industrial activities impact water quality, while climate change may reduce water availability through shrinking glaciers, more frequent dry seasons, increased evapotranspiration and water quality deterioration (Bates *et al.*, 2008). In 2006, 1.2 billion people experienced physical water scarcity, and another 1.6 billion people economic water scarcity (Comprehensive Assessment of

Water Management in Agriculture, 2007). Physical water scarcity mainly results in competition between different uses and is a major concern for future agricultural opportunities in the *resource poor* and *poor water, better soils* patterns for North Africa, the Middle East and Central Asia. Economic water scarcity includes underdeveloped water infrastructure with high vulnerability to droughts that not only impact on agriculture but also on access to save drinking water. People that will be most affected are those living in already water-stressed basins, while problems will be most critical in economically depressed areas, where water stress is increased by socio-economic factors. This mainly includes the *resource poor* and *poor water, better soils* patterns for Sub-Saharan Africa and the *extremely overused* pattern for South Asia.

Although fertility rates are dropping all over the world, population increase is projected to remain high in underdeveloped areas in Sub-Saharan Africa and South Asia (UNDESA, 2007). This is especially a challenge for the *extremely overused* pattern for South Asia where population pressure is already very high. In the *resource poor* and *poor water, better soils* patterns for Sub-Saharan Africa resource availabilities seem to limit opportunities for improving human well-being on the basis of agricultural production, let alone with an increasing population. Combined population and environmental pressures may force people to migrate to better endowed areas, impacting on the development opportunities there, or drive them towards cities that not necessarily provide good income opportunities for these migrants.

Entry points to strengthen human security in drylands

The variety in vulnerability patterns suggests that there is no single set of options for improving human security in dryland areas. Unsustainable use of natural resources, increasing competition over fertile land and water resources, and unmitigated global changes increase the pressure on already stressed livelihoods, as is shown in the previous section. Some of these pressures will only become apparent in the long term – after 2015 or even 2030 – potentially having a backlash on development progress. Therefore, policies fostering increased access to food, water and improved energy sources should address the human-environment system as a whole, while also taking environmental changes into account in development strategies.

The *resource poor, severe poverty* pattern requires attention in almost all vulnerability dimensions. As people in these areas are extremely poor, agriculture is the single most important source of income. Improving water-use efficiency could create positive effects in agricultural production, although immediate interventions to improve basic livelihood conditions are also required to reduce the very high incidence of poverty.

To relieve the pressure in the *poor water, better soils* patterns, land-use intensity should be decreased through more sustainable resource management, but when resources are more constrained, intensities can also be decreased through diversified livelihoods. This requires moving water-intensive agriculture to more water rich areas, as well as creating off-farm income opportunities. It should be noted that decreasing poverty through more sustainable resource management is more realistic in cases where population growth is limited. The same holds for areas with the *extremely overused* pattern, where only a very small income can be generated from relatively good natural resources due to very high population densities. The pressure is unlikely to be relieved by more sustainable agricultural practices alone. Therefore, pressure on production also has to be reduced.

The most promising situation for development is in the *rivers* pattern, where the natural resource situation is good, population pressure is intermediate, and average income is the highest of all patterns. Here, institutional processes, such as land reform and measures to improve the position of smallholder farmers, could relieve pressures from unequal access to natural resources.

Conclusions

In this chapter we have addressed future developments in human security on an aggregated regional scale, focusing on the impact on child mortality caused by inadequate access to food, safe drinking water, basic sanitation, and improved energy sources. In addition, we have analyzed specific vulnerability creating mechanisms that impose potential risks to human security for small-holder farming in dryland areas, due to global change, on a more local scale. Dryland areas are hot spots with respect to the challenges and trade-offs in improving human security in a fragile environment with limited natural resources and high risks of overexploitation.

The regional analysis concludes that, although access to food, water and energy is expected to increase significantly, related health risks remain substantial; even by 2030. If current trends continue, the MDG 4 target for reducing child mortality is far out of reach, even with higher economic growth and greatly improved agricultural productivity. Furthermore, the sub-national scale analysis of drylands concludes that there is no single set of options for improving human security in dryland areas. Unsustainable use of natural resources, increasing competition over land and water resources, and unmitigated global environmental changes increase the pressure on already stressed livelihoods. As some of these pressures will only become apparent in the long term – after 2015 or even 2030 – they may backlash on development progress.

Therefore, to improve human security, and with child mortality as our main indicator, accelerated growth in access to food, improved drinking water and basic sanitation, and improved energy sources is key. Furthermore, policies should address the human-environment system as a whole, while also taking expected global changes into account. In this way, the MDG4 child mortality target may not only be achieved, but the achievement may also become more sustainable beyond this target year.

Note

1 See Kok *et al.* (2010) for a more detailed discussion on the vulnerability creating mechanism and the data used, and Sietz *et al.* (2011) for a similar dryland analysis using a slightly different data set.

References

Alkire, S. (2010) *Human Development: Definitions, Critiques, and Related Concepts*, OPHI Working Papers, Oxford: University of Oxford.

Barnett, J., Matthew, R.A. and O'Brien, K.L. (2010) 'Global environmental change and human security: An introduction', in R.A. Matthew, J. Barnett, B. McDonald and K.L. O'Brien (eds) *Global Environmental Change and Human Security*, Cambridge, MA: The MIT Press.

Bates, B., Kundzewicz, Z.W., Shaohong, W. and Palutikof, J. (2008) *Climate Change and Water*, Technical Paper of the Intergovernmental Panel on Climate Change, Geneva: IPCC Secretariat.

Black, R.E., Allen, L.H., Bhutta, Z.A., Caulfield, L.E., de Onis, M., Ezzati, M., Mathers, C. and Rivera, J. (2008) 'Maternal and child undernutrition: Global and regional exposures and health consequences', *The Lancet*, 371 (9608): 243–260.

Black, R.E., Cousens, S., Johnson, H.L., Lawn, J.E., Rudan, I., Bassani, D.G., Jha, P., Campbell, H., Walker, C.F., Cibulskis, R., Eisele, T., Liu, L. and Mathers, C. (2010) 'Global, regional, and national causes of child mortality in 2008: A systematic analysis', *The Lancet*, 375 (9730): 1969–1987.

Bourguignon, F., Bénassy-Quéré, A., Dercon, S., Estache, A., Gunning, J.W., Kanbur, R., Klasen, S., Maxwell, S., Platteau, J.-P. and Spadaro, A. (2008) *Millennium Development Goals at Midpoint: Where Do We Stand and Where Do We Need to Go?*, Background report for the European Report on Development, 2010, European Commission.

Bouwman, A.F., Kram, T. and Goldewijk, K. (2006) *Integrated Modelling of Global Environmental Change. An Overview of IMAGE 2.4*, Bilthoven: Netherlands Environmental Assessment Agency (MNP).

Cairncross, S. and Valdmanis, V. (2006) *Water Supply, Sanitation, and Hygiene Promotion*, New York: Oxford University Press.

CIESIN (2005) *Poverty Mapping Project: Global Subnational Infant Mortality Rates*, Palisades, NY: Center for International Earth Science Information Network (CIESIN), Socioeconomic Data and Applications Center (SEDAC), Columbia University.

Comprehensive Assessment of Water Management in Agriculture (2007) *Water for Food, Water for Life: A Comprehensive Assessment of Water Management in Agriculture*: London Earthscan, and Colombo: International Water Management Institute.

Dobie, P. (2001) *Poverty and the Drylands*, The Global Drylands Development Partnership, United Nations Development Programme.

FAO (2006) *World Agriculture: Towards 2030/2050 – Prospects for Food, Nutrition, Agriculture and Major Commodity Groups*, Rome: Food and Agriculture Organisation of the United Nations (FAO).

Geist, H.J. and Lambin, E.F. (2004) 'Dynamic causal patterns of desertification', *BioSience*, 54: 9.

Hilderink, H.B.M., Lucas, P.L. ten Hove, A., Kok, M.T.J., de Vos, M.G., Janssen, P.H.M., Meijer, J.R., Faber, A., Ignaciuk, A., Petersen, A.C. and de Vries, H.J.M. (eds) (2008) *Towards a Global Integrated Sustainability Model: GISMO 1.0 status report*, Bilthoven: Netherlands Environmental Assessment Agency (PBL).

IAASTD (2009) *International Assessment of Agricultural Science and Technology for Development. Summary for Decision Makers of the Global Report*, Washington, DC: Island Press.

IPCC (2007) *Impacts, Adaptation and Vulnerability*, Contribution of Working Group II to the Fourth Assesment Report of the Intergovernmental Panel on Climate Change, Parry, M.L., Canziani, O.F., Palutikof, J.P., van der Linden, P.J., Hanson, C.E. (eds), Cambridge: Cambridge University Press.

Jäger, J. and Kok, M.T.J. (2007) 'Vulnerability of people and the environment: Challenges and opportunities', in UNEP (ed.) *Global Environmental Outlook-4: Environment for Development*, Nairobi: Earthscan.

Janssen, P., Walther, C. and Luedeke, M.K.B. (2012) *Cluster analysis to Understand Socio-ecological Systems: A Guideline*, PIK report No. 126: Potsdam Institute for Climate Impact Research (PIK) and Netherlands Environmental Assessment Agency (PBL).

Kassahun, A., Snyman, H.A. and Smit, G.N. (2008) 'Impact of Rangeland degradation on the pastoral production systems, livelihoods and perceptions of the Somali pastoralists in Eastern Ethiopia', *Journal of Arid Environments*, 72 (7): 1265–1281.

Kok, M.T.J. and Jager, J. (2009) *Vulnerability of People and the Environment – Challenges and Opportunities*, Background Report on Chapter 7 of GEO-4, Bilthoven: The Netherlands Environmental Assessment Agency (PBL).

Kok, M.T.J., Lüdeke, M.K.B., Sterzel, T., Lucas, P.L., Walter, C., Janssen, P. and de Soysa, I. (2010) *Quantitative analysis of Patterns of Vulnerability to Global Environmental Change*, PBL Netherlands Environmental Assessment Agency, Potsdam Institute for climate change research (PIK), Norwegian University of Science and Technology (NTNU).

Le Sage, A. and Majid, N. (2002) 'The livelihoods gap: Responding to economic dynamics of vulnerability in Somalia', *Disasters*, 26 (1): 10–27.

Levy, M., Babu, S., Hamilton, K., Rhoe, V., Catenazzi, A., Chen, M., Reid, W.V., Sengupta, D., Ximing, C., Balmford, A. and Bond, W. (2005) 'Ecosystem conditions and human well-being', in Hassan, R., Scholes, R., Ash, N. (eds) *Ecosystems and Human Well-being: Current State and Trends, Volume 1*, Washington, DC: Island Press.

MA (2005) *Ecosystems and Human Well-being: Synthesis*, Washington, DC: Island Press.

Mehta, S., Gore, F., Prüss-Üstün, A., Rehfuess, E. and Smith, K. (2006) 'Modeling household solid fuel use towards reporting of the Millennium Development Goal Indicator', *Energy for sustainable development*, 10 (3): 36–45.

MNP/OECD (2008) *Background report to the OECD Environmental Outlook to 2030: Overviews, Eetails en Methodology of Model-based analysis*, Bakkes, J.A., Bosch, P.R. (eds), Bilthoven, the Netherlands: Netherlands Environmental Assessment Agency (MNP) and Organisation for Economic Co-operation and Development (OECD).

Mustafa, D. and Qazi, M.U. (2007) 'Transition from Karez to Tubewell Irrigation: Development, modernization, and social capital in Balochistan, Pakistan', *World Development*, 35 (10): 1796–1813.

O'Brien, K.L. (2006) 'Are we missing the point? Global environmental change as an issue of human security', *Global Environmental Change*, 16: 1–3.

Oxfam (2006) *Causing Hunger: An Overview of the Food Crisis in Africa*, Oxfam.

PBL (2009) *Beyond 2015: Long-term Development and the Millennium Development Goals*, Bilthoven: PBL Netherlands Environmental Assessment Agency.

PBL (2011) *Scarcities in a Sea of Plenty: Global Resource Scarcities and Policies in the European Union and the Netherlands*, Bilthoven: PBL Netherlands Environmental Assessment Agency.

Ram, K.A., Tsunekawa, A., Saha, D.K. and Miyazaki, T. (1999) 'Subdivision and fragmentation of land holdings and their implication in desertification in the Thar Desert, India', *Journal of Arid Environments*, 41 (4): 463–477.

Reynolds, J.F., Mark Stafford Smith, D., Lambin, E.F., Turner II, B.L., Mortimore, M., Batterbury, S.P.J., Downing, T.E., Dowlatabadi, H., Fernández, R.J., Herrick, J.E., Huber-Sannwald, E., Jiang, H., Leemans, R., Lynam, T., Maestre, F.T., Ayarza, M. and Walker, B. (2007) 'Global desertification: Building a science for dryland development', *Science* 316: 847.

Safriel, U. and Adeel, Z. (2008) 'Development paths of drylands: Thresholds and sustainability', *Sustainability Science*, 3 (1): 117–123.

Safriel, U., Adeel, Z., Niemeijer, D., Puigdefabres, J., White, R., Lal, R., Winslow, M., Ziedler, J., Prince, S., Archer, E. and King, C. (2005) 'Dryland Systems', in R. Hassan, R. Scholes and N. Ash (eds) *Ecosystems and Human Well-being: Current State and Trends. Vol. 1*, Washington, Covelo, London: Island Press: 623–662.

Shiferaw, B., Obare, G. and Muricho, G. (2008) 'Rural market imperfections and the role of institutions in collective action to improve markets for the poor', *Natural Resources Forum*, 32 (1): 25–38.

Sietz, D., Lüdeke, M.K.B. and Walther, C. (2011) 'Categorisation of typical vulnerability patterns in global drylands', *Global Environmental Change*, 21 (2): 431–440.

Smith, L. and Haddad, L. (2000) *Explaining Child Malnutrition in Developing Countries: A Cross-Country Analysis*, Washington, DC: International Food Policy Research Institute (IFPRI).

Thomas, R.J. (2008) 'Opportunities to reduce the vulnerability of dryland farmers in Central and West Asia and North Africa to climate change', *Agriculture, Ecosystems and Environment*, 126 (1–2): 36–45.

Turner, B.L., Kasperson, R.E., Matson, P.A., McCarthy, J.J., Corell, R.W., Christensen, L., Eckley, N., Kasperson, J.X., Luers, A., Martello, M.L., Polsky, C., Pulsipher, A. and Schiller, A. (2003) 'A framework for vulnerability analysis in sustainability science', *Proceedings of the National Academy of Sciences of the United States of America*, 100 (14): 8074–8079.

Twomolow, S.J., Riches, C., O'Neill, D.H., Brookes, P. and Ellis-Jones, J. (1999) 'Sustainable dryland smallholder farming in Sub-Saharan Africa', *Annals of Arid Zone*, 38 (2): 95–136.

UNDESA (2007) *World Population Prospects: The 2006 Revision*, United Nations, Department for Economic and Social Information and Policy Analysis, New York.

UNDP (2010) *Human Development Report 2010 – The Real Wealth of Nations: Pathways to Human Development*, New York: United Nations Development Programme.

UNEP (2007) *Global Environment Outlook IV: Environment for Development*, London, UK: United Nations Environment Programme: Earthscan.

WHO (2002) *The World Health Report 2002, Reducing Risks, Promoting Healthy Life*, Geneva: World Health Organization.

World Bank (2005) *Global Economic Prospects: Trade, Regionalism and Development*, Washington, DC: World Bank.

World Bank (2008) *Global Monitoring Report 2008 – MDGs and the Environment: Agenda for Inclusive and Sustainable Development*, DC: World Bank.

Part II

Breakthrough conditions

Perspectives

9 Social ecological resilience and human security

Katrina Brown

Introduction: what is resilience?

This chapter discusses how resilience concepts are used in contemporary discussions of global change. It explores the areas of overlap, contradiction and tension, and highlights how resilience ideas might inform and enrich human security understandings and analyses. The discussion starts by introducing the term and its origins, and current usage.

Resilience is a term which is widely used in common parlance and which is associated with a number of fields of (social and natural) science. The Oxford English Dictionary definition of resilience is: 1) the ability to recoil or spring back into shape after bending, stretching, or being compressed; 2) (of a person) ability to withstand or recover quickly from difficult conditions. The word originates from the Latin *resilire* 'to leap back'. As I show here, it is used in quite specific ways in different fields. Most of the discussion presented here centers on how resilience is used in studying linked social ecological systems and applied to environmental change.

The concept of resilience is presently infusing policy and scientific debates. In the field of global change, resilience is gaining popularity and traction in fields such as climate change, environmental change, economic development and national security, in addition to more "traditional" fields such as Disaster Response and Recovery. Resilience ideas are evident in international science and policy statements such as the Millennium Ecosystem Assessment, and in the recent Human Development Report (UNDP, 2008), Commission of Climate Change and Development (2009), and initiatives such as the World Bank's Program for Climate Resilience. Non-governmental organizations including Christian Aid and Oxfam, and think-tanks such as UK Institute for Public Policy Research (IPPR) and the World Resources Institute (WRI) have used the term to frame their policy documents and analysis. Thus the concept is rapidly gaining salience within environmental and development policy and within research at the interface of natural and social sciences.

Resilience is also prominent in a number of distinct fields including engineering, mental health and child and developmental psychology. In each of these fields there are a different evolution of concepts and different applications of resilience. For example, Masten and Obradovic (2008) provide a helpful overview of four decades of research and theory on resilience in human developmental sciences. Originating in the 1970s these perspectives have explored how individuals overcome adversity, aiming to identify the factors which promote or protect individuals. In human developmental science, individual resilience refers to "processes of, capacity for or patterns of positive adaptation during or following adverse experiences that have the potential to disrupt or destroy the successful functioning of the person" (Masten and Obradovic, 2008:2). The authors show how the strands of resilience thinking from the human development and social-ecological systems fields come together to inform policy on disaster preparation and recovery and making the link between human individuals and the complex dynamic systems within which they live.

How is resilience applied to environmental change?

Resilience perspectives are increasingly used as an approach for understanding the dynamics of social ecological systems and how they respond to a range of different perturbations (see the definitions in Table 9.1). In this context, resilience is understood as the capacity of a system to absorb recurrent disturbances so as to retain its essential structures, processes and feedbacks. The term is becoming widely used in analyzing the way in which environmental change (including climate change, erosion of biological diversity, land use change) affects ecosystem services and the people and societies who depend on them. Originating in ecological science and closely linked to Holling's concept of the adaptive cycle (Holling 1973; Gunderson, 2000), resilience is now used in interdisciplinary analysis of the interactions of people and nature, applied to the notion of a linked social ecological system, understood as a complex adaptive system comprising a bio-geophysical unit and its associated social actors and institutions (e.g. Berkes and Folke, 1998).

Resilience in this literature is fundamentally about how *a system* can deal with disturbance and surprise. For social ecological systems examined as a set of interactions between people and the ecosystems they depend on, resilience involves three properties: the amount of change a system can undergo and retain the same structure and functions; the degree to which it can re-organize; the degree to which the system can build capacity to learn and to adapt. The present literature encompasses a range of concepts; complexity, transformability and thresholds, dynamics and disequilibria, adaptation, renewal, re-organization and learning (e.g. Carpenter *et al.*, 2001; Walker *et al.*, 2006).

Janssen and colleagues' analysis (2006) shows how research on resilience has spread from ecological sciences, how scholarly networks have become established and how it links with work on vulnerability and adaptation. Central concepts for social sciences in this area of research on resilience of social ecological systems are adaptive capacity, self-organization and learning. These ideas have been used particularly in studies of adaptive governance (Folke *et al.*, 2005) and adaptive institutions (Brown, 2006) that examine how systems can be managed to promote resilience and support learning and adaptation. This area of research brings together ideas from ecology on adaptive management (Berkes and Folke, 1998) and from social sciences on adaptive co-management and collaborative or social learning (Lee, 1993; Ruitenbeek and Cartier, 2001). Lebel *et al.* (2006) have

Table 9.1 Policy prescriptions on resilience

How resilience is defined	*Prescription/focus*	*Reference*
Community resilience is the capacity to adapt to and influence the course of environmental, social and economic change.	Coastal resilient communities take deliberate action to reduce risk from coastal hazards with the goal of avoiding disaster and accelerating recovery in the event of a disaster. They adapt to changes through experience and applying lessons learned (pp.3–1).	US Indian Ocean Tsunami Warning System Programme, 2007 *How resilient is your coastal community?*
Resilience is the ability to withstand the impact of shocks and crises. It is determined by people's assets and their ability to access services provided by external infrastructure and institutions.	The Building Disaster resilient Communities project supports local partner organizations …to strengthen communities' abilities to manage and recover from crises and to prepare for and reduce the risk of future disasters (p.8).	Christian Aid, nd. *Overexposed: Building disaster-resilient communities in a changing climate*
Resilience is the ability of a joint social and ecological system – such as a farm – to withstand shocks, coupled with the capacity to learn from them and evolve in response to changing conditions. Building resilience involves creating strength, flexibility and adaptability.	Makes the case for investing in building up the resilience of vulnerable farming communities as a critical stepping stone to addressing the global challenges of food security; climate change adaptation; and climate change mitigation (p.7).	Oxfam, 2009 *People-centred Resilience*
Resilience is not specifically defined – it is the opposite of vulnerability and is a character of critical infrastructure; enterprise and communities.	A commitment to building national resilience, especially in our infrastructure, by measures including educating and increasing the self-reliance of communities is an integral part of security policy (summary p4).	Commission on National security in the twenty-first century, 2008 *Shared Responsibility*
Resilience not specifically defined, but aim to "integrated climate risk and resilience into core development planning, while complementing other on-going activities".	Aim to provide incentives for scaled-up action and transformational change in integrating consideration of climate resilience in national development planning consistent with poverty reduction and sustainable development goals (section B paragraph 4).	World Bank, 2008 *Pilot Program for Climate Resilience under the Strategic Climate Fund*

summarized a series of empirical studies to demonstrate how various attributes of governance relate to capacities to manage resilience, thus linking issues such as participation, deliberation, accountability, and scale to knowledge, uncertainty, ability to learn and to self-organize. Institutional analysis has been led by political scientists, with anthropologists contributing important insights from the study of local level management of natural resources by specific ethnic groups or in particular cultural contexts. Elinor Ostrom's analysis of common property resource management and institutional diversity in managing complex systems has been at the forefront of institutional analysis of resilience (e.g. Ostrom, 2005). Box 9.1 highlights some of the key features of resilience thinking most pertinent to understanding global and environmental change. Relevant aspects are the distinction between general and specific resilience; fast and slow variables; and recognizing uncertainty surprise and thresholds as "normal" characteristics of a system and how it changes.

But of course there are significant shortcomings and gaps in the social scientific conceptualization of resilience and fundamental difficulties in transferring concepts originating in the ecological realm to social analysis. Reusswig (2007), in reviewing a major interdisciplinary effort in this area (Walker *et al.*, 2006), has drawn attention to some key weaknesses in the applicability of resilience theory for social systems: there is little evidence from history that societies evolve in an "adaptive cycle" proposed in ecology (Gunderson and Holling, 2002); and social resilience theory is bereft of the analysis of structural factors in society that promote resilience for some in the face of lack of resilience for many.

Box 9.1 What does a resilience approach highlight?

- Expect change, manage for change – leads to a prescriptive focus on adaptive management
- Expect the unexpected – uncertainty and surprise as features of systems
- Recognizes different types of change; slow and fast changes and the interactions between them
- Crises may be seen as providing windows of opportunity – to move to a new regime which may be either better or worse than the existing one
- Thresholds are a feature of change and most change is not uniform or regular or predictable – thresholds are ecological and social and may be manifest as "tipping points"
- Multi- and cross-scale issues are important – understanding the interplay and links and interactions is a challenge and has led to examination of polycentric institutions, and the concept of "panarchy"
- Interactions with other stressors – climate change, livelihoods, health, markets, migration and settlement – are recognized by the systems lens – but there may be both general and specific forms of resilience
- Resilience can be "good" or "bad" – it can lead to rigidity or "traps".

In a review of the contribution of a resilience framework to understanding how societies adapt to environmental change Nelson *et al.* (2007) highlight the respective strengths and weaknesses of a resilience approach compared to conventional analysis of adaptation. Conventional social science approaches (originating in human geography, sociology, and development studies) examine the agency of social actors and the capacity of institutions to respond to specific environmental stimuli. This yields insights into vulnerability of different actors to specific stressors. But a resilience approach takes a systems view, giving a more dynamic perspective and enabling a more process-orientated analysis across scales. These aspects, I maintain, could also inform human security approaches to global change.

Where do security and resilience meet?

What then is the relationship between ideas of security and resilience, applied specifically in the field of understanding human dimensions of environmental change – i.e. human security approaches?

First we need to briefly review how resilience ideas permeate conventional security debates and conceptualizations. Resilience and security are often each interpreted as counterparts of vulnerability, such that enhancing resilience and enhancing security are offered as means of overcoming vulnerability – of people and places. The use of resilience in conventional security approaches is quite distinctive, in terms of how resilience is interpreted, from how it applies to and could inform a human security approach.

Resilience and national security

In conventional international relations understandings of security, resilience is used in particular ways which are quite distinct from the social ecological systems perspective. Thus, for national security, resilience is about strengthening responses, resisting change and defending and maintaining stability. In many respects this interpretation of resilience under conventional security approaches (as opposed to human security approaches) is closer to engineering resilience in infrastructure (see Table 9.1). For instance, the IPPR Commission on National Security in the Twenty-first Century (2008) final report devotes a whole chapter to resilience. Here it is framed as the need to protect critical infrastructure in the face of intensified vulnerabilities. The approach requires minimum resilience standards and technical, regulatory and effective planning and information to maintain services. The UK government's perspective is encapsulated in the Cabinet Office's UK Resilience homepage where it declares its aim to "reduce the risk from emergencies so that people can go about their business freely and with confidence ... provide a resource for civil protection practitioners, supporting the work that goes on across the UK to improve emergency preparedness" (see www.cabinetoffice.gov.uk/ukresilience.aspx). This approach can also be seen to echo elements of Dalby's understanding of securitization (2009: 46). As such it does not conform to a more dynamic view of resilience that emerges from the social ecological systems literature. Furthermore, Evans and Steven (2009) show how the interpretation and adoption of resilience ideas in national security approaches is inherently political, and that for conservatives, resilience's appeal to tradition and identity is strong. However, they argue, the conservative instinct to resist change of all kinds is a clear threat to a system's ability to adapt to change, and goes against a social ecological systems view of resilience. Thus there are many contradictions and tensions in how the term is used.

Resilience and human security

Resilience is interpreted quite differently from the human security perspective, and is becoming more widely applied in writing on vulnerability and adaptation to environmental change. The phrase and notion most prominent is that of "building resilience" as a means of ameliorating vulnerability or strengthening disaster preparedness, reduction and responses. For example, Oxfam's recent briefing paper (2009) talks about "people centred resilience" which can help vulnerable farmers address climate change and enhance food security. This mirrors the WRI's definition of "Roots of Resilience" and "routes to resilience" as a means of increasing capacity and wealth of the poor in developing countries through improved natural resource management efforts, payments for ecosystem services and supportive policy and infrastructure measures (WRI, 2008). A number of NGOs have also developed strategies to build "disaster-resilient communities" to address climate change and natural resource related events such as tsunamis, tropical storms and floods (e.g. Christian Aid, nd; US Indian Ocean Tsunami Warning System Programme, 2007) (see Box 9.1). A human security approach on building resilience also leads towards an explicit or implicit targeting of the most vulnerable groups, taking the differentiated capacities of individuals, households and communities as a fundamental starting point.

I suggest that there are two very important differences between the two concepts (human security and resilience from a social ecological systems perspective) and how they are applied, and how in turn they are related to linked terms such as vulnerability and adaptation. These two differences concern first, how they emphasize stability and dynamics; and second, how they view individuals and their agency.

A resilience approach is based on the idea of multiple equilibria or stable states, meaning that it recognizes that change is an integral part of how systems function and that they should be managed for change and in ways to encourage beneficial change, rather than resist change (see Box 9.1). A systems approach also sees dynamic linkages as important ways in which systems function – through feedbacks and interplay across scales or levels. This is somewhat different from many human security approaches which emphasize stability in coping with change and disturbance.

Second, resilience in the social ecological systems literature is focused on systems and how they operate. This approach does not give such prominence to understanding the role of individuals and particularly individual agency in responding to changes. This is an area which is more strongly theorized and empirically studied in the human security approach. The social ecological resilience literature addresses issues such as social learning, leadership and innovation and in the past decade analyses of social networks, institutions and knowledge have made important contributions to broadening the field. However, the resilience approach is criticized for under-emphasizing the role of actors, their agency and issues such as power, social relations, as well as the social construction of scientific knowledge and resource governance (see, for example, Nadasdy, 2007, 2010; Leach 2008).

How can resilience inform human security?

Having drawn these important distinctions between how security and resilience concepts – in selected fields of human dimensions of environmental change – this concluding section discusses how some of the ideas from resilience thinking may inform and enhance research on human security (and indeed vice-versa). I suggest there are a number of insights which can be gained across these fields.

The first insight concerns an understanding of the dynamics of change and how different systems – be they ecological systems or social systems such as a community, a city or a household – can respond to different types of disturbance. A resilience approach provides useful insights into the different types of changes and the different responses. First, it distinguishes between fast and slow controlling variables; slow variables are those which determine the dynamics whereas fast variables tend to be those of direct interest to managers. We might think of culture and worldviews as slow variables for example, whereas prices are fast variables. Making this distinction aids in understanding the linkages between different types of change, the impacts and time scale of change and the points of traction for responses. The case of so-called "land grabs" (see Cotula *et al.*, 2009; von Braun and Meinzen-Dick, 2009) is often presented as both a national and human security issue. Yet it also provides an excellent example through which to distinguish the interactions between fast and slow variables, such as the on-going conversion of natural habitats for agriculture and land use and the evolution of tenure and property rights systems over time, versus the sudden price rises in fuel and food. It also highlights the trade-offs between security and resilience approaches occurring at different scales and for different social actors. Thus a resilience lens brings additional dimensions to the human security understanding of land grabs.

The second insight is that resilience thinking recognizes that change in complex adaptive systems is not uniform or predictable; it is characterized by uncertainty and by feedback and abrupt changes often manifest as thresholds or even "tipping points." Tipping points have come to focus recently in the study of earth systems and the possibility of climate change precipitating shifts in how components of the system function, where a tipping point is defined as the critical point at which a transition is triggered, often by a small incremental change. Tipping points have also been identified for social phenomena – as in the well-known book by

Malcolm Gladwell (2000), and indeed interdisciplinary scientists have started to discuss tipping points in relation to climate change and the possible social limits to adaptation (see Adger *et al.*, 2009). A resilience approach also raises the possibility of crises as providing "windows of opportunity" to trigger latent adaptive capacity and perhaps lead to beneficial system change as described by McSweeney and Coomes (2011) in Honduras following Hurricane Mitch.

A third set of insights stems from the view of adaptive capacity which emerges from resilience conceptualization and is supportive of efforts to build capacity and enhance human security. But resilience thinking also takes us beyond adaptation and adaptive capacity to consider the possibility for transformation and transformative capacity. Thus in a resilient social ecological system, disturbance has the potential to create opportunity for doing new things, for innovation and development (Folke, 2006). According to this view, a resilience perspective *potentially* shifts policies from those that aspire to control change in systems which are assumed to be stable, to managing the capacity of social ecological systems to cope with, adapt to and shape change.

When these ideas are coupled with the insights on multiple stressors and "double exposure," which have come from human security analysis (e.g. Leichenko and O'Brien, 2008) and insights from adaptation studies which characterize different responses (see Ensor and Berger, 2009), then the understanding of how different societies or systems can respond to change might be considerably expanded. In conclusion, there are areas where social ecological resilience concepts and human security approaches together can add significantly to current analysis of the human dimensions of environmental change.

Acknowledgment

Katrina Brown is grateful for support from ESRC Professorial Fellowship 'Resilient development in social ecological systems' (RES-051-27-0263).

References

Adger, W.N. (2000) 'Social and ecological resilience: Are they related?' *Progress in Human Geography*, 24 (3): 347–364.

Adger, W.N., Lorenzoni, I. and O'Brien. K.L. (eds) (2009) *Adapting to Climate Change: Thresholds, Values and Governance*, Cambridge: Cambridge University Press.

Berkes F. and Folke, C. (eds) (1998) *Linking Social and Ecological Systems: Management Practices and Social Mechanisms for Building Resilience*, Cambridge: Cambridge University Press.

Brown, K. (2006) 'Adaptive institutions for coral reef conservation', in I. Cote and J. Reynolds (eds) *Coral Reefs Conservation*, Cambridge: Cambridge University Press.

Carpenter, S., Walker, B., Anderies, J.M. and Abel, N. (2001) From metaphor to measurement: Resilience of what to what? *Ecosystems*, 4: 765–781.

Christian Aid (nd) 'Overexposed: Building disaster-resilient communities in a changing climate', London: Christian Aid.

Commission on Climate Change and Development (2009) *Closing the Gaps*, Stockholm, Sweden: Ministry for Foreign Affairs.

Commission on National Security in the 21st Century (2008) *Shared Destinies: Security in a Globalised World*, London: Institute for Public Policy Research.

Cotula, L., Vermeulen, S., Leonard, R. and Keeley, J. (2009) *Land Grab or Development Opportunity? Agricultural Investment and International Land Deals in Africa*, London/Rome: IIED/FAO/IFAD.

Dalby, S. (2009) *Security and Environmental Change*, Cambridge: Polity Press.

Ensor, J. and Berger, R. (2009) *Understanding Climate Change Adaptation*, London: Practical Action.

Evans, A. and Steven, D. (2009) 'Risks and resilience in the new global era', *Renewal*, 17 (1): 44–52.

Folke, C. (2006) 'Resilience: The emergence of a perspective for social ecological systems analyses', *Global Environmental Change*, 16: 253–267.

Folke, C., Hahn, T., Olsson, P. and Norberg, J. (2005) 'Adaptive governance of social-ecological systems', *Annual Review of Environment and Resources*, 30: 441–473.

Gladwell, M. (2000) *The Tipping Point*, New York: Back Bay Books, Little, Brown and Company.

Gunderson, L.H. (2000) 'Ecological resilience in theory and practice', *Annual Review of Ecology and Systematics*, 31: 425–439.

Gunderson, L.H. and Holling, C.S. (eds) (2002) *Panarchy: Understanding Transformations in Human and Natural Systems*, Washington, DC: Island Press.

Holling, C.S. (1973) 'Resilience and stability of ecological systems', *Annual Review of Ecology and Systematics*, 4: 1–23.

Holling, C.S. (1986) 'Resilience of ecosystems; local surprise and global change', in W.C. Clark and R.E. Munn (eds) *Sustainable Development of the Biosphere*, Cambridge: Cambridge University Press.

Janssen, M., Schoon, M., Ke, W. and Börner, K. (2006) 'Scholarly networks on resilience, vulnerability and adaptation within the human dimensions of global environmental change', *Global Environmental Change*, 16: 240–252.

Leach, M. (ed.) (2008) *Re-framing resilience: Trans-disciplinarity, reflexivity and progressive sustainability – a symposium report*, STEPS Working Paper 13, Sussex: IDS.

Lebel, L., Anderies, J. M., Campbell, B., Folke, C., Hatfield-Dodds, S., Hughes, T.P. and Wilson, J. (2006) 'Governance and the capacity to manage resilience in regional social-ecological systems', *Ecology and Society*, 11 (1): 19.

Lee, K. (1993) *Compass and Gyroscope. Integrating Science and Politics for the Environment*, Washington, DC: Island Press.

Leichenko. R.M. and O'Brien, K.L. (2008) *Environmental Change and Globalization: Double Exposures*, Oxford: Oxford University Press.

Ludwig, D., Walker, B. and Holling, C.S. (1997) 'Sustainability, stability, and resilience', *Conservation Ecology*, 1 (1): 7.

Masten, A.S. and Obradovic, J. (2008) 'Disaster preparation and recovery: Lessons from research on resilience in human development', *Ecology and Society*, 13 (1): 9.

McSweeney, D. and Coomes, O.T. (2011) 'Climate-related disaster opens a window of opportunity for rural poor in northeastern Honduras', *PNAS*, 108 (13): 5203–5208.

Nadasdy, P. (2007) 'Adaptive co-management and the gospel of resilience', in D. Armitage, F. Berkes and N. Doubleday (eds) *Adaptive Co-Management: Collaboration, Learning and Multi-Level Governance*, Canada: UBC Press.

Nadasdy, P. (2010) 'Resilience and truth: A response to Berkes', *MAST*, 9 (1): 41–45.

Nelson, D., Adger, N. and Brown, K. (2007) 'Resilience and adaptation to climate change: Linkages and a new agenda', *Annual Review of Environment and Resources*, 32: 395–419.

Ostrom, E. (2005) *Understanding Institutional Diversity*, Bloomington, IN: Indiana University Press.

Oxfam (2009) 'People-centred resilience', Briefing Paper 135 November 2009, Oxfam International.

Reusswig, F. (2007) 'Book review: Exploring resilience in social-ecological systems. Comparative studies and theory development', *Regional Environmental Change*, 7: 121–122.

Ruitenbeek, J. and Cartier, C. (2001) *The Invisible Wand: Adaptive Co-management as an Emergent Strategy in Complex Bio-economic Systems*, CIFOR Occasional Paper no.34, Bogor, Indonesia: Centre for International Forestry Research.

UNDP (2008) *World Development Report*, New York: UNDP.

US Indian Ocean Tsunami Warning System Programme (2007) *How Resilient is your Coastal Community*? Washington, D.C.: USAID.

von Braun J. and Meinzen-Dick, R. (2009) *Land Grabbing By Foreign Investors in Developing Countries: Risks and Ppportunities*, IFPRI Policy Brief 13 April 2009, IFPRI: Washington, D.C.

Walker, B. and Meyers, J.A. (2004) 'Thresholds in ecological and social–ecological systems: a developing database', *Ecology and Society*, 9 (2): 3.

Walker, B., Holling, C.S., Carpenter, S.R. and Kinzig, A. (2004) 'Resilience, adaptability and transformability in social-ecological systems', *Ecology and Society*, 9 (2): 5.

Walker, B. H., Kinzig, A.P., Gunderson, L.H., Folke, C., Carpenter, S.R. and Schultz, L. (2006) 'A handful of heuristics: propositions for understanding resilience in social-ecological systems', *Ecology and Society*, 11 (1): 13.

World Bank (2008) *Pilot Program for Climate Resilience under the Strategic Climate Fund*, Washington, D.C.: World Bank.

World Resources Institute (2008) *World Resources 2008 – Roots of Resilience – Growing the Wealth of the Poor*, Washington, D.C.: WRI.

10 Climate change and the global financial crisis

Stresses, synergies, and challenges for human security

Robin M. Leichenko, Karen O'Brien and William D. Solecki

Introduction

Climate change and globalization weave together the fates of individuals, households and communities across all regions of the globe. While both processes contribute to growing risks and increasing uncertainties, both also create new opportunities for transformative change. Researchers across the natural and social sciences have paid much attention to climate change and globalization as separate and distinct processes, and growing attention is now being been directed to the interactions between the processes (e.g., Liverman and Vilas, 2006; Leichenko and O'Brien, 2008; Keskitalo, 2008; Adger *et al.*, 2009; Silva *et al.*, 2009; Jeffers, 2013). Yet many questions remain regarding potential feedbacks between processes of globalization and climate change, adaptation to climate change under conditions of rapid socio-economic change, and resilience to the risks and uncertainties associated with both processes.

In this chapter, we build upon the "double exposure" framework developed by Leichenko and O'Brien (2008) to explore the connections and between climate change, globalization and human security. The double exposure framework represents a generalized approach for analysis of the interactions between global environmental and economic changes, paying particular attention to the ways that the two interacting processes spread risk and vulnerability over both space and time (Leichenko *et al.*, 2010). As such, the framework contributes to broader efforts by social scientists to enhance understanding of factors influencing the vulnerability, resilience, and adaptive capacity of communities, regions, and socio-ecological systems to shocks and stresses associated with processes of global change (e.g., Turner *et al.*, 2003; Eakin and Luers, 2006; Berkes, 2007; Nelson *et al.*, 2007; Polsky *et al.*, 2007; Acosta-Michlik *et al.*, 2008; Eakin and Wehbe, 2009; Eriksen and Silva, 2009).

In the next part of the chapter, we present the basic elements of the double exposure framework and consider how the framework may be applied to assess interactions between the global financial crisis and climate change. Focusing on the 2007–2009 drought in California and the simultaneous collapse of the housing market in California's Central Valley, we illustrate connections between the impacts and adaptation to climate change and the long term fall-out from the global financial crisis and subsequent economic slowdown. We show how globalization processes increased vulnerability of households, workers, and communities in the region to climate-related risks, while also influencing capacity to adapt. We conclude by arguing that the emergence of new types of intersections and interactions between globalization and climate change calls for new approaches to policy responses that take into consideration both processes.

The double exposure framework

The double exposure framework, as developed by Leichenko and O'Brien (2008) and summarized in Leichenko *et al.* (2010) provides a general approach for analysis of many types of interactions between environmental change and globalization. One important difference between the double exposure framework and other vulnerability approaches is that other frameworks do not typically take into account the full extent of the potential interactions between climate change and other global change processes, particularly concerning how the outcomes of multiple processes interact across space and over time. There is also relatively little recognition within other frameworks of how the two processes can together transform the context in which people and places experience and respond to changes of many types. Many vulnerability frameworks stress the importance of social, political, economic and environmental context for explaining both differential outcomes and vulnerabilities, yet the frameworks seldom recognize the extent to which the context itself is dynamic, dramatically changing as the result of both global environmental change and globalization. Within the double exposure framework, changing contextual conditions affect exposure and responses to future global change processes, resulting in new patterns of vulnerability and new challenges for social and ecological resilience (Leichenko *et al.*, 2010).

The double exposure framework's point of departure is that multiple global change processes are occurring both simultaneously and sequentially, creating either negative or positive outcomes for individuals, households, communities, and social groups. Within the framework, global environmental change and globalization manifest as either gradual or sudden changes (i.e., stressors or shocks) that have differential effects across a particular exposure frame. Depending on the focus of the research, an exposure frame might be a spatial, political or ecological region, an economic sector or a network of institutions. Exposure results in measurable outcomes, which may, in turn, affect the processes as well as the context in which future changes are experienced (Leichenko *et al.* 2010).

In each case, exposure to global change processes is influenced by the characteristics of the change (e.g., direction, rate, magnitude, intensity, and spatial extent) and by factors in the contextual environment (e.g., institutional, economic, social, political, biophysical, cultural and technological conditions). Responses, which may include actions taken either in anticipation of or following from exposure, are conditioned by factors in the contextual environment, as

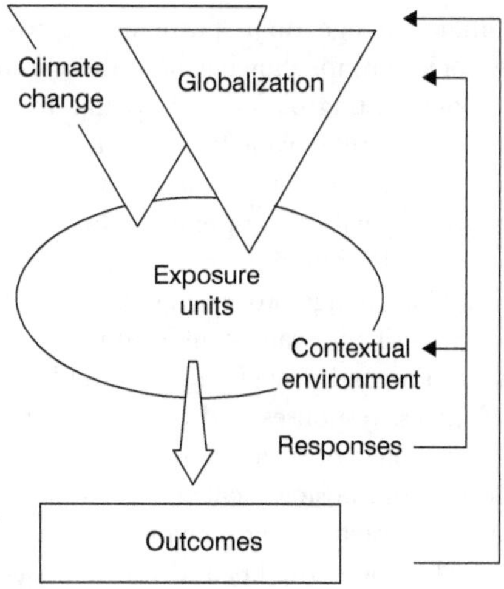

Figure 10.1 A schematic of the double exposure framework. Source: Adapted from Leichenko and O'Brien (2008)

well as by the individual attributes of each affected actor (e.g., education, values, beliefs, cognition). Outcomes depend on both the degree of exposure to each global process and on the actions taken by the affected individuals or other actors (Leichenko *et al.*, 2010).

Figure 10.1 provides a simple illustration of the main components of the double exposure framework. Processes of global environmental change and globalization are represented as partially overlapping triangles. These processes manifest in a specific contextual environment, portrayed as an oval. The extent or magnitude of exposure to the processes is depicted as the intersection between the triangles and the oval. An arrow leading from the contextual environment to a square representing outcomes symbolizes responses to the processes. Outcomes are depicted as separate from the contextual environment to emphasize that any outcome reflects measurable conditions at a specific point in time.

The framework incorporates dynamic linkages between the components. Processes may alter the contextual environment; responses may affect the processes; outcomes may affect responses, and so forth. Dynamics are also incorporated in the framework through recognition that processes and outcomes are often reflexive. Within the figure, the arrow leading from responses and outcomes back to the process triangles depicts these types of circular linkages, which are termed "feedbacks." Although the figure focuses on a single exposure frame, it is important to note that outcomes and responses that occur within one exposure frame may have widespread influence on other exposure frames both across space and over time (Leichenko *et al.*, 2010).

The double exposure framework articulates a number of potential pathways of interaction between the two processes (Leichenko and O'Brien, 2008). The pathway of *outcome* double exposure highlights overlapping impacts of both globalization and climate change on a particular exposure unit, whether it is a region, sector or social group, showing how the combined effects of both processes often exacerbate existing patterns of spatial and social inequality and vulnerability. This pathway identifies what may be referred to as "double winners" and "double losers." The pathway of *context* double exposure shows how one process can influence the capacity to respond to shocks and stresses associated with the other process, often leading to negative outcomes. By incorporating the temporal dynamics of global change processes, context double exposure provides insights on how long-term resilience can be undermined by current changes to the contextual environment. The pathway of *feedback* double exposure demonstrates how the contextual changes, responses and outcomes associated with either or both processes may contribute to drivers of the processes, thereby perpetuating cycles of double exposure and posing challenges to long-term sustainability. By emphasizing the dynamic interactions between processes, responses and outcomes, the framework aims to elicit new insights and research questions, beyond those associated with separate framings of each global change process (Leichenko and O'Brien, 2008). As illustrated in the case example below, the double exposure framework can also provide insights on system-wide economic shocks and how interconnections between multiple processes spread and magnify risks and vulnerabilities (Leichenko *et al.*, 2010).

The global financial crisis and climatic risk: California's Central Valley

Numerous connections can be identified between the global financial crisis and climate change. For example, the globalization of finance contributed to increased availability of low-interest rate loans, which in turn helped spawn energy-intensive, automobile-oriented suburban developments throughout the world (Leichenko and Solecki, 2005). These new developments, which are often concentrated in environmentally sensitive, "amenity" landscapes, such as coastal zones, forested hill slopes, and exurban desert regions, not only contribute to increased greenhouse gas emissions, but also increase vulnerabilities to many types of extreme climatic events, including hurricanes, wildfires, and droughts. Although climate variability, in itself, is considered a "normal" risk, climate change is contributing to increased climate variability and more extreme events, including less predictability of weather, storms and floods, and droughts.

In this section, we illustrate how the double exposure framework can contribute to a better understanding of the complex inter-relationships between drought, the collapse of the housing market, and the fiscal crisis in the California. Drawing on the case of California's Central Valley, this application of the double exposure framework demonstrates how interactions between economic and climatic change have exacerbated social vulnerabilities and undermined efforts to respond and adapt.

California is the largest food producer in the United States, supplying nearly half of the country's fruits, nuts and vegetables. Yet this sector – itself a hallmark of the globalization of the agricultural sector – has long relied on cheap and flexible immigrant labor to produce specialty crops, contributing to persistent poverty in agricultural areas (Mitchell, 1996; Martin *et al.*, 2006; Guthman, 2008). High-value agriculture in California also largely depends on irrigation, creating risks of salinization of soil and groundwater contamination and depletion (Schoups *et al.*, 2005).

California's agriculture is also highly vulnerable to drought (Roland-Holst and Kahrl, 2008). Between 2007 and 2009, a persistent drought devastated many agricultural communities throughout the state. Environmental policies aimed at the preservation of threatened species exacerbated conditions for farmers by reducing the export of water from the Delta region of California and limiting options for additional water allocations in response to the drought (Howitt *et al.*, 2009). The drought contributed to lower profits, increased unemployment and rising food prices, with many of these adverse effects clustered in the Central Valley.

Under climate change, the agricultural sector in the Central Valley will have to adapt to new and more complex conditions (Roland-Holst and Kahrl, 2008). These conditions include reduced water availability in California and the adjoining U.S. Southwest region (e.g., from increased evapotranspiration rates, reduced rainfall, or changes in runoff from snowmelt), changes in temperature regimes, and more frequent outbreaks of pests and disease. Winter warming alone could dramatically reduce the area's fruit and nut production. While California's agriculture is often seen as having a high capacity to adapt to climate change through engineering and technology, this capacity comes at a significant environmental and economic cost, including costs associated with California's groundwater storage and extensive water transfers among regions and users (Tanaka *et al.*, 2006). As discussed by Moser (2009), there are also many barriers to adaptation in California beyond technological and financial constraints, including, for example, lack of institutional capacity and political cooperation among jurisdictions, and social and cultural values that emphasize private property and individual rights (Leichenko *et al.*, 2010).

During the drought, Central Valley farmers responded to the lack of rain and to limited availability of other state or federal supplied water by fallowing fields and planting low-maintenance and low-labor crops (McKinley, 2009). As a stopgap measure to avoid agricultural catastrophe, the state's governor ordered the release of 100,000 acre-feet of water to Central Valley farmers in July of 2009 from the State Water Project with the proviso that it be repaid by the end of November of that same year. This approach countered a "zero allocation" policy for farmers who purchase water from the Central Valley Project, a federally managed irrigation project.

Drought and climate change are not the only crises facing farmers in this region. Cutter and Finch (2008) highlight the Central Valley as one of the U.S. regions with the highest social vulnerability to all types of natural hazards. During the housing boom that ended in 2007, the region experienced some of the most rapid residential housing growth in the country. Real estate speculation had long been a key driver of California's economic growth (Guthman, 2008), and investments in the Central Valley grew dramatically between the mid-1990s and the mid-2000s, such that the region bore the full brunt of the housing crisis that began in 2008. The median sales price of homes in the Central Valley plummeted, and there were thousands of foreclosures on the market. Several of the major cities in the Valley experienced some of the greatest housing price declines of any city in the United States, with losses in some cases exceeding 40 percent. The tax-assessed value of homes in the Valley declined by 11 percent during the first half of 2009 (Borenstein, 2009), and local housing markets were not expected to recover until foreclosures have been absorbed (Streitfeld, 2008). The collapse of the housing sector in the Central Valley, as well as other parts of the state, directly impacted credit markets and local economies. This made capital resources to respond more difficult to acquire. Collapse of the local housing market also devastated the tax revenues for municipalities, further restricting response capacity.

The already limited capacities of local governments were further hampered by a state budget resolution in the summer of 2009, which was built around the state borrowing money from localities with a process of paying back these funds as California's economy improved. Moreover, the State of California had limited capacity to intervene in the Valley, both because it could not raise property taxes due to institutional and legal constraints and because there were few funds available for social services that might benefit vulnerable populations. The state had also instituted "new public management" approaches, commonly associated with neoliberalism and globalization, which entail devolution of resource management responsibilities to local levels (Christensen and Lægreid, 2001; Moser, 2009). In the case of drought management, devolution shifted some of the responsibility to the local level, even though the financial crisis meant that local institutions had limited capacity to manage regional water systems.

Pathways of double exposure

The three pathways of double exposure help to untangle this complex situation, and offer insights on response strategies that may alleviate long-term vulnerability to multiple stressors. For example, the pathway of outcome double exposure highlights the fact that households impacted by drought-related income and livelihood changes may also be subjected to instability and to a loss of shelter or equity linked to the housing and financial crises. Those individuals at the economic and employment margins are particularly vulnerable to the overlapping consequences of multiple crises. Within the Central Valley, this includes those working in industries directly linked to agriculture, such as field workers, processing handlers, food packers, truckers, as well as the local, small businesses that serve these various groups (Leichenko *et al.*, 2010).

The pathway of context double exposure explores transformations in the contextual environment as the result of either or both processes of global change. This interaction between the two processes may sometimes create conditions that undermine adaptive capacity, thereby limiting future responses to one or both crises. In California, as in other parts of United States and throughout the world, the financial and economic crisis affected credit markets such that farmers were unable to borrow money from banks for investments in high value, climate risk-tolerant crops, or for more efficient water supply and irrigation systems. The financial crisis also undermined California's efforts to implement measures to promote adaptation to climate change, many of which would require financial, technical and institutional resources that went beyond "normal" management and maintenance expenses (Moser, 2009). This was especially true in the case of water infrastructure, where adaptation might entail increasing the resilience of water supply systems to extreme climate events via costly new water collection and conveyance infrastructure, or through changes in allocation via new regulatory approaches, review of systems of legacy entitlement and public/private cost sharing (Roland-Holst and Kahrl, 2008).

The pathway of feedback double exposure draws attention to the ways that housing development and urbanization processes in the Central Valley interact with patterns of energy consumption. While suburban expansion has been an important driver of growth in greenhouse gas emissions, the continuing contraction of the housing sector amidst the financial crises and subsequent economic downturn present an opportunity to alter this trend. Such a response might include an emphasis on construction of green buildings, expansion of alternative energy production, and water conservation. In order to be effective and sustainable at the local level, however, such efforts must incorporate adaptation planning and attention to the social and institutional supports needed to ensure resilience to both climatic and economic shocks and stresses.

Conclusion

Globalization and climate changes are well underway and individuals, households, communities, sectors, and regions are confronted with impacts of both of these processes, whether it is through shifts in investments and sectoral upheavals, through more extreme weather events, or through some combination of both. The double exposure framework provides a way to move beyond descriptive statements about the scope of change, and to more fully analyze and explain how interactions between the processes together influence human security.

The particular case of the Central Valley reveals that efforts both to adapt to climate change and to mitigate its drivers need take into account interactions with other types of socially created risks. In addition to using regulations and incentives to manage global financial markets and to limit greenhouse gas emissions, there is a simultaneous need to address shifts in the underlying context that make people vulnerable to shocks, and to respond to uneven outcomes that contribute to increased inequality and future vulnerability. Careful analysis of interactions between processes, responses and the outcomes of multiple stressors is needed to ensure that policy solutions also respond to the local-level, human security challenges of global change.

References

Acosta-Michlik, L., Kumar, K., Klein, R. and Campe, S. (2008) 'Application of fuzzy models to assess susceptibility to droughts from a socio-economic perspective', *Regional Environmental Change*, 8: 150–161.

Adger, W.N., Eakin, H. and Winkels, A. (2009) 'Nested and tele-connected vulnerabilities to environmental change', *Frontiers in Ecology and the Environment*, 7: 150–157.

Berkes, F. (2007) 'Understanding uncertainty and reducing vulnerability: Lessons from resilience thinking', *Natural Hazards*, 41: 283–295.

Borenstein, D. (2009) 'Property tax revenue loss hits cities', Bay Area News Group, 26 July 2009.

Christensen, T. and Lægreid, P. (eds) (2001) *New Public Management: The Transformation of Ideas and Practice*, Farnham: Ashgate.

Cutter, S. and Finch, C. (2008) 'Temporal and spatial changes in social vulnerability to natural hazards', *PNAS*, 105: 2301–2306.

Eakin, H. (2006) *Weathering Risk in Rural Mexico: Climatic, Institutional, and Economic Change*, Tucson, AZ: The University of Arizona Press.

Eakin, H. and Luers, A. (2006) 'Assessing the vulnerability of social-environmental systems', *Annual Review of Environment and Resources*, 31: 365–394.

Eakin, H. and Wehbe, M. (2009) 'Linking local vulnerability to system sustainability in a resilience framework: Two cases from Latin America', *Climatic Change*, 93: 355–377.

Eriksen, S. and Silva, J. (2009) 'The vulnerability context of a savanna area in Mozambique: Household drought coping strategies and responses to economic change', *Environmental Science and Policy*, 12: 33–52.

Guthman, J. (2008) 'Neoliberalism and the making of food politics in California', *Geoforum* 39: 1171–1183.

Howitt, R., MacEwan, C. and Medellin-Azuara, J. (2009) 'Economic impacts of reductions in delta exports on Central Valley agriculture', *Agricultural and Resource Economics Update*, 12: 1–4.

Jeffers, J.M. (2013) 'Double exposures and decision-making: Adaptation policy and planning in Ireland's coastal cities during a boom-bust cycle', *Environment and Planning A* (forthcoming).

Keskitalo, C.H. (2008) *Climate Change and Globalization in the Arctic: An Integrated Approach to Vulnerability Assessment*, London: Earthscan.

Leichenko, R. and O'Brien, K. (2008) *Environmental Change and Globalization: Double Exposures*, New York: Oxford University Press.

Leichenko, R. and Solecki, W. (2005) 'Exporting the American dream: The globalization of suburban consumption landscapes', *Regional Studies*, 39: 241–253.

Leichenko, R., O'Brien, K. and Solecki, W. (2010) 'Climate change and the global financial crisis: A case of double exposure', *Annals of the Association of American Geographers*, 100: 963–972.

Liverman, D.M. and Vilas, S. (2006) 'Neoliberalism and the environment in Latin America', *Annual Review of Environment and Resources*, 31: 2.1–2.37.

Martin, P., Fix, M. and Taylor, J.E. (2006) *The New Rural Poverty: Agriculture and Immigration in California*, Baltimore, MD: Urban Institute Press.

McKinley, J. (2009) 'Drought adds hardships to California,' *New York Times*, 21 February 2009: A1.

Mitchell, D. (1996) *The Lie of the Land: Migrant Workers and the California Landscape*, Minneapolis, MN: University of Minnesota Press.

Moser, S. (2009) 'Adaptation planning in California: Process, progress, challenges and opportunities,' Paper presented at the Global Environmental Change and Human Security (GECHS) Synthesis Conference, Oslo, 24 June.

Nelson, D., Adger, W.N. and Brown, K. (2007) 'Adaptation to environmental change: Contributions of a resilience framework', *Annual Review of Environment and Resources*, 32: 395–419.

O'Brien, K., Hayward, B. and Berkes, F. (2009) 'Rethinking social contracts: Building resilience in a changing climate,' *Ecology & Society*, 14: 12.

Polsky, C., Neff, R. and Yarnal, B. (2007) 'Building comparable global change vulnerability assessments: The vulnerability scoping diagram,' *Global Environmental Change*, 17: 472–485.

Roland-Holst, D. and Kahrl, F. (2008) *California Climate Risk and Response: Executive Summary*, Berkeley, CA: University of California.

Schoups, G., Hopmans, J.W., Young, C.A., Vrugt, J.A., Wallender, W.W., Tanji, K.K. and Panday, S. (2005) 'Sustainability of irrigated agriculture in the San Joaquin Valley, California', *PNAS*, 102: 15352–15356.

Silva, J.A., Eriksen, S. and Ombe, Z.A. (2009) 'Double exposure in Mozambique's Limpopo River Basin', *Geographical Journal*, 176: 6–24.

Streitfeld, D. (2008) 'In the Central Valley, the ruins of the housing bust', *New York Times*, 24 August 2008: A25.

Tanaka, S.K., Zhu., T., Lund, J.R., Howitt, R.E., Jenkins, M.W., Pulido, M.A., Tauber, M., Ritzema, R.S. and Ferreira, I.C. (2006) 'Climate warming and water management adaptation for California', *Climatic Change*, 76: 361–387.

Turner II, B.L., Kasperson, R.E., Matson, P.A., McCarthy, J., Corell, R.W., Christensen, L., Eckley, N., Kasperson, J.X., Luers, A., Martello, M.L., Polsky, C., Pulsipher, A. and Schiller, A. (2003) 'A framework for vulnerability analysis in sustainability science,' *PNAS*, 100: 8074–8079.

11 National level policies versus local level realities – can the two be reconciled to promote sustainable adaptation?

Katharine Vincent, Lars Otto Naess and Marisa Goulden

Introduction

Climate change poses multiple threats to economic growth, poverty reduction, food security and, crucially, the achievement of the Millennium Development Goals in developing countries (Agrawala, 2005; DFID; 2006, Parry *et al.*, 2007a; Stern, 2007). The need for adaptation to reduce the potential adverse impacts of climate change is now widely recognized. At the same time, there is a growing body of evidence pointing to the range of adaptation strategies to climate impacts being employed by households and communities. However, there is so far little evidence of whether and how these agendas are linked, in particular to what extent national policies are supporting poor and vulnerable groups' ability to adapt. We investigate this issue using three case studies in Africa: South Africa, Tanzania and Uganda.

The next section outlines the background literature around adaptation definitions and approaches to adaptation in practice at national and local levels; then the third section describes the methods. The fourth and fifth sections outline the results of research, concentrating first on case study evidence for local level adaptation, and then documentary analysis around national level adaptation priorities. The sixth section discusses the implications of the discord between these two levels, and the chapter finishes with some conclusions on next steps.

Adaptation at national and local levels

Within the academic community, adaptation is variously defined and classified: such as the distinction between planned adaptation, where purposive actions are taken in advance of a climatic event, and autonomous adaptation, where actions after a climatic event serve to reduce future vulnerability (Parry *et al.*, 2007b). The development community tends to define it more generally, for example: "adaptation is about reducing the risks posed by climate change to people's lives and livelihoods" (DFID, 2006: 97–98).

Whilst autonomous adaptation can be observed – in terms of responses to past climatic events, planned adaptation is more problematic as it involves prior responses to a potential future event. As climate change also refers to an increase in variability, as well as gradual change, how this will unfold in reality is often unknown, particularly since climate models are unable to project anticipated exposures to changing variability and extreme climate events at high resolution. There are many case studies that document local level adaptations (e.g. Seo *et al.*, 2008; the UNFCCC database on local coping strategies[1]) and the contribution of local knowledge and skills to understandings of vulnerability and adaptation is increasingly acknowledged (Ford *et al.*, 2006; Parry *et al.*, 2007a; Naess, 2013). At the same time, there is a growing call for an "upscaling" of lessons from individual studies to make them more relevant for national level policy (Burton *et al.*, 2008), with suggestions that adaptation responses must be integrated into national development priorities, whilst being flexible enough to build local level resilience (Prowse *et al.*, 2009).

At the time of research in our case study countries, national level policies for adaptation were enshrined in National Adaptation Programs of Action (NAPAs) (in the case of Tanzania and Uganda), and/or in other policies and frameworks (South Africa[2]).[3] At the time that NAPAs were proposed, case study evidence from Bangladesh showed that in order to ensure equity of outcome, and promote sustainable adaptation, assessment of immediate needs and concerns of the most affected communities at the local level would be required (Huq and Khan, 2006). In reality, however, despite UNFCCC guidelines and best practice promoting the inclusion of local consultation in the process of designing NAPAs, the degree to which this has happened in practice has varied.[4] Osman-Elasha and Downing (2007) identified a number of constraints faced by NAPA teams in east and southern African countries including unhelpful bureaucracy, communication problems, lack of technical capacity and inadequate finance.

As a result of the difficulty of assessing the evidence at the local level, national level policies on adaptation to climate change tend to prioritize large-scale infrastructural projects and interventions. A review of 18 NAPAs showed that the majority of prioritized adaptation projects were on sector-based natural resources activities and the development of infrastructure and disaster relief (Agrawal and Perrin, 2009). They also found that the priority projects were more aimed at capacity building of national governments and agencies than strengthening adaptation capacities of local actors and institutions, and that there was little engagement with local experiences.

We take the notion that adaptation refers to considered long-term changes to livelihood activities that reduce the vulnerability to future climate change exposure, and distinguish this from short-term coping strategies that help to ensure survival through adversity (after Smit and Skinner, 2002). The question remains, do national policies on adaptation take into account local level realities, and is it possible to balance the two? In the next sections we present our observations from three case studies in Africa to highlight the differences in the above approaches.

Methods

We compare findings from research on coping and adaptation responses made by households in five communities; one located in South Africa, two in Tanzania and two in Uganda. Data collection methods included surveys of households (85 in South Africa, 140 in Tanzania and 80 in Uganda), semi-structured interviews and participatory-style group exercises conducted from 2003 to 2006 (Goulden, 2006; Vincent, 2007; Naess, 2008), combined with document and literature analysis of policies and NAPAs.

Crop and livestock farming are common livelihood activities in all of the case studies, and in the Tanzanian case forest resources are also important, whilst in Uganda lakes provide fishing-related livelihood opportunities. Non-natural resource-based activities, such as formal sector employment and informal trading of packaged goods, have become more important in the South African village. All of the villages have suffered droughts in recent years, and the villages in South Africa and Uganda have also experienced floods. In the following sections the results of the research are presented, with special focus on the evidence for local level adaptation, and then national adaptation support through NAPAs and policy.

Evidence for local level adaptation

A range of local level coping strategies and adaptations were observed in South Africa, Tanzania and Uganda, where all communities showed evidence of including their perceptions of climate risks in their livelihood decisions. We classified these local realities into three main types: livelihood diversification, local knowledge of traditional resource management, and the use of social networks (see Table 11.1). Livelihood diversification ensures that in case of risk exposure affecting one source of income, others may still be maintained (Ellis, 1998). Local knowledge of traditional resource management has often been passed down through generations orally and is not otherwise documented. Households use their social networks by capitalizing on their relationships with others, in order to adapt to risks and cope with shocks (Goulden *et al.*, 2009).

Table 11.1 Adaptations observed in rural case studies in Uganda, Tanzania and South Africa. Source: based on Goulden *et al.* (2009)

Livelihood diversification	*Local knowledge based adaptation*	*Social links based adaptation*
Concurrent diversity	Knowledge of historical variability	Loans, exchange, informal credit, remittances
Temporal diversity	Traditional skills	Laboring for other households
Spatial diversity	Traditional practices	Collective action groups, government assistance

Note: All of these strategies are predicated on the recognition of risk and the decision that a change in activities, i.e. an adaptation, is required to sustain livelihoods.

Livelihood diversification can further be divided into three main types: concurrent, temporal and spatial (Goulden, 2006; Goulden *et al.*, 2013). Concurrent diversification refers to multiple activities taking place at any one time: for example many households in the South African case study farmed maize at a subsistence level, but would also try to engage in formal sector labor participation, including piece jobs, in order to diversify their livelihoods. Temporal diversity refers to swapping activities over time, often as a response to the failure of a livelihood activity, for example crop failure following a drought, and is evident in all three case studies. When this no longer becomes tenable they may then have to decide whether it is worth a household member migrating to the town in search of paid employment. Spatial diversity refers to exactly this type of relocation: where the areal extent over which livelihoods are gained must change/be expanded in order to reduce livelihood risk.

Also notable in the identification of local level realities was the fact that access to various adaptation strategies differs between households, depending on their human and financial capital, and often involves trade-offs and hard choices to overcome barriers (Goulden *et al.*, 2009). Furthermore, national level policy interventions designed for purposes other than climate adaptation can hinder the immediate resilience of households to climate impacts. For example, strengthening of fisheries management institutions and harsher enforcement of fishing regulations in the Ugandan case study caused a shock to household incomes and a reduction in assets and livelihood diversity (Goulden, 2006).

National adaptation support through NAPAS and policy

Tanzania's NAPA prioritizes agriculture, water and energy among key sectors. It has been heralded as one of the better NAPAs, with linkages to the National Strategy for Growth and Reduction of Poverty (NSGRP)[5] (Prowse *et al.*, 2009). However, the NAPA document does not specify in any detail how the priority areas were arrived at (e.g. number of consultations, with whom, and where), beyond noting that ranking exercises were done by NAPA teams in consultation with stakeholders, and that "consultations were done at the community level, especially with farmers" (United Republic of Tanzania, 2007: 2). Dietz (2009) argues that participation in the Tanzanian NAPA formulation process was largely limited to a small number of elite actors, with only token local consultations. The NAPA does make reference to local coping strategies and indigenous knowledge, but there is a notable absence of recognition of local people's agency in the selected priority projects, and little sense of whether and how local skills, capacities and knowledge may contribute in either project design or implementation.

Uganda's NAPA also has a strong focus on natural resource management. Indigenous knowledge creation and awareness was given top priority by participating communities but reduced to sixth priority after the NAPA team reviewed national development goals. Many local coping strategies were documented, some of which are specific to certain locations and others that are more widely used. However, they were separated into two categories: those that should be encouraged; and those that should be discouraged. The document also advises that "unplanned coping strategies should be discouraged" (Government of Uganda, 2007: 44), which implies a desire to control local level adaptation rather than leaving people the flexibility to adapt in the most appropriate way for them.

South Africa launched its National Climate Change Strategy (NCCS) in 2004. It was aimed mainly at government departments who were meant to use it to integrate consideration of climate change (related to energy, agriculture, water and waste management) into their programs and policies. However, it has subsequently been shown that there is poor institutional understanding of the implications of adaptation for government departments' own roles and responsibilities (Koch *et al.*, 2006).

In terms of adaptation priorities, the NCCS recognizes that the sectors with the highest vulnerability to climate change in South Africa are: health, maize production, plant and animal diversity, water resources, and rangelands (DEAT, 2004). Whilst outline plans for adaptation in these sectors often go down to the local level, they still reflect top-down priorities. With regard to water resources, for example, the proposed adaptation strategies center on water resource management and contingency planning at the national level, with no recognition of existing local level water conservation/management measures that may be functioning as effective adaptation strategies. The NCCS does mention "creation of sustainable and adaptable livelihoods" as an objective (DEAT, 2004: 17), but contains no reference to the use of local knowledge or explicit guidelines on incorporating local adaptation strategies. A National Climate Change Summit was held in 2009 with the purpose of formally launching a stakeholder-inclusive policy process that would translate Cabinet's climate change policy decisions and directives into fiscal, regulatory and legislative packages as well as sectoral implementation plans (DEAT, 2008). Although adaptation is included here as one of six themes, it is currently poorly elucidated. This reflects the fact that adaptation has received far less attention at the national level than mitigation (although there is recognition that more research and work are required, and that this should include recognition of local coping and adaptation mechanisms).

Discussion

It is clear from the evidence presented here that for the most part, local level adaptation strategies observed in South Africa, Tanzania and Uganda are not recognized in the national policies for adaptation outlined in NAPAs and other relevant policy frameworks. This is expressed in different ways; for South Africa, there is recognition that local adaptation strategies exist, but as yet they have not been systematically charted and thus do not (yet) inform national level adaptation policy. In Tanzania, the selection of priorities implies a focus on transfer of technology and local communities as recipients, not agents. Finally, in Uganda, a number of local coping strategies are documented, but the government wishes to define what is desirable and undesirable adaptation and this may lead to the hindrance of valuable local adaptation strategies.

At the local level, various adaptation realities are observed, including concurrent, temporal and spatial diversification, local knowledge and traditional resource management, and the use of social networks. National policies, however, tend to focus on macro level and sector-based projects with very little recognition of the existence and further potential for local adaptation. Whilst it is not necessarily a problem *per se* for the approaches to differ, it does raise questions about the nature of interactions between the two, and in particular whether a focus on large national adaptation projects might inadvertently undermine existing local strategies.

Thus, despite the intentions in processes such as the development of the NAPAs, the Nairobi Work Program[6] and the diverse funding mechanisms now being set up, it is unclear whether the lessons from local realities will be taken up in nationally-driven adaptation efforts. Already potential conflicts between national level strategies and local level realities are visible. In Tanzania, forest regulations are impeding community-level management of natural resources. Furthermore, drought-resistant crop varieties supported by national programs require labor and capital inputs that put extra burdens on the poorest households (Naess, 2008). At best, people on the ground may adapt to regulations, but derive little benefit from them, and they may in fact promote maladaptation, thus reducing the resilience of communities. In South Africa, for example, national policies that enshrine the right to water mean that people in water-stressed areas are cushioned from variability as government funds the installation of boreholes. As well as being environmentally unsustainable in the long run, it also discourages traditional resource management and conservation practices, which should rather be encouraged as a form of local adaptation.

Similarly in both Tanzania and South Africa there are structural constraints from institutions that actually provide little support to those that need it the most. In South Africa, for example, the national Department of Agriculture provides drought and flood assistance to affected parties, but in reality access to it is scanty (Vincent, 2007). Fisheries management in Uganda has been undergoing changes since 2003 with attempts to introduce co-management institutions at multiple levels of governance. However, lack of financial resources from central government, inadequate local capacity and a difference in interests between national government and fishing communities has meant that the ability of the co-management institutions to contribute to resilience to climate impacts and other stresses is substantially weakened (Goulden, 2006; Goulden *et al.*, 2013).[7] Thus rather than having a neutral or positive effect on local level reality, national level decisions can actively obstruct the development of climate-resilient livelihoods.

In light of this evidence, we suggest that critical questions are raised by the discord between the national level focus on large-scale adaptation projects and sectoral policy interventions, and the local reality, which shows a promising array of adaptation responses that need to be supported rather than hindered by government interventions. When moving from practice to policy, there is a need to activate local level adaptive capacity, and design adaptation interventions that take account of this reality, whilst ensuring adaptation is sustainable so that it provides resilience to climate change impacts.

Conclusion

We suggest that there is a definite discord between evidence from case studies in South Africa, Tanzania and Uganda, which show that at the community level a number of adaptation strategies are being employed; and the focus at the national level which, whilst paying lip service to the existence of such strategies, tends to favor technical interventions by sectors. What needs to be considered in adaptation policy, therefore, is an explicit recognition that local level adaptation is occurring, and that it is the responsibility of each country to compile the research outlining such local level adaptation in their countries. Whilst there is a place for macro scale adaptation projects and sectoral policy interventions, special attention must be paid to ensure that national adaptation policies do not inadvertently inhibit the development of local climate-resilient livelihoods. In reality, this means that policy frameworks need to be flexible and supportive of local mechanisms that are conducive to adaptation. In terms of practice, this means that new and innovative ways of supporting adaptation are needed that facilitate adaptation whilst leaving the final choice over which adaptation actions are most appropriate to those who are affected.

Acknowledgments

The research presented in this paper was supported by PhD studentships funded by NERC and ESRC in the UK and the Research Council of Norway. We thank our colleagues at the Tyndall Centre for Climate Change Research, UEA and our research assistants and participants in South Africa, Tanzania and Uganda.

Notes

1 http://maindb.unfccc.int/public/adaptation/
2 South Africa is not classified as a Least Developed Country and thus does not produce a NAPA.
3 Since this research took place, South Africa has introduced its National Climate Change Response policy (white paper), Uganda plans to release a climate change strategy and implementation and communication plan before the end of 2013, and Tanzania is in the process of finalizing a National Climate Change Response Strategy. These documents and the process of compiling them are not analyzed in this chapter, but the findings presented here can inform the criteria by which these and future policy documents can be assessed.
4 For more information on the process of NAPA development in several countries, see *Tiempo* issue 65, 2007, available online at http://www.tiempocyberclimate.org/portal/archive/pdf/tiempo65high.pdf
5 NSGRP is a successor to the Poverty Reduction Strategy Paper in Tanzania.
6 Set up under the UNFCCC to assist developing countries, in particular, to improve their understanding of impacts, vulnerability and adaptation to climate change and improve planning on practical adaptation actions and measures.
7 Funding problems continue to hinder the functioning of fisheries co-management institutions in Uganda (Government of Uganda, 2011. Department of Fisheries Resources Annual Report 2010/11, URL: http://www.agriculture.go.ug/userfiles/DFR%20ANNUAL%20REPORT%202012.pdf accessed 21/11/12).

References

Agrawala, S. (2005) 'Putting climate change in the development mainstream: introduction and framework', in S. Agrawala (ed.) *Bridge over Troubled Waters: Linking Climate Change and Development*, Paris: Organization for Economic Co-operation and Development.

Agrawal, A. and Perrin, N. (2009) 'Climate adaptation, local institutions and rural livelihoods', in W.N. Adger, I. Lorenzoni and K.L. O'Brien (eds) *Adapting to Climate Change: Thresholds, Values, Governance*, Cambridge: Cambridge University Press.

Burton, I., Dickinson, T. and Howard, Y. (2008) 'Upscaling adaptation studies to inform policy at the global level', *The Integrated Assessment Journal*, 8 (2): 25–37.

DEAT (2004) *A National Climate Change Response Strategy for South Africa*, Pretoria: Department for Environmental Affairs and Tourism.

DEAT (2008) 'Government's vision, strategic direction and framework for climate policy', Presentation made at the launch of the LTMS, August 2008. Online. Available HTTP: http://www.environment.gov.za/HotIssues/2008/LTMS/Media%20LTMS%20 29July2008.ppt (accessed 19 December 2009).

DFID (2006) *Eliminating World Poverty: Making Governance Work for the Poor*, White Paper on International Development, London: Department for International Development.

Dietz, K. (2009) 'Prima Klima in den Nord-Süd-Beziehungen? Die Antinomien globaler Klimapolitik: Diskurse, Politiken und Prozesse', in H.-J. Burchardt (ed.) *Nord-Süd-Beziehungen im Umbruch: Neue Perspektiven auf Staat und Demokratie in der Weltpolitik*, Frankfurt: Campus, pp. 183–218.

Ellis, F. (1998) 'Household strategies and rural livelihood diversification', *Journal of Development Studies*, 35 (1): 1–38.

Ford, J.D., Smit, B. and Wandel, J. (2006) 'Vulnerability to climate change in the Arctic: A case study from Arctic Bay, Canada', *Global Environmental Change-Human and Policy Dimensions*, 16 (2): 145–160.

Goulden, M. (2006) 'Livelihood diversification, social capital and resilience to climate variability amongst natural resource dependent societies in Uganda', PhD thesis, Norwich, UK: School of Environmental Sciences, University of East Anglia.

Goulden, M., Naess, L.O., Vincent, K. and Adger, N. (2009) 'Diversification, networks and traditional resource management as adaptations to climate extremes in rural Africa: opportunities and barriers', in W.N. Adger, I. Lorenzoni and K. O'Brien (eds) *Adapting to Climate Change: Thresholds, Values, Governance*, Cambridge: Cambridge University Press.

Goulden, M.C., Adger, W.N., Allison, E.H. and Conway, D. (2013) 'Limits to resilience from livelihood diversification and social capital in lake social-ecological systems in Uganda'. *Annals of the Association of American Geographers*, DOI:10.1080/00045608.201 3.765771.

Government of Uganda (2007) *Climate Change: Uganda National Adaptation Programmes of Action*, Kampala, Uganda. Online. Available HTTP: http://unfccc.int/resource/docs/ napa/uga01.pdf (accessed 10 October 2012).

Huq, S. and Khan, M.R. (2006) 'Equity in national adaptation programmes of action (NAPAs): the case of Bangladesh', in W.N. Adger, J. Paavola, S. Huq, and M.J. Mace (eds) *Fairness in Adaptation to Climate Change*, Cambridge, MA: MIT Press.

Koch, I., Vogel, C. and Patel, Z. (2006) 'Institutional dynamics and climate change adaptation in South Africa', *Mitigation and Adaptation Strategies for Global Change*, 12 (8): 1323–1329.

Naess, L.O. (2008) 'Local knowledge, institutions and climate adaptation in Tanzania', PhD thesis, Norwich: School of Environmental Sciences, University of East Anglia.

Naess, L.O. (2013) 'The role of local knowledge in adaptation to climate change', *Wiley Interdisciplinary Reviews: Climate Change*, 4 (2): 99–106.

Osman-Elasha, B. and Downing, T. (2007) 'Lessons learned in preparing National Adaptation Programmes of Action in Eastern and Southern Africa', unpublished paper, Stockholm: Stockholm Environment Institute. Online. Available HTTP: http://www.eurocapacity.net/downloads/ecbi_NAPA_PA_Project_2007.pdf (accessed 10 October 2012).

Parry, M.L., Canziani, O.F., Palutikof, J.P., van der Linden, P.J. and Hanson, C.E. (eds) (2007a) 'Cross-chapter case study', in *Climate Change 2007: Impacts, Adaptation and Vulnerability*, Contribution of Working Group II to the Fourth Assessment Report of the Intergovernmental Panel on Climate Change, Cambridge: Cambridge University Press.

Parry, M.L., Canziani, O.F., Palutikof, J.P., van der Linden, P.J. and Hanson, C.E. (eds) (2007b) 'Appendix 1: glossary', in *Climate Change 2007: Impacts, Adaptation and Vulnerability*, Contribution of Working Group II to the Fourth Assessment Report of the Intergovernmental Panel on Climate Change, Cambridge: Cambridge University Press.

Prowse, M., Grist, N. and Sourang, C. (2009) *Closing the Gap between Climate Adaptation and Poverty Reduction Frameworks*, ODI Project Briefing 29, June 2009.

Seo, N., Mendelsohn, R., Kurukulasuriya, P., Dinar, A. and Hassan, R. (2008) *Differential Adaptation Strategies to Climate Change in African Cropland by Agro-ecological Zones*, Policy Research Working Paper 4600, Washington, D.C.: World Bank.

Smit, B. and Skinner, M.W. (2002) 'Adaptation options in agriculture to climate change: a typology', *Mitigation and Adaptation Strategies for Global Change*, 7: 85–114.

Stern, N. (2007) *The Economics of Climate Change: The Stern Review*, Cambridge: Cambridge University Press.

United Republic of Tanzania (2007) National Adaptation Programme of Action, Dar es Salaam, Tanzania. Online. Available HTTP: http://unfccc.int/resource/docs/napa/tza01.pdf (accessed 10 October 2012).

Vincent, K. (2007) 'Gendered vulnerability to climate change in Limpopo Province, South Africa', PhD thesis, Norwich: School of Environmental Sciences, University of East Anglia.

12 Fisheries, resource management and climate change

Local perspectives of change in coastal communities in Northern Norway

Grete K. Hovelsrud, Jennifer West and Halvor Dannevig

Introduction

In this chapter we present empirical findings from case studies in Northern Norway that describe the complex environmental and societal context that shapes vulnerability and adaptation to climate change in coastal fishing communities. The case studies incorporate local knowledge in an approach that provides a means for researchers and community members to interactively identify local vulnerability factors, and which empowers communities to reflect on local adaptation needs. Our research is founded on the recognition that local communities should be involved in defining research questions with local relevance in order to understand how vulnerability is shaped at the local level and what future adaptation strategies will be required to meet multiple changing conditions. We illustrate this point with case studies in which we examine the multiple linkages between coastal fisheries, fishing regulations, changing environmental, social and economic conditions, and local knowledge.

Northern Norway has experienced an increase in annual mean air temperature over the last 100 years, with the warmest period having occurred in the last 20 years (Førland *et al.*, 2009). Ocean temperature increases and a shift in the frequency and intensity of storms and extreme weather events in the region are expected to occur over the coming century (ibid.). Changes in the climate of Northern Norway will have consequences for fish stocks and in turn for coastal fishing activities. Such changes are not necessarily anticipated in current fisheries management practices and regulations, and will also challenge fishers' experience-based knowledge of local sea, weather and fishing conditions.

We start by describing the case study context, and in turn explain our methods and approach. Next we discuss the empirical findings from the cases, highlighting the complex social, environmental and climatic interlinkages that underpin vulnerability and adaptation at the local level. We conclude that local knowledge is essential for capturing salient research questions that can help coastal fishing communities meet future challenges and opportunities.

Case studies context

Coastal fisheries have provided the basis for settlement in Northern Norway from historical times until the present, and remain important for employment, income and identity in many small communities in the region (e.g. Hovelsrud and West, 2008; Keskitalo and Kulyasova, 2009; West and Hovelsrud, 2010). Our case studies in Northern Norway are located near some of the world's richest cod fisheries; historically the region has played a vital role for the Norwegian economy with its 1,000-year-old tradition of producing stock fish for national and international markets (e.g. Berge, 1996). While the number of fishermen has declined substantially over the last few decades due to technological, regulatory, economic and wider societal changes, fishing activities remain important in the case communities. In addition to fisheries and related processing activities such as stock fish production, agriculture, aquaculture, small-scale tourism and the service sector provide jobs locally.

Because economic, cultural, livelihood and recreational activities are closely connected to the natural environment, the communities are particularly exposed to environmental changes, including those that are driven or reinforced by climate change (Hovelsrud *et al.*, 2010). Although fewer people hold jobs in the primary sector today compared to previously, these jobs still serve as the economic cornerstone for many northern, coastal communities. In the context of small and declining populations and a range of societal changes with which the region as a whole is dealing, climate-driven change is an added factor shaping vulnerability in the case communities.

Fishermen have, through their livelihood activities and experience, developed extensive knowledge of their environment, which has made them highly adaptable to natural variations in the region's climate and fish stocks. Climate variability and change are key factors affecting the cod industry; ocean temperature determines where the cod will spawn and the production of stock fish requires a particular set of climate conditions at a specific time of year. A recent increase in ocean temperature in the Barents Sea has been linked to changes in the spatial distribution, abundance, and growth rates of key commercial fish stocks, including North Atlantic cod (Sundby and Nakken, 2008).

Norwegian fisheries are regulated and managed at a variety of levels, including individual fishing vessels, through quota and fishing ground assignments, and particular species (Jentoft and Mikalsen, 2004). With the introduction of the quota system in the early 1990s, the coastal fishery sector has adapted to fisheries management reforms and higher efficiency requirements that have effectively reduced the total number of fishers and fishing vessels in operation (ibid.). The fishing industry is also exposed to major fluctuations in fish prices due to variability in demand for fish products nationally and internationally.

Method and approach

A crucial aspect of research that aims to support community adaptation and reduce vulnerability to changing environmental and societal conditions is to involve local

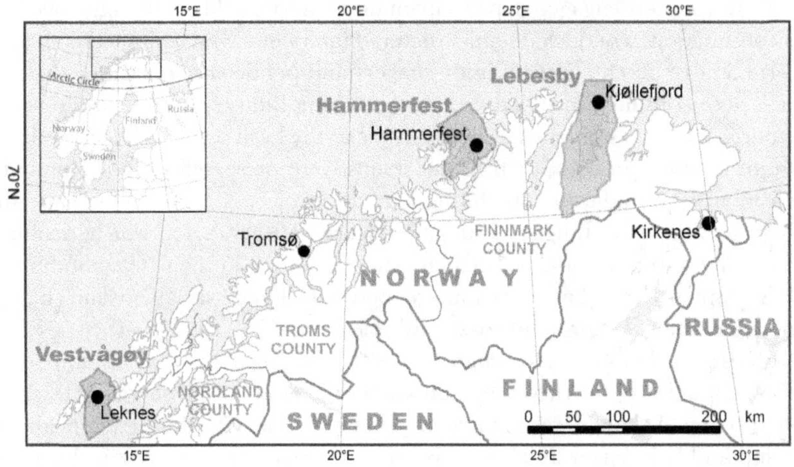

Figure 12.1 The case study sites in Northern Norway: Vestvågøy in Nordland County and Hammerfest, Lebesby and Unjárga/Nesseby in Finnmark County

Table 12.1 Development in population and number of fishermen in the case communities from 1970 to 2008. Source: Norwegian Directorate of Fisheries (2010)

	Population development			*Number of fishermen*		
	1970	*2008*	*% change*	*1970*	*1990*[1]	*% change*
Vestvågøy	11526	10710	-7.1	629	482	-23.3
Hammerfest	7057	9407[2]	33.3	391	206	-46.8
Lebesby	2261	1304	-42.3	203	89	-56.2

Notes
1) Not available at municipal level after 1990.
2) The municipality of Sørøysund was merged with the municipality of Hammerfest in 1992

actors early in the process to identify locally relevant and useful research questions (see Berkes and Jolly, 2001; Turner *et al.*, 2003; Ford and Smit, 2004; Lim *et al.*, 2004; Keskitalo, 2004, 2008; Tyler *et al.*, 2007; Smit, *et al.*, 2008; and West and Hovelsrud, 2010; Hovelsrud *et al.*, 2010). One of the fundamental principles of the research, therefore, is that it is relevant to the communities. This demands close collaboration with local actors (see also Gearheard *et al.*, 2006; Hovelsrud and Winsnes, 2006). The case specific research questions driving the studies reported here were framed in cooperation with local actors. Fieldwork in the communities then helped to determine significant local conditions that shape vulnerability and require adaptation. Local involvement in the research design and process has contributed to increased awareness by researchers of local conditions and has in turn empowered the communities to reflect on their specific adaptation needs.

The researchers engaged in participant observation, held community meetings and consultations, conducted semi-structured interviews, and consulted with local government representatives (decision-makers and politicians) at each of the case sites. Those consulted included fishers, the fishing industry, local entrepreneurs, the tourist industry, store owners and "people at the local pubs." The research has been an iterative process in which the results were presented to the community for feedback and adjustments throughout the project. Climate variability and change are linked to a range of issues of concern to fisheries, tourism, agriculture, aquaculture and municipal maintenance and planning in the case communities. In this chapter we emphasize the linkages between climate variability and change, local vulnerability factors, and coastal fisheries.

To ensure a holistic understanding of the broad changes that are important to the adaptations undertaken by and vulnerability of local fisheries actors, the topic of climate change was not a focal point of discussions during the fieldwork (Marino and Schweitzer 2009; Hovelsrud *et al.*, 2010). In some cases, climate or simply topics about the weather were raised by our local partners in connection with fisheries activities. In other cases we asked deliberate questions about climate and weather in connection with the topics previously identified by the municipalities, fishers and the fishing industry. Once the topic of climate change was raised, discussions ensued about broader climatic conditions and specific local observations of changes in climatic variability and weather that affect coastal fishing activities.

Incorporating local knowledge in scientific research projects has received significant attention over the past two decades, and the value of such knowledge for understanding linkages between social and environmental changes is clear (e.g. Berkes *et al.*, 1995). While the specific aspects of local knowledge (also termed traditional ecological knowledge) that is rooted in indigenous culture may not be directly transferrable to non-indigenous communities, both indigenous and non-indigenous people who depend directly on renewable natural resources, such as fisheries, rely on their extensive knowledge about the local environment and resources in question (e.g. Hovelsrud and Winsnes, 2006). Broad local knowledge of all aspects of coastal fisheries is gained through cumulative experience acquired by individual fishers over many years, and in turn constitutes a source of collective knowledge upon which fishers can draw to respond to changes in the resource base. Local knowledge is dynamic, and depends on active practice and transfer of knowledge and experience between generations, and from older to younger fishermen, so as to remain relevant and useful. Coastal fishermen in Northern Norway rely on their own knowledge about weather, oceanographic, and fish interactions, and adjust their fishing activities according to their own oceanographic and weather observations combined with knowledge obtained from local meteorological forecasts and fishing technologies, and within the constraints of particular social, economic and regulatory circumstances. The weather in Northern Norway is often described as harsh, extreme, highly variable and unpredictable, and coastal communities pride themselves on being adaptable to such conditions. As one fisher noted: "Nowadays some of the fishers even compete about being able to fish in the harshest weather." This perception of "adaptability" can be argued to be a defining feature of the culture of coastal fishing activities in Northern Norway.

Findings and discussion

The fishing sector is considered by scientists to be sensitive to the consequences of climate change, in particular to the effects of altered marine environments on northern fish stocks. In this section we present and discuss a selection of locally-identified climatic conditions to which coastal fishing activities, defined in Norway as fishing activities that take place within 12 miles of the continental shelf, are considered sensitive (see West and Hovelsrud, 2010; Hovelsrud *et al.*, 2010 for more details). We find that *combined climate elements* are of critical importance locally and should inspire further community relevant climatological research. Thereafter, we discuss the social factors that our local partners identified as shaping the flexibility of their coastal fishing activities and which we argue are important for understanding vulnerability and adaptation needs locally. In particular we find that fishing regulations may increase fishers' vulnerability in the context of climate change if they do not consider changing environmental conditions and responses that may be required locally to deal with these. Successful regulatory adjustments will require that policy makers have a better understanding of the consequences of climate change for coastal fisheries. Lastly, we argue that there are increasingly complex social and economic reasons for why the coastal fisheries in Northern Norway may be vulnerable to the impacts of future climate change. These include an aging fisherman population, outmigration, and current income tax regulations. In view of the findings, we argue that climatic, social, and economic factors together shape coastal fishing activities, and need to be considered holistically to arrive at a better understanding of local vulnerability and adaptation challenges.

Local climate elements

Ocean temperature: Although coastal fishers are accustomed to dealing with high natural variability in weather, climate and the resource base, local observations of changes in ocean temperature and in the distribution of fish species are consistent with recent scientific measurements and data. This data shows that even though the ocean temperature in the Nordic Seas fluctuates between cold and warm periods, the last 30 years show a rapid increase to levels higher than the long-term median (Hanssen-Bauer, 2009). Increased ocean temperature affects the production and distribution of biomass (Loeng and Drinkwater, 2007), and a linear relationship has been established between ocean temperature and the distribution of spawning cod and herring (Sundby and Nakken, 2008). In addition to changes in the distribution and abundance of commercially important fish stocks such as Norwegian cod (*Gadus morhua*), herring (*Clupea harengus*), and capelin (*Mallotus villosus*), an increased influx of southerly species such as mackerel (*Scomber scombrus*) monk/goose fish (*Lopus piscatorius*) and blue whiting (*Micromesistius poutassou*) to

Finnmark, in Northern Norway, is anticipated (ACIA, 2005; Drinkwater, 2005; Loeng and Drinkwater, 2007; Sundby and Nakken, 2008). Moreover, the ocean is warming both horizontally and vertically with the result that the fish feed deeper in the water column (Sundby, personal communication). Fishers' responses to these changes are two-tiered: fishing farther out to sea, and employing fishing gear that can reach fish at greater depths.

Polar lows, storms and extreme weather: Coastal fishermen are accustomed to storms and extreme weather, as such conditions affect when, where and with what risk they operate. Whether polar lows are identified by coastal fishers as being of particular concern depends on the proximity of their fishing activities to the polar ice in the Barents Sea. A polar low is a small cyclone forming over open sea during the cold season, is highly localized and commonly not forecast, forming quickly and with hurricane force winds (Hamilton, 2004). Although the drivers of polar lows and Arctic fronts are not yet well understood, we know that a polar low forms when a pocket of cold polar air moves across relatively warmer water (Ibid.). According to our local partners in Finnmark, polar lows represent a significant hazard to coastal fishers during winter. The small coastal fishing vessels (*juksa*, less then 10m in length) are particularly vulnerable to polar lows and storm activity because of their size and because they are typically manned by a one-person crew (West and Hovelsrud, 2010). Local knowledge in and of itself is insufficient for detecting polar lows.

Across the sites, storms and extreme weather are generally highlighted locally as affecting coastal fishing activities, and in turn the supply of fish to local fish processors and the subsequent shipment to market. Results from recent climate modeling suggest an increase in strong wind events in the Nordic Seas (Barstad *et al.*, 2009). When storm conditions prevent fishers from either leaving harbor to fish, or from hauling in catch from traps or nets left out over night, fishing gear may become tangled and fish quality may suffer, costing both time and money (West and Hovelsrud, 2010). Stormy weather often leads to closure of roads, ports and airports, which in turn interrupts the transportation of processed fish out of the region.

Coastal fishers explain that during winter, the main cod fishing season, strong winds from the south and inland are often cold and, when meeting the warm, moist maritime air, may cause ice to accumulate on fishing vessels and gear. Accumulation of ice on vessels and equipment may, in rare cases, cause a vessel to capsize, and fishers are acutely aware of this hazard. The combination of wind, temperature and moisture together create hazardous conditions which are difficult to forecast. Instead the fishers rely on their own experience and knowledge for assessing local conditions and whether it is safe to fish.

Conditions at sea in the study region have already started to change beyond what has previously been adapted to, creating challenges for the coastal fishing fleet. Increased ocean temperatures are occurring, and storm surges and wave heights and wind are all expected to undergo changes in the future (Loeng, 2008; Førland *et al.*, 2009). With increasing ocean temperatures the sea ice will continue to retreat and the distributional range of commercially important fish stocks will continue to shift and may even increase in the Barents Sea. Current fishing regulations delimit access to fish stocks by different vessel size classes and fishing zones and these classifications may not be optimal or appropriate under future marine environmental conditions. In the absence of adapted fishing regulations, coastal fishers may be prevented from traveling farther out to sea to catch distant fish stocks, or if regulations encourage them to do so, fishing vessels currently in use may not be suitable for traveling long distances or withstanding increased storm activity.

Wider social context

In addition to environmental and meteorological observations, actors identify a range of social, cultural, institutional and economic factors that shape coastal fishing activities, and which they perceive to be of more immediate concern for their livelihoods than climate change (West and Hovelsrud, 2010; Hovelsrud *et al.*, 2010). Outmigration to larger centers and an aging fisher population are two ongoing trends of concern in our case-communities. Outmigration of young people reduces recruitment to fisheries while also decreasing the tax base in the municipality, which is needed to support and renew fishing infrastructure such as landing facilities, and to provide social and economic support to fishing families. An aging fisher population also threatens the traditional basis for income and employment, and according to respondents is interrupting the transfer of fishing knowledge and skills between generations. The high cost of gaining access to and participating in the current limited-entry, quota-based management regime is a main reason cited by local fishers for why few young people choose to become fishermen. This trend is taking place despite the current profitability of fishing. Explaining the full socio-cultural and economic context of coastal fisheries requires a detailed discussion, beyond the scope of this chapter (see instead West and Hovelsrud, 2010). However, it is clear that the viability of coastal fishing communities is closely connected to local and regional developments within the fisheries.

Actors in our case studies consider fisheries regulations and management to be major determinants of both changes in fish stocks and in fishermen's potential adaptation to these changes. Reform of Norwegian fishing regulations over the past four decades has involved a transition from an open to a limited entry quota-based fishery founded on the precautionary principle, and total allowable catch (TAC) limits (MoFC, 2007). This has increased the monetary value of fishing quotas and led to a decrease in the total number of coastal fishing vessels and employment in the sector. Structural changes to the fishing fleet over the past two decades therefore reflect the changing management regime and resulting cost structure of participating in a limited access fishery. National fisheries management goals include broad concerns for economic efficiency, social welfare and ecological sustainability. Societal goals are in line with Norwegian district policy, which emphasizes an equitable balance of fisheries participation and rights in different regions in order to maintain a dispersed coastal settlement pattern.

While traditional coastal fishing activities were flexible and adaptive, in terms of fish species, fishing gear and fishing grounds in response to a highly variable and migratory resource (see Jentoft, 1998; Coulthard, 2009), today Norwegian fisheries are highly regulated, institutionalized and politicized (Jentoft, 1998; Jentoft and Mikalsen, 2004; Keskitalo 2009). It is perhaps not surprising, when looking at the Norwegian example, that fisheries management and governance decisions are considered by many to play a more important role than climate change in determining the fate of northern fish stocks (Vilhjálmsson *et al.*, 2005; Browman, 2008; Eide, 2008). In partial support of this position we argue that climate change is an added stressor to the current resource management regimes, and may in fact serve to increase local vulnerability. For example if regulations expand current fishing zones farther out to sea to enable coastal fishers to reach more distant fish stocks, and this coincides with more frequent extreme weather conditions in these same areas, safety hazards may increase for small fishing vessels. Climate change may increase the need for adaptation and flexibility in the fisheries because of a shift in fish species and interactions, and changing weather and navigation conditions at sea (Hovelsrud and West, 2008).

Fishing seasons, quotas and management of individual fish stocks do not take into account the variable weather patterns and longer-term climate trends which often determine when, where and under what conditions coastal fishing activities can take place. These are conditions that, according to fishers, scientists and managers, are already being affected by climate change (Browman, 2008; Hovelsrud *et al.*, 2010). In the past coastal fishers were also farmers or held part-time jobs to ensure a stable year-round income. This was a successful adaptation to a seasonal and variable resource base (Brox, 1966). However, the current national income tax system and quota maintenance requirements for fishers restrict multiple incomes, creating barriers for livelihood diversification and potential obstacles for maintaining vibrant coastal fishing communities.

Economically, coastal fisheries are increasingly integrated and exposed to global market mechanisms (e.g. Keskitalo, 2008; Kaskitalo and Kulyasova, 2009). The current global financial crisis has led to a reduced demand for the exclusive Norwegian cod that has coincided with excellent fishing conditions in many parts of Norway. At the same time, a reduction in the number of landing facilities, along with a weakened economy, have led to reduced storage capacity at remaining landing sites, resulting in cases where coastal fishers have been denied the opportunity to land their catch. The linkages between small coastal fishing communities and the broader international and economic contexts are clear, and illustrate the complexities of understanding local vulnerability and adaptation concerns. Supply of fish depends on environmental, including climatic conditions, while access to fish depends on fishing regulations, technology choice, weather conditions, and price and cost factors (West and Hovelsrud, 2010).

Conclusions

Coastal fishing communities are accustomed to dealing with variability and change in local climate, weather, fish stock abundance and distribution, and, more recently, changing fishing regulations and management. Our research suggests that three interlinked factors are currently of concern for coastal fisheries: changes in bio-physical conditions (ocean and fish); fisheries management and regulations (vessel size, species, quotas) and societal conditions (outmigration, economics, and transfer of knowledge, among others). The interlinkages between these elements in shaping coastal fishers' vulnerability and adaptation have first and foremost been captured and understood through local involvement in our research. Climatic, social, and economic factors together shape coastal fishing activities, and must be considered holistically in order to understand local vulnerability and adaptation challenges.

Local observations of increasing ocean temperatures and shifting fish distribution across our case sites corroborate scientific data and projections, despite the fact that there remain uncertainties about future changes in both knowledge fields. The method by which the knowledge we have presented in this chapter has been obtained differs from more traditional scientific approaches, and is driven by different needs for understanding (see Hovelsrud and Winsnes, 2006). Based on our findings we suggest that considering both local and scientific knowledge systems together leads to a broader understanding of the complexities inherent in fisheries' vulnerability and adaptation issues. Incorporating local knowledge into the research design of projects will likely improve the relevance of questions asked during the research for the communities in which research takes place. The latter statement is supported by our empirical findings that show that a combination of climate elements, rather than individual climate parameters are of critical importance locally and should spur further community relevant climatological research, and that current fishing regulations may increase future vulnerability in the context of climate change if they do not consider changing environmental conditions. Adjustments in these regulations will require policy makers to better understand the consequences of climate change for coastal fisheries. There are increasingly complex social and economic reasons for why coastal fisheries in Northern Norway may be vulnerable to the impacts of climate change. The issue of poor recruitment of young fishers to the coastal fisheries shows that adaptation challenges arise due to a range of factors, including inflexible tax regulations and high entry costs. The consequences are a breakdown in the transfer of local fisheries knowledge, and an uncertain future for coastal fisheries, which remain a cornerstone for local culture and identity in the case communities.

Local interest in and trust of scientific information may improve with an exchange of observations and knowledge. Fishers express dissatisfaction with what they experience as lack of interest in their knowledge from the marine science community: there is no tradition for broad consultations between scientists and individual local fishers about the fisheries resources and their harvests, and needs based on social and economic premises (e.g. Jentoft and Mikalsen, 2004). However, involving local actors in the research design and research process and thereby making the research relevant and useful locally will increase overall awareness and provide the communities with more tools for increasing their adaptive capacity. Our empirical findings show a high level of detailed local knowledge about relationships between fisheries activities, changing climatic conditions, fisheries management and regulations, and economics not previously addressed holistically in research on fisheries and climate change. We suggest that it is of critical importance for community empowerment – an element of adaptive capacity – that local actors are consulted early on, and throughout the research, and that they and their knowledge be treated as equal partners in scientific investigation. Only in this way can research on and for adaptation lead to successful and legitimate actions and policy interventions.

References

ACIA (2005) *Impacts of a Warming Arctic: Arctic Climate Impact Assessment*, Cambridge: Cambridge University Press.

Barstad, I., Sorteberg, A., Flatøy, F. and Déqueet, M. (2009) 'Precipitation, temperature and wind in Norway: dynamical downscaling of ERA40', *Climate Dynamics*, 33: 769–776.

Berge, G. (1996) *Tørrfisk: thi handlet du red'lig, og tørket din fisk: en bok om tørrfiskkultur, Nord-Norge og Bergen*, Norway: Orkana.

Berkes, F. and Jolly, D. (2001) 'Adapting to climate change: social-ecological resilience in a Canadian Western Arctic community', *Conservation Ecology*, 5 (2): 18–31.

Berkes, F., Folke, C. and Gadgil, M. (1995) 'Traditional ecological knowledge, biodiversity, resilience and sustainability', in C. Perring, K-G. Mäler, C. Folke, C. S. Holling, and B.O Janson (eds) *Biodiversity Conservation: Policy Issues and Options*, Dordrecht: Kluwer Academic Publishers.

Browman, H.I. (ed.) (2008) *Report on the Conference: Fisheries Management and Climate Change in the Northeast Atlantic Ocean and the Baltic Sea. Implications for Resource Management Policy*, Copenhagen: Tema Nord, Nordic Council of Ministers.

Brox, O. (1966) *Hva Skjer i Nord-Norge? En Studie i Norsk Utkantpolitikk*, Oslo: Pax Forlag.

Drinkwater, K.F. (2005) 'The response of Atlantic cod (Gadus morhua) to future climate change', *ICES Journal of Marine Science*, 62: 1327–1337.

Coulthard, S. (2009) 'Adaptation and conflict within fisheries: insights for living with climate change', in W.N. Adger, I. Lorenzoni, and K. O'Brien (eds) *Adapting to Climate Change: Thresholds, Values, Governance*, Cambridge: Cambridge University Press.

Eide, A. (2008) 'An integrated study of economic effects of and vulnerabilities to global warming on the Barents Sea cod fisheries', *Climatic Change*, 87: 251–262.

Ford, J. and Smit, B. (2004) 'A framework for assessing the vulnerability of communities in the Canadian Arctic to risks associated with climate change', *Arctic*, 57: 389– 400.

Førland, E. J. (ed.), Benestad, R.E., Flatøy, F., Hanssen-Bauer, I., Haugen, J.E., Isaksen, K., Sorteberg, A. and Ådlandsvik, B. (2009) *Climate Development in North Norway and the Svalbard Region During 1900–2100. Temarapport for NorACIA*, Polarinstituttets rapportserie, 128, Tromsø: Norwegian Polar Institute.

Gearheard, S., Matumeak, W., Angutikjuaq, I., Maslanik, J., Huntington, H.P., Leavitt, J., Kagak, D.M., Tigullaraq, G. and Barry, R. G. (2006) 'It's not that simple: a collaborative comparison of sea ice environments, their uses, observed changes, and adaptations in Barrow, Alaska, USA, and Clyde River, Nunavut, Canada', *Ambio*, 35 (4): 203–211.

Hamilton, L. (2004) 'The Polar Low phenomenon', *GEO Quarterly*, 1: 10–12.

Hanssen-Bauer, I. (ed.) (2009) *Klima i Norge 2100. Bakgrunnsmateriale til NOU Klimatilpasning*, Oslo and Bergen: Norsk Klimasenter.

Hovelsrud, G.K. and West, J.J. (2008) 'Socioeconomic consequences of climate change in fisheries: a progress report of ongoing research' in H.I. Browman (ed.) *Report on the Conference Fisheries Management and Climate Change in the Northeast Atlantic Ocean and the Baltic*

Sea. Implications for Resource Management Policy, Bergen, Norway, Copenhagen: TemaNord, Nordic Council of Ministers.

Hovelsrud, G.K. and Winsnes, C. (eds) (2006) *Conference Proceedings: NAMMCO Conference on User Knowledge and Scientific Knowledge in Management Decision Making, Reykjavik, Iceland*, Tromsø: The North Atlantic Marine Mammal Commission.

Hovelsrud G.K. and Smit, B. (eds) (2010) *Community Adaptation and Vulnerability in the Arctic Regions*, Heidelberg: Springer.

Hovelsrud, G. K., Dannevig, H., West, J.J. and Amundsen, H. (2010) 'Adaptation in fisheries and municipalities: three communities in Northern Norway', in G.K. Hovelsrud and B. Smit (eds) *Community Adaptation and Vulnerability in the Arctic Regions*, Heidelberg: Springer.

Jentoft, S. (ed.) (1998) *Commons in a Cold Climate: Coastal Fisheries and Reindeer Pastoralism in Norway: The Co-management Approach*, Man and the Biosphere Series 22, Paris and New York: UNESCO Parthenon.

Jentoft, S. and Mikalsen, K.H. (2004) 'A vicious circle? The dynamics of rule making in Norwegian fisheries', *Marine Policy*, 28: 127–135.

Keskitalo, E.C.H. (2004) 'A framework for multi-level stakeholder studies in response to global change', *Local Environment*, 9 (5): 425–435.

Keskitalo, E.C.H. (2008) *Climate Change and Globalization in the Arctic: An Integrated Approach to Vulnerability Assessment*, London: Earthscan Publications.

Keskitalo, E.C.H. and Kulyasova, A.A. (2009) 'The role of governance in community adaptation to climate change', *Polar Research*, 28 (1): 60–70.

Lim, B., Spanger-Siegfried, E., Burton, I., Malone, E. and Huq, S. (2004) *Adaptation Policy Frameworks for Climate Change: Developing Strategies, Policies and Measures*, New York: United Nations Development Programme.

Loeng, H. (ed.) (2008) *Klimaendringer i Barentshavet - Konsekvenser av økte CO2-nivåer i atmosfæren og havet*, Report No. 126, The Norwegian Polar Institute (in Norwegian).

Loeng, H. and Drinkwater, K. (2007) 'An overview of the ecosystems of the Barents and Norwegian Seas and their response to climate variability', *Deep-Sea Research II*, 54: 2478 –2500.

Loeng, H., Brander, K., Carmack, E., Denisenko, S., Drinkwater, K., Hansen, B., Kovacs, K., Livingston, P., and McLaughlin, E.S. (2005) 'Marine systems', Chapter 9, *Arctic Climate Impact Assessment*, Scientific Report, New York: Cambridge University Press.

Marino, E. and Schweitzer, P. (2009) 'Talking and not talking about climate change' in S. Crate and M. Nuttall (eds) *Anthropology and Climate Change*, Walnut Creek, CA: Left Coast Press.

Ministry of Fisheries and Coastal Affairs (MoFC) (2007) *Norwegian Fisheries Management*, Oslo: Ministry of Fisheries and Coastal Affairs. Online. Available HTTP: http://www.regjeringen.no/en/dep/fkd/Documents/Legal-handbooks-and-guidelines/2007/norwegian-fisheries-management.html?id=464195 (accessed 30 January 2009).

Norwegian Directorate of Fisheries (2010) 'Registered main occupation fishermen by age in the years 1983–2012 by municipality'. Available online: http://www.fiskeridir.no/english/statistics/norwegian-fisheries/norwegian-fishing-vessels-fishermen-and-licenses (accessed 10 April 2010).

Sundby, S. and Nakken, O. (2008) 'Spatial shifts in spawning habitats of Arcto-Norwegian cod related 828 to multi-decadal climate oscillations and climate change', *ICES Journal of Marine Science*, 65: 953–962.

Smit, B., Hovelsrud, G. and Wandel, J. (2008) *CAVIAR: Community Adaptation and Vulnerability in Arctic Regions*, University of Guelph, Department of Geography, Occasional Paper No. 28.

Turner, B.L., Kasperson, R.E., Matson, P.A., McCarthy, J.J., Corell, R.W., Christensen, L., Eckley, N., Kasperson, J.X., Luers, A., Martello, M.L., Polsky, C., Pulsipher, A. and Schiller, A. (2003) 'A framework for vulnerability analysis in sustainability science', *Proceedings of the National Academy of Sciences*, 100 (14): 8074–8079.

Tyler, N., Turi, J.M., Sundset, M.A., Strøm Bull, S.K., Sara, M.N., Reinert, E., Oskal, N., Nellemann, C., McCarthy, J.J., Mathieson, S.D., Martello, M.L., Magga, O.H., Hovelsrud, G.K., Hanssen-Bauer, I., Eira, N.I., Eira, M.G. and Corell, R.W. (2007) 'Saami reindeer pastoralism under climate change: applying a generalized framework for vulnerability studies to a sub-Arctic social-ecological system', *Global Environmental Change*, 17: 191–206.

Vilhjálmsson, H., Hoel, A.H., Agnarsson, S., Arnason, R., Carscadden, J.E., Eide, A. and Fluharty, D. (2005) 'Fisheries and aquaculture', in *Arctic Climate Impact Assessment*, Scientific report, Cambridge: Cambridge University Press.

West, J. J. and Hovelsrud, G.K. (2010) 'Cross-scale adaptation challenges in the coastal fisheries: findings from Lebesby, Northern Norway', *Arctic*, 63 (3): 338–354.

13 Stuck in the twilight zone or moving towards sustainable climate adaptation?

Integrated Coastal Zone Management in Coastby

Sofie Storbjörk and Johan Hedrén

It is widely accepted today that the societal consequences of our changing climate need to be dealt with in parallel to emission reductions. Climate adaptation embraces attempts to respond to current climate vulnerability, extremes and future climate change and is currently recognised as a major challenge to both developing and developed countries (Adger *et al.*, 2007; Gagnon-Lebrun and Agrawala, 2007). Adaptation to climate variability and extremes is often seen as an important starting point for targeting future climate change and represents the most common approach taken so far (Burton, 2004; Næss *et al.*, 2005; Tompkins, 2005; Schipper and Pelling, 2006; Wall and Marzall, 2006; Adger *et al.*, 2007). In the climate policy literature, a broadening of the current policy-agenda, explicitly linking climate change to sustainable development, is suggested to avoid negative trade-offs and identify potential synergies (Tompkins and Adger, 2005; Linnér, 2006; Burch and Robinson, 2007; Swart and Raes, 2007).

For adaptation to be sustainable, the long-term social, economic and environmental consequences inherent in and emanating from current adaptation practices need to be clarified (Eriksen and O´Brien, 2007; Ulsrud *et al.*, 2008). Such clarifications are equally important for developing and developed countries. At the same time, current research indicates a predominant focus on short-term reactive considerations in climate adaptation that risk furthering business-as-usual, rather than triggering reflections on our current development in a way that would make possible a more proactive, integrated and long-term sustainable adaptation (Brown, 2011). In this context it is important to also be attentive to the conflicting interpretations, tensions and contentious priorities inherent in the sustainable development policy-framework (Owens, 2003; Stevenson and Richardson, 2003; Hedrén, 2009; Hedrén and Linnér, 2009; Storbjörk *et al.*, 2009; Brown, 2011).

One approach with the potential to broaden the current agenda for climate adaptation in coastal zones is the principles of Integrated Coastal Zone-Management (ICZM), developed at the EU-level in the last decade. ICZM is framed as an integrated and strategic approach aimed at informed stakeholder involvement in addressing social/cultural, economic and environmental sustainability in coastal zones. It takes a long-term perspective based on the precautionary principle and the needs of present and future generations to make human activities more "environmentally friendly, socially responsible and economically sound" (EC 2002/413/EC; EC COM/2000/547). An evaluation completed in 2006 however concluded that no country had yet implemented an ICZM National Strategy (Rupprecht Consult, 2006). Later studies have shown evidence of the challenges in implementing practices of ICZM (e.g. Deboudt *et al.*, 2008; McKenna *et al.*, 2008). Still, a number of municipalities have chosen to adopt an ICZM-approach in managing their coastal zones. In a pioneer Swedish municipality with the assumed name of Coastby (Storbjörk and Hedrén, 2011), a policy-shift can today be witnessed where the former reactive approach with its reliance on technical fixes, hard shoreline protection, and business-as-usual regarding waterfront development has – at least on paper – been challenged by a more integrated, long-term and potentially sustainable approach of ICZM.

By taking the Coastby case as a starting point, the aim of this chapter is to discuss the current policy-shift as well as highlight a few critical points that seem to hinder the process of moving towards a more proactive climate adaptation in the form of ICZM. The chapter is empirically based on local documentation on ICZM and climate change as well as interviews with key officials and politicians.

In transition towards local ICZM?

Coastby is a small municipality populated by 27,000 inhabitants located in the south-east of Sweden with 40 kilometres of coastline. The coastline holds natural reserves, pastures, beaches, and infrastructure, such as summer houses, harbours and sewage treatment works, and is valuable in terms of recreation, employment and residential areas. Over the years, large areas of land have eroded and houses have disappeared into the sea. These problems are expected to increase with climate change and rising sea levels (Coastby, 2008). According to interviews, the awareness of eroding land in Coastby can be dated back to the 1820s, but the local administration has only worked actively with erosion-management since 1958.

In the 1980s Coastby was involved in two conferences with engineers at a nearby university, one in 1986 on the use and protection of coastal areas and the other in 1989 on the impacts of climate change and coastal planning. As a result of the first conference, an expert committee consisting of two leading Swedish scientists and two local officials was formed in 1987 to assess conditions and prioritise measures for local erosion-management (Coastby, 1988). The inventory also triggered an explicit local position in erosion-management, revealing strong disagreement between experts and local politicians. The experts explicitly stated how unsatisfactory it was that their recommendations were not politically accepted and warned that the politically preferred approach, relying strongly on hard shoreline protection rather than working with nature by, for example, removing summer cottages most at risk, would aggravate the situation and move erosion downstream. With reference to future sea level rise and the need for long-term coastal planning, problems with allowing attractive waterfront buildings were stressed (Coastby, 1988). These words of caution did not, however, influence local erosion-management. The technical adaptation undertaken in the coming years instead was based on a predominant trust in the effectiveness of hard shoreline protection. Over the years, erosion kept pushing the coastline further inland, resulting in the loss of camping sites, houses and land in unprotected areas, of which some are part of the EU biodiversity network of protected areas, Natura 2000. The solid constructions, together with the building of harbours, have tended to limit public access to beaches and aesthetic values while also drastically reducing the availability of sand and creating new problems by moving erosion downstream (Coastby, 2004; 2007a).

From local interviews we learnt that in the last few years, several concurrent factors have triggered the call for a new strategy for ICZM in Coastby. For example, a Centre for Erosion-Management was formed in 2004 as a joint venture between coastal municipalities, expert authorities and scientists aimed at developing flexible, cost-efficient and environmentally adjusted erosion-management techniques. Further, Coastby was directly involved in several EU-funded projects aimed at providing tools for ICZM, increased knowledge-exchange and learning between countries. Erosion-management has also experienced a heightened public and political priority due to the climate debate. Finally, the need for a more long-term and holistic approach was identified when processing environmental action plans of the Swedish Environmental Goal 'A Balanced Marine Environment, Flourishing Coastal Areas and Archipelagos' (Coastby, 2005).

A new policy based on ICZM was processed, mandated by the local environmental drafting committee in 2007 and accepted by the local government council in 2008. It is stated in the policy that Coastby needs to be adapted to meet future threats such as climate change within the framework of sustainable coastal zone management, with the long-term aim of balancing economic, social and cultural goals with environmental and recreational interests (Coastby, 2007b; 2008). A change from a reactive to a proactive approach is warranted in areas where further erosion would have severe consequences for the public, or unacceptable loss of valuable land such as settlements, buildings, infrastructure and sites for nature protection, recreation and tourism. Further the beaches need to be preserved.

The basic idea behind the policy is that nature should take its course and that erosion is a part of natural coastal dynamics. It is stated that in areas where existing settlements, facilities and infrastructure are threatened by erosion, re-localisation needs to be considered as a first measure. As a secondary measure soft methods such as beach nourishment should be evaluated. Solid constructions are only to be considered 'after alternatives with less environmental impact have been evaluated and eliminated' (Coastby, 2007b, 2008: 12). The policy is thus rather strict and brings back suggestions made in the 1980s by the expert committee. If properly implemented, the policy would involve a clear shift in how erosion is approached with the potential to achieve a more long-term, integrated and sustainable coastal zone management. However, the path forward in turning words to action is far from uncomplicated.

Critical points

A number of critical points can be raised based on the Coastby case that are likely to influence the ability to achieve long-term, integrated and sustainable adaptation, as outlined in the new policy.

The technical fix of waterfront planning

The new policy states that considerations to climate change are to be made in waterfront planning by – as far as possible – minimising the risk of damage. Moreover it states that development is to be *avoided* in areas exposed to erosion, flooding and increased groundwater levels thus following a precautionary principle (Coastby, 2008). The practice in Coastby and many other Swedish municipalities, however, reveals a predominant reliance on technological support systems and attempts to assess what is secure or not rather than employing a precautionary approach. Such technical fixes and dependence on scenarios may in fact increase vulnerability, as it does not deal with the causes of the problems such as where we choose to live in the first place (Changnon, 2000; Sarewitz *et al.*, 2003). Strategies to manage flood-risks have led to the justification of further exploitation and development in flood-prone areas which in fact have increased social vulnerability (Bass and Etkin, 2004).

To follow the new policy, there is a need for tough decisions counteracting the otherwise strong political emphasis on attractive waterfront planning. Recent media reporting in the local newspaper highlights disagreements between the municipality and private landowners, where the latter, call for unrestricted building-permits and hard shoreline protection. Such collision of values and priorities related to the safety vs. scenery divide (Storbjörk, 2007) need to be brought into the open and mediated. Similar problems have been identified in studies in the United States and the United Kingdom, where dilemmas over appropriate planning for future settlement and infrastructure, as well as short-sighted pro-development policies hamper forward-looking coastal management (Moser, 2005; Few *et al.*, 2007; Nicholson-Cole and O'Riordan, 2009).

Erosion-management for whom?

In the British context, one of the key challenges of adaptive coastal governance is accounting for different expectations and understandings in defining what a future sustainable coast is (Nicholson-Cole and O'Riordan, 2009). In the Coastby ICZM-policy, the involvement of citizens is emphasised but nowhere are the divergent interests of citizens in coastal zone management acknowledged. Instead of assuming that the public speaks with one voice, the question 'sustainable for whom and in what perspective' needs to be posed. As a way to broadly sort out different citizen stakes, Pilkey states that 'the erosion problem is caused by people who have purchased beachfront property adjacent to a retaining shoreline, a relatively small number of people compared to beach users' and that beachfront property owners tend to be comparatively influential people with the ability and will to defend their houses (Pilkey, 2000: 159). This is easily seen in Coastby, where actors from the local landowners' association play a strong lobbying role towards national authorities and politicians (Storbjörk and Hedrén, 2011). According to local interviews, private landowners have stacked piles of rock on the beaches outside their houses which has transformed waterfront areas, limited public access and moved erosion downstream.

In contrast to private property-defenders, public investments are specifically targeted at preserving public recreational beaches and waterfront areas attractive for citizens and tourists, due to their important monetary and recreational values. The difficulty in achieving citizen-engagement in erosion-management is, however, a local reality and, according to local interviews, the reason behind a local exhibition on coastal erosion. Local interviews indicate that comments in the local media show a divergent public, where some acknowledge the measures taken by the local administration and others speak of them as literally throwing taxpayers' money into the sea as nature cannot be tamed complaining that taxpayers' money is invested in preserving the interests of private landowners rather than citizens at large. Following this there is no unified view among the public of what a sustainable future coast is. The lack of focus on social sustainability and justice appears to be symptomatic in the Swedish context where justice between the global North and South appears to be more easily accommodated in the sustainable development framework than justice among different groups within Swedish society, at least on a rhetorical level (Bradley *et al.*, 2008).

Colliding institutional frameworks and policy agendas

ICZM specifically aims at improved coordination among policies pertaining to the coastal zone (EC 2002/413/EC). The Coastby case instead reveals that conflicting formal institutional frameworks and lack of sufficient vertical administrative interplay between authorities at national, regional and local scales obstruct current erosion-management. For example, the soft methods for erosion-management outlined in the ICZM-framework generally mean working with rather than against nature by restoring sand, strengthening sandbanks and planting vegetation (Coastby, 2007b). Interviews show that after being permitted to extract 500 000 m^3 of sand as beach nourishment in 2001, local officials have written several applications for permission to nourish beaches at a larger scale, which has been done in several erosion-burdened countries, but for a long time the applications have been rejected and passed between different national authorities and legislative settings. Further, the lack of coherent coastal zone policies at regional and national levels means that the views, interpretations and knowledge of individual administrative officials largely affect the standpoint of authorities in practical erosion-management. What is allowed or not varies from time to time, which is seen as highly problematic by the local administration (Storbjörk and Hedrén, 2011). Similar challenges in vertical administrative interplay have also been noted in a British context (Few *et al.*, 2007).

There is also a difficulty emanating from conflicting policy agendas, where ambitions in coastal natural reserves and Natura 2000-areas of seeing erosion as part of the natural dynamics and letting nature take its course clash with ambitions in local erosion management to protect different societal values at risk. As a further conflicting agenda in the coastal-zone, there is a continuous pressure on politicians and planners to keep planning and developing attractive waterfront areas despite identified climate risks. Such tensions between policy-agendas, values and priorities among environmental officials, erosion-managers and planners and their respective implications for long-term social, economic and environmental sustainability are crucial to acknowledge and openly deal with in trying to live up to the new proactive ICZM policy.

Horizontal integration, ownership and learning

The new integrated policy in Coastby calls for more cross-sectoral engagement, interaction and knowledge-exchange in erosion management (Coastby, 2008). But in interviews, however, erosion management is described as 'a one-man show' with a strong and committed key official with erosion as his chief professional concern (Storbjörk and Hedrén, 2011). Because erosion has been so firmly anchored in the work of the technical unit with a few specialised actors building internal competence, networking with external actors and pushing for change has meant a corresponding lack of horizontal integration and knowledge-exchange across important administrative boundaries. Managing erosion has been seen as a predominant technical problem not influencing the agendas or practical activities of other sectors such as planning and environment. Differing professional cultures and interests as well as professional integrity and interdepartmental rivalry (e.g. not having the head of one administrative unit telling the other what to do) seems to stand in the way of more integrated approaches, leading to lack of mutual ownership and learning between administrative units and officials in managing the coastal zone (Storbjörk, 2010). Criticism has also been directed at local politicians for not being sufficiently involved in coastal zone management in Coastby (Ernst and Young, 2007).

If the new policy is to succeed in better balancing present and future economic, social and environmental interests, the sectoral tendencies that so far appear to have characterised the practice of adapting to coastal erosion need to be overcome and local key actors need to work together on how to manage coastal erosion. That limited internal coordination has had a negative influence on the outcome of local adaptation processes has been noted in several parallel studies both in Sweden and other countries (Moser, 2005; Lidskog and Uggla, 2009; Nicholson-Cole and O'Riordan, 2009; Glaas *et al.*, 2010).

Concluding remarks

Attempts at taking a long-term and broad approach to the coastal zone problematic are essential in light of climate change. Here the policy-framework of Integrated Coastal Zone Management (ICZM) is promising as it strives to do just that. However it also needs to be able to account for tensions, conflicting values, interests and priorities that are a critical part of both current and future coastal zone management in practical planning and decision-making. Contrary to the consensual policy framing of sustainable development, where environmental, economic and social perspectives are framed as being easily reconciled, sustainability is in practice inherently political and contested (Hedrén, 2002; Stevenson and Richardson, 2003; Storbjörk *et al.*, 2009; Brown, 2011).

Adopting the new policy on paper represents an important first step towards a more integrated, proactive and long-term approach to coastal zone management but, the real challenges are expected to appear when transforming the words into concrete action. In this chapter we have highlighted and briefly discussed a few critical points that are likely to influence the ability of actually implementing the ICZM policy. Specifically, we suggest there is a need to overcome the technical fix of waterfront planning, account for different interests, expectations and understandings of how to best manage the eroding coastlines and what a future sustainable coast is, openly dealing with conflicting institutional frameworks and policy-agendas and, finally, finding ways of facilitating cross-sectoral integration, ownership and learning. Besides bringing the different interests, values and priorities at hand into the open to be explicitly acknowledged, and ideally even reconciled, a more sustainable adaptation would also require increased awareness of the social, economic and environmental linkages of our current way of approaching coastal zones.

Acknowledgements

We are grateful to the following: the Swedish Research Council for Environment, Agricultural Sciences and Spatial Planning for funding the research project 'Planning for climate adaptation – adaptive capacity and conditions for change'; the anonymous reviewers and editors for providing valuable comments, reflections and suggestions on how to improve the manuscript; the interviewees in Coastby for generously sharing their viewpoints and experiences of coastal zone management and climate change.

References

Adger, W.N., Agrawala, S., Mirza, M.M.Q., Conde, C., O'Brien, K., Pulhin, J., Pulwarty, R., Smit B., and Takahashi, K. (2007) 'Assessment of adaptation practices, options, constraints and capacity', in M.L. Parry, O.F. Canziani, J.P. Palutikof, P.J. van der Linden and C.E. Hanson (eds) *Climate Change 2007: Impacts, Adaptation and Vulnerability, Contribution of Working Group II to the Fourth Assessment Report of the Intergovernmental Panel on Climate Change*, Cambridge, UK: Cambridge University Press.

Bass, B. and Etkin, D. (2004) 'How should Canada adapt to climate change? Debating the merits of a technological fix versus a behavioural approach', in A. Fenech, D. Maciver, H. Auld, R. Bing Rong, and Y. Yin (eds) *Climate Change: Building the Adaptive Capacity*, Toronto, Canada: Ministry of Public Works and Government Services.

Bradley, K., Gunnarsson-Östling, U. and Isaksson, K. (2008) 'Exploring environmental justice in Sweden – how to improve planning for environmental sustainability and social equity in an "eco-friendly" context', *Projections*, 8: 69–78.

Brown, K. (2011) 'Sustainable adaptation: An oxymoron?', *Climate and Development*, 3: 21–31.

Burch, S. and Robinson, J. (2007) 'A framework for explaining the links between capacity and action in response to global climate change', *Climate Policy*, 7: 304–316.

Burton, I. (2004) 'Climate change and the adaptation deficit', in A. Fenech, R. Bing Rong, D. MacIver and H. Auld (eds) *Climate Change: Building the Adaptive Capacity. An International Conference on Adaptation Science, Management and Policy Options*, Toronto: Meteorological Service of Canada.

Changnon, S.A. (2000) 'Flood prediction: Immersed in the quagmire of national flood mitigation strategy', in D. Sarewitz, R.A. Pielke Jr. and R. Byerly Jr. (eds) *Prediction. Science, Decision-Making, and the Future of Nature*, Washington DC: Island Press.

Coastby (1988) *Arbetsgruppen för strandskydd. Resultatet av ett års arbete*, Coastby: Coastby Municipality (in Swedish).

Coastby (2004) *Erosionsrapport over kusten i Coastby kommun. En nulägesbeskrivning*, Coastby: Coastby Municipality (in Swedish).

Coastby (2005) *Agenda 21 med miljöhandlingsprogram*, Coastby: Coastby Municipality (in Swedish).

Coastby (2007a) *Erosion – Hav och stad i en evig kamp*, Coastby: Coastby Municipality (in Swedish).

Coastby (2007b) *Policy för förvaltning och skydd av kusten. Bakgrundsmaterial*, Coastby: Coastby Municipality (in Swedish).

Coastby (2008) *Policy för förvaltning och skydd av kusten*, Coastby: Coastby Municipality (in Swedish).

Deboudt, P., Dauvin, J. and Lozachmeur, O. (2008) 'Recent developments in coastal zone management in France: The transition towards integrated coastal zone management (1973–2007)', *Ocean & Coastal Management*, 51: 212–228.

EUR-Lex (2000) *Communication on ICZM: A Strategy for Europe*. EC COM/2000/547 Online. Available HTTP: http://eur-lex.europa.eu/LexUriServ/LexUriServ.do?uri=CELEX:52000DC0547:EN:NOT accessed 7 October 2012.

EUR-Lex (2002) *Recommendation concerning the implementation of ICZM in Europe*. EC 2002/413/EC Online. Available HTTP: http://eur-lex.europa.eu/LexUriServ/LexUriServ.do?uri=CELEX:32002H0413:EN:NOT accessed 7 October 2012.

Eriksen, S.H. and O'Brien, K. (2007) 'Vulnerability, poverty and the need for sustainable adaptation measures', *Climate Policy*, 7: 337–352.

Ernst & Young (2007) *Revisionsrapport nr 4 2007*, Coastby: Coastby Municipality (in Swedish).

Few, R., Brown, K. and Tompkins, E.L. (2007) 'Climate change and coastal management decisions: Insights from Christchurch Bay, UK', *Coastal Management*, 35: 255–270.

Gagnon-Lebrun, F. and Agrawala, S. (2007) 'Implementing adaptation in developed countries: An analysis of progress and trends', *Climate Policy*, 7: 392–408.

Glaas, E., Jonsson, A., Hjerpe, M. and Andersson-Sköld, Y. (2010) 'Managing climate change vulnerabilities: formal institutions and knowledge use as determinants of adaptive capacity at the local level in Sweden', *Local Environment*, 15: 525–539.

Hedrén, J. (2002) 'Critical notes on sustainability and democracy', in U. Svedin, and B. Hägerhäll Aniansson (eds) *Sustainability, Local Democracy and the Future: The Swedish Model*, Dordrecht: Kluwer Academic Publishers.

Hedrén, J. (2009) 'Shaping sustainability: Is there an unleashed potential in utopian thought?' *Futures*, 41: 220–225.

Hedrén, J. and Linnér, B-O (2009) 'Utopian thought and the politics of sustainable development', *Futures*, 41: 210–219.

Interview with former head of the technical unit/ Director of strategic affairs.

Interview with official with strategic responsibility for environment and climate.

Interview with official formerly in charge of erosion at the County Administrative Board.

Interview with official in charge of erosion at the State Geotechnical Institute. Interview with politician currently holding the chair of the local environmental drafting committee.

Interview with politician formerly holding the chair of the local environmental drafting committee.

Interview with Professor at a recent University with a long professional involvement in erosion-management.

Interview with leader of the local action-group and former landowners association.

IPCC (2007) 'Summary for policymakers', in M.L. Parry, O.F. Canziani, J.P. Palutikof, P.J. van der Linden and C.E. Hanson (eds) *Climate Change 2007: Impacts, Adaptation and Vulnerability, Contribution of Working Group II to the Fourth Assessment Report of the Intergovernmental Panel on Climate Change*, Cambridge: Cambridge University Press.

Lidskog, R. and Uggla, Y. (2009) 'Lokalt klimatarbete: kommunen som lärande organisation', in Y. Uggla, Y. and I. Elander (eds) *Global Uppvärmning och Lokal Politik*, Stockholm: Santérus Academic Press.

Linnér, B-O. (2006) 'Authority through synergism: The roles of climate change linkages', *European Environment*, 16: 278–289.

McKenna, J., Cooper, A. and Ohagan, A. (2008) 'Managing by principle: A critical analysis of the European principles of Integrated Coastal Zone Management (ICZM)', *Marine Policy*, 21: 941–955.

Moser, S. (2005) 'Impact assessments and policy responses to sea-level rise in three US states: An exploration of human-dimension uncertainties', *Global Environmental Change*, 15: 353–369.

Næss, L-O., Bang, G., Eriksen, S. and Vevatne, J. (2005) 'Institutional adaptation to climate change: Flood responses at the municipal level in Norway', *Global Environmental Change*, 15: 125–138.

Nicholson-Cole, S. and O'Riordan, T. (2009) 'Adaptive governance for a changing coastline: Science, policy and publics in search of a sustainable future', in W.N. Adger, I. Lorenzoni and K.L. O'Brien (eds) *Adapting to Climate Change. Thresholds, Values, Governance*, Cambridge: Cambridge University Press.

Owens, S. (2003) 'Is there a meaningful definition of sustainability?' *Plant Genetic Resources*, 1 (1): 5–9.

Pilkey, O.H. (2000) 'What you know can hurt you: Predicting the behaviour of nourished beaches', in D. Sarewitz, R.A. Pielke Jr. and R. Byerly Jr (eds) *Prediction. Science, Decision Making and the Future of Nature*, Washington, DC: Island Press.

Rupprecht Consult (2006) *Evaluation of Integrated Coastal Zone Management (ICZM) in Europe. Final Report.* Online. Available HTTP: http://ec.europa.eu/environment/iczm/pdf/evaluation_iczm_report.pdf (accessed 7 October 2012).

Sarewitz, D. Pielke Jr, R. and Keykhah, M. (2003) 'Vulnerability and risk: Some thoughts from a political science perspective', *Risk Analysis*, 23: 805–810.

Schipper, L. and Pelling, M. (2006) 'Disaster risk, climate change and international development: Scope for, and challenges to, integration', *Disasters*, 30: 19–38.

Stevenson, R. and Richardson, T. (2003) 'Policy integration for sustainable development', *Journal of Environmental Policy & Planning*: 95–118.

Storbjörk, S. (2007) 'Governing climate adaptation in the local arena: Challenges of risk-management and planning in Sweden', *Local Environment*, 12: 457–469.

Storbjörk, S. (2010) 'It takes more to get a ship to change course. Barriers for organisational learning and local climate adaptation in Sweden', *Journal of Environmental Policy and Planning*, 12: 235–254.

Storbjörk, S. and Hedrén, J. (2011) 'Institutional capacity-building for targeting sea level rise in the climate adaptation of Swedish coastal zone management. Lessons from Coastby', *Ocean & Coastal Management*, 54: 265–273.

Storbjörk, S., Lähteenmäki-Smith, K. and Hilding-Rydevik, T. (2009) 'Conflict or consensus: The challenge of integrating environmental sustainability into regional development programming', *European Journal of Spatial Development*, 34: 1–22.

Swart, R. and Raes, F. (2007) 'Making integration of adaptation and mitigation work', *Climate Policy*, 7: 288–303.

Tegner Anker, H., Nellemann, V. and Sverdrup-Jensen, S. (2004) 'Coastal zone management in Denmark: Ways and means for further integration', *Ocean & Coastal Management*, 47: 495–513.

Tompkins, E.L. (2005) 'Planning for climate change in small islands: Insights from national hurricane preparedness in the Cayman Islands', *Global Environmental Change*, 15: 139–149.

Tompkins, E. and Adger, N.W. (2005) 'Defining response capacity to enhance climate change policy', *Environmental Science & Policy*, 8: 562–571.

Uggla, Y. (2009) 'Framtiden är inte längre vad den varit: anpassning till ett förändrat klimat', in Y. Uggla, and I. Elander (eds) *Global Uppvärmning och Lokal Politik*, Stockholm, Santérus Academic Press.

Ulsrud, K., Sygna, L. and O'Brien, K. (2008) 'More than rain: Identifying sustainable pathways for climate adaptation and poverty reduction', Oslo: The Development Fund.

Wall, E. and Marzall, K. (2006) 'Adaptive capacity for climate change in Canadian rural communities', *Local Environment*, 11: 373–397.

14 The rights of internally displaced persons in connection with natural hazard-related disasters

Vikram Kolmannskog

Introduction

There is an increasing understanding of the importance of climate change as a risk multiplier (IPCC, 2012), also in the context of human mobility. It is recognised that people move for a complex set of reasons. The majority of displaced persons remain within their country of origin (IASC, 2008a; IDMC, 2012). This chapter addresses internal displacement and rights in the context of climate change and natural hazard-related disasters. Using secondary sources such as UN reports, media reports and research papers, it draws upon some experiences relating to the 26 December 2004 tsunami and Hurricane Katrina in August 2005. The tsunami affected several Asian countries; approximately 1.7 million people were displaced in India, Sri Lanka and Indonesia alone (Inderfurth *et al.*, 2005). Katrina hit New Orleans and the Missisippi Gulf Coast region where estimates of displaced people ranged from 600 000 to 1 million (Lewis, 2009).

Developing countries are particularly vulnerable to climate change and disasters and have high social and economic losses (O'Brien *et al.*, 2012: 441). The example of Katrina illustrates that also developed countries are vulnerable, and degrees of vulnerability vary among people within a country. Furthermore, a premise behind the references to these two disasters is that the experiences, needs and rights are similar both in climate-related disasters (hurricane) and non-climate-related disasters (tsunami).

The chapter begins by presenting the 1998 Guiding Principles on Internal Displacement, the rights framework for all internally displaced persons. Then it considers what major issues arose in the context of the tsunami and Katrina and how they were framed as human rights concepts such as non-discrimination, participation and durable solutions to displacement. Next the chapter highlights some international developments in the aftermath of the two disasters. The chapter concludes with a short consideration of how rights can be further promoted.

Internally displaced persons and rights

Several rights and systems of law, in particular human rights, have been identified as relevant for internally displaced persons (IDPs). Rather than creating a separate convention, the 1998 UN Guiding Principles on Internal Displacement (UN Doc. E/CN.4/1998/53/Add.2; hereafter: Guiding Principles or GP) is a synthesis of such law, specified and applied to the context of IDPs. The Guiding Principles cover three different phases of internal displacement: protection against displacement, protection during displacement, and durable solutions to displacement. Included are elements from the 1966 International Covenants on Civil and Political Rights (ICCPR) and Economic, Social and Cultural Rights (IESCR).

Strictly speaking, the principles are so-called 'soft law' and not legally binding. One can, however, refer to the 'hard law' behind them, such as rights in the International Covenants (Kälin, 2001), and the Guiding Principles have been recognized in several UN General Assembly Resolutions, including at the World Summit in 2005 (G.A. Res. 60/L.1, §132, U.N. Doc. A/60/L.1). There are also several examples of national and regional IDP legislation and policy inspired by the Guiding Principles, such as the 2009 African Union Convention for the Protection and Assistance of Internally Displaced Persons in Africa.

A major challenge for IDPs is that states, which have the primary responsibility for people on their territory, are sometimes unwilling or unable to protect them, and in some cases even deny them international assistance by referring to the principle of national sovereignty and non-interference. The introduction to the Guiding Principles clarifies that the principles presented aim to guide non-state actors, including NGOs, as well as states. Where states are not willing or able to protect, the role of the non-state actors becomes crucial both in advocacy and providing actual protection on the ground.

'Internally displaced persons' according to the broad and descriptive definition in the Guiding Principles are:

> persons or groups of persons who have been forced or obliged to flee or to leave their homes or places of habitual residence, in particular as a result of or in order to avoid the effects of armed conflict, situations of generalized violence, violations of human rights or natural or human-made disasters, and who have not crossed an internationally recognized State border.
>
> (GP, Introduction: Scope and Purpose, 2)

When the Guiding Principles were drafted some argued that only persons fleeing persecution and violence should be included, but the majority favoured the broader definition, that also included people fleeing due to disasters (Cohen, 2009). Still, many states, UN agencies, NGOs and academics have continued to neglect people displaced by disasters. Protection in natural hazard-related disasters is still poorly understood, especially by state agencies (Albuja and Cavelier, 2009), and an approach that treats people as objects of state care rather than with dignity is still prevalent (Kälin, 2009). The UN leadership role is decided upon by the UN High Commissioner for Refugees (UNHCR), the Office of the High Commissioner of Human Rights (OHCHR) and UNICEF, on a case-by-case basis, an arrangement that does not provide necessary predictability or rapidity of response. Furthermore, many organisations provide support to target groups identified in their own mandates and have historically focused on conflict IDPs (Birkeland, 2003; Kolmannskog, 2008). In Aceh, Indonesia, the government preferred labelling those uprooted by the tsunami as 'homeless', while the US government described those displaced by Katrina as 'refugees', 'evacuees' and 'disaster victims'; IDPs in their view were people displaced by conflict elsewhere (Cohen, 2009). In the following, some particular issues that arose during the two disasters are discussed and framed in rights language.

An opportunity for (non-)discrimination

From a sociological point of view, de Sousa Santos (2002: 487) speaks of citizenship as the 'sum total of rights effectively exercised' and a matter of degree. He argues that 'super-citizens' have the right connections and cash resources to be on top. The remainder of citizens make up what he calls 'the strange civil society' which comprise second, third and fourth class citizens. Many IDPs find themselves somewhere in this latter group, and within the group of IDPs some are less citizens than others in this sociological sense. At the same time, non-discrimination (and equality) is a paramount principle in human rights (GP 4, GP 22, ICCPR arts. 2(1) and 26, IESCR art. 2(2)) and modern law and justice in general (Eng, 2007: 186).

The three states where Katrina had the most impact were among the poorest in the United States with more than one out of six living in poverty prior to the hurricane (Liu *et al.*, 2011). The majority of the poor were Afro-Americans. While Afro-Americans constitute approximately 12 percent of the US population, they make up some 50 percent of the homeless population (Human Rights Committee, 2006: para 22). Due to entrenched and historic racism and classism, Katrina had a disproportionately negative impact on poor Afro-Americans (Lewis, 2009).

Evacuation is among the first possible disaster response measures. In connection with Katrina, rescue and evacuation plans were based on the assumption that people would use their private vehicles (Edwards, 2006; Human Rights Committee, 2006: para 26; Kälin, 2009). This was not discriminatory on the face, but since everyone knows that poverty is more rampant in the Afro-American communities, the latter were in effect discriminated against (Edwards, 2006). Many poor Afro-Americans did not have their own big American car that they could drive away in.

The state's duty to protect people entails an obligation to help *all* people move from zones where they face a danger (see for example the European Court of Human Rights, 2008). Effective non-discrimination may require that distinctions are made to take into account special protection needs. The Human Rights Committee (2006) was concerned about reports that the poor, and in particular Afro-Americans, were disadvantaged during Katrina. The non-existence of an adequate evacuation plan could also contribute to a breach of right to life (Edwards, 2006: 378). When prisoners were left in flooding jails, the basic rights to life and the ban on torture and inhuman treatment were compromised (Lewis, 2009: 241).

In Sri Lanka, evacuation and relocation after the tsunami resulted in discrimination of another kind. 'Buffer zones' were established along the coast in a discriminatory manner allowing construction of tourism facilities while local residents were not allowed to return and reconstruct their homes (Fletcher *et al.*, 2005; Kälin, 2009). Both the tsunami and Katrina show how issues of discrimination can arise in the context of disasters and how they can be framed in human rights language such as the principle of non-discrimination.

During displacement, pre-existing patterns of discrimination are often exacerbated (Fletcher *et al.*, 2005; IASC, 2008b). The tsunami resulted in much international aid, both in terms of money and NGO engagement. In India, Dalits ('the oppressed'; official term 'scheduled castes') were discriminated against in the distribution of aid (Asian Human Rights Commission, 2005). In Sri Lanka, aid and reconstruction was focused on the Sinhalese-dominated South, neglecting the equally hard hit, Tamil-dominated East (Fletcher *et al.*, 2005: 63). Aid was sometimes distributed first to men with the unfounded assumption that males would then automatically provide it to women and children (Fletcher *et al.*, 2005). In Thailand, Burmese migrant workers were excluded from relief efforts and many were deported back to Burma (Asian Human Rights Commission, 2005). According to General Comment 3 (1990) by the UN Committee on Economic, Social, and Cultural Rights, even developing countries with limited 'available resources' are under an immediate obligation to distribute the domestic and international aid available to them in an even-handed manner and to protect survivors from discrimination by others.

After Katrina poor Afro-Americans were portrayed as 'looting' food and other materials by the media while whites as having 'found' food (Edwards, 2006: 374). When the government issued 'shoot on sight' orders, Afro-Americans may have been particularly targeted (Ibid.). Latinos and other immigrants who did not have formal U.S. citizenship status were once again reminded that they could not even depend on basic rights to food, water, a living wage, housing, and health care (Lewis, 2009). Mississippi Governor Haley Barbour, repeatedly sought, and obtained, waivers from the Department of Housing and Urban Development (HUD) allowing the state to spend federal funds that would normally be reserved for low- to moderate-income residents on other projects (Maloy, 2007). Therefore, it can be argued that the relief and recovery efforts did not reach all people equally. The Human Rights Committee (2006: para 26) has highlighted the importance of ensuring that the rights of the poor, in particular Afro-Americans, are fully taken into consideration in the recovery and reconstruction plans.

Who gets to participate and who is pushed around?

It is often those most vulnerable to hazards that are left out of decision-making processes. Within households women and children may be ignored; within communities or within nations marginalised groups may not be heard, and social division and political power influence the agenda. As evidenced during the tsunami as well as other disasters, limited-term response and reconstruction budgets, and an understandable desire for rapid action meant that central governments and international organisations imposed their own values and visions (O'Brien *et al.*, 2012; Fletcher *et al.*, 2005). Yet, participation is a key element in dignity (Kant, 1785/1993), human rights and modern law in general. IDPs have a right to be informed, consulted and to participate in decisions affecting them (GP 7(3)c and d, GP 18(3) and GP 28(2)).

Consultation and participation can also promote learning, a key component for living with uncertainty and disasters (O'Brien *et al.*, 2012). In sociology of law and legal theory, 'reflexive legal rationality' is seen as a modern development that combines forms of material regulation with procedural regulation and thereby accommodates the complex and constantly changing processes that are to be regulated (Sand, 2008). Participation and involving key actors are characteristics of reflexive legal rationality. Reflexive arrangements are particularly useful in dealing with climate change and disasters as they are highly complex phenomena.

While participation is important, we should not romanticise local actors and their views, which may be maladaptive (O'Brien *et al.*, 2012). At the same time, there is a real risk that states and other non-local actors use climate change and disasters as a pretext while in reality they have other motives for relocating certain groups from certain areas; the 'buffer zones' in Sri Lanka where land was cleared of people to accommodate commercial and tourist industry, provide one example of this (see also Barnett and Webber, 2009).

Freedom of movement (ICCPR art 12) includes the right to remain at the place of habitual residence. Based on this right and other law, the Guiding Principles expressly state a right not to be arbitrarily displaced (GP 6). In the context of climate change and disasters, forced movement is considered legitimate only if it is required by the safety and health of those affected (GP 6(2)d; see also Human Rights Council, 2012: para 4h). The evacuation or relocation must be a measure of last resort; all feasible alternatives must have been explored in order to avoid displacement altogether (GP 7(1); see also Barnett and Webber, 2009; Human Rights Council, 2012: para 4h). This means that authorities must make real efforts to implement other adaptation measures to make the areas safe enough so people can choose to stay. Involuntary relocation and resettlement rarely leads to improvements in the quality of life of those who are moved, so moving communities in anticipation of climate change and disasters may precipitate vulnerability more than it avoids it (Barnett and Webber, 2009).

While consultation and participation may be difficult during the emergency phase of a disaster, it is particularly important, with regard to return, resettlement and recovery (GP 28(2); see also Human Rights Council, 2012: para 4g). The tsunami experience showed that IDPs are often isolated in camps or otherwise ignored, while government officials in distant towns and cities formulate and implement programs, sometimes in favour of special interests (Fletcher *et al.*, 2005). There were similar issues after Katrina, and the Committee on the Elimination of Racial Discrimination eventually recommended 'that every effort is made to ensure genuine consultation and participation of persons displaced [...] in the design and implementation of all decisions affecting them' (2008: para 31).

Creating a new home

Many people have not been able to return home even years after Katrina. Sufficient affordable housing is still not available. In fact, New Orleans still has the second highest rate of homelessness in the United States (Bradley, 2012). On the basis of freedom of movement, Guiding Principle 28 spells out three durable solutions. IDPs have the right to choose between returning to the place of origin, local integration in the place of displacement, and resettlement in another part of the country. The affected should receive compensation for property and land lost as well as assistance in resettling and re-establishing their livelihoods and residence (GP 28 and 29).

During the tsunami some villages became part of the sea (Fletcher *et al.*, 2005: 40). For conflict displaced people, returning to the place of origin remains an option, but in some disaster situations people simply cannot return. As in the case of initial evacuation or relocation, it can only be for the safety and health of those affected that a state prohibits return to an area (GP 6(2)d).

As a corollary to Guiding Principle 28, Guiding Principle 15 d stipulates the right of IDPs to be protected against forcible return or resettlement to places where their life, safety, liberty or health would be at risk. In human rights law, there is a ban on sending a person, independent of conduct or status, to places where they risk certain rights violations, in particular torture and inhuman or degrading treatment (see for example the landmark case European Court of Human Rights, 1989). In connection with the 'buffer zones', fishing communities were moved inland and given no retraining or resources to survive there (Kälin, 2009). One could question whether this is inhuman treatment and therefore illegal. It is also possible that some Thai deportations of Burmese migrants back to Burma may have been in violation of the ban.

Experience indicates that there is not sufficient focus on the creation of livelihood opportunities and the provision of basic services such as water, sanitation, education, and health at the relocation and resettlement sites (EACH-FOR, 2009; Kälin, 2009; O'Brien *et al.*, 2012). After the tsunami, fishing communities that were moved inland regularly ended up destitute since no adequate livelihood alternatives were available or made available. On the American Gulf Coast, the absence of jobs for those with skills that were not suited to the rising casino industry, such as the shrimpers, have had little assistance from the federal government (Williams and Bennett, 2008). A successful example of integration and livelihood support is that of Houston, USA, where city authorities took active measures to integrate displaced people into the local labour market by providing them with the necessary information and individual counselling (Kälin, 2009). Where people are unable to return to previous sources of livelihood, appropriate measures including provision of re-training opportunities should be taken (IASC, 2008b: C 4.2).

According to the previous Representative of the UN Secretary General on the Human Rights of IDPs, problems relating to housing, property, and livelihoods in most cases result from an inadequate legal and budgetary framework, and the fact that the affected often come from marginalised sections of society that continue to be marginalised (Kälin, 2009).

Recognition of IDPs, (un)natural disasters and rights

Climate change represents a system-scale transformation that will have widespread consequences. Responses to climate change and changes in disaster risk can in turn be both incremental and transformational (O'Brien *et al.*, 2012). 'Transformational responses are not always radical or monumental – sometimes they simply involve a questioning of assumptions or viewing a problem from a new perspective' (O'Brien *et al.*, 2012: 466). A transformation may be that we start seeing disasters not merely as natural, but that we see how human rights (violations) play a role in vulnerability and that human rights can be a framework for addressing displacement as well as disaster risk reduction, response, and recovery more generally. Rights may be violated in disaster situations not because of conscious intention but because of a lack of rights-based planning (IASC, 2008b).

Case studies of social-ecological systems suggest that there are three phases involved in systems transformations, including navigating the transition by making use of a sudden crisis as an opportunity for change, whether the crisis is real or perceived (O'Brien *et al.*, 2012). The tsunami and Katrina were two such crises. Although the contexts were very different, this chapter has shown how major human rights principles such as non-discrimination and participation were seen as relevant in a developed, Western country and developing, Asian countries. De Sousa Santos (2002) argues that,

> once cross-culturally reconstructed, human rights are one of the most powerful factors in bringing about the unthinking of modern law and politics, thereby generating the emancipatory energies necessary to face the challenges of the new times ahead.
>
> (282)

The Sri Lankan National Human Rights Commission took up hundreds if not thousands of cases of persons with human rights problems (Cohen, 2009), and in the aftermath of the tsunami there was increased international recognition of the importance of a human rights approach in dealing with those affected by disasters (Cohen, 2009). After visiting the region, Walter Kälin, Representative of the UN Secretary-General on the Human Rights of IDPs at the time, concluded that displaced people share many common types of vulnerability regardless of the reasons for their displacement and that 'it is no less important in the context of natural disasters than it is in cases of displacement by conflict to examine and address situations of displacement through a "protection lens"' (Cohen, 2009: 58). Kälin added disaster IDPs to the concerns of his mandate. The Human Rights Committee (2006) also stressed the need to review relevant,

> practices and policies to ensure the full implementation of [the] obligation to protect life and of the prohibition of discrimination, whether direct or indirect, as well as of the United Nations Guiding Principles on Internal Displacement, in matters related to disaster prevention and preparedness, emergency assistance and relief measures.
>
> (para 26)

The tsunami also triggered the development of the Operational Guidelines and Field Manual on Human Rights Protection in Situations of Natural Disaster (IASC, 2008b). The Human Rights Council (2012: para 4 g) seems to acknowledge them as a soft law instrument along with the Guiding Principles.

In 2006 the International Law Commission included the topic 'Protection of persons in the event of disasters' in its programme of work (ILC, 2012). Since then, they have drafted articles stressing the application of non-discrimination and human rights more generally in natural hazard-related disasters.

Katrina was also seen as a human rights crisis. Among the goals of the organisers of the 'International Tribunal on Hurricanes Katrina and Rita' was to expose to the world the human rights abuses committed by the US government, and to attain national and international recognition as IDPs (Lewis, 2009). Shadow reports to, for example, the UN Committee on the Elimination of Racial Discrimination resulted in official reactions from the Committee. Several reforms have followed the disaster in the areas of evacuation planning, schooling, health, government ethics and accountability, and broad participation in public meetings and processes as well as grassroots organization (Liu *et al.*, 2011). Seven years later, when Hurricane Isaac arrived on the same day as Katrina, New Orleans was better prepared (Bradley, 2012).

Increased focus on climate change has also led to important international developments in this field. The OHCHR has clarified that climate change and disasters potentially impinge upon the enjoyment of the full range of internationally protected human rights, including the right to not be arbitrarily displaced (OHCHR, 2009). A special report from the Intergovernmental Panel on Climate Change (IPCC, 2012) describes how a changing climate is influencing natural hazards such as storms, floods and droughts in terms of frequency, severity, timing and/or spatial distribution. Hazards combined with human vulnerability can result in disasters (ISDR, 2004). The concept of social-ecological systems recognizes the interdependence of social and ecological factors (O'Brien *et al.*, 2012). The social or 'human' factor is highly relevant in what has been called 'natural disasters'. First, human-made climate change can influence the hazard side. Second, human factors, including human rights realisation or violation, influence vulnerability.

In 2009 the Internal Displacement Monitoring Centre, which up until then had focused on conflict displacement, started monitoring displacement due to natural hazard-induced disasters. According to the centre, millions of people are forced to flee every year due to sudden-onset disasters such as floods and storms (IDMC, 2012). At the Nansen Conference on Climate Change and Displacement in the 21st Century, held in Oslo June 2011, the participants emphasized that '[r]esponses to climate and environmentally-related displacement need to be [...] guided by the fundamental principles of humanity, human dignity, human rights and international cooperation' (Wahlström, 2011: 5).

To dispel all doubt, the UN, nation states and NGOs have clarified that the IDP definition covers all climate change-related and disaster-related displacement occurring within a country (IASC, 2008a; Beyani, 2011; Wahlström, 2011). In its 2011 resolution on IDPs (A/C.3/66/L.45/Rev.1) the General Assembly included climate change and disaster displacement and encouraged the Special Rapporteur, in close collaboration with states, IGOs and NGOs, to continue to explore the human rights implications and dimensions in order to support states to build local resilience and capacity for prevention and protection.

Humanisation and beyond

This chapter has explored internal displacement and rights in the context of climate change and natural disasters through experiences from the tsunami and Hurricane Katrina and later developments. States and other actors should now take a rights-based and inclusive approach to IDPs. This would be a transformative approach to disasters and displacement. Adopting a human rights framework will help actors identify the most vulnerable and deliver assistance in a manner that decreases vulnerabilities or at least does not increase them. Moreover, human rights activism is one tool among other social justice efforts. As argued by Lewis (2009: 204) in the context of Katrina, the human rights approach has the advantage that it is '...broad enough to take in the amazing variety of complex abuses that occurred, as well as to set them in global context.'

A recurrent theme in this chapter is that disasters can highlight existing discrimination and marginalisation in society. Disasters can be an opportunity to address these issues. Disasters can also exacerbate these issues. A major obstacle to human rights realisation, closely related to the concept of non-discrimination and de Sousa Santos' concept of citizenship, is that while we may recognise that all humans should have certain rights, we dehumanise some people so much that they are no longer seen as humans that can claim such rights. A comment attributed to Mrs Barbara Bush regarding those housed temporarily in awful conditions in the Houston Astrodome is illustrative: 'Many of the people in the arena here, you know, were underprivileged anyway so this is working very well for them' (Edwards, 2006: 395). One way of addressing this dehumanisation is through insisting on a human rights approach and storytelling that creates empathy and serves as 'sentimental education' (Rorty, 1998). The stories that the IDPs from the tsunami and Katrina insisted on telling both domestic and international audiences form an important part of this. The storytelling may help consolidate new common identities, for example the superordinate identity of human being. This in turn can change whom we consider ingroup and outgroup and our behaviour towards each other (Hogg and Vaughan, 2011).

The internationalisation of the issues is important because who is viewed as human in the Human Rights Council or Committee on Elimination of Racial Discrimination is hopefully different from that in the US White House or Colombo at the time; sometimes distance works well. While the international human rights system has weak enforcement mechanisms, a significant source of pressure has been naming, shaming, and the public outrage and 'interest convergence' resulting from an 'embarrassment effect' (Lewis, 2009: 237).

Considering the universal obligations of human dignity and that climate change is a global process, the international community also has a responsibility to support and strengthen different states' ability to provide protection. Paragraph 14(f) of the Cancun Agreements now invites parties to enhance adaptation by undertaking '[m]easures to enhance understanding, coordination and cooperation with regard to climate change induced displacement, migration and planned relocation, where appropriate, at the national, regional and international levels' (UNFCCC, 2011). Importantly, the reference calls for cooperation and funding for actions related to displacement. When it comes to states' (lacking) will to protect IDPs, the international community must be prepared for 'political mobilization of international human rights or [. . .] humanitarian intervention in situation of extreme, life-threatening forms of social exclusion' (de Sousa Santos, 2002: 488).

Finally, the climate change and global environmental change framing of internal displacement, along with the principle of common heritage of human kind – which is increasingly referred to in International Law – may point towards a *jus humanitatis*,

> a law of and for humanity as a whole, the law of a decent condition in a non-dualist, but rather mutualist, interaction with nature [...] Indeed, from the perspective of outer space, we are enabled to see the arbitrary borders that separate nations, and we become keenly aware of a single delicate ecological system.
>
> (de Sousa Santos, 2002: 309)

In addition to finding and feeling and identifying with a superordinate identity, a sense of interdependence and shared goals can also help shape group dynamics and new social identities (Hogg and Vaughan, 2011). Rilke (1903/2005) writes, 'I live my life in widening circles / that reach out across the world. / I may not complete this last one / but I give myself to it.'

References

Albuja, S. and Cavelier, I. (2009) 'Work in progress towards protection in natural disasters: challenges from a conflict/disaster context', presented at Protecting People in Conflict and Crisis: Responding to the Challenges of a Changing World, University of Oxford: Refugee Studies Centre/Humanitarian Policy Group.

Asian Human Rights Commission (2005) 'In the wake of the tsunami, deception and discrimination', *Article 2*, 4, 1.

Barnett, J. and Webber, M. (2009) *Accommodating Migration to Promote Adaptation to Climate Change*, Stockholm: Swedish Commission on Climate Change and Development.

Beyani, C. (2011) 'Human rights-based approach to protection of environmentally displaced persons', presented at The Nansen Conference on Climate Change and Displacement in the 21st Century, Oslo, 6–7 June 2011. Online. Available HTTP: www.nansenconference.no (accessed 10 October 2012).

Birkeland, N. (2003) '"Last time I fled because of war, this time because of hunger" – environmental change and internal displacement in the Huambo Province, Angola', in N. Shanmugaratnam, R. Lund and K. A. Stølen (eds) *In the Maze of Displacement. Conflict, Migration and Change*, Kristiansand: Høyskoleforlaget.

Bradley, M. (2012) 'Watching Isaac, reflecting on Katrina', *Ottawa Citizen*, 29 August 2012. Online. Available HTTP: http://www.ottawacitizen.com/opinion/op-ed/Watching+Isaac+reflecting+Katrina/7163783/story.html (accessed 13 October 2012).

Cohen, R. (2009) 'An institutional gap for disaster IDPs', *Forced Migration Review* 32.

Committee on the Elimination of Racial Discrimination (2008) *Concluding Observation of the Committee on the Eliminatio of Racial Discrimination, United States of America,* CERD/C/USA/CO/6. Online. Available HTTP: http://www2.ohchr.org/english/bodies/cerd/cerds72.htm (accessed 10 October 2012).

de Sousa Santos, B. (2002) *Toward a New Legal Common Sense*, East Kilbride: Butterworths Lexis Nexis.

Dworkin, R. (1993) *Life's Dominion: An Argument About Abortion, Euthanasia, and Individual Freedom,* New York: Vintage.

EACH-FOR (2009) *Environmental Change and Forced Migration Scenarios Project*, Synthesis Report. Online. Available HTTP: http://www.each-for.eu/documents/EACH-FOR_Synthesis_Report_090515.pdf (accessed 10 October 2012).

Edwards, G.E. (2006) 'International human rights law violations before, during, and after hurricane Katrina: An international law framework for analysis', 31 *Thurgood Marshall Review*, 353.

Eng, S. (2007) *Rettsfilosofi*, Oslo: Universitetsforlaget.

European Court of Human Rights (1989) Soering v. United Kingdom, 11 EHRR 439.

European Court of Human Rights (2008) Budayeva and others v. Russia, app.nos. 15339/02, 21166/02, 20058/02, 11673/02 and 15343/02, 30 March.

Fletcher, L.E. Stover, E. and Weinstein, H.M. (2005) *After the Tsunami: Human Rights of Vulnerable Populations,* Berkeley: Human Rights Center, University of California, Berkeley.

Hogg, M.A. and Vaughan, G.M. (2011) *Social Psychology*, Harlow: Pearson Education Limited.

Human Rights Committee (2006) *Concluding Observations of the Human Rights Committee: United States of America,* CCPR/C/USA/CO/3. Online. Available HTTP: http://www.unhchr.ch/tbs/doc.nsf/%28Symbol%29/CCPR.C.USA.CO.3.En?Opendocument (accessed 10 October 2012).

Human Rights Council (2012) *Adequate housing as a Component of the Right to an Adequate Standard of Living in the Context of Disaster Settings*, A/HRC/19/L.4, 15 March 2012.

IASC (2008a) 'Climate change, migration and displacement: Who will be affected?' Submission to UNFCCC, 31 October 2008, Inter-Agency Standing Committee. Online. Available HTTP: http://unfccc.int/resource/docs/2008/smsn/igo/022.pdf (accessed 10 October 2012).

IASC (2008b) *Operational Guidelines and Field Manual on Human Rights Protection in Situations of Natural Disaster,* Inter-Agency Standing Committee, Washington DC: Brookings-Bern Project on Internal Displacement.

IDMC (2012*) Global Estimates 2011: People Displaced by Natural Hazard-inducedD,* Internal Displacement Monitoring Centre. Online. Available HTTP: http://www.internal-displacement.org/natural-disasters (accessed 10 October 2012).

ILC (2012) *Protection of Persons in the Event of Disasters*, International Law Commission. Online. Available HTTP: http://untreaty.un.org/ilc/summaries/6_3.htm (accessed 6 November 2012).

Inderfurth, K.F., Fabrycky, D. and Cohen, S. (2005) *The 2004 Indian Ocean Tsunami: Six Month Report,* Washington DC: The Sigur Center Asia Papers.

IPCC (2012) 'Summary for policymakers', in C.B. Field, V. Barros, T.F. Stocker, D. Qin, D. Dokken, K.L. Ebi, M.D. Mastrandrea, K.J. Mach, G-K. Plattner, S.K. Allen, M.

Tignor and P.M. Midgley (eds) *Intergovernmental Panel on Climate Change Special Report on Managing the Risks of Extreme Events and Disasters to Advance Climate Change Adaptation*, Intergovernmental Panel on Climate Change, Cambridge: Cambridge University Press.

ISDR (2004) *Terminology: Basic Terms of Disaster Risk Reduction*, International Strategy for Disaster Reduction. Online. Available HTTP: http://www.unisdr.org/eng/library/lib-terminology-eng%20home.htm (accessed 10 October 2012).

Kälin, W. (2001) 'How hard is soft law? The guiding principles on internal displacement and the need for a normative Framework', Presentation at the *Roundtable Meeting Ralph Bunche Institute for International Studies*, CUNY Graduate Center, 19 December 2001. Online. Available HTTP: http://www.brookings.edu/fp/projects/idp/articles/Kaelin12-19-01.pdf (accessed 10 October 2012).

Kälin, W. (2008) 'Guiding principles on internal displacement: annotations', *Studies in Transnational Legal Policy*, No. 38, Washington, DC: The American Society of International Law and tBrookings Institution – University of Bern Project on Internal Displacement.

Kälin, W. (2009) *Protection of Internally Displaced Persons in Situations of Natural Disasters*, Report of the Representative of the Secretary-General on the Human Rights of Internally Displaced Persons, Walter Kälin, to the Human Rights Council, tenth session, A/HRC/10/13/Add.1, 5 March 2009.

Kant, I. (1785/1993), translated by J. W. Ellington, *Grounding for the Metaphysics of Morals*, 3rd ed., Cambridge: Hackett.

Kolmannskog, V. (2008) 'Climates of displacement', *Nordic Journal of Human Rights*, 26 (4): 302–320.

Lewis, H. (2009) 'Race, class, and Katrina HR and (un)natural disaster', in F.C. Steady (ed.) *Environmental Justice in the New Millenium*, New York: Palgrave Macmillan.

Liu, A., Anglin, R.V. Mizelle, M. and Plyer, A. (2011) 'Introduction', in A. Liu, R. V. Anglin, R. M. Mizelle and A. Plyer (eds) *Resilience and Opportunity: Lessons from the U.S. Gulf Coast after Katrina and Rita*, Washington, DC: Brookings Institution Press.

Maloy, S. (2007) 'Media ignored Mississippi's use of waivers to redirect funds designated for low-income Katrina victims', *Media matters for America*, 30 August 2007. Online. Available HTTP: http://mediamatters.org/research/2007/08/30/media-ignored-mississippis-use-of-waivers-to-re/139727 (accessed on 11 October 2012).

O'Brien, K., Pelling, M., Patwardhan, A. Hallegatte, S. Maskrey, A., Oki, T., Oswald-Spring, U., Wilbanks, T. and Yanda, P.Z. (2012) 'Toward a sustainable and resilient future', in C.B. Field, V. Barros, T.F. Stocker, D. Qin, D. Dokken, K.L. Ebi, M.D. Mastrandrea, K.J. Mach, G-K. Plattner, S.K. Allen, M. Tignor, M. and P.M. Midgley (eds) *Intergovernmental Panel on Climate Change Special Report on Managing the Risks of Extreme Events and Disasters to Advance Climate Change Adaptation*, Cambridge: Cambridge University Press.

OHCHR (2009) *Report of the Office of the United Nations High Commissioner for Human Rights on the Relationship between Climate Change and Human Rights*, UN Doc A/HRC/10/61, Office of the United Nations High Commissioner for Human Rights, New York: United Nations. Online. Available HTTP: http://ochaonline.un.org/OchaLinkClick.aspx?link=ocha&docId=1116540 (accessed 10 October 2012).

Rilke, R.M. (1903/2005) *Book of Hours: Love Poems to God*, translated by Anita Barrows and Joanna Macy, New York: Riverhead Trade.

Rorty, R. (1998) 'Human rights, rationality, and sentimentality', in R. Rorty (ed.) *Truth and Progress Philosophical Papers* Vol. 3, Cambridge: Cambridge University Press.

Sand, I.J. (2008) 'The interaction of society, politics and law: The legal and communicative theories of Habermas, Luhmann and Teubner', *Scandinavian Studies in Law* 53.

Türk, V. (2011) 'Can protection of environmentally displaced persons be found in existing protection regimes? What are the next steps, from a protection perspective?' *The Nansen Conference on Climate Change and Displacement in the 21st Century*, Oslo, 6–7 June 2011. Online. Available HTTP: www.nansenconference.no (accessed 10 October 2012).

UNFCCC (2011) *Cancun Agreements*, FCCC/CP/2010/7/Add.1, United Nations Framework Convention on Climate Change. Online. Available HTTP: http://cancun.unfccc.int/ (accessed 10 October 2012).

Wahlström, M. (2011) 'Chairperson's summary', *The Nansen Conference on Climate Change and Displacement in the 21st Century*, Oslo, 6–7 June 2011. Online. Available HTTP: www.nansenconference.no (accessed 10 October 2012).

Williams, J.S. and Bennett, K. (2008) 'Impact of international human rights law on internally displaced Gulf Coast citizens', 77 *Miss. L.J.* 853.

15 Cities, human security and global environmental change

David Simon and Hayley Leck

Introduction

This chapter contributes to nascent efforts to integrate human security perspectives into the hitherto largely neglected urban (as opposed to global, national and rural) dimensions of global environmental change. Key urban challenges for human security in the context of environmental change relate to overall ecological footprints, maintaining institutional and infrastructural integrity, and safeguarding shelter, utilities, economic activities and livelihoods. These take on somewhat different complexions in coastal and inland urban areas. While in wealthier parts of the world, urbanization levels have generally been stable or declining, they are much more diverse and variable in poorer countries and regions due to differing and complex interacting forces and processes. These increasingly include rural impoverishment and hunger through the adverse effects of environmental change on rural livelihoods. Moreover, megacities, which face complex human security challenges in hosting often predominantly poor populations, are increasing rapidly in number and overall concentration in developing contexts (Bohle and Warner, 2008). Providing for large in-migrant populations in expanding cities with inadequate resourcing whilst simultaneously seeking to mitigate and adapt to the effects of environmental change poses severe challenges that are likely to threaten political stability and human security in many contexts.

The Global Environmental Change and Human Security (GECHS) Science Plan (Lonergan *et al.*, 1999) was produced in the late 1990s, at a time when concern for urban issues was beginning to rise up the international development agenda again after a period dominated by rural concerns. Global environmental change research up to that time had also been dominated by issues of global governance and essentially rural dynamics like land use and cover change that were amenable to remote sensing analysis. Hence it is perhaps unsurprising that the GECHS Science Plan makes only passing reference to urban issues in the summary of research foci. Focus 3 refers to "urbanization and human security" (Lonergan *et al.*, 1999: 29, 30). Despite initial insubstantial attention (GECHS, 2003; Leitmann, 2003; Gutberlet, 2003), GECHS has engaged increasingly over time with urban issues, especially through several Associates and collaboration with the International Human Dimensions Programme on Global Environmental

Change's (IHDP) Urbanization and Global Environmental Change (UGEC) core project. The first concrete evidence of this was the joint UGEC/GECHS publication in 2007, which contends that insufficient attention has been paid to the tight linkages between global environmental change, human security and urbanization to date. The contributing authors argue for the adoption of multidimensional perspectives, more appropriate for providing integrated approaches for understanding the implications of these links and associated events (IHDP Update, 2007). Recently, urban research has featured more prominently in GECHS publications (e.g. Matthew *et al.*, 2010) while the GECHS synthesis conference (June, 2009) included several urban sessions, signaling growing awareness of the importance of urban dimensions of human security.

Against this background and as a contribution to the assessment of GECHS' achievements that this volume represents, we now assess the extent of urban global environmental change – human security engagement more broadly and outline some directions for future research.

Urban global environmental change research into human security

The UGEC core project's Science Plan (Sánchez-Rodríguez *et al.*, 2005: 34) stipulates that research on adaptation strategies to the adverse effects of environmental change necessitates "a better understanding of the structural causes behind the vulnerability of urban systems to those impacts." Here, there is considerable potential for UGEC and GECHS collaboration as GECHS research is also centrally concerned with understanding the underlying social, economic and political processes contributing to injustices and vulnerabilities with respect to adverse environmental change effects (Lonergan *et al.*, 1999). Moreover, vulnerability analysis constitutes a core focus of both projects. Incorporating an urban focus into GECHS research does not warrant a completely new toolkit as such, but rather the adaption of existing methodologies, conceptual frameworks and tools to the urban context. Indeed, various approaches and methodological tools predominantly associated with the "rural" have more recently been applied successfully to urban contexts. For instance, urban disaster risk reduction is now well established (e.g. Pelling and Wisner, 2009) while a combination of political ecological and sustainable livelihoods analyses appears valuable for developing crucial new insights into urban areas because "[v]ulnerability to disaster and the impacts of gradual GEC erode the rights and opportunities of urban poor" (Parnell *et al.*, 2007: 9).

Much urban research into environmental change risks, vulnerabilities and coping mechanisms inevitably addresses aspects of human security but without either doing so explicitly or invoking a human security framework. In other words, using explicit coverage of human security within urban environmental change research as an indicator underestimates the degree of integration or commonality between these two agendas. One aspect of this relates to work on disaster risk reduction, which has addressed human security and risk (e.g. Pelling and Wisner,

2009), and which has become increasingly integrated with environmental change research over recent years, even if not explicitly acknowledged (Romieu *et al.*, 2010; Simon, 2010; Manuel-Navarrete *et al.*, 2011; see also the contributions to *Environment and Urbanization*, 2011).

Urbanization and global environmental change

Urbanization, globalization and accelerated environmental change are perhaps the three most powerful anthropogenic forces on earth. *Homo sapiens* has now become a predominantly urban species. The actual significance of this urban shift and associated consequences, particularly in terms of the adverse effects of environmental change and human security implications, is yet to be adequately contemplated and acted upon. Humankind's urban locus is an unprecedented phenomenon, which brings with it unprecedented implications and challenges, particularly in terms of environmental change and human security issues. This urban shift is expected to persist over the coming decades, particularly in poorer countries where rising aspirations and widespread deterioration of rural livelihoods caused by many interacting forces are driving large scale rural-urban migration.[1] As Martine *et al.* (2008: 3) remind us, "cities are also at the heart of local, national and global environmental change." Yet, remarkably, "discussions about global environmental problems and the possibilities for a 'sustainable' future customarily ignore the urban origin of many problems" (Swyngedouw and Heynen, 2003: 899). Helm and Hepburn (2009) represent a recent example of this tendency.

Fundamentally, while cities contribute greatly to environmental problems and also concentrate them spatially, cities must also be recognized as crucial to most solutions (Bohle and Warner, 2008; Martine *et al.*, 2008; Pelling and Wisner, 2009). Cities contain vast potential in terms of economies of scale and the potential of urbanization to contribute to environmental sustainability needs to be further recognized and exploited. Crucially, therefore, urbanization and environmental change have a bidirectional relationship; urbanization should be understood as both a cause and an effect of environmental change (Sánchez-Rodríguez *et al.*, 2005; Simon, 2007; Sánchez-Rodríguez, 2008; Dawson *et al.*, 2009). Urban areas worldwide constitute engines of economic growth, concentrating a multitude of economic activities within often densely populated built-up areas as part of ongoing globalization processes.

Moreover, both environmental change and economic globalization result in "winners" and "losers" (admittedly problematic terms in themselves) and the combination of the two phenomena can result in double winners or double losers. Attention to socio-environmental justice is thus essential (Barnett, 2006; Leichenko and O'Brien, 2006; Parnell *et al.*, 2007). Hence, the impacts of globalization and environmental change need to be considered simultaneously in all development plans since the interaction of these two evolving phenomena will modify and create new complex vulnerabilities to their combined effects. Such a holistic focus has been somewhat lacking in the urban literature to date since, while both phenomena are receiving unprecedented attention, they have not been considered together to a sufficient degree.

Vulnerability to adverse environmental change effects in urban areas is caused by various interconnected and multiscalar social, economic, political and environmental factors. Each factor contributes to vulnerability or, conversely, to resilience to the effects of environmental change (Dawson *et al.*, 2009). Moser *et al.* (1994, cited in Pelling, 2003: 57–58) usefully identify three features of urban life and livelihoods which distinguish rural and urban experiences of vulnerability as follows: "urban life is more commodified than rural life"; the "complexity of environmental risk is greater"; and "there is greater social fragmentation because of high residential mobility and the loss of support networks."

This notwithstanding, it is also critical not to lose sight of the close interconnectedness and bidirectional interactions characterizing the rural-urban continuum. Environmental pressures and degradation in rural areas also contribute to increased hazard and environmental degradation in urban areas and vice-versa, particularly when coastal systems are affected. As has been emphasized elsewhere (e.g. Pelling, 2003; Sánchez-Rodríguez *et al.*, 2005; Parnell *et al.*, 2007; Moser and Satterthwaite, 2008) urban-rural linkages are vital for mitigation and adaptation to environmental change, for instance, in terms of providing greater flexibility for risk spreading and opportunity enhancement across different activity spaces. Therefore, efforts at enhancing human security in urban areas are closely linked to activities in rural areas. As such, horizontal collaboration between rural and urban authorities is crucial to enhancing regional human security.

Most urban areas are periodically subject to dramatic crises, caused by interconnected biophysical, socio-economic and political factors. As Fragkias (2007: 25) notes, individual cities have unique "stress bundles" which can be best identified through integrative research approaches. "Stress bundles" consist of environmental factors (such as the combined effect of sea level rise and storm surge), that interact with physical characteristics (such as flood-prone or landslide prone areas) and socio-economic characteristics (such as increasing inequalities and polarization), which all combine to create different vulnerability contexts for urban inhabitants. Furthermore, urban environmental change risks and associated human security threats arise from macro-scale shortfalls in tackling global and national inequality, coupled with failures at local scales in implementing sustainable urban development (Barnett and Adger, 2005; Parnell *et al.*, 2007; Bohle and Warner, 2008). Underlying this is the more profound issue of how risk is conceptualized and communicated among different epistemic communities and social groups, and how their respective values affect the responses and realistic extents of behavioral change to enhance their (human) security in respect of disaster risk and environmental change (Hulme, 2009: 211–247; O'Brien, 2009; Brauch, 2011) – in this case in urban contexts, where there is considerable scope for further research.

The vulnerabilities of coastal urban settlements to the effects of environmental change are significant and could potentially lead to widespread erosion of human security. The *2008/9 State of the World's Cities* report (UN-Habitat, 2008: 141) cautions that failure to adapt port cities and urban coastal settlements risks creating devastating local and international consequences and "any rise in sea level is therefore potentially catastrophic for millions of urban dwellers and the global economy." Coastal zones vulnerable to the combined effects of sea level rise and storm surge are, of course, home to rural populations as well, but urban populations predominate. Almost half of the world's largest cities are situated within 50 kilometers of the coast and coastal population densities are 2.6 times greater than inland population densities (Millennium Ecosystem Assessment, 2005). This situation is likely to remain largely unchanged in the foreseeable future given that urbanization rates and levels in coastal areas are higher than in any other environment in the world (UN-Habitat, 2008: 142; see also Millennium Ecosystem Assessment, 2005; McGranahan *et al.*, 2007; Simon, 2010).

Urban human security and meeting basic needs

As alluded to above, environmental change and human security are linked in complex and multifaceted ways across diverse sectors in urban contexts and vary both temporally and spatially. Building on this, the following section highlights some of the key connections between predicted climate change impacts, meeting basic needs (with particular reference to the Millennium Development Goals (MDGs)) and human security.

The most explicitly urban MDG is Target 11 within MDG 7 (Environmental Sustainability). As modified by UN-Habitat in May 2003, this target reads:

> We commit ourselves to allocating the necessary financial, human and technical resources to improve significantly the lives of at least 100 million slum dwellers by 2020, while providing adequate alternatives to new slum formation, and to developing and implementing national and local action plans for slum upgrading.
>
> (UN-Habitat, 2005: 5)

At current rates of progress, there is little prospect of this target being achieved by the deadline, especially in countries with large low-income urban populations. Since shelter is one of the most fundamental basic human needs, and inadequate shelter is often associated with deficiencies in relation to other basic needs, this situation raises the very real prospect of an ongoing threat to urban human security by virtue of the failure of many of the world's poorest urban residents adequately to meet their basic needs, even in the absence of significant environmental change impacts. Climate change is already having adverse impacts on urban areas worldwide and will, in many cases, affect the poorest and most vulnerable sectors of urban populations most severely (McGranahan *et al.*, 2007; Parnell *et al.*, 2007; Satterthwaite *et al.*, 2007; Simon, 2007, 2010; Douglas *et al.*, 2008; Hardoy and Pandiella, 2009; Little and Cocklin, 2010; Ahammad, 2011; Banks *et al.*, 2011). Since these groups are often already at acute risk from other forms of disaster (Pelling and Wisner, 2009), their vulnerability will be exacerbated by urban displacement and threats to their existing livelihoods, however inadequate these are.

In this context, it is useful to explore more fully the likely climate change implications for urban human security in the short-, medium- and long-term time horizons of the projections in the Intergovernmental Panel on Climate Change's (IPCC's) Fourth Assessment Report (IPCC, 2007), namely 2009–2020, 2021–2050 and 2051–2100 respectively (Wisner *et al.*, 2007).

The short-term urban implications for many of the basic needs will be mutually reinforcing. For example, food shortages leading to malnutrition will probably exacerbate urban unrest of the sort already seen during "bread riots" when food is scarce and prices rise in poorer countries, reduce labor productivity and contribute to greater vulnerability to diseases (health). Similarly, failure to meet the urban water MDG will impact upon health, morbidity and labor productivity (Wisner *et al.*, 2007). The impacts in respect of mega-projects, disasters and gradual increases in urban displacement through sea level rise will exacerbate unemployment and are likely to increase political pressure for mitigation and adaptation interventions (Wisner *et al.*, 2007).

In the medium term (2021–2050), increased rural displacements as a result of food insecurity, water shortages in many areas, mega-projects and disasters will reinforce peri-urban and intra-urban pressures under all the basic needs headings. There is a real prospect that vulnerable urban residents will seek to leave cities in search of greater security elsewhere just as the rural displacees head to town. The consequences will be contingent on the balance of these countervailing processes and other local dynamics in each case (Wisner *et al.*, 2007).

In the longer term (2051–2100), sea level rise and heat island effects coupled with salinization of freshwater in coastal aquifers from which a significant proportion of coastal urban water supply is derived, will greatly exacerbate the urban sustainability challenge (Wisner *et al.*, 2007). Furthermore, long distance water transfer to cities will become increasingly precarious in areas experiencing more variable and declining rainfall and increased ambient temperatures because of greater evapotranspiration, reduced supply and increasing rural demand. By mid-century sea level rise in many low- and mid-latitude regions is likely to be at least 20–30 centimeters, submerging or threatening most low-lying parts of coastal urban areas. While wealthier residents may have the resources and resilience to cope, many poorer, more vulnerable residents are likely to lose their homes and livelihoods, requiring large-scale public intervention and international aid. Although there is considerable socio-economic and spatial overlap, poverty and vulnerability are rarely synonymous, even in urban areas (Eriksen and O'Brien, 2007; Parnell *et al.*, 2007; Douglas *et al.*, 2008; Hardoy and Pandiella, 2009). The combined effect of all these short- to long-term impacts will probably be increased political and social upheaval, threatening the stability and survival of urban local authorities as well as regional and national governments. Because of the concentration of environmental change-related impacts in cities, the human security threat posed by environmental change impacts in urban areas could become even more profound than that from changes in rural areas. Added to this is the key role that the integrity of urban infrastructure plays in determining levels of human security, something which has only recently emerged as a prominent concern in human security debates (e.g. Pelling and Wisner, 2009; Dalby, 2009; Monstadt, 2009). Indeed, water, waste, energy and transport systems infrastructure is central to the promotion of sustainable urban development and human security (Monstadt, 2009).

From the perspective of political leaders, the implications of the culmination of all the above-mentioned impacts, together with inadequate and inappropriate urban infrastructure and urban governance will quite literally make cities increasingly ungovernable through direct protest action. Thus, adopting a three-pronged strategy that simultaneously accounts for immediate, medium- and long-term concerns is crucial for progressing along sustainable development paths in the context of unprecedented environmental change and linked human security threats. However, politicians and political leaders are often predisposed to adopting short-term developmental outlooks based on election periods.

The way forward

Environmental change poses significant threats to human security. Such threats are likely to escalate in the near and long term future as a result of failure to meet MDGs in the context of increasing environmental change impacts. It is in urban areas that these impacts are likely to be most acutely felt, threatening the lives and livelihood security of millions of urban residents, particularly in poorer countries. Thus, the rising awareness of the urban-specific issues associated with environmental change and human security needs to be more urgently acted upon. Accordingly, key urban challenges in the context of environmental change and GECHS need to be recognized explicitly and incorporated into research endeavors as well as policy and practical action on the ground.

In research terms, much progress has been made in the last few years as a result of greater synergies between the respective GECHS and UGEC agendas but also in highlighting the complementarities between disaster risk reduction and environmental change perspectives, for instance. Seeking to understand how risk is conceived, communicated and responded to by different actors and stakeholder groups in the face of changing urban conditions is one good example of where potential compatibility among these perspectives exist. Taking these agendas forward will require further holistic and coordinated research linking up various approaches from the urban, disaster risk reduction, GECHS, UGEC and other literatures. Such collaborative and inclusive research offers the greatest potential to formulate crucial new insights to guide the research community, practitioners, policy makers, NGOs and civil society alike in meeting the imperative of enhancing urban human security.

Programmatically, the MDG period ends in 2015 and there is currently much debate about their successors, likely to be styled Sustainable Development Goals (SDGs). If the MDG shortcomings in relation to urban and human security issues identified above can be addressed by giving them greater coherence in one or more SDGs, the prospects for achieving measurable action will be enhanced because of the way in which such targets become embedded in international development assistance and national government expenditure negotiations. Although the instrumentality of such apparent accountability needs to be recognized, greater progress is likely than without such goals and the associated commitments. Some of these commitments apply at the subnational, in this case urban, scale. More broadly, since urban local authorities are the most important state-sector actors in driving urban environmental change mitigation, adaptation and transformation, it is also vital that more holistic understandings of urban human security issues are embedded in their development/planning and implementation agendas.

Acknowledgments

This is an updated and revised precis of the paper presented in the session on Cities and Climate Change: Human Security Challenges, at the GECHS Synthesis Conference, "Human Security in an Era of Global Change," Oslo, 23 June 2009. Thanks to the organizers for the invitation to participate and for comments on the draft; session participants also provided helpful feedback. A different version of the full paper appeared as "Urbanizing the Global Environmental Change and Human Security Agendas" in *Climate and Development* 2, 2010: 263–275.

Note

1 Although there is still considerable uncertainty about their relationship, inter and intra-regional migration have been identified as key risk-driving factors for human security, and Barnett and Adger (2005, 2007) attest that political and institutional responses to migrants are a principal concern for human security and violent conflict in the context of GEC.

References

Ahammad, R. (2011) 'Constraints of pro-poor climate change adaptation in Chittagong,' *Environment and Urbanization*, 23: 503–515.

Banks, N., Roy, M. and Hulme, D. (2011) 'Neglecting the poor in Bangladesh: research, policy and action in the context of climate change,' *Environment and Urbanization*, 23: 487–502.

Barnett, J. (2006) 'Climate change, insecurity and injustice,' in N. Adger, J. Paavola, S. Huq and M.J. Mace (eds) *Fairness in Adaptation to Climate Change*, London: MIT Press: 115–129.

Barnett, J. and Adger, W.N. (2005) *Security and Climate Change: Towards an Improved Understanding*, Paper presented at Human Security and Climate Change: An International Workshop, Oslo, 20–21 June 2005. Online. Available HTTP: http://www.auswaertiges-amt.de/cae/servlet/contentblob/389714/publicationFile/4506/17-GF-Barnett.pdf (accessed 10 October 2011).

Barnett, J. and Adger, W.N. (2007) 'Climate change, human security and violent conflict,' *Political Geography*, 26: 639–655.

Bohle, H.G. and Warner, K. (eds) (2008) *Megacities, Resilience and Social Vulnerability*, SOURCE: Studies of the University: Research, Council, Education, Publication Series of UNU-EHS, No. 10/2008.

Brauch, H.G. (2011) *Implementing Climate change Commitments: Sustainability Revolution, Changes in Worldviews and Mindsets*, Presentation made at the Third World International Studies

Conference, The World in Crisis: Revolution or Evolution in the International Community Porto, Portugal, 17–20 August 2011. Online. Available HTTP: http://www.afes-press.de/html/pdf/Brauch_Porto_WISC_110817.pdf (accessed 2 October 2011)

Dalby, S. (2009) *Security and Environmental Change*, Cambridge: Polity Press.

Dawson, R.J., Hall, J.W., Barr, S.L., Batty, M., Bristow, A.L., Carney, S., Dagoumas, A., Evans, S., Ford, A., Harwatt, H., Köhler, J., Tight, M.R., Walsh, C.L. and Zanni, A.M. (2009) *A Blueprint for the Integrated Assessment of Climate Change in Cities*, Tyndall Centre for Climate Change Research, Working Paper 129.

Douglas, I., Alam, K., Maghenda, M., McDonnell, Y., McLean, L. and Campbell, J. (2008) 'Unjust waters: climate change, flooding and the urban poor in Africa', *Environment and Urbanization*, 20: 187–205.

Environment and Urbanization (2011) 'Community-driven disaster risk reduction and climate change adaptation in urban areas,' *Environment and Urbanization*, 23 (2).

Eriksen, S. and O'Brien, K. (2007) 'Vulnerability, poverty and the need for sustainable adaptation measures', *Climate Policy*, 7: 337–352.

Fragkias, M. (2007) *Interactions and Responses to Global Environmental Change (GEC) and Their Implications for Human Security in Urbanized Coastal Zones – a synthesis*, in International Human Dimensions Programme (IHDP) on Global Environmental Change Update, October 2007, Issue No. 2.

GECHS (2003). *Aviso* no. 11 (double issue), Global Environmental Change and Human Security Project. University of Victoria, BC, Canada.

Gutberlet, J. (2003) 'Cities, consumption, and the generation of waste,' *Aviso*, 11: 13–22.

Hardoy, J. and Pandiella, G. (2009) 'Urban poverty and vulnerability to climate change in Latin America,' *Environment and Urbanization*, 21: 203–224 (republished in J. Bicknell, D. Dodman and D. Satterthwaite (eds) (2009) *Adapting Cities to Climate Change; Understanding and Addressing the Development Challenges*, Earthscan, London).

Helm, D. and Hepburn, C. (eds) (2009) *The Economics and Politics of Climate Change*, Oxford: Oxford University Press.

Hulme, M. (2009) *Why We Disagree About Climate Change*, Cambridge: Cambridge University Press.

IPCC (2007) 'Summary for policymakers, in climate change 2007: impacts, adaptation and vulnerability', contribution of Working Group II to the Fourth Assessment Report of the Climate Change Intergovernmental Panel on Climate Change, Cambridge: Cambridge University Press.

International Human Dimensions Programme (IHDP) (2007) *Global Environmental Change Update*, October 2007. Issue No. 2, Bonn.

Leichenko, R. and O'Brien, K. (2006) 'Is it appropriate to identify winners and losers?', in N. Adger, J. Paavola, S. Huq, and M.J. Mace (eds) *Fairness in Adaptation to Climate Change*, London: MIT Press, London: 97–114.

Leitmann, J. (2003) 'Urbanisation and environmental change: issues and options for human security,' *Aviso*, 11: 2–13.

Little, L. and Cocklin, C. (2010) 'The vulnerability of urban slum dwellers to global environmental change', in R.A. Matthew, J. Barnett, B. McDonald and K.L. O'Brien (eds) *Global Environmental Change and Human Security*, Cambridge, MA and London: MIT Press: 77–95.

Lonergan, S.C., Brklacich, M., Cocklin, C., Petter Gleditsch, N., Gutierrez-Espeleta, E., Langeweg, F., Matthew, R., Narain, S. and Soroos, N. (1999) *Global Environmental Change and Human Security: GECHS Science Plan*, International Human Dimensions Programme on GEC (IHDP) Report No 11, Bonn.

Manuel-Navarrete, D., Pelling, M. and Redclift, M. (2011) 'Critical adaptation to hurricanes in the Mexican Caribbean: development visions, governance structures, and coping strategies,' *Global Environmental Change*, 21: 249–258. Martine, G., McGranahan, G., Montgomery, M. and Fernandez-Castilla, R. (eds) (2008) *The New Global Frontier: Urbanization, Poverty and the Environment in the 21st Century*, London: Earthscan.

Matthew, R.A., Barnett, J., McDonald, B. and O'Brien, K.L. (eds) (2010) *Global Environmental Change and Human Security*, Cambridge MA and London: MIT Press.

McGranahan, G., Balk, D. and Anderson, B. (2007) 'The rising tide: assessing the risks of climate change and human settlements in low-elevation coastal zones', *Environment and Urbanization*, 19: 17–37 (republished in J. Bicknell, D. Dodman and D. Satterthwaite (eds) (2009) *Adapting Cities to Climate Change; Understanding and Addressing the Development Challenges*, London: Earthscan).

Millennium Ecosystem Assessment (2005) *Ecosystems and Human Well-being: A Framework for Assessment*, Washington, DC: Island Press.

Monstadt, J. (2009) 'Conceptualizing the political ecology of urban infrastructures: insights from technology and urban studies', *Environment and Planning A*, 41: 1924–1942.

Moser, C. and Satterthwaite, D. (2008) *Towards Pro-poor Adaptation to Climate Change in Urban Centres of Low- and Middle-income Countries*, Human Settlements Discussion Paper Series, Human Settlements Group and Climate Change Group, London: International Institute for Environment and Development (IIED).

Moser, C., Gauhurts, M. and Ginhan, H. (1994) 'Urban poverty research sourcebook: sub-city level research', in M. Pelling (ed.) *The Vulnerability of Cities: Natural Disasters and Social Resilience*. London: Earthscan.

O'Brien, K.L. (2009) 'Do values subjectively define the limits to climate change adaptation?' in W.N. Adger, I. Lorenzoni and K.L. O'Brien (eds) *Adapting to Climate Change: Thresholds, Values and Governance,* Cambridge: Cambridge University Press: 164–180.

Parnell, S., Simon, D. and Vogel, C. (2007) 'Global environmental change: conceptualising the growing challenges for cities in poor countries', *Area*, 39: 357–369.

Pelling, M. (2003) *The Vulnerability of Cities: Natural Disasters and Social Resilience*, London: Earthscan.

Pelling, M. and Wisner, B. (eds) (2009) *Disaster Risk Reduction: Cases From Urban Africa*, London: Earthscan.

Romieu, E., Welle, T., Schneiderbauer, S., Pelling, M. and Vinchon, C. (2010) 'Vulnerability assessment within climate change and natural hazard contexts: revealing gaps and synergies through coastal applications', *Sustainability Science*, 5: 159–170.

Sánchez-Rodríguez, R. (2008) 'Urban sustainability and global environmental change: reflections from an urban agenda', in G. Martine, G. McGranahan, M. Montgomery, and R. Fernandez-Castilla (eds) *The New Global Frontier: Urbanization, Poverty and the Environment in the 21st Century*, London: Earthscan: 149–163.

Sánchez-Rodríguez, R., Seto, K.C., Simon, D., Solecki, W.D., Kraas, F. and Laumann, G. (2005) *Science Plan, Urbanization and Global Environmental Change*, IHDP Report 15, International Human Dimensions Programme on Global Environmental Change, Bonn.

Satterthwaite, D., Huq, S., Pelling, M., Reid, H. and Romero Lankao, P. (2007) *Adapting to Climate Change in Urban Areas – The Possibilities and Constraints in Low and Middle-income Nations*, Human Settlements Discussion Paper Series, Human Settlements Group and Climate Change Group, London: International Institute for Environment and Development (IIED).

Simon, D. (2007) 'Cities and global environmental change: exploring the links. Urbanization and global environmental change: 21st century challenges', *The Geographical Journal,* 173: 75–92.

Simon, D. (2010) 'The challenges of global environmental change for urban Africa', *Urban Forum,* 21(3): 235–248.

Swyngedouw, E. and Heynen, N.C. (2003) 'Urban political ecology, justice and the politics of scale', *Antipode*, Special Issue, 35 (5): 898–918.

UN-Habitat (2005) *The Millennium Declaration: An Urban Perspective*, Nairobi and Washington, DC: UN-Habitat and the World Bank.

UN-Habitat (2008) *State of the World's Cities 2008/9 – Harmonious Cities*, London: Earthscan.

Wisner, B., Fordham, M., Kelman, I., Johnston, B.R., Simon, D., Lavel, A., Brauch, H.G., Spring, U.O., Wilches-Chaux, G., Moench, M. and Weiner, D. (2007) *Climate Change and Human Security, Radix-Radical Interpretations of Disasters*. Online. Available HTTP: http://www.radixonline.org/cchs.html (accessed 16 April 2007).

16 Suburbia on fire

Human security, climate change and emergency management

Jason Flanagan and John Handmer

Introduction

The observation by the US Department of Homeland Security that wildfires may have more impact than terrorism is no surprise to those working on climatic disasters – whether we think of the devastation resulting from floods on nearly every continent, evacuation of nearly a million people in California and the loss of thousands of homes to wildfire, or the loss of tens of thousands of lives in the European heatwave of 2003. In the wake of the "Black Saturday" fires that swept through the state of Victoria in February 2009, resulting in 173 deaths and the destruction of over 2,000 homes, there have been growing calls in Australia to treat wildfires as a security issue. Such calls reflect trends in other parts of the world, and are consistent with a long history of wildfire securitization in the United States. Securitization is a tempting option, given both the catastrophic impact of recent wildfires and the extraordinary response securitization would imply. This chapter will argue, however, that while there would be benefits to securitizing wildfires, these benefits would be outweighed by the problems that would logically flow from securitization.

Wildfires and security

Modern fire and emergency management have dual origins. In many countries, modern emergency management organizations have their origins in the "civil defense" or "home guard" units developed during World War II. After the war, it was not long before another military threat, the Cold War and the possibility of nuclear attack, created the imperative for maintaining civil defense capacities. These war-related organizations found themselves increasingly busy with more "everyday" emergencies and crises, and during the 1970s most civil defense organizations in western countries formally shifted to an emergency management rather than war-related focus. This was a significant shift, as civil defense was not about risk management as such – it did not attempt to reduce threat of war, but rather sought to protect the state and to a lesser extent the people. The second origin is completely different and concerns fire services, which evolved largely from a desire to protect property and for commercial reasons associated with insurance.

In many parts of the world we are seeing a reversal of this historical shift from war-related security logic to an emergency management focus on protecting people. This reversal is connected to a much wider effort to broaden and reconceptualize security. The end of the Cold War was seen by many as an opportunity to address both longstanding threats, overlooked during the superpower standoff, and newly emergent threats to people's lives and welfare. It was recognized that citizens of states that were considered "secure" according to the traditional concept of security, were often quite insecure in terms of their daily existence due to non-military threats. Recognizing that for most people "a feeling of insecurity arises more from worries about daily life than from the dread of a cataclysmic world event," the concept of human security extended the security label to economic, food, health, environmental, personal, community and political issues (UNDP, 1994; Tadjbakhsh and Chenoy, 2007).

Wildfire is one of many non-military dangers that threaten communities and populations around the world, with local wildfire risk and occurrence on the increase due to a combination of global climate change and social and demographic trends. Climate change is likely to more frequently create those weather conditions that produce large-scale high-intensity wildfires, called "megafires" in the US lexicon (Goldammer and Price, 1998; Kitzberger *et al.*, 2001; Flannigan *et al.*, 2005; Hennessy *et al.*, 2006; Moriondo *et al.*, 2006; Karoly, 2009). Pyne (2007), among others, argues that the modern era of wildfire records, dating from the 1930s, occurred during a period of relatively benign climate, with the past no longer serving as a good guide to future fire weather conditions. The Californian fire season, for example, is perceived as longer today than a decade ago, with fire chiefs now speaking of low and high risk periods rather than fire and non-fire seasons (Sullivan, 2008). European Mediterranean countries are in a similar situation, and are likely to see longer fire seasons with increases in both the number of days of fire risk and the number of extreme fire weather events (Reinhard *et al.*, 2005; Moriondo *et al.*, 2006).

Wildfires are generally only considered a threat when they intersect with people and things we value. Unfortunately, these intersections are being rapidly expanded in scale by the growth of urban settlements and recreation into fire prone ecosystems. In the US and Australia we see relentless suburban sprawl into fire-prone rural and wildland areas, with growth in the most rapidly expanding cities occurring at the urban-rural fringe. For example, according to a study by the US Forest Service and the University of Wisconsin, between 1990 and 2000, 61 percent of the new housing in the states of California, Oregon and Washington was built at the urban-rural fringe (Boxall, 2008: para. 20). In Australia this strong urban sprawl is accompanied by an influx of people into smaller regional and coastal centers, driven by a mix of environmental amenity, lifestyle change, and the attraction of affordable housing. Such "sea-" and "tree-changers" often move into fire prone areas with no local livelihood basis and a lack of anything but the most basic of understandings of what fire risks mean for them. In other parts of the globe, socio-economic changes exacerbate such trends. In European Mediterranean countries, for example, changes from traditional land use have led to the abandonment of large areas of farmland and the regrowth and accumulation of vegetation as fuel for wildfires. Combined with the expanding urban edge and increasing tourist use, such factors have dramatically increased fire frequency (Dimitrakopoulos and Mitsopoulos, 2006).

Is wildfire a "security" issue?

On 7 February 2009, devastating wildfires swept through the state of Victoria, Australia. These fires – the so-called "Black Saturday" fires – had a catastrophic impact upon local communities, destroying over 2,000 homes and resulting in 173 deaths (Victorian Bushfires Royal Commission, 2009). Due to their destructive magnitude, some have since called for wildfires to be treated as a national security issue. This logic is part a larger trend where securitization is justified in quantitative terms. For example, at the January 2000 meeting of the United Nations Security Council on the issue of HIV/AIDS, the then United States vice president Al Gore declared:

> We tend to think of a threat to security in terms of war and peace. Yet no one can doubt that the havoc wreaked and the toll exacted by HIV/AIDS do threaten our security. The heart of the security agenda is protecting lives – and we now know that the number of people who will die of AIDS in the first decade of the 21st Century will rival the number that died in all the wars in all the decades of the 20th Century.
>
> (Gore, 2000: para.4)

Gore's remarks reflect not only the central argument for securitizing AIDS, but the underlying logic of most securitization efforts – that the enormity of the issue's real or potential death toll warrants its elevation to the realm of security.

The use of the security label, however, is not a simple and direct reflection of the magnitude of a given danger. The so-called Copenhagen School of security studies has convincingly argued that securitization – the framing of an issue as a security concern – can best be understood as a level of politicization. Ole Waever, Barry Buzan and Jaap de Wilde have asserted: "It is a *choice* to phrase things in security . . . terms, not an objective feature of the issue" (Buzan *et al.*, 1998: 211). Security, in this view, is an extreme form of politicization, with the Copenhagen School suggesting a three level spectrum of politicization. On the first level are non-politicized issues, which are not issues of state action or of public debate or decision. On the next level are politicized issues, which are part of public policy, requiring government decision and the allocation of resources. Finally there are securitized issues, which are "presented as an existential threat requiring emergency measures and justifying actions outside the normal bounds of political procedure" (Buzan *et al.*, 1998: 23–24). Securitization theory helps us pose the question as to whether "it is a good idea to make this issue a security issue – to transfer it to the agenda of panic politics – or whether it is better handled within normal politics" (Buzan *et al.*, 1998: 34; Waever, 1999).

There are obvious benefits to securitizing an issue like wildfire risk. First and foremost it serves to generate attention and mobilize national and international economic, social and political resources. Along with the inherent benefits, however, securitization brings with it a number of pitfalls. It can remove an issue from cosmopolitan and altruistic frameworks and place it within a state-centric one, thus relieving states of their obligation to address issues remote from their own national borders and interests (Caballero-Anthony, 2006; Elbe, 2006). Defining an issue in security terms may also be counterproductive when its security implications are far from compelling. For example, in her study of epidemic disease and national security, Susan Peterson, has rightly argued that from a national security viewpoint "AIDS poses a far smaller threat to most states than it does from almost any other viewpoint" (Peterson 2002/3: 79).

Securitization can also bring with it zero-sum thinking and an "us vs. them" mentality. The dangers of such logic have been evident in the widespread securitization of migration, where an "us vs. them" logic has, among other things, served to block social integration processes and thus threaten the social cohesion of host countries (Huysmans, 2000; Ceyhan and Tsoukala, 2002). In a slightly different vein, Daniel Deudney has argued that conceptualizing climate change as a security concern "risks undercutting both the globalist and common fate understanding of the situation and the sense of world community that may be necessary to solve the problem" (Deudney, 1990: 486). Even when strict "us vs. them" thinking is not evident, securitization necessarily involves the "threat-defense" logic of traditional security thinking (Buzan *et al.*, 1998: 29). While securitization can give issues greater profile, defining issues as a "threat" that must be "defended" against is not always conducive to finding solutions.

Securitization can also transfer the responsibility for addressing an issue from civil society to security institutions. This can be a tempting option. The securitization of HIV/AIDS in Africa, for example, has served to move the issue away from struggling health ministries to better resourced and more influential security-related ministries (Altman, 2000; Piot, 2000). Security institutions, however, are poorly equipped to address such "non-traditional" threats. They are prepared, through very specialized equipment and training, to fight conflicts and protect the nation from external violence. It is very difficult to successfully transfer such equipment and skills to non-military dangers. Moreover, at least in the developed world, military threats are generally delegated to the military and other organs of national security in totality, with nothing being required of the average citizen, except in those rare incidents of total war. In contrast, issues such as climate change require *all* people to change their daily behavior (Deudney, 1990), and for wildfire safety all those at risk need to change. Such issues cannot simply be delegated to the national security apparatus, leaving citizens to carry on as before. Issues such as climate change are not crises in the foreign policy sense of the word. Crises, like conflicts, generally only mobilize short-term action and support. Crises call for resolution, just as wars call for victory. Issues such as climate change cannot be resolved in the short term, but rather require long-term changes in lifestyle and patterns of behavior. While crisis rhetoric can motivate people in the short term, failure to bring about a relatively speedy solution to a declared crisis generally leads to the evaporation of initial support (Deudney, 1990; Bostdorff, 1994).

Securitizing wildfire

Wildfires pose a significant threat to numerous rural and peri-urban communities around Australia, and in the wake of the Black Saturday fires we have seen what the Copenhagen School calls "securitizing moves" on the issue (Buzan *et al.*, 1998: 25). Anthony Bergin of the Australian Strategic Policy Institute (ASPI), published a piece in *The Age* newspaper entitled "Defending the home front is top priority," in which he declared: "In the aftermath of Black Saturday, it's time to rethink the protection of the Australian people . . . We now need, as the Prime Minister noted last December, to face up to climate change as a fundamental national security challenge" (Bergin, 2009). Bergin went on to discuss how "defending" against climate change at home might require different "battle tactic and skills," outlining how Australian intelligence and military equipment and personnel should play a direct role in addressing natural disasters. An August 2009 ASPI report similarly recommended that climate change be "acknowledged by Australian security planners as a significant homeland security threat," thus giving "security agencies ... a stronger mandate to contribute to addressing climate change impacts" (Yates and Bergin, 2009: 2–3).

Even without conscious efforts to securitize the issue of wildfire, the language used to describe them is already militarized. Most obvious are the terms "fire-fighting" and "fire-fighter." Moreover, we hear of fire-fighters "battling" and "fighting" fire "fronts," which "destroy" and "devastate" land and homes. In 2003, the then Prime Minister John Howard, with deliberate allusion to terrorism, spoke of the "terror from the bush." At least figuratively, Australia is already fighting a war on wildfire. This in part represents the projection of urban concepts onto rural and wild settings. In an urban landscape fire is indeed threatening and dangerous, whereas in rural and wild settings fire is often a vital force of ecological maintenance and renewal (Dimitrakopoulos and Mitsopoulos, 2006; Smith, 2006; Pyne, 2007).

Limited knowledge and cultural conditioning, linked to a 24-hour news cycle increasingly driven by sensationalism, entertainment and opinion, serve to reinforce the threatening image of wildfires. Stories of terrible fires destroying pristine wilderness and threatening lives are very reportable, whereas discussion of the vital ecological role of wildfires lacks drama, is generally outside the expertise of reporters, has less cultural resonance, and cannot be easily reduced to brief sound bites. The public's tendency to think of wildfires as "threats" that need to be "fought" and defeated is thus reinforced by the media's desire to report them as such (Smith 1992, 2006). In turn, this creates political pressure for an emphasis on fire control and suppression, rather than on the protection of assets, livelihoods and in some cases, individual safety. Politicians and other actors cannot afford to be perceived as failing to directly confront the threat of destructive wildfires.

Les AuCoin, examining the American example, describes what he calls the "Mr. Goodwrench" syndrome, in which "pressured legislators feel compelled to act as problem solvers even though they may be making matters worse" (AuCoin, 2006).

The United States, and particularly California, provide an instructive point of comparison with the Australian experience. Like south-east Australia, California has regular fire seasons, as well as significant property destruction and mobilization of fire fighting forces. While Australia employs similarly militarized language and media framing as the United States, its fire suppression efforts have not yet been likewise securitized. The American "war" on wildfire – originally declared by the US Forest Service in the early 1900s (Dombeck *et al.*, 2004) – reveals many of the elements feared by the Copenhagen School. For example, large fire suppression incidents are often declared as a "state of emergency," which gives the US Forest Service sweeping emergency powers, allowing it to suspend or ignore its own conservation strategies and forest plans, as well as fundamental national conservation laws (e.g. see Barnard, 2005). Note that while we draw on US Forest Service examples – as it is the National US fire fighting agency with significant global influence – we recognize that the situation varies greatly by Forest Service Region and by state, and that our argument probably applies best to California. Moreover, just as it is fundamentally unpatriotic to reduce expenditures on soldiers in battle, so too vast sums of money continue to be spent on fire-fighting. This long-standing blank cheque approach is, however, now being critically scrutinized with attempts to control the spiraling costs of fire fighting in the US (Cabán, 2007; O'Toole, 2007; US-GAO, 2007).

In parts of the United States wildfire policy is not only framed as a "war" on fire, but is actually fought with military equipment and tactics. In some cases fire-fighters, for reasons of access, fuel reduction and attempts at containment, destroy some of what is being protected. Containment lines in particular may have a number of adverse environmental impacts, such as killing and removing vegetation, compacting and eroding soil, degrading water quality, and fragmenting forest stands (Ingalsbee, 2005; Ingalsbee, 2006). More dramatically, fire-fighters, like soldiers on the battlefield, often call in air support – in this case planes and helicopters dropping water and fire retardants. While aerial fire-fighting appliances are promoted by the media as a solution – and have likely become politically unavoidable – Australian research has shown that aerial suppression is generally no more effective in stopping a fire's forward spread than ground crews with tankers and bulldozers (Loane and Gould, 1985). The use of aerial suppression measures is often considered to be as much a public relations exercise as an effective fire-fighting measure, with fire officials in the United States referring to them as "CNN drops" (Cart and Boxall, 2008). Fire retardants have a fleeting effect on wildfire, but a lasting effect on ecosystems, being particularly dangerous to aquatic wildlife (Kalabokidis, 2000).

In addition to water bombing aircraft, fire-fighters rely on a wide arsenal of incendiary weapons in their war on fire. Such equipment is not only used in backburning operations, aimed at strengthening containment lines and changing the fire's direction, but also in burning unburned "green islands" in the interior of fires. The use of fire can be the most effective and environmentally sound way of managing wildfires. Such effectiveness and environmental benefits, however, are negated when combined with a war-like mindset. At this point the use of fire becomes merely one tactic in the single-minded pursuit of fire suppression. While the burning of "green islands," can shorten the time period needed to declare a wildfire under control, it also destroys refuges for flora and fauna that are vital to natural post-fire recovery processes (Ingalsbee, 2005; Ingalsbee, 2006).

Wildfire security or safety?

The "war on wildfire" approach has its winners: large agencies use this approach; it creates heroes, helps enlarge budgets, and creates a clear mission. While individual fire fighters are celebrated as heroes, the organizations have also recently been subject to intense negative scrutiny as a result of escalating cost, issues of effectiveness and perceived commercial imperatives (for example see Boxall and Cart, 2008). The approach may create significant commercial opportunities, especially in aerial fire fighting. In the United States we have seen the emergence of what has been labeled a "fire-military-industrial-complex" (Ingalsbee, 2006; Wuerthner, 2006: 261–282), with suggestions that commercial interests are driving approaches that return profit as a first priority rather than address fire risk management objectives. Profits are best generated by the use of expensive technologies. They are not as easily generated by low-cost community resilience and risk reduction activities. When seen from the perspectives of cost, practicality and use of science, such an approach may not be well aligned with a policy objective of community safety. The emphasis is all too often only on fighting fire, despite the fact that the limits of fire suppression are well known to wildfire scientists and fire agencies (Luke and McArthur, 1978). Thus the threat-defense logic inherent in securitization triumphs at the cost of what we would argue should be the central objective of fire policy: improving safety and reducing all types of loss.

A better approach to wildfire risk management is one where the objective is the protection of people and key assets – with an emphasis on individual safety. The focus is explicitly on the people at risk, who can themselves become part of the solution instead of being part of the problem. This focus on individual safety can be addressed through the concept of human security, and indeed we have seen a growing use of the "security" label with explicit reference to this concept. Use of the concept of human security, however, does not address the problems associated with securitization, but rather merely complicates the issue. One must remain cognizant of the dangers of the "security" label, which are not resolved merely by adding the term "human." As Ole Waever has discussed, the term "security," at least in the public sphere, "carries with it a history and a set of connotations that it cannot escape. At the heart of the concept we find something to do with defense and the state. As a result, addressing an issue in security terms still evokes an image of threat-defense, allocating to the state an important role in addressing it" (Waever, 1995: 47). Looking at the lessons of the American example, it is clearly counterproductive to brand wildfires a security issue in Australia. Using the Copenhagen School's spectrum, wildfires have already achieved a very high level of politicization both in Australia and the United States. This high level of politicization delivers all the benefits of securitization – public attention, political and economic resources, etc. – without the associated problems of militarization. We must recognize wildfires as a significant and indeed growing danger, while at the same time adopting an approach defined by the concepts of resilience, risk sharing and a focus on the safety of the people at risk. While securitization is a tempting option, we cannot hope to win a "war on wildfire" like that being fought in the United States. Wildfires are not an enemy to be defeated, but rather a risk to be reduced and a set of consequences to be managed.

References

Altman, D. (2000) 'Understanding HIV/AIDS as a global security issue', in L. Kelly (ed.) *Health Impacts of Globalization: Towards Global Governance*, London: Palgrave.

AuCoin, L. (2006) 'Don't get hosed: How political framing influences fire policy', in G. Wuerthner (ed.) *Wildfire: A Century of Failed Forest Policy*, Washington, DC: Island Press.

Barnard, G. (2005) 'Judge finds wildfire retardant drops violate environmental law', *Casper Star-Tribune*, 25 October. Online. Available HTTP: http://trib.com/news/state-and-regional/article_b80fbb52-48f4-565c-8fca-8311f899dbef.html (accessed 10 May 2009).

Bergin, A. (2009) 'Defending the home front is top priority', *The Age*, 21 February. Online. Available HTTP: http://www.theage.com.au/opinion/defending-the-home-front-is-top-priority-20090220-8dl2.html (accessed 14 May 2009).

Bostdorff, D.M. (1994) *The Presidency and the Rhetoric of Foreign Crisis*, Columbia: University of South Carolina Press.

Boxall, B. (2008) 'A Santa Barbara area canyon's residents are among many Californians living in harm's way in fire-prone areas', *Los Angeles Times*, 31 July. Online. Available HTTP: http://www.latimes.com/news/science/environment/la-me-wildfires31-2008jul31,0,4933797.story (accessed 14 August 2009).

Boxall, B. and Cart, J. (2008) 'As wildfires get wilder, the costs of fighting them are untamed, *Los Angeles Times*', 27 July. Online. Available HTTP: http://www.latimes.com/news/local/la-me-wildfires27-2008jul27,0,4093174.story (accessed 17 May 2009).

Buzan, B., Waever, O. and De Wilde, J. (1998) *Security: A New Framework for Analysis*, Boulder: Lynne Rienner.

Caballero-Anthony, M. (2006) 'Combating infectious disease in East Asia: Securitization and global public goods for health and human security', *Journal of International Affairs*, 59: 105–127.

Cabán, A.G. (2007) 'Wildland fire management policy and fire management economic efficiency in the USDA Forest Service', keynote presented at the IV International Wildland Fire Conference: 13–17 May 2007, Seville, Spain. Online. Available HTTP: http://www.fire.uni-freiburg.de/sevilla-2007/Keynote-Gonzales-Caban.pdf (accessed 20 May 2009).

Cart, J. and Boxall, B. (2008) 'Air tanker drops in wildfires are often just for show', *Los Angeles Times*, 29 July. Online. Available HTTP: http://www.latimes.com/news/local/la-me-wildfires29-2008jul29,0,5666042.story (accessed 23 August 2009).

Ceyhan, A. and Tsoukala, A. (2002) 'The securitization of migration in Western Societies: Ambivalent discourses and policies, *Alternatives: Global, Local, Political*, 21: 21–39.

Dimitrakopoulos, A.P. and Mitsopoulos, I.D. (2006) *Global Forest Resources Assessment 2005 – Report on Fires in the Mediterranean Region*, Fire Management Working Papers, Rome: Food and Agriculture Organization of the United Nation.

Dombeck, M.P., Williams, J.E. and Wood, C.A. (2004) 'Wildfire policy and public lands: Integrating scientific understanding with social concerns across landscapes', *Conservation Biology*, 18 (4): 883–889.

Deudney, D. (1990) 'The case against linking environmental degradation and national security', *Millennium: Journal of International Studies*, 19 (3): 461–476.

Elbe, S. (2006) 'Should HIV/AIDS be securitized? The ethical dilemmas of linking HIV/AIDS and security', *International Studies Quarterly*, 50: 119–144.

Flannigan, M.D., Amiro, B.D., Logan, K.A., Stocks, B.J. and Wotton, B.M. (2005) 'Forest fires and climate change in the 21st century', *Mitigation and Adaptation Strategies for Global Change*, 11 (4): 847–859.

Goldammer, J. G. and Price, C. (1998) 'Potential impacts of climate change on fire regimes in the tropics based on MAGICC and a GISS GCM-derived lightning model', *Climatic Change*, 39: 273–296.

Gore, A. (2000) 'Remarks as prepared for delivery by Vice President Al Gore United Nations Security Council Opening Session', 10 January 2000. Online. Available HTTP: http://clinton4.nara.gov/textonly/WH/EOP/OVP/speeches/unopen_fp.html (accessed 5 May 2009)

Handmer, J. and Dovers, S. (2007) *The Handbook of Disaster and Emergency Policy and Institutions*, London: Earthscan.

Hennessy, K., Lucas, C., Nicholls, N., Bathols, J., Suppiah, R. and Ricketts, J. (2006) *Climate Change Impacts on Fire-weather in South-east Australia*, CSIRO Marine and Atmospheric Research, Bushfire CRC and Australia Bureau of Meterology.

Huysmans, J. (2000) 'The European Union and the securitization of migration', *Journal of Common Market Studies*, 38: 751–777.

Ingalsbee, T. (2005) *Collateral Damage: The Environmental Effects of Firefighting*, The 2002 Biscuit Fire Suppression Actions and Impacts, Eugene, OR: Firefighters United for Safety, Ethics, and Ecology, May 2005. Online. Available HTTP: <http://www.fusee.org/content_pages/docs/biscuit_suppression_impacts.html> (accessed 12 May 2009).

Ingalsbee, T. (2006) 'The war on wildfire: Firefighting and the militarization of forest fire management', in G. Wuerthner (ed.) *Wildfire: A Century of Failed Forest Policy*, Washington, DC: Island Press.

Karoly, D. (2009) 'The recent bushfires and extreme heat wave in southeast Australia', *Bulletin of the Australian Meteorological and Oceanographic Society*, 22: 10–13.

Kitzberger, T., Swetnam, T.W. and Veblen, T.T. (2001) 'Inter-hemispheric synchrony of forest fires and the El Niño-Southern Oscillation', *Global Ecology and Biogeography*, 10: 315–326.

Kalabokidis, K.D. (2000) 'Effects of wildfire suppression chemicals on people and the environment – a review', *Global Nest: The International Journal*, 2 (2): 129–137.

Loane, I.T. and Gould, J.S. (1985) *Aerial Suppression of bushfires, Cost-benefit Study for Victoria*, Canberra: National Bush Fire Research Unit CSIRO.

Luke, R.H. and McArthur, A.G. (1978) *Bushfires in Australia*, Canberra: Australian Government Publishing Services.

McCarthy, G.J. (2003) *Effectiveness of Aircraft Operations by the Department of Natural Resources and Environment and the Country Fire Authority 1997–1998*, Department of Sustainability and Environment, Victoria, Research Report 52.

Moriondo, M., Good, P., Durao, R., Bindi, M., Giannakopoulos, C. and Corte-Real. J. (2006) 'Potential impact of climate change on fire risk in the Mediterranean area', *Climate Research*, 31 (1): 85–95.

O'Toole, R. (2006) 'Money to burn: Wildfire and the budget', in G. Wuerthner (ed.) *Wildfire: A Century of Failed Forest Policy*, Washington, DC: Island Press.

O'Toole, R. (2007) 'The perfect firestorm: Bringing forest service wildfire costs under control', *Policy Analysis*, 591, 30 April. Online. Available HTTP: http://www.scribd.com/doc/13673534/-The-Perfect-Firestorm-Bringing-Forest-Service-Wildfire-Costs-under-Control-Cato-Policy-Analysis-No-591- (accessed 3 June 2008).

Peterson, S. (2002/3) 'Epidemic disease and national security', *Security Studies*, 12 (2): 43–81.

Piot, P. (2000) 'Global AIDS pandemic: Time to turn the tide', *Science*, 288: 2176–2178.

Pyne, S.J. (2007) 'Megaburning: The meaning of megafires and the means of their management', Fourth International Conference on Wildland Fire, Seville, Spain, May 2007.

Reinhard, M., Rebetez, M. and Schlaepfer, R. (2005) 'Recent climate change: Rethinking drought in the context of forest fire research in Ticino, south of Switzerland', *Theoretical and Applied Climatology*, 82 (1): 17–25.

Singer, P.W. (2002) 'AIDS and international security', *Survival*, 44 (1): 145–158.

Smith, C. (1992) *Media and the Apocalypse: News Coverage of the Yellowstone Forest Fires, Exxon Valdez Oil Spill, and Loma Prieta Earthquake*, Westport, CT: Greenwood.

Smith, C. (2006) 'Hot news: Media coverage of wildfire', in G. Wuerthner (ed.) *Wildfire: A Century of Failed Forest Policy*, Washington: Island Press.

Sullivan, B.K. (2008) 'California fire season now year-round in era of "mega" blazes', *Bloomberg*, 18 November. Online. Available HTTP: http://www.bloomberg.com/apps/news?pid=newsarchive&sid=a8pdA6Wy9FOA (accessed 15 November 2012).

Tadjbakhsh, S. and Chenoy, A. (2007) *Human Security: Concept and Implications*, London: Routledge.

UNAIDS (2008) *Report on the Global AIDS Epidemic*, Geneva: UNAIDS.

UNDP (1994) *Human Development Report 1994*, United Nations Development Programme, New York: Oxford University Press.

US-GAO (2007) *Wildland Fire Management: A Cohesive strategy and Clear Cost-containment Goals are Needed*, Government Accountability Office 07-1017T, June 19, 2007. Online. Available HTTP: http://www.gao.gov/new.items/d071017t.pdf (accessed 16 November 2012).

Victorian Bushfires Royal Commission (2009) *Victorian Bushfires Royal Commission interim report*. Online. Available HTTP: http://www.royalcommission.vic.gov.au/Commission-Reports/Interim-Report (accessed 20 October 2012).

Waever, O. (1995) 'Securitization and desecuritization', in R.D. Lipschutz (ed.) *On Security*, New York: Columbia University Press.

Waever, O. (1999) 'Securitizing sectors? reply to Eriksson', *Cooperation and Conflict*, 34 (3): 334–340.

Weaver, D.H., Beam, R.A., Brownlee, B.J., Voakes, P.S. and Wilhoit, G.C. (2006) *The American Journalist in the 21st Century: U.S. News People at the Dawn of a New Millennium*, New York: Routledge.

Wuerthner, G. (ed.) (2006) *Wildfire: A Century of Failed Forest Policy*, Sausalito, CA: Island Press.

Yates, A. and Bergin, A. (2009) *Hardening Australia: Climate Change and National Disaster Resilience*, Special Report 24, Australian Strategic Policy Institute, August 2009.

17 Ethics, environmental change and human security

Roberto P. Guimarães

Introduction: sustainable development, from oxymoron to paradigm shift

Humankind has given the impression, especially in the past decade, that it is acquiring a clear understanding of the challenges that modern civilization will have to overcome in order to mitigate and adapt to environmental stress. Yet, the issues that seem to permeate the debate within and between nation-states elude the nature of the environmental crisis. The naiveté of many world leaders today recalls the example offered by Alvin Toffler (1974) about the simplemindedness of the elders of an Indian tribe that for centuries lived off the produce of a river at its doorstep. Its culture and economy are based upon fishing, boat building, and harvesting from the soil fertilized by the river, so that the future of this community merely repeats its past. But what happens when this tribe pursues its traditional style of development unaware that a dam is being built upstream? Its image of the future is misled, dangerously misled, for the river will soon dry up or become a trickle (Guimarães, 2003).

In effect, never before have we been so close not to disaster – in the way many zealot environmentalists believe – but rather to experiencing, on a planetary basis, the limitations of our fragile life support systems. Activists and intellectuals alike tell us that nothing short of a planetary ethics must arise if we are to survive as a species (Guimarães, 1991a). This transition to sustainability implies a profound change in the prevailing archetype of civilization, particularly in what it refers to as its cultural pattern of articulating humans and nature. An adequate understanding about this transition imposes the recognition that humankind is facing the exhaustion of a style of development that is *ecologically wasteful* (depletes the natural resource base), *socially perverse* (generates poverty and inequality), *politically unjust* (freezes absolute and relative scarcity of access to resources), *ethically objectionable* (disrespects human and non-human life forms), and *culturally alienated* (estranges fellow human beings and subjugates nature) (Guimarães, 2001). Thus, the often repeated statement that human beings must constitute the center and the *raison d'être* of development calls for a new development style that must be:

- environmentally sustainable in the access and use of natural resources and in the preservation of biodiversity;
- socially sustainable in the reduction of poverty and inequality and in promoting social justice;
- culturally sustainable in the conservation of the system of values, practices and symbols of identity that, in spite of their permanent evolution, determine national integration through time;
- politically sustainable by increasing democracy and guaranteeing access and participation of all sectors in public decision-making; and
- guided by a new development ethics, one in which the economic objectives of growth are subordinated to the laws governing the operation of natural systems, subordinated as well to the criteria of respect for human dignity and of improvement in the quality of life for current and future generations; in short, a development paradigm that respects the integrity of the life support systems of the planet.

(Guimarães, 2001)

To further complicate matters, most scenarios discussed today about the new millennium emphasize the process of globalization, often defined exclusively in economic, financial or commercial aspects. Not a few, including myself, consider it more revealing to unveil globalization from a sustainable development perspective. These questions include, for instance, the economic rationality of globalization in its many dimensions vis-à-vis the dynamics and the pace of natural processes. Indeed, the prospects of a process of globalization founded upon an upward, unlimited and unchecked economic growth model are also seriously disputed, particularly in view of the reality of exhaustion of many natural resources (e.g., fauna, flora, non-renewable sources of energy) as well as of deterioration of natural processes which are crucial for the ecosystem viability of life on the planet (ozone layer, climate, etc.) (e.g., MEA, 2005; IPCC, 2007). As the poignant remark attributed to Kenneth Boulding (Uldrich, 2008:184; RSA, 2011), a leading pioneer of sustainability, indicates "anyone who believes that exponential growth can go on forever in a finite world is either a madman or an economist."

Environmental change as an ethical challenge

The sudden centrality of an ethical dimension in social life poses hitherto unforeseen challenges to the social sciences, and to the everyday concerns of citizens, government and private enterprises as well. Compounding these challenges is the current ideological discourse on ethics, markets, and governmental regulation. In this respect, what we confront nowadays no longer refers solely to the situation pictured in the report of the Club of Rome, *The Limits to Growth* (Meadows *et al.*, 1972), when attention was drawn to the exhaustion of natural resources. After all, society may confront such challenges, albeit in an imperfect and limited fashion, via substitution of natural capital for physical capital, be it through the invention of new products that substitute near-exhausted resources (e.g., hydrogen for petroleum to fuel transportation), be it by new technologies (more fuel-efficient engines) that enlarge existing reserves. What we face today are radically different circumstances. Present institutions must cope with the frailty of environmental processes that cannot be simply substituted by others. One cannot substitute the ozone layer, neither can climate stability be substituted, unless one accepts as valid the search of another planet to where earthlings may migrate once the cycles and natural processes which support life on Earth are ultimately drained. It characterizes the uniqueness of the contemporary world that the higher we rise in advancing our technological society, the more intimate and demanding the interconnections between ourselves and our forgotten nature become, as some of environmental services and natural resources for satisfying them become exhausted by global processes such as climate change. As the competition to grasp a greater share of resources goes unchecked, we place increasing strains upon the stability of our social and political institutions.

As already pointed out, this new (sustainability) paradigm emphasizes that development must produce qualitative changes in the quality of life of human beings. More than the mercantile goods and services exchanged in the market, these aspects include the social, cultural and aesthetic dimensions of meeting material and spiritual needs. After all,

> in situations of extreme poverty, the impoverished individual, marginalized or excluded from society and from the national economy, does not have any commitment to avoid environmental degradation if society is not able to thwart his or her own deterioration as a human being.
>
> (Guimarães, 1991b: 24)

Klaus Töpfer (2001) could not have been more poignant, condemning the 11 September 2001 terrorist attacks against the United States, when he warned that:

> When people are denied access to clean water, soil, and air to meet their basic human needs, we see the rise of poverty, ill-health and a sense of hopelessness. Desperate people can resort to desperate solutions. They may care little about themselves and the people they hurt [...] I am not suggesting for a moment that poverty and environmental degradation are factors on their own. Intolerance also has its role. But it can fan the flames of hate and ignite a belief that terrorism is the only solution to a community's or nation's ills.
>
> (Töpfer, 2001: 1)

What bonds and upholds this specific understanding of sustainability is the urgent need for a new ethics of development that incorporates current concerns for the environment. This follows an already established tradition in development ethics that dates back to Adam Smith (1976), has among its most important modern intellectuals Joseph Lebret and René Moreux (1942), Gunnar Myrdal (1968) and Dudley Seers (1969), and has been further expanded by the seminal works of Amartya Sen (1979, 1987, 1999). A good summary of the main tenets of development ethics can be found, among others, by Dennis Goulet (1971, 1995) and David Crocker (1991).

This subfield of development studies places a particular focus on the so-called agency approach to the traditional capabilities approach, introduced mainly by Amartya Sen in the 1980s (for more on the uses of the capabilities approach, see Des Gasper, 2002 and Martha Nussbaum, 2003). Both, agency and capabilities attempt to overcome the shortcomings of development theory and its derived policy-oriented models that aim for economic growth for its own sake to the detriment of social goals and individual freedom and happiness. Thus, for an ethics of development, aspects such as affluent consumption, hunger, inequality and other dimensions of discrimination based on gender or race assume the forefront of concerns. Not surprisingly, development ethics has grown in interest as the world becomes more unequal and more polarized between rich and poor nations. Likewise, for this brand of ethics the distribution of welfare within and among societies is central to guaranteeing development as much as differences in capabilities to transform resources into valuable activities for individual and social welfare. In short, development is not synonymous to economic growth but to enlargement of freedoms intra- and internationally.

The modern day concerns for the environment have extended development ethics even further precisely because they highlight the unequal distribution of resources, of access to resources and, most importantly, the negative impacts of environmental degradation brought about by unsustainable patterns of consumption mostly in affluent societies and minorities in poor countries. Hence, what is referred to here as the need for "new" ethics of development requires that economics rescue its roots, identity and initial purposes as *oikonomy*, or the material supply of the *oikos*, the household where humans live in. In a most fortunate coincidence, economics shares the same semantic root of ecology, the study of the laws that govern this home. Unfortunately, with the acceleration of modern times, economics has ceased to study ways and means to achieve well being for people. It has turned into an end in itself, a science in which anything that does not have monetary value is not worthy or valuable. This is turning into one of the most pernicious fetishes of modern times, despite the warnings of economists of the stature of the already mentioned 1999 Nobel economics laureate, Amartya Sen:

> Economics assigns an order of preferences to a man and when it is needed it assumes that this utility function reflects his own interests, represents his well being, summarizes his idea of what should be done and describes his choices … In effect, the purely economic man is almost mentally retarded from a social point of view. Economic theory has dedicated much attention to this rational fool comfortably relaxed behind his unique order of preferences for all purposes.
> (Sen, 1979: 202; see also Sen, 1987)

Empirical reality indicates also that wealth accumulation or economic growth does not constitute and has never been a requirement or precondition for development. Human welfare options are projected well beyond economic welfare

(UNDP, 1999; UNDESA, 2005). Wealth in itself is not the decisive factor, rather it is the use a community gives to its wealth. Numbers clearly indicate that countries with equivalent levels of economic riches have radically different welfare levels. This combined reality of wealth inequalities and power concentration has led the UNDP to state that:

> the new globalization rules – and the actors that write them – are aimed at integrating global markets, neglecting the needs of the people that the markets cannot satisfy. This process is concentrating power and is marginalizing poor countries and people.
>
> (UNDP, 1999: 30)

Human security as a public good, the policy crossroads of ethics and global environmental change

The sad reality today is that thirty years after the Stockholm conference, the warnings voiced by Margaret Mead are still waiting to be echoed by world leaders. In her wise words,

> we can recognize that the ways of our forebears are ways to which we can never return, but that the more we can recapture of this earlier wisdom, in a form we can understand, the better we can understand what is happening today, when a generation almost innocent of a sense of history has to learn how to cope with an unknown future, one for which they were not reared.
>
> (Mead, 1970: 70)

Until now, what one can see are only cosmetic changes to "green" current growth, without in fact enforcing those changes that governments committed themselves to in Rio in 1992.

As more scientific knowledge is available the limits for negotiating environmental conflicts that emerge can be summarized as follows: National leaders do not acknowledge that the security of the nation depends upon an environmentally sound development strategy; instead, environmental criteria are subsumed either by national security interests or economic criteria. Furthermore, rapid economic growth has high priority over conservation. On top of that, the techno-bureaucracy and the corporate elite share an ideological orientation towards the private allocation of natural resources and of the "commons" in general (Guimarães, 1991a).

Economic elites in general, but particularly technocrats, have also learned the lessons of coping with innovations. The need to address global change problems constitutes a fairly recent public issue and, thus, represents an innovation, almost a revolutionary innovation to the policy process as well as to strategic development decisions in general. Faced with this new challenge, the governmental bureaucracy has continually adopted what Donald Schon (1973) calls dynamic conservatism. First, one accepts a discourse that incorporates the new issue, something that has been successfully demonstrated from Stockholm 1972 to Johannesburg 2002.

Then follows the institutional stage of "containment and isolation," when one literally throws the discourse into a bureaucratic box in the governmental structure, or into an internationally adopted agreement. Containment and isolation also have another important, beneficial side effect for dynamic conservatism. These processes lead to compartmentalization. Now that adequate institutions are in place, busily tilting against the windmills of global change, there comes the phase of "selective inattention." In other words, environmental agencies must be at a bus stop where the bus of power does not stop. In short, in dynamic conservatism one should promote the minimum change possible so as to guarantee that nothing will actually change. It is dynamic because it is not the result of a carefully conceived scheme of overt resistance. There is no conspiracy theory at work here. The individual, group, or class is able to establish a connection between their special interests and the (inertial) interests of the social system as a whole (known not to be very fond of dramatic or profound changes). Because everyone is bound to be affected by hard policy choices in response to global change, there is no need to connive in accord against taking them seriously. It is just a question of letting the bureaucratic process run its course.

The foregoing remarks require also an attempt to rescue the sociopolitical aspects of the conceptual quandary embedded in much of the debate surrounding the new configurations of power brought about by globalization and the "new" international order. In light of the fact that these processes gained momentum after the fall of the Berlin Wall, not few scholars were quick to declare "the end of History" (e.g., Fukuyama, 1990), yet history seems to suggest with enough clarity that the market has never been a foundational principle of social organization, even though it may condition the economic behavior of social actors as producers and consumers. History indicates as well that the market economy is an excellent generator of wealth but it is also a most sure generator of profound social asymmetries (see e.g., Guimarães, 1990; UNDP, 1999; UNDESA, 2005).

Sustainable development, understood as the maintenance of the stock of natural resources and the environmental viability of societies to satisfy the basic needs and improve the quality of life of current and future generations, requires the democratization of the state, not its abolition, as well as the empowerment of civil society. As a matter of fact, there is a growing recognition that despite the ideological seesaw of the past decade, the state still holds a distinctive role in fostering development. First of all, the state still provides a contribution to development, which is unique, necessary and indispensable. It is unique because its logic transcends the logic of market forces, particularly in dimensions such as equity and social justice, which are foreign to market mechanisms and institutions. It is necessary because the very logic of capital accumulation requires the production of "public goods" that cannot be produced by competitive actors in the marketplace, particularly in the imperfect markets of emerging countries. It is indispensable because it addresses issues such as climate change, biodiversity depletion and many other "variables" not amenable to the micro-economic calculus of discount rates and rates of return, especially when future generations (who, by definition, cannot participate in today's market) are brought into the forefront of environmental regimes (Guimarães, 2003).

The current debate over climate change offers a good illustration of the shortsightedness of using solely economic criteria to overcome the crisis (e.g., Beckerman and Hepburn, 2007; Broome, 2008). Equally misguided is the attempt to replace the State for market forces to foster adaptation instead of mitigation to face impending climate change. As has been forcefully argued in an analysis germane to the approach advocated here, limits to adaptation are endogenous to society, hence dependent upon the ultimate ethics, knowledge and cultural aspects, such as attitudes to risk, those different societies uphold (Adger *et al.*, 2009).

Moreover, the challenges posed by poverty and social inequality aggravated by environmental degradation cannot be defined as individuals' problems, embodying instead social, collective problems (Guimarães, 1990). It is definitely not a case of guaranteeing access, via the marketplace, to education, housing, and health or to an environment free of pollutants. Instead, these encompass the duty of recovering the collective (solidarity driven) practices of fulfilling material and spiritual needs of human well being. Against the backdrop of this reality, the internationalization of markets poses the greatest threat to the ability of state and public actors to maintain social cohesion and national identity, producing the fragmentation of power and strengthening the transnational articulation of partial, private segments of national societies.

Having been subjected to intense attack and having barely evaded "extinction" at the hands of the apostles of neoliberal economics and politics, the state is undoubtedly wounded and weaker now. If this situation is not reversed, be it in the arena of environmental regimes or in human rights and trade regimes, our societies will be confronted with public policies that, albeit their best intentions, will amount to ambulances that run around collecting the wounded and discarded people of a globalization riding the horses of neoliberal economics. All this takes place in a context in which greater portions of the decisions that are crucial for a nations' future, and for social cohesion as well, are arrived at outside its territory and implemented by actors completely foreign to the social, political and economic realities of the country (Guimarães, 2004).

Concluding remarks: ethics and human security in the context of global environmental change

The single most important challenge facing the world in the New Millennium is in fact the quality of growth (i.e. the increase in the levels of well-being and the reduction of the socioeconomic inequalities), much more than its quantity (i.e. the simple increment of material output). One must also pay close attention to the implications of globalization for governance at all levels, planetary, regional, national, and sub-national. Many global dynamics simply ignore national boundaries. The erosion of nation states brings with it powerless governments and may lead to the end of governance. As suggested in a book commissioned by the UN University and introduced by Kofi Annan, then UN Secretary General:

> [M]any applaud this erosion of governance – indeed, many see it as the main attraction of globalization. *These are the true anarchists, perhaps more so than the masked youth that smashed windows at the WTO meeting of Seattle in 1999.*
>
> (Grunberg and Khan, 2000: 18, emphasis added)

There is also a growing consensus about the social unsustainability of the current styles of development, a reality of globalization in the midst of increasing poverty, social inequality and exclusion, a reality which certainly precedes but has been exacerbated by the very process of globalization. Addressing poverty and inequality requires efforts to achieve a balance between many complex, countervailing socio-economic forces. Although economic growth is necessary, it is not a sufficient condition to reduce poverty. Reforms are required in a number of different areas to increase the opportunities for and capabilities of the poor and other marginalized groups in order to spur inclusive growth and development and thereby reduce inequality. A healthy, well-educated, adequately employed, socially and environmentally protected citizenry contributes to social cohesion. Improved access by the poor to public assets and services (social and environmental alike) and income transfer programs to sustain the poorest families are essential to changing the structure of opportunities and are key to reducing the intergenerational transmission of poverty and inequality (UNDESA, 2005).

Hence, the future of sustainable development is intimately associated with the future of politics. Clive Lewis perhaps best captured the socio-political nature of the environmental crisis when he stated that "what we call the power Man over Nature is in fact the power of some men over other men, using nature as their instrument" (Lewis, 1947: 69). This implies that situations of environmental degradation reveal nothing else than inequalities of social and political power (patterns of relations among human beings and the form in which society is organized), as well as structural distortions of the economy (patterns of consumption of society and patterns of production organized to satisfy those). If this represents indeed an accurate description of environmental problems, it becomes rather evident that possible solutions to the current crisis of civilization via sustainable development will have to be found within the social system itself and not in some technological magic or market gimmick.

Projecting the realities of power between human beings in the long run, with the consequent implications about how these will bring about the incorporation of nature, the scenario turns even more delicate and somber. Indeed, just as the relationships of power are synchronically determined, there is also diachronic, inter-generational asymmetry of power among humans. In other words, each generation exercises power on subsequent generations (in making use of nature); while the latter, when modifying the inherited natural endowment, resist and attempt to limit the power of their predecessors. This process, repeated indefinitely, ends up not representing more power over the natural world, just the opposite, rendering human society more fragile in nature's hands. The later in time a generation lives, and, by definition, the closer it lives to the extinction of species, the less power it will have at its disposal to adapt nature to its needs and wants and to dominate other humans. As Lewis (1947) concludes poignantly, albeit so appropriately (in a period when sustainability was not yet fashionable), *human* nature will be the last part of Nature to surrender to Man. Those subjugated will not be "men at all; they will become artifacts" (1947: 77). Man's final conquest will prove to be in fact "the abolition of Man."

References

Adger, W.N., Dessai, S., Goulden, M., Hulme, M., Lorenzoni, I., Nelson, D.R., Naess, L.O., Wolf, J. and Wreford, A. (2009) 'Are there social limits to adaptation to climate change?', *Climatic Change*, 93 (3–4): 335–354.

Beckerman, W. and Hepburn, C. (2007) 'Ethics of the discount rate in the Stern Review on the economics of climate change', *World Economics*, 8 (1): 187–210.

Broome, J. (2008) The ethics of climate change: Pay now or pay more later? *Scientific American*, May 19, 2008. Online. Available HTTP: http://www.scientificamerican.com/article.cfm?id=the-ethics-of-climate-change (accessed 5 August 2010).

Crocker, D.A. (1991) 'Toward development ethics', *World Development* 19 (5): 457–483.

Fukuyama, F. (1990) 'El fin de la historia?', *Estudios Públicos*, 37, Summer: 5–31.

Gasper, D. (2002) 'Is Sen's capability approach an adequate basis for a theory of human welfare?', *Review of Political Economy* ,14 (4): 435–461.

Goulet, D. (1971) *The Cruel Choice: A New Concept in the Theory of Development*, New York: Athenaeum.

Goulet, D. (1995) *Development Ethics: A Guide to Theory and Practice*, New York: The Apex Press.

Grunberg, I. and Khan, K. (eds) (2000) *Globalization: The United Nations Development Dialogue: Finance, Trade, Poverty, Peace-Building*, New York: United Nations University Press.

Guimarães, R.P. (1990) 'Desarrollo con equidad ¿Un nuevo cuento de hadas para los años noventa?', *SINTESIS: Revista Documental de Ciencias Sociales Iberoamericanas*, 10, January–April: 15–68.

Guimarães, R.P. (1991a) *The Ecopolitics of Development in the Third World: Politics and Environment in Brazil*, Boulder, CO and London: Lynne Rienner Publishers.

Guimarães, R.P (1991b) 'O Brasil e o desafio do desenvolvimento sustentável', in R.P. Guimarães (ed.) *O Desafio do Desenvolvimento Sustentável: Relatório Nacional do Brasil para a Conferência das Nações Unidas sobre Meio Ambiente e Desenvolvimento*, Brasília: Secretaria de Imprensa da Presidência da República: 13–24.

Guimarães, R.P (2001) 'The politics and ethics of "sustainability" as a new paradigm for public policy formation and development planning', *International Journal of Economic Development*, 3 (3): 1–54.

Guimarães, R.P (2003) 'The political and institutional dilemmas of sustainable development', in S. Nagel and G. Mudacumura (eds) *Handbook of Developmental Policy Studies*, New York, Marcel Drekker: 447–469.

Guimarães, R.P (2004) 'Waiting for Godot: Sustainable development, international trade and governance in environmental policies', *Contemporary Politics*, 10(3–4): 203–225.

Guimarães, R.P (2006) 'Politics, the missing link in the science-policy interface', *IHDP UPDATE*, 3–4: 8–13.

IPCC (2007) *Climate Change 2007: Impacts, Adaptation and Vulnerability*, Summary for Policymakers, Contribution of Working Group II to the Fourth Assessment Report of the Intergovernmental Panel on Climate Change.

Lebret, J. and Moreux, R. (1942) 'Manifeste d'économi et humanisme', *Économie et Humanisme* (February–March): 6.

Lewis, C.S. (1947) *The Abolition of Man: How Education Develops Man's Sense of Morality*, New York: Macmillan Publishing Company.

Mead, M. (1970) *Culture and Commitment*, New York: Doubleday.

Meadows, D.H., Meadows, D.L., Randers, J. and Behrens, W.W. (1972) *The Limits to Growth: A Report to the Club of Rome's Project on the Predicaments of Mankind*, New York: Universe Books.

Millennium Ecosystem Assessment (2005) *Ecosystems and Human Well-being: Synthesis*, Washington, DC: Island Press.

Myrdal, G. (1968) *Asian Drama: An Inquiry into the Poverty of Nations*, New York: Pantheon.

Nussbaum, M. (2003) 'Capabilities as fundamental entitlements: Sen and social justice', *Feminist Economics*, 9 (2–3): 33–60.

Royal Society for the encouragement of Arts, Manufactures and Commerce (RSA) (2011) *The Understandable Madness of Economic Growth: 2011 RSA President's Lecture*. Online. Available HTTP: http://www.youtube.com/watch?v=1sP291B7SCw (accessed 13 March 2011).

Schon, D.A. (1973) *Beyond the Stable State*, New York: W.W. Norton.

Seers, D. (1969) 'The new meaning of development', *International Development Review*, 11 (4): 3–4.

Sen, A. (1979) 'Rational fools: A critique of the behavioural foundations of economic theory', in F. Hahn and M. Holis (eds) *Philosophy and Economic Theory*, Oxford: Oxford University Press: 87–109.

Sen, A. (1987) *On Ethics and Economics*, Oxford: Blackwell.

Sen, A. (1999) *Development as Freedom*, New York: Oxford University Press.

Smith, A. (1976) *The Theory of Moral Sentiments*, Oxford: Oxford University Press, reprint of 1759 edition.

Toffler, A. (1974) 'The psychology of the future', in A. Toffler (ed.) *Learning for Tomorrow: The Role of the Future in Education*, New York: Random House: 31–47.

Töpfer, K. (2001) *UN Environment Chief Urges World to Fight Root Causes of Civil Unrest that Can Lead to Terrorism*, Nairobi: UNEP Press Release, 21 September.

Uldrich, J. (2008) *Jump the Curve*, Avon, MA: The Platinum Press.

UNDESA – United Nations Department of Economic and Social Affairs (2005) *The Inequality Predicament: Report on the World Social Situation*, New York: UNDESA.

UNDP (1999) *Human Development Report, 1999: Globalization with a Human Face*, New York: UNDP.

Wilson, J.O. (1992) 'Socio-economic justice', in P. Ekins and M. Max-Neef (eds) *Real-Life Economics: Understanding Wealth Creation*, London and New York: Routledge: 54–59.

Paradigms

18 From poverty to prosperity

Addressing growth, equity and ethics in a changing environment[1]

Asunción Lera St. Clair and Victoria Lawson

Introduction

The Global Environmental Change and Human Security Programme (GECHS) defines human security as an integral concept, as a state that is achieved when and where individuals and communities have the options necessary to end, mitigate or adapt to threats to their human, environmental and social rights; have the capacity and freedom to exercise these options; and actively participate in pursuing these options (GECHS, 1999). As discussed in several chapters in this volume, this emphasis on agency, and on people's and communities' existing capacities to respond to any threats – whether social, human or environmental – presumes that there is no fundamental difference between threats and harms caused by entitlement failures or pervasive conditions of poverty, and threats and harms caused by environmental changes, including climate change. This focus on freedom to act and recognition of people's own capacities to respond to structural and environmental conditions implies that poverty and environmental change are more than linked; they are both part of a single frame that includes both symptoms and results of unjust and unsustainable development pathways.

Furthering this line of thinking, we argue that the dominant approach to poverty studies is ill-prepared to contribute insights on the human dimensions of environmental change. This dominant approach presumes that the reduction of poverty is dependent upon continuing a vision of development that is based on increased production and consumption, unending economic growth, externalization of environmental impacts, and the underlying view that "more is better." This is, not surprisingly one of the central arguments used by energy companies to promote the need for more oil and carbon-based energy sources. The outcomes of this approach are both destructive to nature *and* the source of massive inequalities; empirical evidence shows that increases in material wealth (including the wealth produced by the exploitation of oil) have ended up in the hands of a few, leaving a large number of people powerless and unable to exercise their rights, and without access to basic needs (including access to energy). What we call in this chapter "dominant poverty knowledge" (hereafter DPK) is a widespread discourse that sees the reduction of poverty as contingent upon unending growth and consumption.[2]

DPK ignores 1) the interconnections between poverty and prosperity and 2) the consequences of the growth agenda for the environment. The causes of poverty and processes of wealth creation are seen as unrelated, or with diverse causal linkages. Poverty and prosperity are often considered two separate categories, corresponding to two distinct social groups. Social inequality is taken as a given, and hence receives insufficient scholarly attention. Meanwhile, unending, inequitable growth remains the dominant answer to solving poverty, and it is the source of massive global environmental change all over the planet. At times when climate change and extreme weather events are making their impacts more and more visible, this dichotomy is more pronounced than ever. In a world locked in to corporate and carbon-based growth strategies, processes of mass production, financial speculation, and increasing consumption of luxuries are the prime characteristics of prosperity. When disassociated from processes of wealth creation, the poor and their poverty are depicted as outcomes of insufficient growth, rather than an unequal distribution of wealth, or the result of processes of production or governance that do not provide decently paid jobs and social services. Reducing poverty is then used to justify more of this type of growth, and thus more of these processes that associate prosperity with production of carbon-based energy sources.

Even more alarming, a continuation of current development trends means that the world will be heading towards dangerous climate change which will severely affect ecological and human systems and lead to more poverty. Impacts include accelerated sea level rise, more intense heat waves, melting of glaciers affecting river flows, droughts and floods threatening food production, or changing disease vectors that are very likely to increase poverty (Eriksen and O'Brien, 2007; Stern, 2007; Skoufias *et al.*, 2011). Impacts will lead to displacement of affected populations and disrupted access to food, water and shelter, which in some cases could fuel conflict, more exposure to illness, and less preparedness for disasters (see IPCC, 2012). We know that all these impacts already affect the poor first and more intensely. We also know that solutions to climate change can also have negative consequences on poor people, from land grabbing for biofuels and the reordering of land for reforestation programs such as REDD, to Clean Development Mechanism (CDM) programs, which have proved detrimental to poor communities all over the world (Lohman, 2006; McMichael, 2009; Marino and Ribot, 2012). Nonetheless, such programs are often referred to as "win-win solutions" that will enable the transition to low-carbon economies and facilitate the reduction of poverty. In addition, the dominant discourse on climate change already assumes that we understand what poverty is and what the poor need in order to escape poverty. This discourse frames the relations between climate change and poverty as if climate change threatens our ability to reduce poverty, as if we already were on the path towards a just world, were it not for the "intruding" threat of climate change.

These circular narratives represent the real poverty trap, and it is a vicious one, for it diverts our attention from the actual causes of global environmental change and ecological and social vulnerability. The poor are instead used as the trump cards in arguments that resist transforming dominant socio-economic and political landscapes. A focus on "the poor" as a separate category ignores the underlying causes of poverty, those that both produce and maintain poverty and environmental destruction. It is sadly ironic that these synergistic processes (poverty and prosperity; economic growth and environmental degradation) are presented as separate issues by both dominant discourses on climate change and dominant discourses on poverty, while at the same time the continued existence of poverty serves as an excuse to continue with current development models and dependency on carbon economies.

In this chapter we explore "win-win" climate and poverty reduction strategies, including the emergent concept of "green growth" (UNRISD, 2010, 2011; Resnick *et al.*, 2012). We argue that these ignore the structural processes that produce both poverty and negative impacts on the environment, and that they are unlikely to produce positive results or lead to what often is referred to as a "no-regrets" solution (Heltberg *et al.*, 2009). Dominant forms of economic development have produced and are perpetuating poverty in the majority world, while also creating the current climate change crisis, despite the reality that alternatives exist (Mishan, 1967; Daly, 1991; Martinez-Alier, 2002; Jackson, 2011; Schneider *et al.*, 2010). We suggest that a critical understanding of poverty, as relational, historically grounded, and embedded in global and local political economies would contribute to clarity about the processes and structures that underlie the problems. A critical poverty studies research agenda is needed both to prevent further environmental damage in the name of the poor, as well as to protect the poor from further socio-economic shocks and negative climate change impacts.

Dominant poverty knowledge

DPK embodies the following fundamental characteristics: it is economistic and based on neoliberal premises; it emphasizes economic over ethical concerns; and it disassociates the majority and minority worlds, depicting poverty as outside social and geographical relations, in fact obscuring the forces that produce poverty in the first place, also in advanced economies (Hickey, 2009; Mosse, 2010; Lawson, 2012; O'Connor, 2001).

First, DPK is economistic, arguing for the primacy of economic causes of poverty and for narrow quantitative and aggregative definitions of poverty, primarily in terms of income. Indeed, econometrics has become "the science" of poverty and has been influential in discounting alternatives. DPK is also methodologically individualistic, reductionist, constructing "poor people" as a category that stands outside social relations. Such perspectives disregard history, presenting both ideas and data in an ahistorical manner. They ignore the role of economic and political power differentials in creating and perpetuating poverty, limiting rather than empowering the agency of people and collectivities. DPK is based on neoliberal premises to the point of even privatizing poverty and possible policy solutions, and does so very often by situating the task of poverty reduction under the mandates of global development agencies largely driven by neoliberal politics of "aid" and consistent with the goals of global capitalism. One of the most pervasive and perverse consequences of this economistic and neoliberal view of the poor is that it ignores processes of accumulation and wealth creation that have produced and perpetuated situations of poverty across the globe.

The DPK also shifts the burden of responsibility for poverty to individuals. Poverty is explained in terms of personal deficiencies, inappropriate or even immoral behavior, or as a result of insufficient engagement with markets, rather than the result of structural socio-economic and political conditions (Schram, 2000; O'Connor, 2001; Roy, 2010; Lawson, 2012). As argued by Lawson (2012), this shifting of responsibility applies to places as well as people. Failing places are viewed as responsible for producing poverty through deficiencies of governance, corruption and/or an overall cultural disposition that is seen as preventing market forces taking hold or preventing people from engaging their (presumably inherent) entrepreneurial spirit. Poor places are framed as isolated, disconnected and inherently lacking the wherewithal (capital, free and functioning markets *inter alia*) to raise themselves out of poverty as recently expressed in the 2009 World Development Report.

Second, dominant poverty knowledge blinds us to the ethical and justice aspects of poverty and the poor by treating poverty issues as technical matters with managerial solutions, removed from social relations. This view leads analysts to ignore the fundamental importance of social relations (for example, in work, governance, representation of the other) between poorer sectors, middle classes and elites as sites of intervention for building lasting solutions to poverty (Deacon and Cohen, 2011; Haarstad and St. Clair, 2011; Lawson, 2012; O'Connor, 2001). The most influential global ideas about poverty follow the ethics of the market. That is, the presumption that market relations are the most fair way to allocate resources, condemning those in structural traps to their own destinies and leading to ineffectual policy solutions that have later been reframed as market failures. The space that the ethics of the market occupies has further consequences, as it displaces a serious reflection about the extent to which poverty is, in fact, not a market failure but a moral failure. Treating poverty as a technical issue prevents ethical thinking, while the dominant perception of economics as the science of poverty leads to the rejection of political science and philosophy as sources of relevant knowledge on ethics and global justice. Moreover, this technocratic understanding of poverty makes completely irrelevant the role of non-poor people's agency, and their actual capacity to act and change the situation of the poor, while also ignoring the many ways in which poor people themselves are able to cope and to creatively act in seeking alternatives (UNRISD, 2010). It also neglects the enormous importance of global norms, such as, for example, the relations between the Treaty on Intellectual Property Rights (TRIPS) and the lack of entitlement to medicines for the poor, or the role that global trade agreements crafted by the WTO have in perpetuating unfair global relations that benefit elites at the expense of the majority (Pogge, 2010).

Third, DPK disassociates the majority and the minority worlds by representing poverty as a problem that occurs only in the global south and ignoring the role of global and historical economic and political relations across countries and peoples. It is in this sense a methodologically territorialist framing that presents poverty as spatially static and contained, i.e., something that can be mapped. Poverty is conceived of as a problem of and in specific places, despite the fact that places are integrally connected through a multiplicity of processes that are continuously produced and reconfigured. Methodological territorialism enhances the misguided idea that poverty is not relational, while also hiding the role of the non-poor in either producing or helping to eradicate poverty. This disassociation also serves to make invisible the millions of people in advanced economies who also experience conditions of relative poverty and in some cases live close to absolute poverty. Perhaps the most perverse consequence of this dislocation is the blindness towards the very important lessons from advanced economies in minimizing poverty and protecting the poor from the negative consequences of markets and ruthless capitalism.

Importantly, dominant poverty knowledge avoids a very serious factual question. How feasible is it, after all, to eradicate poverty? At times of unprecedented wealth the abolition of severe poverty is less costly than the bailouts most western countries' governments made of selected private sector companies during the financial crisis in 2008/2009. In Europe, bailouts to companies are accompanied by structural adjustment policies that have passed on the crises to average citizens, leading to increasing poverty and insecurity particularly for children and poor households (Ortiz and Cummings, 2012).

These broad characteristics of DPK are incompatible with the search for win-win strategies that address environmental issues, including climate change. DPK is antithetical to the framing and vision of human security proposed in this volume. If poverty is viewed through an economistic lens and action is based on neoliberal premises, then this diffuses rather than promotes collective responsibility, and negates rather than embraces social relations (Eriksen and Silva, 2009; O'Brien *et al.*, 2010). Very importantly, employing a similar frame for environmental change will lead to viewing climate change along similar lines.

A rich history of problematizing development

The critique of dominant models of development and their negative social and environmental consequences is nothing new. In this section we summarize literature from three complementary fields: critical development studies, development ethics, and pioneers of alternative economic thinking to illustrate that there is a rich history of problematizing mainstream development and DPK. Climate change is a development issue. Today, we know climate change is the result of using fossil fuels to pursue unsustainable models of progress and development associated with unending economic growth, land and natural resources use, and mass consumption (IPCC, 2007; Parry, 2009; O'Brien *et al.*, 2010; Dryzek *et al.*, 2011). In order to build models of development that change direction and help address global environmental change and climate change we first and foremost have to revisit the critique of dominant models and the many alternatives that have been put forward. Equally important, fighting poverty does not require destructive growth. While levels of poverty are associated with limited consumption and absence of basic needs and opportunities, the relationship between poverty reduction and unending economic growth has always been contentious. This carbon-driven economic growth, promoted through capitalist economic liberalization policies worldwide, large-scale development projects such as the construction of dams, mines, or massive transformations in agricultural production through industrial agriculture schemes has led to unequal harms and benefits empirically documented in a large body of development research literature that needs to be revisited when addressing climate impacts. Just to take one leading theory, the capabilities approach, the message is very clear: prosperity is not the same as economic growth (Sen, 1999; Nussbaum, 2011).

The development models that have led to the climate crises are the same processes of rapid capitalist transformation identified by Karl Marx. As Karl Polanyi puts it in his brilliant book the *Great Transformation*, the emergence of the market economy means primarily one thing: the disembedding of economic action from nature and the social (Polanyi, 1944). While the term development is used to convey the planned social change coordinated by international and national development aid agencies that emerged after World War II, this obscures the fact that since the nineteenth century there have been competing views about the direction of socio-economic transformation (Cowen and Shenton, 1995). A particular view became successful and dominant: progress as the pathways of societies that are commercialized, individualized, industrialized and urbanized. What emerges is a system that creates poverty discursively and materially, while it perpetuates existing inequalities and depends on structural conditions that keep poor people and communities subsumed to the functioning of a world economy in the making and unmaking of the "third world" (Escobar, 1995). As Denis Goulet (1980) argued, development is a cruel choice, leading to winners and losers. The professionals of development, the aid agencies, act like one-eyed giants, asking poor people in the global south to follow certain development pathways while not seeing the negative consequences of such pathways (Goulet, 1980). Modernity led to dreamed lives for some, especially in colonizer countries, but much poverty existed and still exists within modernities, bringing "disenchantment, alienation and loss of meaning, and from concern for the vulnerable and suffering" (Gasper and St. Clair, 2010: 20). Some get gains and some pains, argues one of the first social scientists to populate the World Bank, Michael Cernea, referring to the socially and environmentally harmful consequences of large development projects such as building large dams, as the result of ignoring both people and nature (Cernea, 1985).

Decades of this purposeful development have led to a present pathway marked exactly by these two ills: massive inequalities and extensive environmental damage. Inequality, although very well documented, remains an externality in most processes of planning and social transformation and is a defining characteristic of contemporary globalization (Munck, 2004; Milanovik, 2007; Gills, 2008). Much of the critical literature on development and inequality has one common characteristic: it proposes the thesis that any plausible conception of what may constitute fair globalization and social transformation processes needs to address the power relations that produce and reproduce global poverty (St. Clair, 2006; 2007). Current data on poverty trends present a very sad picture. There are almost 1 billion hungry people on the planet, and a similar number lack suitable housing, access to safe water or to basic education. More than two billion lack access to sanitation and essential medicines and health care. Over 200 million children work for wages, and more than 1.6 billion people do not have access to electricity or any energy service (CROP, 2010). As Branko Milanovik (2007) argues, the 6.4 percent of global household income that has gone into expanding the share of the top tenth, would be sufficient to double all incomes for the bottom seven tenth of humankind.

Equally important, unending economic growth models of development have been the main cause of pervasive environmental degradation, ecosystem degradation and biodiversity loss. While growth is often depicted as the driver for better environmental regulation and thus environmental quality, the so-called U-shaped curve linking more growth and more environmental protection is a theoretical construct rather than a "real fact." Economic liberalization and other policies that promote gross national product growth are not substitutes for environmental policy (Arrow *et al.*, 1995).There are limits to the carrying capacity of the planet, but these are not central to contemporary growth practices that presumably are the main driver of poverty reduction. Economic theory driving the modern project of prosperity has always treated the environment as a sink and a never-ending source of resources to be turned into products. There is a large body of environmental economics and alternative economic thinking that has warned against unending growth and its negative effects. In 1968, Ezra Mishan argued that the dream of unending economic prosperity and rising living standards was deeply dangerous because of its social and environmental costs. Mishan (1967) contended we may be heading toward chaotic environmental damage. Later, steady state economics, proposed by Herman Daly (1991), warned of the flaws in traditional economic theory for improperly valuing natural resources. Environmental economists such as Joan Martinez-Alier demonstrated how environmentalisms are numerous, and that an environmentalism for the poor requires valuation of natural systems and accounting of ecological debt to the poor (Martinez-Alier, 2002).

More recently a de-growth movement has developed in response to climate change impacts and widespread environmental degradation. The term de-growth was invented by the French economist Sergio Latouche, as part of his critique of dominant models of development social and environmental impacts (see http://www.degrowth.eu/). Recently Tim Jackson (2011) has argued for shifting gears in relation to unending growth by changing the ways we conceive of and practice economic transactions and investments. What we see emerging is a new conception of prosperity and progress that takes social and environmental consequences as an internal part of their definition. Amongst those alternatives, we must also add a new way of seeing the causes of poverty, who are the poor and what to do in order to eradicate poverty while protecting the natural environment through valuing differently. It is doable and feasible, we suggest, if power shifts were to enable such transitions to new development pathways. Together, these alternatives offer a way out of the vicious cycle of promoting destructive growth in the name of the poor.

Alternative perspectives on poverty: connecting human security and the climate crisis

A reframing of both poverty and climate change must be part of building alternatives visions that no longer view climate change as a threat to our ability to reduce poverty. Such visions do not present the current efforts to reduce poverty as an ideal state, as if we already were on the path towards a just world were it not for the threat of climate change. Alternative perspectives on poverty challenge this discourse and strengthen the possibility of alliances across critical fields of development studies, globalization studies and environmental economics. We propose a perspective more suited to an integrated, relational, agency-based and equitable joint framing of both challenges: poverty and the environment.

An alternative framing of poverty starts with the proposition that poverty occurs in all parts of the globe (both majority and minority worlds, or Global South and Global North) and poses questions about the interconnected global material and discursive dimensions of poverty processes in their relation to wealth and prosperity. Poverty, we argue, is best understood as relational, produced through material social relations and discursive processes in the context of specific histories, geographies and ecologies (O'Connor, 2001; Hickey, 2009; Mosse, 2010; Lawson, 2012). Poverty is produced by the changing political-economy of globalized capitalism across both majority and minority worlds. Ecological connections between human beings and the places they inhabit are key to understanding poverty and situating humans as part of their environment. The ways in which poverty is pictured and addressed in different places are connected: for example, how the poor are depicted in the majority world helps us to understand constructions of the poor and poverty policy in the minority world. And this is relevant also in relation to the current debate on poverty and climate change. For example, the presumption that those with more resources are less likely to be affected by climate and also that they will be more able to adapt, reinforces the distance between the poor and the well-off in the face of dangerous climate change. We need to rejoin this artificial dislocation between the poor and the non-poor, the vulnerable to climate and the non-vulnerable. A relational view focuses our attention on the interlinkages between the poor and non-poor, rather than eliding these interconnections.

We argue for work that addresses simultaneously the contradictions inherent in market-based approaches to poverty reduction as well as market responses to climate and global environmental change. This includes critiques of the extension of market relations into almost everything, and the implications of neo-liberal roll-backs in institutional involvements in poverty reduction (at various scales with different profiles in different places). We suggest that market-based approaches to poverty reduction are deeply political (despite their apolitical claims). DPK produces flawed subjects and flawed places through categorizing "the poor"; by discursively framing who is counted as deserving/undeserving, in their exclusion of care activities and relations, and in how the state is reframed as a neoliberal actor in policy decisions. By contrast, relational analyses of poverty focus attention on the co-production of poverty, attending to how people interpret and challenge both the category of poverty and processes producing impoverishment, and contest and negotiate poverty processes. On the one hand, we know the poor accommodate poverty through adaptive preferences (Sen, 1999). But people also seek to maintain dignity and civility rather than resisting either representations or material productions of poverty/inequality. Co-production also includes attention to how global actors engage in poverty polices, co-produce bureaucratic social orders and institutional power relations at the same time as they produce knowledge (St. Clair, 2006).

Finally, we argue for an ethical perspective and the treatment of poverty as a question of social and global justice, and for attention to the relations to and complementarities with solidarity, virtues, care, and social cohesion and the fundamental role that inequality plays in preventing just processes and outcomes. A relational approach to analyzing poverty can then be formed through including insights from development ethics and global justice; insights from critical care ethics, global social justice, and social inclusion. This includes consideration of special attachments and obligations (Deacon and Cohen, 2011; Haarstad and St. Clair, 2011; Lawson, 2012).

In summary, we argue for a social constructionist political-economy approach to understanding poverty. Poverty is produced through political, economic, and discursive/cultural mechanisms that are connected and recurrent across space and also (co)produced through human actions and places. To adequately address poverty in the context of climate and global environmental change crises, scholars must explore the simultaneity of material processes, the social constructions of poverty and how these depict the relations between poverty and climate change. This leads to placing the central concern on social, global and climate justice. In addition, it leads scholars to engage and co-produce knowledge with those named as poor as creative agents with capacity to define their actions and futures.

Conclusion: reframing poverty and prosperity

This chapter argues that global environmental change, poverty and prosperity are integrally related sets of processes that are linked to specific models of development and particular conceptions of progress. Past visions of social transformation have led to a world that has created prosperity for a few at the expense of the many and the planet. Reversing this trend requires rethinking transformative processes in a holistic way, where egalitarian societies and the health of the planet become central goals. Unlike the dominant poverty discourse and visions of development as unending growth and luxury consumption, the reframing we see in critical social science literature provides a point of departure that may lead not only to a more socially just and pro-poor discourse on environmental and climate change, but also to more feasible, empirically grounded and long-lasting solutions to these problems. There are four fundamental reasons for merging critical agendas for research relating poverty and climate change:

First, it is of fundamental importance to unsettle the DPK and to prevent its uncritical adoption by the rapidly emerging scholarship on the human dimensions of climate change and the worldwide search for easy win-win solutions to the dilemmas posed by climate change to existing poor people and communities. Such uncritical adoption of a discourse that has fundamentally not only failed to eradicate severe poverty or to protect people from vulnerability to poverty conditions, but has actively contributed to them, will lead to maladaptation and will be blind to the now documented negative impacts of mitigation strategies on poor communities. Also, the uncritical acceptance of poverty and development discourses will not encourage crucial self-reflection regarding how processes of production and consumption are linked with the dominant economic development paradigm, and are a primary cause of current climate crises. If we continue along this same path, mitigation will never be possible.

Second, we argue that in the same way that market-based approaches to solving poverty have failed to lead towards the idealized "outcomes" promised by the dream of modernity, we risk similar outcomes for market-based solutions to climate change.

Third, we argue that the reframing of global poverty as a matter of social justice and global solidarity is exactly what is needed in the reframing of climate change. There are no easy answers to what are complex global problems with uncertain outcomes. Precisely because of these characteristics, such problems call for protection, fairness, acceptance of responsibility, and equity.

Fourth, feasible and equitable solutions to both climate change and poverty may be crafted only when climate change and poverty are seen together, guided by solidarity and justice concerns, informed by political economy, linked to consumption and production, and within a joint framework that is critical of dominant discourses on development and prosperity.

It is by seeing both issues in an integral manner, as questions of agency and human action, as moral and ethical challenges, related to human beings with complex and often contradictory values and worldviews that a balanced solution may be found for both mitigation and adaptation. We argue such solutions lie in the capacity to transform the framing of both the causes and solutions to the interconnected crises of global poverty and climate change.

Notes

1 An earlier and much shorter version of some of the arguments presented in this chapter appeared at the International Human Dimensions of Global Environmental Change Program (IHDP) Highlights as an electronic publication under the title "Climate, environment and the poor: The relevance of critical global poverty studies." We thank the editors for their excellent comments and suggestions.

2 Dominant poverty knowledge (in the US and in international development circles) is characterized by a primary focus on income, a methodological individualism and as a

discrete ontological entity: a category or a benchmark defining particular people and/or places (Lawson, 2012: 4; Mosse, 2010; Hickey, 2009; O'Connor, 2001).

References

Alkire, S and Foster, J. (2011) 'Understandings and misunderstandings of multidimensional poverty measurement', *Journal of Economic Inequality*, 9 (2): 289–314.

Arrow, K., Bolin, B., Costanza, R., Dasgupta, P., Folke, C., Holling, C.S., Jansson, B., Levin, S., Maler, K., Perrings, C. and Pimentel, D. (1995) 'Economic growth, carrying capacity and the environment', *Science*, 268 (5210): 520–521.

Cowen, M-P. and Shenton Robert, W. (1995) 'The invention of development', in J. Crush (ed.) *Power of Development*, London: Routledge: 27–43.

Cernea, M. (1985) *Putting People First: Sociological Variables in Development*, Oxford: Oxford University Press.

CROP (2010) 'Status on global poverty-some statistics and sources', Comparative Research Programme on Poverty, Policy Brief Series 1, 2010. Online. Available HTTP: http://www.crop.org/viewfile.aspx?id=208 (assessed 10 November 2012).

Daly, H. (1991) *Steady State Economics*, Washington, DC: Island Press.

Deacon, B. and Cohen, S. (2011) 'From the global politics of poverty alleviation to the global politics of social solidarity', *Global Social Policy*, 11 (3): 233–249.

Dryzek, J., Norgaard, R. and Scholsberg, D. (2011) *Oxford Handbook on Climate and Society*, Oxford: Oxford University Press.

Eriksen, S. and O'Brien, K. (2007) 'Vulnerability, poverty and the need for sustainable adaptation measures', *Climate Policy*, 7 (4): 337–352.

Eriksen, S. and Silva, J.A. (2009) 'The vulnerability context of a savanna area in Mozambique: household drought coping strategies and responses to economic change', *Environmental Science & Policy*, 12 (1), 33–52.

Escobar, A. (1994) *Encountering Development: The Making and Unmaking of the Third World*, Princeton, NJ: Princeton University Press.

Escobar, A. (1995) *Encountering Development: The Making and Unmaking of the Third World Princeton*, Princeton, NJ: Princeton University Press.

Gasper, D. and St. Clair, A.L. (eds) (2010) *Development Ethics: A Reader*, London: Ashgate.

GECHS (1999) *Global Environmental Change and Human Security*, GECHS Science Plan, Bonn: IHDP.

Goulet, D. (1980) 'Development experts: the one eyed giants', *World Development*, 8 (7/8): 481–489.

Gills, B. (2008) 'The swinging of the pendulum: the global crisis and beyond', *Globalizations* 5 (4): 513–522.

Haarstad, H. and A.L. St. Clair (2011) 'Social policy and global poverty: beyond the residual paradigm?', *Global Social Policy*, 11 (2–3): 214–219.

Hayes, T. and Persha, L. (2010) 'Nesting local forestry initiatives: revisiting community forest management in a REDD+ world', *Forest Policy and Economics*, 12 (8): 545–553.

Heltberg, R., Siegel, P.B. and Jorgensen, S.L. (2009) 'Addressing human vulnerability to climate change: toward a "no-regrets" approach', *Global Environmental Change*, 19 (1): 89–99.

Hickey, S. (2009) 'Rethinking poverty analysis from the margins: insights from Northern Uganda', *Afriche e Oriente*, 11 (2): 119–36.

IPCC (2007) *Contribution of Working Group II to the Fourth Assessment Report of the Intergovernmental Panel on Climate Change, 2007*, M.L. Parry, O.F. Canziani, J.P. Palutikof, P.J. van der

Linden and C.E. Hanson (eds) Cambridge, United Kingdom and New York, NY, USA: Cambridge University Press.

IPCC (2012) *Managing the Risks of Extreme Events and Disasters to Advance Climate Change Adaptation, a Special Report of Working Groups I and II of the Intergovernmental Panel on Climate Change*, C.B. Field, V. Barros, T.F. Stocker, D. Qin, D.J. Dokken, K.L. Ebi, M.D. Mastrandrea, K.J. Mach, G.-K. Plattner, S.K. Allen, M. Tignor, and P.M. Midgley (eds), Cambridge, UK, and New York, NY, USA: Cambridge University Press.

Jackson, T. (2011) *Prosperity Without Growth: Economics for a Finite Planet*, London: Routledge.

Lawson, V. (2010) 'Reshaping economic geography? Producing spaces of inclusive development, *Economic Geography*, 86 (4), 351–360.

Lawson, V. with the Middle Class Poverty Politics Research Group (2012) 'De-centering poverty studies: middle class alliances and the social construction of poverty', *Singapore Journal of Tropical Geography*, 33: 1–19.

Lohman, L. (2006) 'Carbon trading: A critical conversation on climate change, privatization and power', *Development Dialogue*, 48, Uppsala: The Dag Hammarskjöld Centre.

Marino, E. and Ribot, J. (2012) 'Adding insult to injury: Climate change and the inequities of climate intervention', *Global Environmental Change*, 22 (2): 323–328.

Martinez-Alier, J. (2002) *Environmentalism of the Poor, A Study of Ecological Conflicts and Valuation*, London: Edward Elgar.

McMichael, P. (2009) 'Banking on agriculture: a review of the World Development Report 2008', *Journal of Agrarian Change*, 9 (2): 235–246.

Milanovik, B. (2007) An even higher global inequality than previously thought: a note on global inequality calculations using the 2005 ICP results', World Bank Policy Research Working Paper Series, Washington DC: World Bank.

Mishan, E. (1967/1993) *The Cost of Economic Growth* (revised edition in 1993), New York: Praeger.

Mosse, D. (2010) 'A relational approach to durable poverty, inequality and power', *Journal of Development Studies*, 46 (7): 1156–1178.

Munck, R. (2004) *Globalization and Social Exclusion: A Transformationalist Perspective*, London: Kumarian Press.

Nussbaum, M. (2011) *Creating Capabilities: The Human Development Approach*, Cambridge, MA: Belknap Press of Harvard University Press.

O' Brien, K.L., St. Clair, A. and Kristoffersen, B. (eds) (2010) *Climate Change, Ethics, and Human Security*, Cambridge: Cambridge University Press.

O'Connor, A. (2001) *Poverty Knowledge: Social Science, Social Policy, and the Poor in Twentieth-Century U.S. History*, Princeton, NJ: Princeton University Press.

Ortiz, I. and Cummings, M. (2011) 'Global inequality: beyond the bottom billion – a rapid review of income distribution in 141 countries', UNICEF Social and Economic Policy Working Paper. Online. Available HTTP: http://dx.doi.org/10.2139/ssrn.1805046 (accessed 10 November 2012).

Ortiz, I. and Cummings, M. (2012) *A Recovery for All: Rethinking Socio-economic Policies for Children and Poor Households*, New York: UNICEF. Online. Available HTTP: http://www.unicef.org/socialpolicy/index_62107.html (accessed 10 November 2012).

Parry, M. (2009) 'Climate change is a development issue, and only sustainable development can confront the challenge', *Climate and Development*, 1 (1): 5–9.

Pogge, T. (2010) *Politics and Usual: What Lies behind the Pro-Poor Rhetoric*, London: Polity Press.

Polanyi, K. (1944) *The Great Transformation: The Political and Economic Origins of our Time*, Chicago: Chicago Press.

Resnick, D., Tarp, F. and Thurlow, J. (2012) *The Political Economy of Green Growth: Illustrations from Southern Africa*, UNU-WIDER Research Paper, World Institute for Development Economic Research (UNU-WIDER).

Roy, A. (2010) *Poverty Capital: Microfinance and the Making of Development*, New York: Routledge.

Schneider, F., Kallisa, G. and Martinez-Alier, J. (2010) 'Crisis or opportunity? Economic degrowth for social equity and ecological sustainability', *Journal of Cleaner Production*, 18 (6): 511–518.

Schram, S. (2000) *After Welfare: The Culture of Postindustrial Social Policy*, New York: New York University Press.

Sen, A. (1999) *Development as Freedom*, New York: Knopf.

Stern, N. (2007) *The Economics of Climate Change – The Stern Review*, Cambridge: Cambridge University Press.

Skoufias, E., Rabassa, M., Olivieri, S. and Brahmbhatt, M. (2011) *The Poverty Impacts of Climate Change*, World Bank Policy Brief, Washington, DC: World Bank.

St. Clair, A.L. (2006) 'Global poverty: the co-production of knowledge and politics', *Global Social Policy* 6 (1): 57–77.

St. Clair, A.L. (2007) 'A methodologically pragmatist approach to development ethics', *Journal of Global Ethics*, 3 (2): 1–22.

St. Clair, A.L. (2010) 'Global poverty and climate change: towards the responsibility to protect', in K. O'Brien, A.L. St. Clair, and B. Kristoffersen (eds) *Climate Change, Ethics and Human Security*, Cambridge: Cambridge University Press: 180–198.

UNRISD (2010) *Poverty Report: Combating Poverty and Inequality: Structural Change, Social Policy and Politics*, Geneva: United Nations Research Institute for Social Development. Online. Available HTTP: http://www.unrisd.org/publications/cpi (accessed 15 August 2012).

UNRISD (2011) *Green Economy and Sustainable Development: Bringing Back the Social Dimension*, Geneva: United Nations Research Institute for Social Development. Online. Available HTTP: http://www.unrisd.org/events/greeneconomy (accessed 30 October 2012).

19 REDD+ and the global discourse on climate change and poverty

Desmond McNeill

Introduction

With global warming the world has encountered the ultimate "global environmental threat," one which undeniably affects all people and requires concerted action by all countries. It is a grim irony that those who will suffer most from climate change are generally those that are already among the poorest in the world, and have done least to cause the problem (Leichenko and O'Brien, 2008). The nations of the world are, however, beginning to recognize the threat and are responding, albeit perhaps too little and too late. Climate change has suddenly and dramatically changed the nature of the long-running and complex relationship between countries of the "North" and "South." It has also changed the global discourse concerning "development" and "environment"; that both shapes and is shaped by these relationships. The aim of this chapter is to examine this discourse; and to assess one specific initiative that has arisen in response to climate change, REDD+ (Reducing Emissions from Deforestation and Forest Degradation in Developing Countries), showing what implications it may have for poverty reduction.

Climate change will necessitate adaptation not only at national level but also at international level: adaptation of the relations among the nations of the world. The old structure, derived from the era of development aid, will have to be replaced by another. The rhetoric of mutual interest – to be found in the Pearson Report of 1969, the Brandt Report of 1980, and even the Brundtland Report – was never very convincing to skeptics. And for those in the rich North it could even encourage a very negative picture of the rest of the world: as a source of disease, terrorism, migrant hordes, etc. But with climate change there is now a more real sense of mutual interest; the people of the world today do truly face a common threat.[1] But the simple North/South dichotomy is now even more questionable than it was half a century ago. Countries such as China, India, and Brazil are rich and powerful in aggregate terms, even if they still have many millions of poor inhabitants. As a result, their rights and responsibilities with regard to the challenge of climate change are very different from those of small, poor countries. As regards development, these countries are now powerful global economic actors, and, in Africa and elsewhere, they are increasingly active, whether as aid donors or as investors. What these countries do – especially with regard to climate change – will be of crucial importance for the world as a whole. In summary, the environment and development discourse is now very different from what is was until quite recently, as I shall briefly show by tracing how first the development, and then the environment, discourse has changed over time.

Discourses

How are these changing international relationships affecting, and affected by, the global discourse concerning "development" and "environment"? I shall argue that the *global environmental discourse* has, from the start, been located within the already established *development discourse*; but that now, with climate change, the situation is changing. But what is a discourse? Rather than quoting standard authors such as Foucault (1972) or, more specifically in relation to development issues, Escobar (1995), I am tempted to say: "you know a discourse when it hits you." In other words, the power of a discourse is all too evident to those who, perhaps unwittingly, seek to escape or challenge it. People who take part in international donor meetings on development or the environment, or submit inputs to influential documents such as the World Bank's annual World Development Report, are very soon confronted by this reality. Certain topics are on the agenda, while others are excluded. And in contributing to the debate certain words can, and even must, be used, while others cannot. As neo-Gramscians would have it, our thoughts are "framed" – and hence, so too are our actions (Bøås and McNeill, 2004). The major effect – and, many would argue, the major function – of such framing is to depoliticize the debate. The study of a discourse is revealing because it tells you where the power lies.

Development – since 1950

Development is a post World War II phenomenon, entirely tied up with development aid (Cowen and Shenton, 1996; Rist, 1997; Degnbol-Martinussen and Engberg-Pedersen, 2003). In the 1950s/1960s, the era of decolonization, aid was seen as a temporary activity, which would be necessary for perhaps a decade or two, involving investment and technology transfer from "developed" to "underdeveloped" countries. Around the 1970s came calls for radical change in North/South relations – a New International Economic Order. But, at least in donor circles, the more reformist agenda of "Redistribution with Growth" (Chenery *et al.*, 1974) predominated – calling for change, not revolution, within countries in the South; and with much less focus on relations between North and South. The 1980s brought reaction, and a neo-liberal agenda – the so-called "Washington consensus"; here the ideology of the market prevailed both at international and national levels. In the 1990s, with the fall of the Berlin Wall, the clash of ideologies between East and West suddenly became largely irrelevant. It is at this point that politics was finally allowed to enter explicitly into the policy debate of development agencies. Though largely invisible in official documents, politics has, of course, always been fundamental in shaping the discourse. Throughout the whole development era, the two key development debates have been around "state or market" and "equity or growth"; both clearly very political issues. But even now, politics is treated largely as a technical issue – of "development management." And the focus of attention has continued to be on intra-national issues; the international dimension still largely excluded from debate.

During the 1980s and 1990s, the aid agenda became increasingly diffuse; every new international conference, commission or report added another new issue to be addressed: gender, population, governance, human rights – and the environment (with the World Commission on Environment and Development: the Brundtland Report, published in 1987). There can be little doubt, however, that it was the last of these that was the most high profile; the 1992 "Conference on Environment and Development" (UNCED), in Rio de Janeiro was the largest such event in history.

Then, in 2000, came a return to poverty as the central issue, with the Millennium Declaration and the Millennium Development Goals (MDGs). This was regarded by many as a necessary, and even overdue, refocusing of attention and resources, to counter what had been a tendency to overload the development agenda by adding yet another "decoration on the Christmas tree" (to quote a staff member of a major multilateral development bank). In this situation, the response of those who still wished to promote a particular issue, such as gender, was often to argue that this was the best means of achieving the now overarching purpose – of poverty reduction. Thus, for example, it was argued that gender planning helps to focus on the needs of women – often the most seriously disadvantaged group; or that good governance is a necessary prerequisite for poverty reduction. But protecting the environment, by contrast, could not easily be reformulated as a means to the end of reducing poverty; indeed some, and in fact many in the South, saw poverty reduction and environmental conservation as conflicting objectives, as noted below. It is therefore relevant to note that in the Millennium Development Goals (MDGs), the emphasis is clearly on poverty reduction and improved welfare; potential conflicts between this and environmental goals are not addressed. Of the eight goals, only one is directly concerned with environmental sustainability, with one sub-goal specifically focused on climate change. And yet again we find, in the MDGs, that the main focus is on the national obligations of poor countries; international obligations receive only very limited mention.

Environment – since 1970

The United Nations Conference on the Human Environment, held in Stockholm in 1972, could be said to mark the beginning of the environment discourse at the international level. But it was mainly concerned with the agenda of countries in the North. Indeed, developing countries were skeptical of supporting the initiative. At a meeting in Founex in Switzerland, before the conference began, Mahbub ul Haq, then at the World Bank, was brought in to persuade the developing country delegates that "there really was good cause, both then and in the future, for them to be concerned about these matters" (Wade, 2004: 74). But their skepticism remained.

The global discourse on environment did not really take off until the Brundtland Report of 1987 (Redclift, 1987; Sachs, 1993). In that report the environmental discourse was intimately linked with the already established development discourse, and debate easily fell into a "North/South" dichotomy. This was not surprising in view of the resistance by the South just noted, which was manifest also in the discussions by the commission members as they traveled around the world. For some of those from the North it was a challenge to have to reconsider their views about the environment in the light of the realities of poverty faced by many in the South. In the Brundtland Report, environment and development were brought together into a single discourse, as manifest in the famous concept of "sustainable development." But the adoption of this term did not in practice resolve the conflicts between environment and development, or between North and South. And this became apparent in subsequent international conferences – from Rio de Janeiro 1992, through to Rio de Janeiro again in 2012. But in many of the central documents concerning sustainable development, the structural inequalities of wealth and power between North and South were largely excluded; and in negotiations the South adopted the well-established position of demanding "new and additional" development assistance. It is also worth noting that the uncomfortable issue of sustainable consumption was avoided.

In brief, environment and development (or poverty reduction) have for many years enjoyed an uneasy relationship: vying for attention as the primary objectives of international cooperation, while at the same time being lumped together under the somewhat unconvincing term "sustainable development." But since about 2005 the world faces a new situation – thanks to climate change.

Climate change and REDD+

The "win-win" vision of sustainable development appeared optimistic when the Brundtland Report came out in 1987. It can be argued on both theoretical and empirical grounds that some environmental problems, notably pollution of various kinds, can be reduced with rising income levels; while others, notably the consumption of non-renewable resources, cannot.[2] But sustainable development seems almost a utopian ambition now that the correlation between high income and consumption, and resulting emissions of greenhouse gases, is so glaringly apparent.

As noted above, the MDGs were predominantly about poverty reduction; the environment received fairly marginal treatment.[3] With climate change the global agenda has changed: for better or worse. Global warming is – incontrovertibly – a global issue. I would even claim, that it is the first (or perhaps second) truly global environmental issue. Most, if not all, other environmental issues, such as urban air pollution, shortage of water, or desertification, even if they are experienced in many countries, are still geographically limited in scope. The depletion of the ozone layer can justifiably claim to be the first truly global issue: in this case, however, the magnitude of the impacts was less, and the problem was simpler to resolve – both technically and politically (Falkner, 2007). With climate change, then, environment and development are more "truly" linked than in the Brundtland Report. The interconnections are much closer: but more in the sense of a potential "lose-lose" situation – by contrast with the (optimistic) "win-win" hopes of the Brundtland Report. Those who will suffer most from climate change are those who are already most disadvantaged.

And this new situation also changes the North/South relationship. The North now needs the South in a way that it did not before. Development assistance has never been merely altruistic; many would justifiably argue that a good deal of aid was, and still is, not charity but payment for services rendered, such as political support in international relations. But payments by rich countries to induce poor countries to reduce emissions are manifestly, and indisputably, payments for services rendered. If development assistance is replaced by mechanisms such as carbon trading and REDD+, what will this mean for the environment-development relationship, and the North-South relationship?

The challenge, and opportunity, involved in seeking to reconcile poverty reduction with combating global warming is well illustrated by REDD+. This initiative, which gained increased international backing at the December 2009 United Nations Convention on Climate Change (UNFCCC) meeting in Copenhagen, is based on the idea that one rewards individuals, communities, or countries that reduce greenhouse gas emissions from forests. A number of countries in the North have indicated their willingness to finance countries in the South which undertake to implement REDD+ projects.

Precisely because they are poor, and have not yet cut down all their forests, countries in the South now have something to sell to the rich North. In some respects this is surely good news. The amount of money involved would far exceed foreign aid (Meridian Institute, 2009). And because these are payments for services they are more akin to automatic transfers, almost like a global income tax – which has long been an apparently unattainable dream (McNeill, 1980). But there is also some bad news. REDD+ may involve massive international transfers, but the money will not go only to poor countries and not necessarily to poor people. The countries which have most to gain include several – such as the Democratic Republic of Congo, or Indonesia – which do not enjoy particularly good governance. In recent years, aid has increasingly been made conditional on its being used for the benefit of the poor; but REDD+ will create huge scope for "misuse" according to standard aid thinking. It is already becoming apparent that reconciling the primary goal of REDD+ – reduction of deforestation and forest degradation – with that of poverty reduction can be difficult in practice. In fact, measures to inhibit or prohibit access to the forest may come largely at the expense of poor and marginalized groups – and especially indigenous communities. A further complicating factor is that REDD+ is also meant to protect biodiversity; but such a "win-win-win" strategy may prove even more elusive.

To quote an authoritative source: "REDD has the potential to deliver large cuts in emissions at a low cost within a short time frame and, at the same time, contribute to reducing poverty and sustainable development" (Angelsen, 2008: viii). The quote continues: "This sounds too good to be true. REDD is based on a simple and appealing idea, but turning the idea into action is much more complex" (ibid.). It is becoming increasingly recognized that REDD+ constitutes a major governance challenge – even in dealing with well-intentioned and honest governments. The problem is not merely the technical difficulty of measuring emission reductions, and attributing them to specific groups. Equally, if not more challenging, is how best to channel funds to these groups – to individuals, villages, districts, or projects – in a way that is fair, efficient and effective (in terms of incentives). It is becoming apparent that local people may not benefit from REDD+. Evidence comes both from NGOs, many of which have been actively mobilizing against REDD+;[4] and also from independent research – including my own, in collaboration with Signe Howell (Howell, 2013). Phelps *et al.* (2010: 326) note that "Many observers have criticized REDD+ projects for their rapid development that has often overlooked the rights, tenure, and engagement of forest-dependent and indigenous peoples." In addition, Börner *et al.* (2010: 280) assert that "Unequal land distribution in the Brazilian Amazon, for example, means that regulatory mandated reductions in deforestation will especially hurt smallholder families struggling to maintain soil productivity in traditional slash-and-burn systems on relatively small plots." And Schroeder (2010) refers to experience from the Noel Kempff National Park in Bolivia launched in 1997, and quotes Griffiths (2008) to conclude that:

> payments for ecosystem services reinforce state and private sector control over forests as benefits are captured largely by state agencies, local governments and international conservation NGOs and not indigenous and forest communities. Design and implementation of such projects are top-down, limit stakeholder participation and exclude indigenous peoples from the benefits arising from such projects.
>
> (Schroeder, 2010: 322)

In conclusion

Climate change has substantial implications not only for international North-South relations but also for the "environment-development" discourse. Climate change is necessarily a "development-and-environment issue." And – if the REDD+ initiative is implemented on a large scale – the South (or at least forest-rich countries) may no longer be dependent on the benevolence of the North for financial transfers, but rather receive payment for services rendered in the form of reduced CO_2 emissions.

But will REDD+ benefit the poor people in those countries? The prospects are not very encouraging. Local communities living in or near forests risk not only missing out on the benefits, but even finding their livelihoods threatened. This is particularly ironic, if this is the price they pay in order that the rich North can avoid having to find other ways of mitigating climate change.

Notes

1 Ironically, the recent financial crisis also constituted a global threat; and in combating this threat the continued economic growth of China and India was very important for the North. It is ironic because it is precisely this continued rapid growth which is exacerbating climate change; the same "emerging middle classes" in these two countries that are saving the North from recession are also generating massive CO_2 emissions.

2 Debate has often been based on the so-called "environmental Kuznets curve" (Stern *et al.*, 1996), which is claimed to be analogous to the, also disputed, Kuznets curve relating to economic growth and inequality (Kuznets, 1955).

3 If a more wide-ranging interpretation of "environment" is adopted, including for example drinking water quality, then more MDGs become relevant.

4 See, for example, the website of the coalition organization "Rights and Resources" at http://www.rightsandresources.org/ and numerous postings on "REDD+ monitor" http://www.REDD+-monitor.org/.

References

Angelsen, A. (ed.) (2008) *Moving Ahead with REDD+: Issues, Options and Implications,* Bogor, Indonesia: CIFOR.

Bøås, M and McNeill, D. (2004) *Global Institutions and Development: Framing the World*, London: Routledge.

Börner, J., Wunder, S., Wertz-Kanounnikoff, S., Tito, M., Pereira, L. and Nascimento, N. (2010) 'Direct conservation payments in the Brazilian Amazon: Scope and equity implications', *Ecological Economics*, 69: 1272–1282.

Chenery, H., Ahluwalia, M.S., Bell, C.L.G., Duloy, J.H. and Jolly, R. (eds) (1974) *Redistribution with Growth,* Oxford: Oxford University Press.

Cowen, M.P and Shenton, R.W. (1996) *Doctrines of Development,* London: Routledge.

Degnbol-Martinussen, J. and Engberg-Pedersen, P. (2003) *Aid: Understanding International Development Cooperation*, London: Zed Books.

Escobar, A. (1995) *Encountering Development: The Making and Unmaking of the Third World,* Princeton: Princeton University Press,

Falkner, R. (2007) *Business Power and Conflict in International Environmental Politics,* London: Palgrave Macmillan,

Foucault, M. (1972) *The Archaeology of Knowledge*, New York: Pantheon Books.

Griffiths, T. (2008) 'Seeing 'REDD+' – avoided deforestation and the rights of indigenous peoples and local communities' *Indigenous Perspectives*, 9 (1–2): 93–118.

Howell, S. (2013) 'Divide and rule: nature and society in a global forest programme' in K. Hastrup (ed.) *Anthropology and Nature*, London: Routledge.

Kuznets, S. (1955) 'Economic growth and income inequality', *American Economic Review*, 45: 1–28.

Leichenko, R. and O'Brien, K. (2008) *Environmental Change and Globalization: Double Exposures*, Oxford: Oxford University Press.

McNeill, D. (1980) *The Contradictions of Foreign Aid*, London: Croom Helm.

Meridian Institute (2009) *Reducing Emissions from Deforestation and forest Degradation: An Options Assessment Report*, Prepared for the Government of Norway by Angelsen, A., Brown, S., Loisel, C., Peskett, L. Streck, C. and Zarin, D. Online. Available HTTP: http://www.REDD+-oar.org/links/REDD+-OAR_en.pdf (accessed 7 December 2012).

Phelps, J. Guerrero, M.C., Dalabajan, D.A., Young, B. and Webb, E.L. (2010) 'What Makes a "REDD+" Country?', *Global Environmental Change*, 20: 322–332.

Redclift, M. (1987) *Sustainable Development: Exploring the Contradictions*, London: Routledge.

Rist, G. (1997) *The History of Development: From Western Origins to Global Faith*, London: Zed Books.

Sachs, W. (ed) (1993) *Global Ecology: A New Arena of Political Conflict*, London: Zed Books.

Schroeder, H. (2010) 'Agency in international climate negotiations: the case of indigenous peoples and avoided deforestation', *International Environmental Agreements*, 10: 317–332.

Stern, D.I., Common, M. S. and Barbier, E.B. (1996) 'Economic growth and environmental degradation: The environmental Kuznets curve and sustainable development', *World Development*, 24 (7): 1151– 1160.

Wade, R. (2004) 'The World Bank and the environment', in M. Bøås and D. McNeill (eds) *Global Institutions and Development: Framing the World*, London: Routledge.

20 Climate change adaptation

Challenging the mainstream

Philip Ireland and Peter Keegan

Introduction

Climate change adaptation (CCA) has rapidly emerged as an important and widely used concept within international development policy and programmatic discourses and practices. Defined by the Intergovernmental Panel on Climate Change (IPCC) as adjustment in natural or human systems in response to actual or expected climatic stimuli or their effects, which moderates harm or exploits beneficial opportunities (IPCC, 2007), the concept has become widely used within the development aid sector.

This chapter explores how CCA may be being absorbed into business as usual approaches to development. We reflect on recent research that has investigated how CCA is being conceptualized and utilized across the development sector, and we show that many development actors may not be changing their approach in light of the new challenges posed by climate change, but rather conceptualizing and utilizing CCA within business-as-usual development aid frameworks. In contrast to these dominant approaches, we draw on postdevelopment approaches to both challenge existing understandings of CCA and provide an impetus for exploring the potential of diverse local understandings and initiatives for adaptation. Such an approach is not embedded in the familiar certainties of development, but is a new effort that calls development actors to move forward uneasily and build upon the diverse practices of local communities.

Three important background considerations inform the argument of the chapter. First, that there is already strong emerging evidence that climate change poses serious and negative threats to communities in developing countries and that some form of adaptation is and will be necessary (McKibben, 1999; Adger *et al.*, 2009; Füssel, 2009; Karbassi *et al.*, 2011; IPCC, 2012). A range of factors contribute to the heightened acuity of the climate change challenge in developing countries. These include geographical (such as in Asian river deltas or low-lying Pacific islands), political (including inadequate urban planning, failures of governance, and the presence of social conflict), and economic factors (such as heavy reliance on climate-sensitive sectors like agriculture and the scarcity of livelihood options for the poor) (World Bank, 2010; Nath and Behera, 2011; IPCC, 2012). These factors cause many scholars and practitioners to conclude that there is a fundamental link between CCA and development, as it is the familiar subjects of development efforts – "the poor" – who also experience the greatest vulnerability to the effects of climate change (Adger *et al.*, 2003; Huq and Ried, 2004; Klein *et al.*, 2007; Leary *et al.*, 2008).

Second, as CCA has emerged as a significant influence within the development sector, it has increasingly been accompanied by an aid-driven response including large and growing volumes of dedicated CCA financing. There are currently over 14 dedicated multilateral CCA funds in addition to a range of other initiatives from bilateral aid programs and international development NGOs (Climate Funds Update, 2012). This trend looks set to increase into the future. Under commitments made at the Copenhagen climate conference in 2009 (UNFCCC, 2009) and re-affirmed at the Cancun climate conference in 2010 (UNFCCC, 2011), climate financing in developing countries, which includes CCA, is expected to be scaled up to around 100bn USD per year by 2020. This is the same order of magnitude as total current annual overseas development assistance (World Bank, 2011).

Finally, in exploring the development sector's conceptualizations and utilizations of CCA, something of the sector's own contested history and nature must also be recognized. Far from being uncontested, there is wide criticism of development aid efforts for their failure to achieve goals and the creation of new and additional problems. Potential perverse outcomes from development initiatives identified in the research literature include environmental damage (Shiva, 1993; Sidaway, 2007), expanding bureaucratic state power (Ferguson, 1990), violent conflict (Verhoeven, 2011), widening social and economic inequality (Sachs 2010) and the creation of an imagined geography of underdevelopment (Escobar 1995). Recent scholarship, characterized by titles such as "The Curse of Aid" (Djankov *et al.*, 2008), "The Trouble with Aid" (Glennie, 2008) and "The Aid Trap" (Hubbard and Duggan, 2009), questions the fundamental logic of aid and development and suggests that many development programs may have a net negative impact on developing countries.

Conceptualizations of climate change adaptation

In seeking to understand the role of development actors in CCA, we begin with the ways in which it is being conceptualized – how it is defined and understood – within the development aid sector. We show that in the literature and amongst development actors, the term is often employed malleably and through the constructs of pre-existing approaches and concepts of development.

Existing development approaches to disaster risk reduction (DRR) have been one of the major influences on contemporary conceptualizations of CCA. There is a wide literature that identifies the similarities between these two approaches and suggesting that DRR should be one of the first steps or key components of adaptation to climate change (cf. Venton and LaTrobe, 2008; Christoplos, 2008; Schipper, 2009; Ireland, 2010; Mercer, 2010). By examining how DRR is being considered in emerging theory and practice around CCA, Ireland (2010) also suggests that the field of CCA is still evolving as stakeholders attempt to understand what CCA is and what actions it involves.

DRR is not the only pre-existing concept that is shaping and influencing understandings of CCA. In the majority of cases, development actors who are engaging with CCA are doing so through the conceptual lenses of their own preferred existing approaches to development aid. A recent study used semi-structured interviews to identify, outline and categorize the conceptualizations of CCA held by 35 development actors (Ireland, 2012b). A range of understandings of CCA emerged from this research that can be categorized into six broad definitions: improving knowledge, reducing risk, improving livelihoods, coping with impacts, planning and infrastructure, and "good development". The majority of participants' conceptualizations of CCA were framed within one of these themes, though it does also need to be acknowledged that there is some degree of interrelationship between the categories.

When Ireland (2012b) further explores the nature of the responses that were included within each of these categories it becomes evident that CCA is principally being conceptualized through the lenses of existing approaches for many of these development actors. Under the theme of improving knowledge, several NGO actors suggested that CCA is about educating the community to understand the potential climatic changes they face. In relation to reducing risk, participants drew upon the pre-existing context of vulnerability and referred to pre-existing approaches such as DRR and sustainable livelihoods. Improving livelihoods also reflects this idea of utilizing pre-existing methods with a focus on specific activities such as crop diversification and resilient agriculture. The theme of "coping with impacts" included responses that focus on actions that are reactive whilst planning and infrastructure focused on pre-emptive actions including activities such as the construction of sea walls and urban planning. Finally, the most dominant theme that emerged is around the assertion that CCA is just "good development". Under this category the participants suggested that CCA is not a new idea or approach, rather it is a component or natural outcome of "good development" which carried different implied meanings for different actors. Some conceptualizations of CCA have attempted to incorporate such a wide range of different activities by presenting adaptation as a continuum of different actions (McGray *et al.*, 2007: Persson *et al.*, 2009; Ireland, 2011). Whilst we can see the usefulness of these conceptualizations, the findings of Ireland (2012b) suggest that some of the actions being proposed as CCA are being driven by an ongoing imperative for certain types of development, more than by a consideration of the unique challenges of climate change.

Ireland and Schipper (in review) also show the tendency for CCA to be conceptualized through pre-existing concepts and linked to pre-existing activities of development organizations such as resilience, coping and risk. In the absence of clearer and more commonly understood conceptual boundaries, this tendency risks making CCA a meaningless term that may be operationalized in such a way that undermines possibilities for adaptation in the future. They argue that:

> Although there has been extensive scholarship on the definition of adaptation and other related terms, these are overlapping and repetitive, do not always translate clearly, and have resulted in a Wild West approach whereby everyone redefines their favorite concepts to fit their arguments. We suggest that there is space for this creativity, but that a shared sentiment about the purpose of adaptation is necessary, in order to avoid endorsement of projects that instead make people more vulnerable to climate change.
>
> (Ireland and Schipper, in review: 1)

The breadth of CCA conceptualizations and the risk that this may lead to projects being funded that do not result in effective adaptation, and at worst make people more vulnerable to climate change, is also reflected in the utilization of the concept as we examine in the next section.

Utilizations of climate change adaptation

In this section we examine the ways in which these different conceptualizations of CCA are being used in the day-to-day practice of development actors. Utilizations of CCA include the way the concept is employed within funding proposals and the actual development projects and programs that are conducted in the name of CCA. We suggest CCA is not only being conceptualized in light of existing development discourses; it is also being utilized as a means to repeat, strengthen and perpetuate these business-as-usual approaches to development aid.

If we consider first the use of CCA concepts in project funding applications and processes, Ireland's (2012b) study of development actors presents an interesting view of the way some actors are utilizing CCA. Several of the interviewed actors explained how they "rebadge" and "rename" current development activities as CCA due to perceptions that this language is more "sexy", can "capture the imagination of the donors" and results in "more money" (Ireland, 2012b). As one NGO manager said, "We call our projects different things to meet the donors' needs … It's not ideal but we need to use the language of the donor" (Ireland, 2012b: 100). The implication of the tendency to simply rebadge existing initiatives as CCA is that despite climate change being recognized as a new and different challenge, there has been little change in the actual development activities that are being implemented including activities implemented in the name of CCA. Indeed, some development actors are taking advantage of the flexible conceptualizations of CCA in order to attain funding for a range of pre-existing work that may or may not actually result in adaptation to climate change.

This is true at many different levels of the development sector, including amongst multi-lateral actors as demonstrated by the World Bank's World Development Report (2010). The World Development Reports are the World Bank's flagship publications and have significant influence upon the shape of policies of governments and development aid organizations (Paalman *et al.*, 1998; Rigg *et al.*, 2009). These reports have been critiqued for being ideologically driven and donor centric (Rigg *et al.*, 2009; Wade, 2001). Ireland and McKinnon (2013) explore the 2010 World Development Report and show that it supports well-worn neoliberal solutions such as privatization, economic growth and removal of subsidies as contributions to CCA. This appropriation of CCA is something that was both recognized and expressed as a concern by some development actors within Ireland's (2012b) study. Several of the participants in this research suggested that the new focus upon CCA is donor driven. In some instances small local organizations are being marginalized and disempowered as they are not able to utilize the concept of CCA in their current work or proposals.

As a result of the tendency by development actors to utilize CCA as a means to further existing development paradigms, we see that it is leading not only to varying, but in some instances directly contradictory, approaches to adaptation. The contrast between the World Bank (2010) and Oxfam (Pettengell, 2010) provides a case in point in relation to developing country agriculture. The two organizations are promoting seemingly opposing approaches to the issue. Whereas the World Bank identifies chemically intensive high yielding approaches associated with the Green Revolution as potential CCA initiatives, Oxfam posits organic agriculture as a more appropriate adaptation strategy. While this example may demonstrate the link between flexible conceptualizations of CCA and divergent actions, it also raises questions around how these actions will ultimately help vulnerable communities cope with and prepare for the impacts of climate change.

When considered critically, the emerging patterns of CCA conceptualization and utilization described in this chapter risk repeating problematic aspects of development aid. At the level of development activities, the patterns of CCA usage described above demonstrate a range of diverse – and in places contradictory – initiatives to support adaptation. With this proliferation and potential contradiction, there is an accompanying risk that at least some activities may drive forms of maladaptation that produce opposite outcomes to the IPCC definition – increasing harm or diminishing beneficial opportunities.

However, beyond these tangible activities, another important set of questions may be raised above the emerging use of CCA in relation to its place within the broader history and discourses of the development sector. There is an underlying tendency within the sector to adopt new foci and discourses in ways that strengthen and perpetuate the pre-existing agendas and ideologies of development actors (Pieterse, 2010). The practice of co-opting and integrating new development ideas with existing business-as-usual approaches to development outlined above is not unique to CCA but has previously been identified in relation to other foci such as the Millennium Development Goals (Langford, 2010; Vandemoortele, 2011). This tendency may serve to obscure more fundamental problems with the broader development aid effort. Indeed, recent critiques have suggested that the development sector continually produces new discourses and fashions precisely in order to conceal the low success rate of development interventions and to maintain itself (Pieterse, 2010). A key question that emerges from this analysis is whether CCA is being appropriated into existing development paradigms or whether it could be something different.

Other critiques suggest that the very foundations of global development aid efforts are predicated on problematic patterns of aid transfers from the "developed" to the "developing" world. The current and emerging conceptualizations of CCA explored in this chapter are often consistent with this underlying paradigm. That is, they tend to regard international donors as benign actors disconnected from the causes of the vulnerability that their adaptation efforts seek to reduce. In this vein, Ireland (2010) explored how the new focus upon adaptation within the development sector raises important questions about the reasons for communities' vulnerability to climate change in the first instance. It is also possible that the effective de-linking of adaptation from mitigation may ultimately obscure the extent to which "[d]ominant Western countries are designing and upholding global institutional arrangements, geared to their domestic elites, that foreseeably and avoidably produce massive deprivations in most of the much poorer regions of Asia, Africa, and Latin America" (Pogge, 2008: 3). Consequently CCA becomes framed as a matter of aid and international altruism rather than justice or international responsibility.

For the concept of CCA, this means on one level there is a risk of becoming yet another development goal that is mal-aligned with the priorities of communities and implemented ineffectively. However, it also means that the increasing emphasis on CCA could be seen through the lens of Pieterse's (2010) critique as having the potential to divert attention from previous development failures, including problematic overarching paradigms. It may simultaneously facilitate funding for new development programs that follow similar patterns. In light of the large volumes of funding that are available for CCA identified in the introductory section, there is a risk that it may not just end up being an avenue for the repetition of well critiqued approaches to development; it may end up strengthening and perpetuating them.

The contribution of postdevelopment perspectives to CCA

Given the evidence to suggest that CCA is at risk of being absorbed into existing business-as-usual approaches and paradigms to development, here we draw upon Ireland and McKinnon's (forthcoming) work to conclude by arguing that postdevelopment perspectives may offer one set of possibilities for challenging the emerging CCA mainstream. These perspectives suggest options "bedded not in familiar certainties but in a new effort to move uneasily, supporting the diverse practices of local communities as they respond in their own particular ways to the universal threat of climate change" (Ireland and McKinnon, 2013).

Postdevelopment theorists call for a critical rethinking of development that is mindful of the long history of critiques (Ferguson, 1990; Escobar, 1995; Saunders, 2002; Sidaway, 2007) and attentive to nascent possibilities (Gibson-Graham, 2005; McKinnon, 2007). They argue that development is always embedded in politics and ideology and warrants critical analysis and mindful interrogation (Ferguson, 1990; McKinnon, 2007). Critics of postdevelopment suggest that it draws upon a narrow version of development and fails to present genuine alternatives or a positive program (Pieterse, 2010). In our view, whilst postdevelopment does not seek to codify universalizing approaches to development, it does offer viable alternatives and does not make the development sector redundant. A postdevelopment approach to CCA requires those engaged in development aid to move with critical uneasiness – without the confidence that any particular approach is the right one – and with openness to new possibilities (Ireland and McKinnon, in review). In this vein, postdevelopment perspectives value diverse practices that are already occurring at the local level in response to a long history of hierarchical logics of scale in development that have obscured the significance of the local (Escobar, 2001; Santos, 2004; Gibson-Graham, 2005).

Ireland and McKinnon (2013) draw on postdevelopment perspectives to identify nascent opportunities and possibilities for adaptation to climate change in the developing world. They draw on field research in Nepal and Bangladesh (including Ireland and Thomalla, 2011; Ireland, 2012a) conducted in communities that are already responding to challenges identified locally as environmental change but that can be linked to broader patterns of climate change. In one example, local women's collectives are responding to challenges associated with the climate in Nepalganj, Nepal by taking specific action at a local level in response to considered reflection on the changes required to living patterns, resource use and collective behavior in the face of environmental changes. It was found that these actions were driven by local actors rather than by aid or the agendas of external development actors (Ireland 2012a). In this context the actions included raising the walkways between their houses to help them cope with flooding and combining some of their savings to purchase land to grow some of their own food in response to a shortage of work and income. In Kohlpur, Nepal, savings and loans collectives have begun to cooperate as disaster assistance networks to help communities prepare for and cope with severe storms and floods (Ireland and Thomalla, 2011; Ireland, 2012a). Similarly, research from Bangladesh showed how after several years of intense storms and cyclones some local NGOs are trialing new methods to better survive in the future. These actions include revegetating roadsides, reducing local deforestation, saving local seeds, developing floating vegetable gardens and building clay pots into the walls of new homes to save basic food and water for times of emergency (Ireland and McKinnon, 2013).

A postdevelopment approach does not lead us to view examples such as these simply as isolated instances that contribute to local adaptation processes. Instead it sees these examples as contributing to "a global network of localised actions […] accumulating towards a diverse set of climate change adaptations that enact important alternatives to the emerging norms of CCA" (Ireland and McKinnon, 2013). Examples like these, and the use of a postdevelopment perspective, point to both the narrowness of many current conceptualizations and utilizations of CCA, and to the possibility of alternatives that take seriously the unique challenges of climate change, are grounded in a critically reflective stance toward dominant development paradigms, and recognize the legitimacy and value of local actions.

Future adaptation efforts by the development sector should be built upon the needs, priorities and capacities of local communities. If the development sector is to contribute to effective adaptation across the world it must move uneasily into the subject of adaptation to climate change. Whilst this approach to adaptation is informed by postdevelopment perspectives we depart from postdevelopment and agree with Ireland and Schipper (in review) who suggest that there is a need for commonly understood boundaries of what adaptation can mean within the development sector. Commonly understood boundaries will inevitably posit some limits on what local actions can be supported as adaptation processes by the development sector, but also minimize the potential for a range of pre-existing and ideologically driven development aid efforts to be labeled and funded as adaptation. We contend that there is a necessary and dynamic tension that development actors need to grapple with between supporting previously unseen opportunities and limiting the potential for adaptation to become a catch-all concept that can be used to mobilize financing for any development aid activity. This approach could limit the incidence of maladaptation and the repetition of problematic aspects of development aid.

References

Adger, W.N., Huq, S., Brown, K., Conway, D. and Hulme, M. (2003) 'Adaptation to climate change in the Developing World', *Progress in Development Studies*, 3 (3): 179–195.

Adger, N.W., Lorenzoni, I. and O'Brien, K. L. (2009) 'Adaptation now', in W.N. Adger, I. Lorenzoni and K.L. O'Brien (eds) *Adapting to Climate Change: Thresholds, Values, Governance*, Cambridge: Cambridge University Press: 1–22.

Christoplos, I. (2008) *Incentives and Constraints to Climate Change Adaptation and Disaster Risk Reduction – A Local Perspective*, Commission on Climate Change and Development, Stockholm: Commission on Climate Change and Development.

Climate Funds Update (2012) *Climate Fund Frofiles*, Online. Available HTTP: http://www.climatefundsupdate.org/listing (accessed 14 February 2012).

Djankov, S., Montalvo, M. and Reynal-Quernol, M. (2008) 'The curse of aid', *Journal of Economic Growth*, 13 (3): 169–194.

Escobar, A. (1995) *Encountering Development: The Making and Unmaking of The Third World*, Princeton, NJ: Princeton University Press.

Escobar, A. (2001) 'Culture sits in places: Reflections on globalism and subaltern strategies of localization', *Political Geography*, 20: 139–174.

Ferguson, J. (1990) *The Anti-Politics Machine: 'Development', Depoliticization and Bureaucratic Power in Lesotho*, Cambridge: Cambridge University Press.

Füssel, M. (2009) 'An updated assessment of the risks from climate change based on research published since the IPCC Fourth Assessment Report', *Climatic Change*, 97: 468–482.

Gibson-Graham, J. K. (2005) 'Surplus possibilities: Postdevelopment and community economies', *Singapore Journal of Tropical Geography*, 26 (1): 4–26.

Glennie, J. (2008) *The Trouble with Aid: Why Less Could Mean More for Africa*, London: Zed Books.

Hubbard, R. and Duggan, W. (2009) *The Aid Trap: Hard Truths About Ending Poverty*, New York: Columbia Business School Publication.

Huq, S. and Reid, H. (2004) 'Mainstreaming adaptation in development', *IDS Bulletin*, 35(3): 15–21.

IPCC (2007) *Climate Change 2007: Impacts, Adaptation and Vulnerability – Contribution of Working Group II to the Fourth Assessment Report of the Intergovermental Panel on Climate Change,* Cambridge: Cambridge University Press.

IPCC (2012) 'Summary for policymakers', in C.B. Field, V. Barros, T.F. Stocker, D. Qin, D. Dokken, K.L. Ebi, M.D. Mastrandrea, K.J. Mach, G.K. Plattner, S.K. Allen, M. Tignor and P.M. Midgley (eds), *Intergovernmental Panel on Climate Change Special Report on Managing Risks of Extreme Events and Disasters to Advance Climate Change Adaptation*, Cambridge: Cambridge University Press: 1–9.

Ireland, P. (2010) 'Climate change adaptation and disaster risk reduction: Contested spaces and emerging opportunities in development theory and practice', *Climate and Development*, 2: 332–345.

Ireland, P. (2011) 'Climate change adaptation exchanges: An exploration of the possibilities and risks', in R.P. Buckley (ed.) *Debt-for-Development Exchanges*, New York: Cambridge University Press: 223–235.

Ireland, P. (2012a) 'Nepalganj, the centre of the world: Local perceptions of environmental change and the roles of climate-change adaptation actors', *Local Environment,* 17 (2): 187–201.

Ireland, P. (2012b) 'Climate change adaptation: Business-as-usual development or an emerging discourse for change?', *International Journal of Development Issues,* 11 (2).

Ireland, P. and McKinnon, K. (2013) 'Strategic localism for an uncertain world: A postdevelopment approach to climate change adaptation', *Geoforum.* Online. Available HTTP: <http://dx.doi.org/10.1016/j.geoforum.2013.01.005> 15th February 2013.

Ireland, P. and Schipper, L. (in review) 'Defining a landscape of adaptation to climate change', *International Journal of Climate Change Strategies and Management.*

Ireland, P. and Thomalla, F. (2011) 'The role of collective action in enhancing communities' adaptive capacity to environmental changes and risks: An exploration of two case studies in Nepal and Thailand', *PLOS Disasters Currents*, 3.

Karbassi, L., Kasten, T., Coleman, H., Putt del Pino, S. and Hopkins, N. (2011) *Adapting for a Green Economy: Companies, Communities and Climate Change*: United Nations Global Compact, United Nations Environmental Programme, Oxfam, World Resources Institute.

Klein, R., Eriksen, S., Naess, L.O., Hammill, A., Tanner, T., Robledo, C. and O'Brien, K. (2007) 'Portfolio screening to support the mainstreaming of adaptation to climate change into development assistance', *Climatic Change*, 84: 23–44.

Langford, M. (2010) 'A poverty of rights: Six ways to fix the MDGs', *IDS Bulletin, 41* (1): 83–91.

Leary, N., Adejuwon, J., Bailey, W., Barros, V., Caffera, M., Chinvanno, S., Conde, C., De Comarmond, A., De Sherbinin, A., Downing, T., Eakin, H., Nyong, A., Opondo, M., Osman, B., Payet, R., Pulhin, F., Pulhin, J., Ratnasiri, J., Sanjak, E., von Maltitz, G., Wehbe, M., Yin, Y. and Ziervogel, G. (2008) 'For whom the bell tolls: Vulnerabilities in a changing climate', in N. Leary, C. Conde, J. Kulkarni, A. Nyong and J. Pulhin (eds) *Climate Change and Vulnerability,* London: Earthscan: 3–30.

McGray, H., Hammill, A. and Bradley, R. (2007) *Weathering the Storm: Options for Framing Adaptation and Development*, Washington DC: World Resources Institute.

McKibben, B. (1999) *The End of Nature* (2nd ed.), London: Penguin.

McKinnon, K. (2007) 'Postdevelopment, professionalism, and the politics of participation', *Annals of the Association of American Geographers,* 97 (4): 772–785.

Mercer, J. (2010) 'Disaster risk reduction or climate change adaptation: Are we reinventing the wheel', *Journal of International Development*, 22: 247–264.

Nath, P. and Behera, B. (2011) 'A critical review of impact of and adaptation to climate change in developed and developing countries', *Environment, Development and Sustainability,* 13: 141–162.

Paalman, M., Bekedam, H., Hawken, L. and Nyhiem, D. (1998) 'A critical review of priority setting in the health sector: The methodology of the 1993 World Development Report', *Health and Policy Planning,* 13 (1): 13–31.

Persson, A., Klein, R., Kehler Siebert, C., Atteridge, A., Muller, B., Hoffmaister, J., Lazarus, M. and Takama, T. (2009) *Adaptation Finance under a Copenhagen Agreed Outcome,* Stockholm: Stockholm Environment Institute.

Pettengell, C. (2010) 'Climate change adaptation: Enabling people living in poverty to adapt', Oxford: Oxfam International.

Pieterse, J.N. (2010) *Development Theory* (2nd ed.), London: SAGE.

Pogge T, (2008) *World Poverty and Human Rights* (2nd edn), Cambridge: Polity Press.

Rigg, J., Bebbington, A., Gough, K.V., Bryceson, D.F., Agergaard, J., Fold, N. and Tacoli, C. (2009) 'The World Development Report 2009 "reshapes economic geography": geographical reflections', *Transactions of the Institute of British Geographers,* 34 (2): 128–136.

Sachs, W. (2010) 'Preface to the new edition', in W. Sachs (ed.) *The Development Dictionary* (2nd ed.), London: Zed Books, vi–xiv.

Santos, B. d. S. (2004) 'The WSF: Toward a counter-hegemonic globalization', in J. Sen, A. Anand, A. Escobar and P. Waterman (eds) *World Social Forum: Challenging Empires,* New Delhi: The Viveka Foundation: 235–245.

Saunders, K. (2002) 'Towards a deconstructive post-development criticism', in K. Saunders (ed.) *Feminist Post-development Thought. Rethinking Modernity, Post-colonialism and Representation,* London: Zed Books: 1–38.

Schipper, L. (2009) 'Meeting at the crossroads?: Exploring the linkages between climate change adaptation and disaster risk reduction', *Climate and Development,* 1 (1): 16–30.

Shiva, V. (1993) *The Violence of the Green Revolution: Ecological Degradation and Political Conflict in Punjab,* London: Zed.

Sidaway, J. (2007) 'Spaces of postdevelopment', *Progress in Human Geography,* 31: 345–361.

UNFCCC (2009) Draft decision/CP15: Copenhagen Accord.

UNFCCC (2011) Decision 1CP16: Cancun Agreements.

Vandemoortele, J. (2011) 'The MDG story: Intention denied', *Development and Change,* 4 (1): 1–21.

Venton, P. and LaTrobe, S. (2008) *Linking Climate Change Adaptation and Disaster Risk Reduction,* London: Tearfund.

Verhoeven, H. (2011) 'Climate change, conflict and development in Sudan: Global neo-Malthusian narratives and local power struggles', *Development and Change,* 42 (3): 679–707.

Wade, R.H. (2001) 'Making the world development report 2000: Attacking poverty' *World Development,* 29 (8): 1435–1441.

World Bank (2010) *World Development Report 2010: Development and Climate Change,* Washington, DC: World Bank.

World Bank (2011) *World Development Indicators,* Washington, DC: World Bank.

21 Why the discursive environment matters

The UK Climate Impacts Programme and adaptation to climate change

Elspeth Oppermann

Introduction

This chapter discusses the UK Climate Impacts Programme's problematization of adaptation to climate change, which both responds to and helps shape a discursive "environment" through which adaptation to climate change is understood. It also discusses the extent to which this might be characterized in terms of human security. Although human security is often said to be threatened by climate change as a biophysical phenomenon (Barnett and Adger, 2007), the understanding of climate change as a threat is discursively produced (Shackley and Wynne, 1995; Demeritt, 2001, 2006; Szerszynski, 2010), as is our understanding of security in relation to it (Campbell, 1998; Dalby, 2002; Feindt and Oels, 2005; O'Brien *et al.*, 2007). As such, rather than directly negotiating the changes in the biophysical environment, humans understand and respond to environmental change, including climate change, through discourse.

These discourses can be understood as themselves creating a "conceptual environment" that reveals some ontological inclusions and obscures others and in doing so exposes particular avenues of visibility and spaces for action, while other paths remain concealed. In doing so, this conceptual environment shapes how current and future actors understand and respond to climate change. As such, understanding climate change requires not just an understanding of the biophysical phenomenon, but also a critical awareness of the structuring and limiting effects of established discourses, including the discursive interventions produced through adaptation policy and advice. Without this awareness, our capacity to conceptualize and respond to the problem of climate change critically and perhaps more effectively will also be limited. This chapter offers a conceptual line of enquiry into the way in which this discursive environment is emerging in regard to climate change adaptation in the United Kingdom through the work of the UK Climate Impacts Programme (UKCIP).

Theoretical basis: why the discursive environment matters for adaptation and human security

Although they use different terminology, O'Brien and colleagues (2007) have described this process of problematization in the field of vulnerability to climate change. They observe that there appear to be two alternative conceptual "framings" of vulnerability and discuss how these shape policy. The first framing sees vulnerability as an outcome of the impacts of climate change. This is based on a scientific framing which has an ontology and epistemology that focuses on the technical and physical aspects of climate change, entailing policies based on empirical, physical and linear responses. The second, human security framing, understands vulnerability through an ontology that is sensitive to the social production of norms and contingency, arising from the context of the community or individual that responds to climate change. This relocates the problem of climate change in the social realm, entailing a policy response that seeks to build social capacity in a broad sense. This framing is linked to a broader critical discourse of environmental change that in turn draws on poststructuralist accounts of the contingency of knowledge and power relations in social systems (O'Brien *et al.*, 2007; Tadjbaksh and Chenoy, 2007).

More recently, Pelling (2011) has accounted for adaptation discourse as divided into three general frames of resilience, transition and transformation. These also have their specific ontologies and epistemologies, but Pelling's insight is that their principal distinction is the difference in their core "intention and action" (Pelling, 2011: 170). With regard to resilience, this intention and action is geared towards ensuring the continuation of the existing state of affairs; for transition it is to enable improvement within the existing framework of social relations; for transformation it is the willingness to critically engage with and potentially transgress the limits of that framework itself. Each of these frames offers consecutive improvements in their traction on social systems, including their political systems, economic structures, and cultural norms. Pelling situates human security within the last of these frames: identifying it as one of the adaptation discourses offering the most critical purchase on social systems, enabling a greater depth to the understanding and practice of adaptation and therefore enabling the possibility of transformation.

The division between the scientific and human security framing of vulnerability, and between different forms of adaptation as resilience, transition and transformation, calls into question the specific ontological and epistemic inclusions in UKCIP's account of adaptation as well as the "intention and action" that forms its political rationale. This is particularly important as these tensions in the wider discursive field appear to have been internalized within UKCIP's discourse (Oppermann, 2011). This calls for an analysis of how UKCIP's problematization of adaptation has responded to and resolved these external discourses, and what this means for its content and coherence.

In order to analyze UKCIP's account of adaptation, it is necessary to establish a theoretical approach. Seeing discourse as an important object of study in its own right is based on the assumption that meaning is not so much referenced by language as produced by it (Potter, 2001). Agreed or general norms about language-use and connected technologies of intervention thus produce an accepted system of knowledge or "system of representation" about a referent object (Foucault, 1998; Hall, 2001: 72). This system can be characterized as a "problematization," where an account of a problem and solution are formed through the use of a particular ontology and epistemology and are linked through a political rationale that sanctions certain practices and politics in order to produce a desired outcome (Campbell, 1998; Foucault, 1998; Flynn, 2005). This creates a particular understanding of a topic that both implicitly and explicitly excludes other, alternative conceptualizations as invisible, irrelevant, false, and even immoral (Dijk, 2001; Jäger, 2001).

A level of linguistic precision is added to the concept of problematization through the account of its structure as the "articulation" of conceptual "elements" as "moments" (Laclau and Mouffe, 2001). The specific manner of their articulation is accounted for through various linguistics-based discourse analysis techniques (including those of Titscher *et al.*, 2000; Wodak and Meyer, 2001). This framework is used here to critically analyze a technical report produced in 2003 by UKCIP, which formed the basis of their account of adaptation. This analysis has formed the starting point for a wider investigation of their discourse between 2003 and 2011, although there is only room to cover the text-based elements here.

At the level of a document, discourse analysis techniques can be used to critically review how this is narratively, rhetorically and semantically constructed (Dijk, 2001; Wetherell *et al.*, 2001). Crucially, this makes visible the content and also the contingency of such articulations, opening the discourse up to critique (Jäger, 2001; Wertsch, 2001), and indeed to the possibility of strategic discursive change should this demonstration of the contingency of content be utilized normatively to further specific parts of the discourse, although this is not the intention of this chapter.

During this research, it became apparent that existing discursive tensions were often resolved through the use of a conceptual "supplement" which was linguistically articulated to the problematization in a manner which appears to extend the discourse, yet which at the same time acts as a "limit point" (Derrida, 1976: 145), through the nature of its articulation as subsidiary or additional to the original or core problematization. However, its very presence nonetheless undermines the coherence of the core problematization and thus intimates the possibility of discursive change (Torfing, 1999; Wertsch, 2001; Hansen, 2006). It is for this reason Derrida refers to the supplement as "dangerous" (Derrida, 1976: 163).

The next section will demonstrate the use of this discourse analysis approach in critically analyzing UKCIP's problematization of adaptation, beginning with a discussion of the discourse's founding moments of articulation in UKCIP's technical report *Climate Adaptation: Risk, Uncertainty and Decision-making* (Willows and Connell, 2003).

The discursive landscape: UKCIP's core problematization of adaptation

In seeking to account for a problematization, the first port of call is often the definition of what is assumed to be the core referent, in this case: adaptation.

In UKCIP's technical report there is no overt account of what substantively constitutes adaptation. Instead, a definition of climate adaptation is offered in the glossary at the end, where adaptation is defined as the "process or outcome of a process that leads to a reduction in harm or risk of harm, or realization of benefits associated with climate variability and climate change …" (Willows and Connell, 2003: 111). What kind of process or outcome ensures a reduction of harm is not explicitly stated, nor is it defined inversely by an account of what constitutes harm. Whilst this abstract openness is common in definition writing, it combines with the absence of a definition in the beginning of the report to leave a gap in substantive meaning. In the context of a technical report, a genre usually characterized by precision and certitude, adaptation as a term is then implicitly and inevitably filled and shaped by association as it is used throughout the text of the report (Johnstone, 2008). It is this process of articulation which will be considered here.

The introductory "Context" section of the report is particularly important in providing adaptation with a substantive meaning by situating it within a persuasive narrative. First, the claim is made that there is certainty that climate change is happening, but that "the exact extent and nature of changes in our climate remains uncertain." This is elaborated a little, then followed by the statement: "Climate therefore represents a changing source of risk. Climate adaptation is about recognizing these altered risks …" (Willows and Connell, 2003: 3). Thus, climate adaptation is immediately articulated to the concept of uncertainty and risk. Crucially, the word choice in these sentences frames climate change as being a manageable problem in a series of subtle ways. The use of the phrase "exact extent" qualifies the meaning of "uncertainty," implying that uncertainty is limited to details rather than substantial uncertainties about the type or magnitude of threat. Similarly, the use of the term "altered," implies that changes to the climate are minor. This diminutive account of the problem is coupled with a lexis that gives the impression that knowledge of climate change through risk is easily accessible: this occurs through the use of the authoritative term "recognizing." Thus we see already that through these opening moments of articulation, climate change is problematized as a threat of limited uncertainty, which is resolved through risk. As the only included epistemology, cast in such accessible terms, risk is produced as the core epistemology in the problematization of adaptation.

The sentence continues: "Climate adaptation is about recognizing these altered risks, and taking decisions that allow the likely impacts to be reduced or managed, and the opportunities to be exploited" (Willows and Connell, 2003: 3). Here, the first "and" between "risks" and "taking decisions" implies a simple transition from knowing the problem through risk to the production of a solution, in the form of a "decision." This introductory sequence thus implicitly establishes a core rationale in which utilizing risk as the means of knowing the uncertainty of climate change claims to provide a straightforward basis for decisions that constitute successful adaptation.

Taken together, these opening statements present a problematization of adaptation that appears to be logical, sufficient and therefore true. Appearing in such a simple and concise form implicitly establishes a definition which leaves no overt space for critical reading, and as such helps to "lock in" this interpretation in the mind of the reader (Johnstone, 2008) before they go on to read the technical detail of the report. This is significant because the coherence and hence validity of this problematization is later undermined in two ways: first, through more tentative and limited claims about the availability of risk-based knowledge, and second, through supplementary concepts that add to but also exceed the risk-based core problematization. Excluding such openness and supplementary concepts, this initial narrative functions to "cover over" the cracks in the conceptual edifice of the problematization of adaptation, allowing it to be resilient in the face of contradictory evidence (Laclau and Mouffe, 2001; Glynos and Howarth, 2007). As we shall see in the following sections, this stabilizing narrative allows the supplement to introduce elements from different framings of adaptation into the discourse without destabilizing the problematization.

The conceptual elements articulated as moments in this problematization of adaptation can be examined through three related facets. First, what ontology or imaginary of the world, its contents and forces, does the report account for or assume? Intimately related to this is the second facet: what is its epistemology, or how does it characterize knowing? Third, based on these foundational claims, what understanding of agency or action is assumed to be necessary to secure adaptation, and what overall political rationale mobilizes these practices, including the values or ethics that explain them? These are discussed below, together with examples of supplementary moments.

Ontology and epistemology

The narrative described above clearly identifies risk as the epistemology of adaptation. In doing so, other types of knowledge are implicitly excluded. This establishment of risk enables the overt exclusion of the social as both a mode of knowing and as an object of knowledge. In fact, the report expends just one sentence to the claim that, comparatively to climate, social change is unknowable (Willows and Connell, 2003: 50) in a manner that dismisses it as a valid object of adaptation discourse or as a basis for a technology of knowledge. This demonstrates significant parallels with the outcome-based scientific framing identified by O'Brien and colleagues (2007) within climate change vulnerability discourse.

However, several supplementary moments disrupt the dominant epistemology, particularly where the report provides various technical means to cover the gaps in knowing uncertainty left by probabilistic risk (see, for example, Willows and Connell, 2003: 70–87). One instance is the use of the concept of "adaptive management" as the constant reassessment of risk (Willows and Connell, 2003: 53), a measure that only makes sense under the ontological and epistemological assumption that there is a limited capacity for prescient and hence preemptive knowledge of how to adapt to climate change. This indicates a supplementary epistemology more akin to a systems-based frame that assumes irreducible uncertainty or ignorance as a condition (Luhmann, 1998). In the report, however, adaptive management calls for the reapplication of the risk-based framework at greater frequencies, rather than changing the epistemology of adaptation per se (Willows and Connell, 2003: 71). As such, it reinstates a risk-based means of knowing as the solution to its own insufficiency in dealing with deep uncertainty, and excludes alternative means of knowing that respond to this irreducible uncertainty, such as the human security approach based on knowing the contextual vulnerability of the social system (O'Brien *et al.*, 2007). Interestingly, since 2003 this supplementary epistemology has come to play an increasing role in UKCIP's approach through the introduction of the Local Climate Impacts Profile methodology (UK Climate Impacts Programme, 2009).

The risk-based epistemology has several ontological implications. First, in order to identify a risk, it is necessary for a potential event be described, and thus not only defined but also rendered as a finite or delimited object as a result. Second, the practice of risk assessment assumes that the identity, norms or values of the organization is consistent or at least predictable, in order to project them into the future to enable the risk assessment process to rank risks in their order of magnitude of effects on the organization (Oppermann, 2011). This outcome-based account of risk thereby centralizes the accounts of actors and agency through the demands of predictive knowledge. This ontology is exacerbated by the almost exclusive use of the singular noun "decision-maker" to define the subject that acts in the problematization produced in the report. The core problematization thus establishes an ontology that delimits and centralizes the social in both time and space, enabling a linear understanding of identity, norms and values, as well as of response capacity. In the assumption of the validity of prescient knowledge and the related occlusion of concern for the observation and utilization of social capacities and networks, the core ontology in operation in UKCIP's problematization again has clear resonances with the scientific framing of responses to climate change as outcome-based, linear, and structured by knowledge of externally originating risk (O'Brien *et al.*, 2007).

However, there are several supplementary moments which appear linguistically as details and caveats in the body of the report which extend the core problematization, yet which also intimate a conflicting ontology of actors and agency. "Decision-makers," for example, are recognized as located amongst "stakeholders" or in groups with shared interests or roles such as "sectors" (Willows and Connell, 2003: 66). This indicates a supposition that an organization's achievement of its objectives is to some extent dependent on relations with other actors, and implies an ontology of networked connections rather than discrete objects populating the ontological imaginary, and as such enables a more distributed account of the agency of decision-making, which resonates with the human security framing of contextual vulnerability. However, the character of these relations is not a focus of knowledge in the report and as such does not disrupt its core problematization, keeping these elements articulated as supplementary moments through the failure to investigate them or present them as central concerns.

Political rationale

The dominance of the predictive, outcome-based epistemology, and the centralized ontology of actors and agency, is reflected in the political rationale of adaptation established by the report. The most direct account of what adaptation "is" lies in the concept of "decision-making" itself, where the basis of those decisions, which we can call its politics, is for the most part left to risk assessment on the basis of the "objectives of the decision-maker" (Willows and Connell, 2003: 53), implying a laissez-faire, neoliberal politics for the report's account of adaptation as a whole. This assumes that the right to act is limited to a single decision-maker. In doing so it centralizes practices and the political in social space and occludes the agency and value of other actors, and the nature of their relations to one another, as both enabling and enacting the decision as part of the problematization of adaptation.

Practices and politics are also represented as centralized in time throughout the document through the collocation of decision with "the" or "a," marking decision-making as occurring in a single instance or moment. This usage obscures the absence of a substantial discussion of the politics or practices of the processes involved before and after this moment. It also creates a political imperative to make the "best" decision possible at the moment of decision, which drives a political ideal of perfect knowledge of the risk, and implies that a single person in a single moment can find a single or right answer. This occlusion of wider, emergent processes tends to support a linear sense of politics and practice, drawing the techno-scientific, risk-based problematization full circle.

Despite the profusion of such examples in the report, there are some attempts to avoid the limitations of a linear approach. These act as a supplementary conceptual basis for the problematization of adaptation. Post-decision review and iterative decision-making, under the rubric of "adaptive management," (Willows and Connell, 2003: 71) suggest that it is not just the moment of decision or the accuracy of what is decided that matters, but instead places value on making decisions frequently. This implicitly responds to the ontology of uncertainty as an irreducible condition rather than as a temporary obstacle that can be controlled by the preemptive technologies of risk-based decision-making.

Moreover, the practice of review as a continual process of making decisions entails that the process itself, rather than the decisions it produces, becomes the central political rationale of adaptation. This resonates with the strong focus on social capacity, rather than precise outcomes, within the human security account of adaptation to climate change and the related account of adaptation as transformation.

However, post-decision review and iterative decision-making have a limited impact on the report's core problematization of adaptation. The iterative review process, although it is noted in the introduction to the report, gets very limited substantiation in the main body of the report and does not appear as an inherent part of the practices it describes. Indeed, its location at the end of the report represents it as post-script both physically and conceptually (Willows and Connell, 2003: 39–40). Moreover, the process of reviewing decisions, although described as "continuous," is also coupled with the phrase "best decision at each decision point" (Willows and Connell, 2003: 30). This implies that the core techno-scientific ontology, which enables certainty, remains in the interim periods of the adaptation process and as such does not dislodge a prescient and preemptive rationale of adaptation.

However, a powerful supplementary moment occurs in the ideal of "keeping options open" (Willows and Connell, 2003: 30). This expresses an ideal of making a decision that maintains flexibility for future decisions and in doing so implies the need for a practice and politics that can respond to an epistemology and temporal ontology of irreducible uncertainty by ensuring flexibility. This sensitivity to systemic links in time is mirrored by a sensitivity to a systemic ontology of social space. Here the supplementary principle of flexibility appears in the request to avoid "adaptation constraining decisions" (Willows and Connell, 2003: 11); that is, to avoid decisions that negatively affect others. As such, both of these supplementary moments place value on ensuring that the process of making decisions is continual, broad and flexible.

Together, these supplementary moments proffer an alternative rationale, where flexibility becomes the basis of enabling good adaptation decisions, which are open and emergent. Although not mentioned in any detail, this would tend to assume the creation and maintenance of strong communication and cooperation abilities, which again has strong parallels with the human security focus on social capacity (Folke *et al.*, 2002; Pelling *et al.*, 2008; Adey and Anderson, 2011; McCormack and Schwanen, 2011). Such supplementary moments thus undermine the centralized politics of individualized "objectives" overtly discussed in the report (Willows and Connell, 2003: 53) and reduce the imperative to create risk-based certainty in order to make the best decision.

However, while "keeping options open" and "avoiding adaptation constraining decisions" are noted relatively strongly in the document, they are not connected into the political rationale as an essential process or outcome, nor to any overt or detailed ontology or epistemology. As such, although they appear to extend the problematization of adaptation, they do not make any substantive changes, particularly at the level of the core political rationale of making the best decision at each decision point as predetermined by the organization.

Implications for human security

Human security is not part of UKCIP's discourse, nor is it the intention of this chapter to use any of the normative claims of human security to critique UKCIP's discourse of adaptation. However, in the spirit of ethical critique (Glynos and Howarth, 2007), it is pointed out that while UKCIP's core problematization occludes key principles of human security, the supplementary moments discussed above do in fact resonate with certain elements of human security's ontology, epistemology, and to a lesser extent, its political rationale.

Emerging relatively recently, human security is characterized as a vast "agglomerate of interconnected concepts" (Tadjbaksh and Chenoy, 2007: 235) that draws on several other disciplinary backgrounds. Although varied, these concepts all link to a central normative principle of protecting human wellbeing through enabling agency and capacity at the community and individual level, rather than at the level of the state. In doing so, the ontology of security, and of adaptation to climate change if understood in this frame, is expanded to include overt ethical engagement with security issues of the human, but most importantly through taking the individual as the referent object of security and making systemic capacity the political rationale for achieving this (Tadjbaksh and Chenoy, 2007).

Crucially this problematization requires an epistemology capable of registering the complex power-relations within political, economic and social structures which enable individual wellbeing, agency, and capacity (O'Brien *et al.*, 2007; O'Brien and Leichenko, 2007). This concern moves the imaginary of the threat of climate change from biophysical change and direct impacts to "the dynamic social context" through which these impacts emerge and come to have meaning (O'Brien and Leichenko, 2007: 2). In short, climate change and adaption emerge as a problem and solution through the degree of individual and wider capacity and agency within that system. This has led to a willingness to critically assess existing norms and governance structures and engage in transformational agendas, bringing into focus the ontology of the system that adapts and its potential rationales for securing such a system (Pelling, 2011), making it both the subject and object of adaptation.

This ontology and epistemology are in stark contrast to the limited account of the social in UKCIP's problematization of adaptation, particularly with regard to the internal nature of organizations and the complete occlusion of the individual. Rather, for UKCIP the extant objectives of administrative and business units determines the referent object of security. At best, a wider account of social systems appears when it notes the need to consider an organization's situation in wider regional networks, but it does not call for internal analysis of organizational structure or organizational change, let alone individual agency. Such an account of the subject is conservative rather than transformative, and in this sense does not fit with human security's willingness to engage with individual agency or norms of governance.

These supplementary moments do however resonate with human security's ontology of time, as it "allows for a continuum from prevention to emergency" (Tadjbaksh and Chenoy, 2007: 235). Here the supplementary moment that calls for keeping options open shares this understanding of time as emergent. Since 2003, UKCIP's increased use of the Local Climate Impacts Profile (UK Climate Impacts Programme, 2009) has increased the presence of this ontology of time through its methodological focus on organizational experience which shifts the time-frame of adaptation to a smaller scale relevant to individual and organizational practices, rather than long-term objectives. From a human security perspective then, it is important to note that UKCIP's supplementary ontological and epistemic moments and their political rationales have become increasingly distinct since 2003, and have begun to develop in a direction more sympathetic to the concerns of human security.

The UKCIP problematization of adaptation appears to have reached a moment of crisis in 2010–2011, when its coherence was undermined and overtly called into question. This was demonstrated in publications by UKCIP in 2010 and 2011 (Lonsdale *et al.*, 2010; Brown *et al.*, 2011) which introduced a much more thorough account of the supplementary moments articulated into an alternative problematization of adaptation with their own political rationale. Importantly, these documents established the centrality of social systems, and particularly internal organizational structure, for adaptation. In doing so they introduced new epistemologies and rationales of communication, decision-making and action that are sensitive to social practices and norms and to irreducible uncertainty, including the critique of the scientific framing by Dessai and Hulme (Hulme and Dessai, 2008; Dessai *et al.*, 2009; Brown *et al.*, 2011). Furthermore, these new concerns overtly connected to existing alternative literatures on adaptive capacity building, specifically at the organizational level. This radically opens up the potential of these supplementary moments, and indeed of the nascent supplementary problematization, to be dangerous in Derrida's sense of supplanting the core problematization and the practices that are called for as a result.

However, while these documents bolstered the presence of the supplementary moments, the precise nature of their articulation nonetheless curtailed their effect on UKCIP's core problematization. The 2010 document offered these alternative accounts to national climate change risk assessment, but did not relate them back to UKCIP's own practices directly. As such they remained supplementary in the sense of being additional, rather than supplanting. The second document, which was UKCIP's final public guidance in 2011 and which is fittingly referred to as "supplementary guidance" (Brown *et al.*, 2011: 9) was held in check in a similar fashion. Nonetheless, it creates a clear alternative to the core problematization from 2003, designating it as a "top-down approach" to be contrasted with a vulnerability-based "bottom-up approach." In doing so, it nonetheless points out the limitations of a vulnerability approach in accounting for adaptation, and ties it back into the need for a techno-scientific approach that limits the appearance of adaptive capacity as an alternative referent of a political rationale for adaptation. In sum, although this last document is beginning to take social conditions on board more generally, it has yet to change UKCIP's political rationale accordingly, thereby retaining the core problematization.

This unresolved status of the emerging supplementary problematization leaves a large question mark over the future of adaptation in the UK as it shifts to a new institutional structure for the provision of adaptation support. October 2011 marked the end of UKCIP's role as the core provider of adaptation advice on behalf of the UK government, which has now been made the responsibility of the Department for Environment, Food and Rural Affairs (Defra) and the Environment Agency (UK Climate Impacts Programme, 2011). The Environment Agency's historical disciplinary focus lay in the techno-scientific or impacts-based problematization, and as such it is a concern that this institutional re-embedding has the potential to undermine the growth of the supplementary problematization. By contrast, the EA's decision to continue UKCIP's work with regional climate change partnerships (Defra Adapting to Climate Change Programme team, 2012; Environment Agency, 2012) provides a potentially sympathetic mode of practice for the supplementary moments identified here. This period of transition has provided a clear opportunity for the reproblematization of adaptation, particularly under the auspices of the Adaptation Sub-Committee to which the 2010 report was submitted. Further research is now needed to see whether and how the supplementary moments emerged, and if they were enabled to produce a transformational account of adaptation in substantive terms, or whether they have receded, to be replaced by a resilient or at best transitional problematization of adaptation that is conservative in its political rationale.

Conclusion

This chapter has tried to demonstrate that far from being a purely biophysical phenomenon, the very understanding and communication of adaptation itself is a discursive "environment." The particular inclusions and exclusions and their articulation as either as substantive or supplementary moments forms a specific problematization of adaptation, a terrain that must be understood and negotiated, and opened up where it is closed off. Indications of these new paths occur in the supplementary moments that offer caveats, alternatives and clarifications to the core problematization. In doing so, these moments are "dangerous" and act as immanent points of potentiality; if the avenues they indicate were followed through, the supplementary problematization they entail would create a very different discourse, one which resonates with the problematization of adaptation put forward by the human security framing.

UKCIP's discourse of adaptation has been explored through the foundation established by the report published in 2003, *Climate Adaptation: Risk, Uncertainty and Decision-making* (Willows and Connell, 2003). This analysis has demonstrated how the report's core problematization was based on a scientific ontology and epistemology, which both shapes and limits the way in which adaptation is problematized. Here we see the threat is understood through outcome-based risk assessment, which is used as enabling preemptive decisions as the solutions, such that adaptation happens in relatively discrete, linear steps by discrete, unchanging and abstract actors.

However, this core problematization has been increasingly marked by supplementary moments, which both extend the core problematization and fracture its coherence. In the case of UKCIP, the articulation of these supplements as moments within the discourse enabled the inclusion of a more comprehensive account of society and social forms of knowledge (Oppermann, 2011), and have gradually brought the social into view as the subject that adapts. Crucially, the 2010 document also made the jump to identifying the social system not only as the subject that adapts but also as the object that is adapted to climate change, through the recognition of internal organizational structures and qualities, and as such indicating alternative political rationales of adaptation based on building flexible capacity in the face of irreducible uncertainty.

These supplementary moments resonate with the concerns of a human security framing of adaptation to climate change, particularly where human security is understood in terms of the shift from an apolitical or post-political techno-scientific account that occludes critical engagement with social systems and norms, to a concern for the shape and values of human systems and particularly the rationales that produce their mode of governance, with the practical focus on social capacity (O'Brien *et al.*, 2007; O'Brien and Leichenko, 2007).

This analysis of UKCIP's discourse has intended to perform the function of ethical critique in opening up this particular discursive "environment" in order to enable critical engagement with an established conceptual terrain, making visible its contents and contours, and its resulting limits and potentialities. It has done this through identifying the ontological and epistemic inclusions and exclusions and political rationale that articulates these. It has also used Derrida's account of the supplement to demonstrate the crucial nuances of articulation that enable a concept to be apparently included but with little substantive effect, thus whilst introducing conceptual potential these also act as limit points. Such critical analysis of the discursive problematization of adaptation is particularly important at a time of change for the actors and institutions involved in the (re) production of the problematization of adaptation to climate change in the UK, as it provides an opportunity for the critical function of the supplement to be overtly and substantively engaged with, and notes that its dangerousness is itself in danger of being lost, due to its lack of codification within UKCIP's core problematization before its role was returned to Defra and the Environment Agency.

References

Adey, P. and Anderson, B. (2011) 'Event and anticipation: UK civil contingencies and the space-times of decision', *Environment and Planning A*, 43: 2878–2899.

Barnett, J. and Adger, W.N. (2007) 'Climate change, human security and violent conflict', *Political Geography*, 26: 639–655.

Brown, A., Gawith, M., Lonsdale, K. and Pringle, P. (2011) *Managing Adaptation: Linking Theory and Practice*, Oxford: UK Climate Impacts Programme.

Campbell, D. (1998) *Writing Security: United States Foreign Policy and the Politics of Identity*, Minneapolis, MN: University of Minnesota Press.

Dalby, S. (2002) *Environmental Security*, Minneapolis, MN: University of Minnesota Press.

Defra Adapting to Climate Change Programme Team (2012) *Environment Agency to Provide Practical Advice to Help England Adapt to Climate Change*. Online. Available HTTP: http://www.defra.gov.uk/news/2011/03/22/ea-climate-change-adaptation/ (accessed 3 August 2012).

Demeritt, D. (2001) 'The construction of global warming and the politics of science', *Annals of the Association of American Geographers*, 91: 307–337.

Demeritt, D. (2006) 'Science studies, climate change and the prospects for constructivist critique', *Economy and Society*, 35: 453–479.

Derrida, J. (1976) *Of Grammatology*, Baltimore, MD: Johns Hopkins University Press.

Dessai, S., Hulme, M., Lempert, R. and Roger Pielke, J. (2009) 'Climate prediction: A limit to adaptation?', in W.N. Adger, I. Lorenzoni, and K.L. O'Brien (eds) *Adapting to Climate Change: Thresholds, Values, Governance*, Cambridge: Cambridge University Press.

Dijk, T.V. (2001) 'Principles of critical discourse analysis', in M. Wetherell, S. Taylor and S.J. Yates (eds) *Discourse Theory and Practice: A Reader*, London: Sage.
Environment Agency (2012) *Adapting to Climate Change Programme*. Online. Available HTTP: http://www.environment-agency.gov.uk/research/132323.aspx (accessed 3 August 2012).
Feindt, P. and Oels, A. (2005) 'Does discourse matter? Discourse analysis in environmental policy making', *Journal of Environmental Policy & Planning*, 7: 1–13.
Flynn, T.R. (2005) *Sartre, Foucault, and Historical Reason: A Poststructuralist Mapping of History*, Chicago, IL: University of Chicago Press.
Folke, C., Carpenter, S., Elmqvist, T., Gunderson, L., Holling, C. and Walker, B. (2002) 'Resilience and sustainable development: Building adaptive capacity in a world of transformations', *Ambio* 31: 437–440.
Foucault, M. (1998) *The Will to Knowledge: The History of Sexuality Volume I*, London: Penguin Books.
Glynos, J. and Howarth, D. (2007) *Logics of Critical Explanation in Social and Political Theory*, Abingdon: Routledge.
Hall, S. (2001) 'Foucault: Power, knowledge and discourse', in M. Wetherell, S. Taylor and S.J. Yates (eds) *Discourse Theory and Practice: A Reader*, London: Sage.
Hansen, L. (2006) *Security as Practice: Discourse Analysis and the Bosnian War*, Abingdon: Routledge.
Hulme, M. and Dessai, S. (2008) 'Predicting, deciding, learning: Can one evaluate the "success" of national climate scenarios', *Environmental Research Letters*, 3: 1–7.
Jäger, S. (2001) 'Discourse and knowledge: Theoretical and methodological aspects of a critical discourse and dispositive analysis', in R. Wodak and M. Meyer (eds) *Methods of Critical Discourse Analysis*, London: Sage.
Johnstone, B. (2008) *Discourse Analysis*, second edition, Oxford: Blackwell.
Laclau, E. and Mouffe. C. (2001) *Hegemony and Socialist Strategy: Towards a radical Democratic Politics*, London: Verso.
Lonsdale, K., Gawith, M., Johnstone, K., Street, R., West, C. and Brown, A. (2010) *Attributes of Well-adapting Organisations: A Report Prepared by UK Climate Impacts Programme for the Adaptation Sub-Committee*, Oxford: UKCIP.
Luhmann, N. (1998) *Observations on Modernity*, Stanford, CA: Stanford University Press.
McCormack, D.P. and Schwanen, T. (2011) 'The space – times of decision making', *Environment and Planning A*, 43: 2801–2818.
O'Brien, K. and Leichenko, R. (2007) *Human Security, Vulnerability, and Sustainable Adaptation*, Human Development Report 2007/2008 Occasional Paper: Fighting climate change: Human solidarity in a divided world. HDROO. Paper.
O'Brien, K., Eriksen, S., Nygaard, L. and Schjolden, A. (2007) 'Why different interpretations of vulnerability matter in climate change', *Climate Policy*, 7: 73–88.
Oppermann, E. (2011) 'The discourse of adaptation to climate change and the UK Climate Impacts Programme: Describing the problematization of adaptation', *Climate and Development*, 3: 71–85.
Pelling, M. (2011) *Adaptation to Climate Change: From Resilience to Transformation (Kindle Edition)*, UK: Taylor & Francis Books.
Pelling, M., High, C., Dearing, J. and Smith, D. (2008) 'Shadow spaces for social learning: A relational understanding of adaptive capacity to climate change within organisations', *Environment and Planning A*, 40: 867–884.
Potter, J. (2001) 'Wittgenstein and Austin', in M. Wetherell, S. Taylor and S.J. Yates (eds) *Discourse Theory and Practice: A Reader*, London: Sage.

Roe, E. (1994) *Narrative Policy Analysis: Theory and Practice*, Durham, NC: Duke University Press.

Shackley, S. and Wynne, B. (1995) 'Global climate change: The mutual construction of an emergent science-policy domain', *Science and Public Policy*, 22: 218–230.

Szerszynski, B. (2010) 'Reading and writing the weather: Climate technics and the moment of responsibility', *Theory Culture Society*, 27: 9–30.

Tadjbaksh, S. and Chenoy, A. (2007) *Human Security: Concepts and Implications*, Abingdon: Routledge.

Titscher, S., Meyer, M., Wodak, R. and Vetter, E. (2000) *Methods of Text and Discourse Analysis*, London: Sage.

Torfing, J. (1999) *New Theories of Discourse: Laclau, Mouffe, and Zizek*, Oxford: Blackwell.

UK Climate Impacts Programme (2009) *A Local Climate Impacts Profile: How to do an LCLIP*, Oxford: UKCIP.

UK Climate Impacts Programme (2011) *6 September 2011: Environment Agency Launch New Phase of Work*. Online. Available HTTP: http://www.ukcip.org.uk/news/transfer-delivery/ (accessed 3 July 2012).

Wertsch, J. (2001) 'The multivoicedness of meaning', in M. Wetherell, S. Taylor and S.J. Yates (eds) *Discourse Theory and Practice: A Reader*, London: Sage.

Wetherell, M., Taylor, S. and Yates, S.J. (2001) *Discourse Theory and Practice: A Reader*, London: Sage.

Willows, R. and Connell, R. (2003) *Climate Adaptation: Risk, Uncertainty and Decision-making, UKCIP Technical Report*, Oxford: UKCIP.

Wodak, R. and Meyer, M. (2001) 'Methods of critical discourse analysis', in D. Silverman (ed.) *Introducing Qualitative Methods*, London: Sage.

22 The institutionalisation of vulnerable conditions and a case study from Germany

Christian Kuhlicke

Introduction: positioning the argument

Vulnerability is no universal theory; it is a middle range theory attempting to integrate conceptual reasoning and empirical research with regard to specific social phenomena (e.g. the occurrence of devastating famines or floods). A meaningful application of the concept should be based on some kind of contextualisation both empirically as well as conceptually (cf. Kuhlicke *et al.*, 2011). Against this background, the current application of the concept of vulnerability and the increasing emphasis on assessment procedures across the globe seems somewhat troubling. Particularly the strong relationship between cause and effect on which most attempts to assess and evaluate vulnerability are based is worth a moment of reflection. Many vulnerability researchers are implicitly or explicitly interested in unravelling the 'causal forces' at work (Watts and Bohle, 1993: 43) that define the relationship between root causes, dynamic pressures and unsafe conditions (Blaikie *et al.*, 1994). Pursuing this view means identifying the root causes that affect the 'allocation and distribution of resources between different groups of people' (1994: 24) and thus shape the vulnerability of people.

While such a research design appears to be an intellectually demanding exercise that requires a thorough understanding of political, economic, and institutional context conditions, it all too often results in simplified and reductionist reasoning based on a stereotyped analysis (cf. also Hewitt, 1998: 82) that assumes that some predefined groups are more vulnerable than others (e.g. the poor, the uneducated, the elderly). However, as plausible such simple rules might appear and as often they are at the core of vulnerability, they should not be taken for granted. Empirical evidence collected in Europe, for example, suggests that such assumptions are not always helpful to understand people's vulnerability (e.g. Kuhlicke *et al.*, 2011; De Marchi and Scolobig, 2012).

This chapter aims to better understand how vulnerable conditions are institutionalised. It is based on research conducted on the 2002 flood affecting large parts of central and Eastern Europe. It concentrates on the city of Eilenburg, population about 17 500, located in Saxony, Germany, that was severely affected by the 2002 Mulde flood. The 2002 flood inundated the historical city centre and further residential districts. The analysis is based on results obtained from a household survey (n=327) and qualitative interviews conducted with decision-makers and affected citizens (n=21). The interview quotes presented below were translated from German into English by the author.

The institutionalisation of vulnerable conditions

This chapter takes an action theoretical approach to vulnerability. The reason for applying such an approach is that action theory makes no a priori assumptions about underlying causalities as it assumes structure and agency to be equally relevant for the analysis. In other words, there is no external reality assumed (i.e. structures), which directly influence the internal reality of actors (i.e. agency) since reality is bound to the possibilities and limitations of actors (Berger and Luckmann, 1967). In this vein, actors not only construct a reality, this reality also affects them; a process Berger and Luckmann describe with the term *realization*: 'Knowledge about society is thus a *realization* in the double sense of the word, in the sense of apprehending the objectivated social reality, and in the sense of continually producing this reality' (1967: 66).

Furthermore, institutionalisation is a process that establishes rules as naturally given and at the same time sanctions alternative meanings and actions and might thus be seen as a social process by which individuals develop a shared definition of social reality (Berger and Luckmann, 1967). By focusing on the process of institutionalisation, the analysis shifts the emphasis from institutions per se (understanding the consequences of a particular regulatory system for a vulnerable population, for instance) to the very process of how institutions take shape, how they are maintained and reproduced in daily life, and how they shape and regulate the interaction of people in certain situations. In this sense, institutions are constantly produced and reproduced and thus are non-static social processes that are subject to re-interpretation, (re-)negotiation and under continuous change at different scales and from different perspectives.

Although institutionalised rules are produced and re-produced in daily interactions, they are not created instantaneously; they are always built up over time. Only if they exist over a certain period of time, they gain their specific quality, that is, their objectivity. To be sure, objectivity does not mean an absolutely given, external reality, quite the contrary applies: Objectivity means a 'relative consensus' (Hitzler, 1999: 473) that members of a society have agreed upon over the course of time to acknowledge specific subjective interpretations of situations as generally valid interpretations. If institutions transcend the life-span of an individual and are transmitted from one generation to the next generation, they start to develop a reality of their own, a reality that seems to be external to the individual and which appears as a coercive fact. At this moment institutions gain stability that resists quick alterations: 'the institutional world "hardens" and "thickens"' (Berger and Luckmann, 1967: 59).

The 2002 summer flood affected large parts of Central and Eastern Europe and was the single most costly flood in German history. One of its decisive characteristics was that it took most villages, cities, authorities and citizens by surprise. In retrospect most citizens view it as simply impossible to have anticipated the extent of the 2002 flood. The flood was far beyond people's and organisations' power of imagination (Kuhlicke, 2010). Why were people and organisations so radically surprised by the flood? Interestingly, most of the study's participants had experienced previous floods and some even considered themselves experts closely connected with the river and the surrounding environment. They furthermore felt well prepared and felt they had developed an enormous contextual knowledge about the river and adaptive routines to cope with floods. One might therefore think that they were able to prepare for the flood, since they felt they knew the river so well. Yet, as the experience of the 2002 flood revealed, they were not. People's knowledge was hardly helpful in 2002 in anticipating and preparing for the impact and the consequences of the flood, as many citizens admitted astoundedly.

The following paragraphs attempt to explain participants' astonishment and which consequences their astonishment had for their vulnerability. To do this, the argument is based on the assumption that it requires a stock of knowledge that is accepted and more or less taken for granted to be radically surprised. One might even say that the more knowledge is considered as valid and taken for granted, the more fundamentally surprised an actor is when this knowledge turns out to be no longer valid. In this sense, radical surprises uncover the limits of the (till then) valid knowledge (Böschen, 2002: 74). However, how is knowledge stabilised to such an extent that it prepares the ground for being radically surprised?

The process of institutionalisation and its unintended consequences

In addition to their controlling character, institutions also relieve individuals from the need to constantly make decisions. People thus save time and effort and establish certain routines eventually established as institutions and more or less taken for granted. According to Berger and Luckmann (1967), this is the background against which division of labour is possible, opening the way for innovations demanding a higher level of attention and leading to specialization and differentiation.

However, the establishment of routines goes hand in hand with unintended and mostly unacknowledged side-effects – the reduction of complexity, eventually resulting in a loss of '"sensibility" of the society as a whole' (Voss, 2006: 66; translation C.K.). If a problem is sufficiently solved and the solution all-embracingly effective, a loss of social attention is an inevitable side-effect. An issue that once stood at the centre of attention is no longer of importance. Investigating the views of experts and residents on social vulnerability to flash floods in an Alpine region of Italy, De Marchi and Scolobig (2012) reveal how this process is defining the vulnerability of a region and their residents. They identify a 'safety paradox' as well as an 'efficiency paradox' that are rooted in the effectiveness of established hazard and disaster management efforts translating into a 'lack of awareness and agency on the part of residents' (2012: 316).

A similar process took place in the city of Eilenburg. Since the 1950s, when large-scale pumps were installed to route surface and ground water from the inner city to the river, and the city centre was systematically protected by dykes, practical knowledge about individual adaptive measures has played virtually no role in the city. Until the 1950s individual households had been trying to protect themselves against fast rising surface water. This had then been necessary as during heavy cloudbursts the surface water inundated neighbourhoods in the lower parts of the city. To prevent the water from penetrating the ground floors, all homeowners living in these areas installed iron splints on the right and left side of their entrance allowing them to fix two rows of boards in case of heavy rainfall. The interspaces were filled with sand to prevent the rising water from inundating the ground floor. An elderly person remembered: 'If a thunderstorm was threatening on a Sunday, a family member had to stay home to install the iron splints'. This form of household adaptation survived for centuries and only became obsolete during the 1950s when a levée was constructed to surround the city and electric pumps were installed to move water from the city towards the Mulde River.

It is Timmermann's modification of the organic-functional model that captures this loss of knowledge pointedly (1981). Generally, this model is based on the historical change from a high-frequency/low amplitude event scenario towards a low-frequency/high-amplitude environment whereas technological interventions such as dykes are the main reasons for this alteration. In the city of Eilenburg floods occur with low frequency and high amplitude and local successes, such as the protection of the city by surrounding levées, have set the stage for a major perturbation like the 2002 flood. As the river seemed to be controlled and therefore the problem of flooding appeared to be adequately solved, household adaptations such as boarding entrances were no longer viewed as necessary. Only the 2002 flood showed that the levée was inadequate as a solution because it promoted certainty and stability during an event where uncertainty and instability were dominant. This process relates to a more fundamental societal process: the uneven distribution of social knowledge by the establishment of an expert-system.

The establishment of an expert-system

With the establishment of an expert system, an important and far-reaching transition with regard to the social distribution of knowledge begins (cf. also Clausen, 2003). Specialists are attributed responsibility for the segment of reality; they are experts for: 'specialist knowledge is regarded as a motive for leaving differentiated stocks of knowledge to the specialists' (Schütz and Luckmann, 2003: 407; translation C.K.). This means on one hand that an established expert system claims to be responsible for a certain section of reality; on the other hand the system is also ascribed responsibility for this segment by people. As a consequence the uneven distribution of skills and practical knowledge begins and eventually becomes solidified. Consequently, people are dispossessed of the 'material culture' of coping and adaptation techniques that had been gained through immediate practical experiences (Clausen, 1994: 30; cited in: Voss, 2006: 67). The material culture is now extracted from a generally relevant stock of knowledge and converted to a specialised stock of knowledge and this opens up and widens the gap between experts and lay-people.

Prior to the 2002 flood, many residents in Eilenburg trusted the ability of experts to manage and control the river and therefore neglected other techniques and measures that constituted the material culture for protecting themselves against floods at the local level. The empirical analysis reveals that it is, above all, one expert system that people refer to – the Reservoir Administration (Landestalsperrenverwaltung, LTV). In the sense of Giddens (1990), the administration has the technical capability and professional expertise to influence large parts of the material and societal environment along the Mulde River. The administration created large retention basins in the upper part of the Mulde River and its tributaries and shaped large parts of the physical environment along the Mulde through technical flood protection measures. Many people expected the experts to be able to reduce the dangers attributed to the river and to hold ready the means for averting life-threatening situations.

This point is explicated by a participant of the study who offers a historical perspective on the development of the retention basins. He makes the case that the river is in principle uncontrollable: 'Our Mulde is an incalculable river.' He recollects that in his youth – that is in the 1920s and 1930s – one might have walked through the river on Wednesday and the water had only reached to the ankles, and by Friday the water would already have risen above the riverbed. He refers to a local fisherman who always told his father that three days of rain in the Ore Mountains would cause the Mulde in Eilenburg to flood. Although the participant introduces a certain regular pattern (strong rain in the mountains three days later results in a water level rise in Eilenburg), he concludes that the river had been unpredictable at that time. This passage contrasts the situation in the 1920s and 1930s with the situation after the 1950s when the expert system was institutionalised and there was large scale construction of reservoirs in the upper course of the river: 'So the river is incalculable – why? Because in those days they had no retention basins at the source of the river'. The participant therefore has always been positive about the ability of the retention basins to help avoid floods: 'For ten or fifteen years, I have always promised people that there will be no floods anymore' and later in the conversation he underlines this view again: 'I have always calmed people and told them, our dams [retention basins, C.K.] we have in the mountains, they catch the water'.

At the same time a strong feeling of dependency existed among participants. Many of them assumed that whenever a situation is not judged appropriately by the experts, they have to bear the consequences in that their properties are flooded. Therefore, many citizens did not believe in the experts' ability to control the river and this not just since the 2002 flood; they also openly mistrusted their expertise. Some participants had expressed their concerns in one form or the other for a long time. However, the symptoms of crisis observed by the population (e.g. floods) were regarded as temporary phenomena: a person did not follow the instructions, an expert misjudged the situation, or – even more shattering – the general institutionalised instructions were simply wrong, since other issues such as the supply of drinking water or the recreational claims of the population to the retention basins are considered as more important than flood protection.

This mismanagement, however, is in principle re-addressable: 'The flood would have come in 2002, I am convinced, because of the heavy rainfall, but not in this manner if one had reacted differently'. This opinion is widely held among the affected participants and also among decision-makers.

Beliefs in the controllability of the river

Underlying this interpretation is a strong belief in the controllability of the river. While the trust in the expert system was substantially shattered by the flood, the faith in the controllability of the river was not questioned. It is only the institutionalised malfunctioning of the experts that resulted in a loss of control over the river. Therefore, paradoxically, although the institutionalised order is challenged by referring to the expert system, its underlying legitimation is maintained unquestioned, whereas the most dominant legitimation resembles what Holling describes as 'nature tolerant' (1978).

Prior to the flood two different beliefs existed about the states of equilibrium of the river: First, ordinary variability of the river resulted in the inundation of the area outside of the levées. This was considered as an unproblematic, ordinary equilibrium and people adapted to these fluctuations. Second, major variability resulted in the flooding of parts of the city in 1932, 1954 and 1974. These variations were considered as extraordinary equilibrium, but they were not perceived as threatening for the community; they were still within an acceptable limit for the community. It was known by most participants that the river could inundate some areas (ordinary equilibrium) and that from time to time the city would be inundated by floods similar to the ones experienced in the twentieth century (extraordinary equilibrium).

The institutionalisation of vulnerable conditions: outline of a framework

This chapter presents some of the findings within an overall framework for better understanding how vulnerable conditions are institutionalised. Being inspired by Blaikie and his colleagues' Pressure and Release (PAR) Model (1994), Figure 22.1 outlines how in a specific location the material culture of handling risks and disasters is shaped by and interdependent with the distribution of knowledge between experts and citizens as well as by underlying myths and legitimations. Before giving a more detailed account on the single components presented in Figure 22.1, a difference to the PAR-model needs to be highlighted. While the PAR-model assumes a causal progression from root causes towards dynamic pressures resulting in specific unsafe conditions (cf. 1994), this study does not assume such a strong relationship between causes and effects. By applying action theory, it is rather interested in how actors assess, take up, reinterpret and hence maintain their reality and therefore does not assume a pre-analytical difference between causes and effects. This is also reflected in Figure 22.1 which emphasizes the interrelation between the different components of the framework.

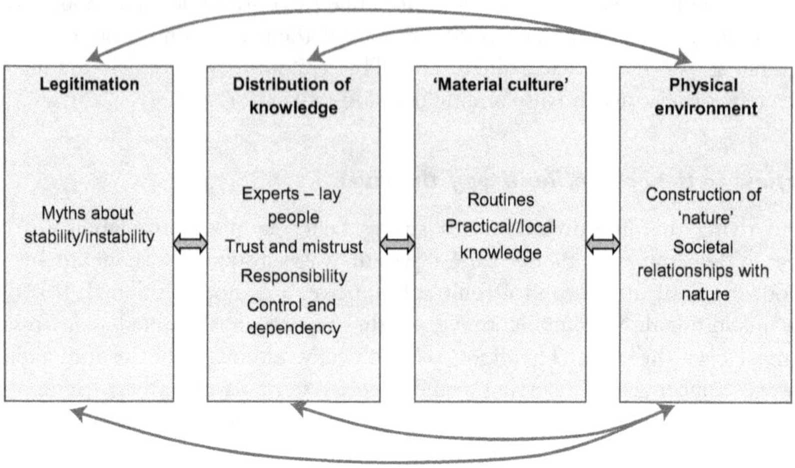

Figure 22.1 The institutionalisation of vulnerable conditions. Source: Adapted from Blaikie *et al.* (1994)

Many citizens in Eilenburg learnt to live with the risk of flooding and developed adaptive routines to cope with flood events, at least in relation to regular, smaller annual floods. They felt quite well prepared for floods and had developed an enormous contextual knowledge about the river. However, many citizens were still radically surprised by the 2002 flood. How to explain this surprise? Paradoxically, many citizens emphasised that because they thought they understand the river and because they had adapted to its annual variations, they could not envision the 2002 flood, since their personally acquired and tested knowledge about the river did not consider the possibility that the 2002 flood would significantly deviate from previous historical flood levels.

At the same time, a deeply seated belief exists in the 'promise of security' attributed to the dikes surrounding Eilenburg. A dominant finding was that citizens as well as decision-makers trust in the ability of dykes to prevent flooding of the city. This view is closely linked to questions of responsibility. Most actors do not consider themselves responsible for flood protection. Rather, certain areas of knowledge are left to experts whilst participants simultaneously feel dependent on the abilities of these experts to regulate the river. Although many citizens mistrust the capabilities of the LTV, because they believe it is engaged in too many contradicting tasks (e.g. supply of drinking water), the underlying faith in the controllability of the river remains unaffected by such doubts. Both developments, a loss of sensitivity on the one hand as well as the delegation of responsibility to an expert system on the other, result in a loss of the 'material culture' in Eilenburg in the long-run.

This process is embedded in a widely shared belief centred on ideas of equilibrium. In Eilenburg, probably similarly to other cities and communities along the Mulde River, knowledge about the river was based on two states of equilibriums of the Mulde. That is, an ordinary equilibrium (considering the annual inundation of the areas located outside the levées), as well as an extraordinary equilibrium (represented by the three historical floods of 1932, 1954 and 1974). Only the 2002 flood demonstrated quite radically that the river is also able to deviate from previous flood levels. This widely held and apparently unquestioned stock of knowledge helps explain the vulnerability of the city: it did not consider the possibility that the river could exceed both equilibriums. However, why did this belief become so powerful? Most importantly, it is not an abstract belief; it is rather based on everyday experiences. The river – apart from single exceptions – appears mostly as controlled and the levées and other technical devices offer a strong sense of security that is deeply interwoven with the reality of everyday life. This experience has been produced and reproduced in the daily interactions with the immediate social and physical environment and was hardly questioned. It was thus, paradoxically, the effective control of the river that prepared the ground for experiencing the 2002 flood as a radical surprise.

The institutionalisation of vulnerability discussed in this chapter presents some challenges to the idea of transformation. The latter aims at going beyond incremental changes that remain within an institutionalised environment; transformation rather aims at fundamentally altering institutionalised characteristics of a society including underlying norms, rules and value systems as well as their specifications in regulatory, legislative, and bureaucratic regimes (IPCC, 2012: 5). As the IPCC SREX report states: 'transformational strategies place emphasis on addressing risk that stems from social structures as well as social behaviour and have a broader scope extending the immediate field of disaster risk management or climate change adaptation' (IPCC, 2012: 37). Any attempt to develop such transformational strategies thus needs to engage with patterns of how institutionalised rules shape practices and interpretations of everyday life and vice versa.

References

Beck, U. and Lau C, (2005) 'Second modernity as a research agenda: theoretical and empirical explorations in the "meta-change" of modern society', *British Journal of Sociology*, 56: 525–557.

Berger, P. and Luckmann, T. (1967) *The Social Construction of Reality: A Treatise in the Sociology of Knowledge*, New York: Anchor Books.

Blaikie, P., Cannon, T., Davis, I. and Wisner B. (1994) *At Risk: Natural Hazards, People's Vulnerability, and Disaster*, London: Routledge.

Böschen, S. (2002) Risikogenese: Metamorphosen von Wissen und Nichtwissen, *Soziale Welt*, 53 (1): 67–86.

Clausen, L. (1994) *Krasser sozialer Wandel*, Opladen: Leske and Budrich.

Clausen, L. (2003) 'Reale Gefahren und katastrophensoziologische Theorie: Soziologischer Rat bei FAKKEL-Licht', in L. Clausen, E. M. Geenen and E. Macamo (eds) *Entsetzliche soziale Prozesse. Theorie und Empirie der Katastrophe,* Münster: LIT Verlag, 51–76.

De Marchi, B. and Scolobig A. (2012) 'The views of experts and residents on social vulnerability to flash floods in an Alpine region of Italy' *Disasters,* 36: 316–337.

Demeritt, D. (2002) 'What is the 'social construction of nature'? A typology and a sympathetic critique', *Progress in Human Geography*, 26: 767–790.

Giddens, A. (1990) *The Consequences of Modernity*, Stanford, CA: Stanford University Press.

Görg, C. (2004) 'The construction of societal relationships with nature', *Poiesis und Praxis,* 3: 22–36.

Hewitt, K. (1998) 'Excluded perspectives in the social Construction of disaster', in E.L. Quarantelli (ed.) *What is a Disaster? Perspectives on the Question*, London: Routledge: 75–91.

Hitzler, R. (1999) 'Welt erkunden – Soziologie als (eine Art) Ethnologie der eigenen Gesellschaft', *Soziale Welt,* 50: 473–482.

Holling, C.S. (1978) 'Myths of ecological stability: Resilience and the problem of failure', in C.F. Smart and W.T. Stanbury (eds) *Studies on Crisis Management*, Toronto: Butterworth & Co. Ltd: 93–106.

IPCC (Intergovernmental Panal on Climate Change) (2012) *Managing the Risks of Extreme Events and Disasters to Advance Climate Change Adaptation. A Special Report of Working Groups I and II of the Intergovernmental Panel on Climate Change.* C.B. Field, V. Barros, T.F. Stocker, D. Qin , D.J. Dokken, K.L. Ebi, M.D. Mastrandrea, K.J. Mach, G-K. Plattner, S.K. Allen , M. Tignor, and P.M. Midgley (eds) Cambridge: Cambridge University Press.

Kuhlicke, C. (2010) 'The dynamics of vulnerability: Some preliminary thoughts about the occurrence of "radical surprises" and a case study on the 2002 flood (Germany)', *Natural Hazards*, 55: 671–688.

Kuhlicke, C., Scolobig, A., Tapsell, S., Steinführer, A. and De Marchi B. (2011) 'Contextualizing social vulnerability: findings from case studies across Europe', *Natural Hazards*, 58: 789–810.

Polanyi, M. (1985) *Implizites Wissen*, Frankfurt: Suhrkamp.

Schütz, A. and Luckmann, T. (2003) *Strukturen der Lebenswelt,* Konstanz: UVK Verlagsgesellschaft mbH, Konstanz.

Timmerman, P. (1981) *Vulnerability, Resilience and the Collapse of Society*, Institute for Environmental Studies, Toronto: University of Toronto.

Voss, M. (2006) *Symbolische Formen: Grundlagen und Elemente einer Soziologie der Katastrophe,* Bielefeld: Transcript.

Watts, M. and Bohle, H-G. (1993) 'The space of vulnerability: The causal structure of hunger and famine', *Progress in Human Geography,* 17: 43–67

23 Disaster risk and vulnerability reduction

Ben Wisner

Introduction

The central focus of this book is global environmental change and human security, with an emphasis on the kinds of transformations (political, economic, cultural and technological) that are required to create human security. The focus of this chapter is on disaster risk and efforts to reduce people's vulnerability to natural hazards that are ever more complicated by environmental changes. Few dispute the salience of such changes, especially when it comes to extreme climate events and their human consequences (IPCC, 2012). Yet to understand risk and vulnerability, it is necessary to think more broadly and consider other global changes, including how these changes interact in dynamic and differentiated local contexts. Such an approach challenges dominant climate change policies, revealing the potential and need for deeper and more transformative actions.

This chapter first discusses some key concepts and processes associated with disaster risk and vulnerability reduction, including capacities, knowledge and skills, and participation. A case study in Northern Tanzania illustrates how these concepts can be differentially interpreted to understand the local context of risk.[1] Next, the chapter uses a systems approach to consider the types of transformations that may be necessary to reduce risk. The final section considers the implications, and proposes an agenda for research, policy and civil society action.

Framing disaster risk: key concepts and processes

Vulnerability

"The focus on vulnerability emphasizes that disaster and crisis are never purely the result of natural processes but are in fact often social products," state Heinrichs *et al.* (2011: 10), who continue, "It essentially refutes 'taken-for-granted beliefs about the essential nature of things'..." (Demeritt, 2002). As White stated more than 60 years ago: "Floods are 'acts of God', but flood losses are largely acts of man [sic]" (White, 1945: 2). What has to be recognized is that "acts of man" are not only affecting loses; people may indeed also affect the flood itself through land use changes and engineering, and potentially also through climate change. Now in the grip of anthropogenic climate change, White's "God" is getting a good deal of help from humanity! What, then, is vulnerability in this context?

Development studies specialist, Robert Chambers introduced the term "vulnerability" into the analysis of rural poverty in his 1983 book, *Rural Development: Putting the Last First.* It came as one of five elements that interlocked with each other, producing what he termed "ratchet effect" or "deprivation trap" (Chambers, 1983: 112): a condition of "integrated rural poverty" that is very difficult to extract oneself from. The other elements were political powerlessness, physical weakness (ill health), isolation, and income poverty. Contemporary users of the term "vulnerability" in the literature of disaster risk reduction as well as climate change might find Chamber's usage odd. Isn't vulnerability precisely a matter of what Blaikie and Brookfield (1987) called "ecological, economic, and political marginality" or what Wisner (1978) termed "eco-demographic marginality"?

It seems from the context that Chambers had in mind social vulnerability when he used the single noun. This has to do with the degree to which one's social status (culturally and socially constructed in terms of roles, responsibilities, rights, expectations concerning behavior) influences differential impact by extreme natural events and the chances or recovery. Thus, depending on the society and situation, social characteristics such as gender, age, occupation (in caste systems), marital status, race, ethnicity, religion may have a bearing on potential loss, injury, life chances in the face of hazards and the nature of one's recovery trajectory. In addition to such socially differentiated vulnerability, there is what I have elsewhere called "generalized vulnerability" (Wisner, 2003). Generalized vulnerability is a characteristic of the poorest of the poor in every society, especially those who not only suffer income poverty and are also politically marginal, spatially marginal (in urban squatter settlements or in remote rural locations), ecologically marginal (livelihoods based on access to meager natural resources or living in degraded environments), and economically marginal (poor access to markets) (Wisner, 1993).

A further important aspect of what I am calling generalized vulnerability is the difficulty households in this situation encounter in recovering from any insult or stress. Again, depending on the specific nature of the stress (flood, spontaneous collapse of an old apartment building, coastal storm, urban arson, or wildfire), it is much more difficult for these households to regenerate their livelihoods and shelter than for the non-poor and non-marginal (in the broad sense of "marginal" that includes political, spatial, ecological, and economic dimensions). The millionaire who lost a USD10 million home in the mountains outside Los Angeles has suffered a greater economic loss than the Mexican migrant farm worker who lost a mobile home. However, without insurance, with few financial reserves, with their extended family living across the border in Mexico perhaps a thousand miles away, afraid of applying for government aid because they are illegal immigrants, the farm worker household will have a much harder time recovering.

The poorest and marginal groups in a society live with chronic indebtedness, malnutrition, ill health, degraded physical surroundings, and often violence. For them any additional stress can have catastrophic results (losing land or meager assets to money lenders, losing children to famine, losing their place in a community as they are forced by circumstances to migrate). In this sense, whatever is on the local geographic menu could have that impact on them – flood somewhere, landslide elsewhere, coastal storm or earthquake in yet other places. If it is possible to identify the poor and the marginal, their numbers and location are a starting point, crude as it might be. Without falling into the trap of assuming that poverty equals vulnerability, one can then zero in on these groups, determining, for example, their initial wellbeing, degree of self-protection, degree of social protection they enjoy, their livelihood resilience, and the existence and extent of social capital accessible to them (Maskrey *et al.*, 1997; Cannon, 2000).

There is ample evidence that in many disasters the elderly, young children, women, people living with disabilities, ethnic minorities, and immigrants (especially illegals) tend to suffer more (Gaillard, 2012). Women died at a higher rate than men in the Asian tsunami (Doocy *et al.*, 2007; Fordham, 2012). Three times as many women died in the Pakistan earthquake in 2005 because they were more likely to be indoors when it hit (Reuters, 2009). The majority of the deaths from Kobe's 1994 earthquake were in the group over 60 years of age (Wisner *et al.*, 2004; Ngo, 2012). It is important as a minimum starting point to recognize this evidence and ensure that the needs of "high risk" groups are planned for. However, emphasizing only needs and not what various individuals or groups can bring to the task of risk reduction represents a limited way of looking at vulnerability. This view emphasizes victimhood rather than capacities.

Capacities

Capacity is a critical component of risk. Most people, even poor people, have capacities (resources, networks, knowledge and skills) that they can put to work in order to prevent, resist, cope with and recover from stresses and shocks such as an extreme natural event (Adger, 2006). Capacity is not the opposite of vulnerability. Someone may have capacities in the form of knowledge and skill but nevertheless have them blocked and thus remain highly vulnerable. Sen (1999) has discussed at length how socio-economic and political systems affect the attainment and use of capabilities. One sub-set of such capabilities is the capacities to cope and to recover from shocks. Clark (2005) provides an overview of Sen's contribution, which considers:

> people as ends in themselves (rather than treating them merely as means to economic activity), recognising human heterogeneity and diversity (through differences in personal conversion functions), drawing attention to group disparities (such as those based on gender, race, class, caste or age), embracing human agency and participation (by emphasising the role of practical reason, deliberative democracy and public action in forging goals, making choices and influencing policy), and acknowledging that different people, cultures and societies *may* have different values and aspirations.
>
> (Clark, 2005: 5)

Capacity to protect oneself, one's family, and assets from extreme events as well as to recover from any damage or loss inflicted by such an event is dependent on access to resources (Wisner *et al.*, 2004: Chapter 4). "Resources" are defined in a broad way in this context to include all bundles of assets that can produce a livelihood, as well as the institutional, political, and social pre-conditions for using those assets. Thus, for example, land and water are assets farmers can use to grow food, but there have to be conventional rules for governing the farmer's access to the land, or, as in many countries, formal legal title to the land. Markets for buying and selling assets, goods, and services are also usually necessary for a household (rural or urban) to realize values that provide a livelihood. Income is often taken as a short cut to identifying "poor" and "less poor" households; yet income is only part of many livelihood systems. Production for own consumption remains important in many parts of the world. Also, many income sources are not usually counted. This is the case with much remittance from family members working abroad or in another part of the country. It is certainly the case where the so-called parallel economy is concerned.

Groups of people – neighbors, extended family members, members of faith communities and other civil society organizations – can be very active in their own interests when facing the possibility or reality of an extreme natural event. Little of this knowledge, skill, sharing of knowledge and material goods, mutual aid, emotional support is visible to government planners and most researchers, and much less is valued in economic terms. Yet the complex interactions among these so-called capitals (social, human, institutional, natural) may have critically positive outcomes for preventing disasters triggered by extreme natural events or at least mitigating the loss they cause, for preparing for such events, for warning and immediate relief. It is well known in disaster relief circles that the vast majority of people rescued during short onset, acute disasters (e.g. earthquakes, landslides, flash floods) are saved by non-professionals: loved ones, neighbors, and passers-by. The complex of social capital and knowledge (human capital) is also critical in the longer-term recovery period, where their interactions with the state's institutional capital may be positive or lead to negative outcomes. In strictly economic terms under most circumstances, investments in strengthening and preserving social capital and in human capital (e.g. education and health care) provide a "two-for" or win-win situation. These investments are likely to increase productivity in normal situations and provide for more resilience at the local level in cases of extreme natural events. Where the locality and local livelihood systems are more resilient, the state is relieved of costly increments of relief and recovery expense.

Knowledge and skills

Local knowledge is important for disaster risk reduction because it is the lens through which people perceive and understand the world and work on the world. All innovation including risk reduction will have to be carried out at the end of the day by *people in places*. But apart from this sociological and geographical reality, there is even a more important reason why local knowledge is critical: People are constantly coping with threats. They share knowledge with neighbors, may draw knowledge in from far away, boil it down and work out ways to apply it locally. Local communities are workshops of knowledge production, not just museums of tradition. Thus for the outside specialist, the village, hamlet, town and city neighborhood are as much sources of new ideas to be tested, refined, and shared as is the outside specialist's skill a resource for local people. There is a broad and deep potential for partnership in knowledge production for reducing disaster risk that is very seldom actually achieved.

Local knowledge comprises the totality of perceptions, beliefs, understandings, and skills that one or more members of a community uses or potentially uses to communicate about and manipulate the world. "World" in this sense is made up of the physical and built environment and also the social, economic and political environment that affect production and consumption at the local scale. This is a formal definition for the sake of clarity. In simpler terms, it is what people living in an urban or rural locality know that is useful to them in their lives. There are many important features of local knowledge that are critical to disaster risk reduction (UNISDR, 2009; 2011). First, local knowledge, like all knowledge, is social. Just as there can be no private language that only one person understands, there can be no knowledge completely separate from what others know and have known in the past. Local knowledge is not entirely "traditional" (passed on by generations). In fact, the notion of "indigenous" or "traditional" knowledge is quite limited and only part of the picture. Local knowledge is more, as it may opportunistically incorporate versions of outside specialist knowledge. For instance, weather or climate forecasts listened to on the radio may be interpreted and modified according to local weather signs and past experience. In fact, the tendency to mix or hybridize knowledges is increasing. Mobility is important to livelihoods, and family members that travel to seek wage employment or trade often send or bring knowledge home as well as money and goods.

Within a community, local knowledge is not uniformly distributed. Not everyone has access to secret knowledge or knowledge associated with local skills such as building, finding water, or midwifery. Occupations and special skills come with sub-sets of local knowledge that may not be widely distributed in a locality. Local knowledge may be a source of power and status. For example, in Sierra Leone, rice farmers have been known to compete with each other in breeding new varieties of rice, some simply for the beauty of plant. This is a source of praise and prestige. Local knowledge may not be explicitly spoken about by those who have it. It is sometimes tacit or implicit in their practices and acts – for instance, where on a slope with different soil characteristics to plant different plants. Western trained or oriented experts find the idea that knowledge is tacit hard to accept, yet it can be made explicit through patient discussion.

Local knowledge is also gendered and age graded, and moreover it varies according to the standpoints of people in different (and to some extent dynamic and changeable) life situations: for example, local knowledge of people living with disabilities, knowledge of people with chronic health problems, or knowledge of people who constitute an ethnic or caste minority in a community. Local knowledge may not appear to Western trained or oriented specialists to have any physical or biological basis or efficacy. This is because local knowledge often bridges physical and social functions and realities. Thus some women in Africa boil stones during a hunger period. "But stones have no nutritional value," the outside observer would object. However, by executing this practice, the women signal a stage at which the community must consider the hunger serious and activate coping measures. The community is also reassured by the ritual which reminds them that they have survived in the past and that control and unity are still possible (Richards, 2009).

In the context of disaster risk reduction, the following insights on knowledge management are significant (Wisner, 2009; cf. Mercer, 2012). First, top-down diffusion of knowledge and practice requires fine tuning to local conditions. Diffusion "by the book" seldom works. Second, at the local scale, people experience threats in a more holistic way, whereas specialists who design practices tend to focused on one hazard or another in isolation from other stressors. Poverty, violence, climate change, and many different natural and other hazards confront people at the scale of 1:1 where they live, work, raise children, celebrate, and suffer. Local efforts to deal with one of these challenges generally involve dealing with the others. Fine tuning takes such experience into account. Finally, there is sometimes a lack of trust between communities and governments or outside/non-local institutions. Trust and partnership must be built; it cannot be assumed. Without trust and mutual respect, the exchange of knowledges and production of a useful hybrid of outside and local knowledge is not possible.

Participation

It is precisely at the local scale that people's own knowledge and capacity can most usefully combine with the outreach of local government (GNDR, 2009; 2011). Over the last decade many NGOs as well as many of the national societies comprising the International Federation of Red Cross and Red Crescent (IFRC) have pioneered participatory action research methods through which communities map local hazards, assess the vulnerabilities they face and capacities they possess (Haghebaert *et al.*, 2008). "Participation" can have many meanings. The so-called "ladder of participation" (Arnstein, 1969; Hart, 1992) includes everything from hosting a meeting for the local official and rubber stamping some proposal or scheme she or he brings along, to local definition of goals, capacities, and needs, and concrete steps by local people to brings these to fruition. These extremes have also been referred to as "weak" or instrumental and "strong" or transformative participation (Wisner, 1988). "Participation" as used commonly by development studies and natural hazards research in such phrases as "participatory action research" and "community participation" is a special case of "inclusion." Büscher and Mascareño (2012) say the term "inclusion":

> conveys the expectation that in principle each individual is treated as a legal entity and may assert his [sic] personal rights in court; that everyone is entitled

> to their own property and can trade with it; that everyone has the right to attend educational institutions and take part in democratic elections . . .
>
> (Büscher and Mascareño, 2012: 57)

Participation is where the untapped potential of faith communities becomes apparent. Religious institutions have the potential of joining with secular civil society in developing local risk awareness and self-assessment of hazards, vulnerabilities and capacities (Anderson and Woodrow, 1989; Wisner *et al.*, 2004). Yet to move from assessment to action only some things can be done in isolation in a village, hamlet or urban neighborhood. More often than not, cooperation with local authorities is needed. This is where trust and development of a common language of risk reduction is required, but the state often demands and enforces control (Scott, 1998) rather than opt for partnerships based on trust (Wisner and Uitto, 2009).

Global change in a local context: insights from Tanzania

Understanding how capacity, local knowledge and participation interact with planning processes to produce or reduce risk and vulnerability calls for a reality check (Wisner *et al.* 2012a). Many processes interact with environmental change to produce diverse outcomes for human security (see Smucker and Wisner, 2008). Working with farmers, herders, local officials and other outside specialists, recent field work carried out for a study on local knowledge and climate change adaptation in northern Tanzania challenges the undifferentiated view of Tanzanian rural society that seems to inform most climate policies.

Small farmers in our Tanzanian study area have been experiencing major changes of many sorts over the past few decades, and climate change is just one of them. Scale is a critical issue. Global climate change manifests differently at regional, sub-regional and local scale. Is there less water in the Ruvu River because there is less snow melt on Mt. Kilimanjaro? Or is it because water users are extracting more upstream? Does it matter locally, at the scale of the village of Mtakuja, where those Maasai households and other people in the village may well be tapping a huge aquifer as yet unmeasured and unmapped?

Consider this example: Maasai pastoralists on the so-called Maasai Steppe (Simanjiro District) narrate increasingly poor and erratic rainfall. The most palatable of grasses for cattle are scarce. Young Maasai men must take herds farther and farther to find grass and water. Across the Ruvu River, a group of Maasai have settled in a village with mixed ethnicities and look forward to taking advantage of borehole irrigation to farm. Meanwhile they have used a small amount of irrigation water to grow water melons for sale and still have livestock on site, while most of the herds are in the distant Tanga Region.

The first part of the example satisfies expectations encouraged by the dominant discourse on climate change, in that it discusses increasing climate variability and growing hardship for people reliant on traditional, resource intensive livelihoods. The rest of the story, however, deviates from that narrative. Some Maasai have taken advantage of other global changes and have begun to adapt spontaneously to rainfall variability. Technological change in the form of irrigation, mobile phones and a motorbike transport system allow perishable goods to reach local markets more efficiently than before. Political and ideological changes banned the burning of pasture lands and contributed to the encroachment of woody species, and they are also responsible for many Maasai's displacement from newly established national parks and hunting reserves – all in the interest of Conservation (with a capital "C") and rational management of resources.

The dominant climate change adaptation narrative does not take into account local contexts. For example, urban environments need to be distinguished from rural ones in terms of impacts of global environmental change, and also in terms of the opportunities for adaptation offered by cultural and economic exchange. Recent field work revealed the vitality of cyclical market sites at various altitudes on gradients from highlands down into the surrounding plains. Fodder for stall feeding cows daily and dried fish move uphill; timber products and bananas move downhill. Preliminary data suggest that increased variability of climate has caused this trading and exchange to increase in volume and frequency (Wangui *et al.*, 2012). Opportunities come along with risks, and rural adaptation sparks urban growth.

Such nuances at the local scale and interconnections among localities at different altitudes are not recognized by national level planners and policy makers in Tanzania. Tanzania was one of the earliest countries to produce a National Adaptation Plan of Action (NAPA), a report to the United Nations Framework Convention on Climate Change (UNFCCC) through which Least Developed Countries identify "urgent and immediate needs – those for which further delay could increase vulnerability or lead to increased costs at a later stage" (UNFCCC, 2012). In the global South, NAPAs provide a unique insight into dominant adaptation framings that serve as the principal blueprint for both national adaptation planning and internationally-financed adaptation programs.

Tanzania's NAPA paints a stark picture of undifferentiated rural poverty and vulnerability exacerbated by climate change. This generalized vulnerability is unmitigated by the dynamism and increasing differentiation of rural livelihoods, nor does it take into account established, local means of coping with climate stress (e.g. Eriksen *et al.*, 2005). Indeed, the NAPA suggests that climate change is a "threat mainly to the agrarian population that still depends on subsistence agriculture for their daily livelihood" (United Republic of Tanzania, 2007: viii). This greatly simplified representation of rural livelihoods is made more problematic by a framing of drought vulnerability as wholly produced by climate variability, with no role for the wider set of social and political "root causes" of vulnerability (Wisner *et al.*, 2004). The NAPA portrays rural communities and their development prospects as wholly at the mercy of the vagaries of climate, as "the major causes of these vulnerabilities at village, district and national levels is climate change associated with prolonged heavy rainfall or drought" (United Republic of Tanzania, 2007: 13).

The exception to this undifferentiated view of Tanzanian rural society is in the document's evocation of climate stress-driven conflict, whereby "shrinkage of rangelands is likely to exacerbate conflicts between livestock keepers and farmers in many areas" (URT, 2007: 7). While no ecological explanation is provided for the imminent decrease in rangeland area, conflict is said to be likely because livestock carrying capacity has already been surpassed. In response, the NAPA recommends controlling movements of livestock, a deeply political proposal in light of ongoing land access struggles of Tanzania's approximately 1.5 million pastoralists (Igoe, 2003; Goldman, 2011; Pearce, 2012).

Nothing in Tanzania's official climate policy acknowledges the pivotal role of public goods and infrastructure as the matrix within which farmer and herder adaptations take place. More importantly, there is little or no reference to issues of governance, poverty or equity in adaptation. Early efforts to draw attention to governance issues (Paavola, 2004) and issues equity and justice (Paavola and Adger, 2002) are ignored in Tanzania's NAPA. In short, the social drivers of vulnerability, the human capacity to adapt, and the value of local knowledge and participation are not adequately considered in national and international climate policies. A reductionist approach is inadequate for dealing with systemic problems.

Risk and vulnerability in linked systems

Systems thinking seems to be rediscovered by every generation. In the 1950s General Systems Theory inspired innovations in geography and human ecology long before "resilience thinking" was a twinkle in anyone's eye. Kenneth Boulding's (1985) systematic approach to nested physical, biological, and human systems pioneered the way long before a marketing niche was carved out for "sustainability science." Whether one uses simple, common sense terms for the complex interactions among humans and their creations and the natural environment or call them socio-ecological systems, the referent is the same. Humans have created many systems, and they affect and are affected by nature in complex ways. These interactions bear heavily on risk and vulnerability, and are important to understand if transformations that support human security are the goal.

Demographic systems

The Tanzanian population, like many others in the global South, has grown, moved about in space, and changed characteristics such as age-sex ratios, health status, longevity, and educational level. Too much emphasis is given in some discourses of global environmental change and human security to mere growth in numbers and population density. There is no clear linear relationship between high density of population and degradation of the environment. In fact, studies in Nigeria and Kenya point in the opposite direction. The name of the now-classic Kenyan study is evocative: "More People, Less Erosion" (Tiffen *et al.*, 1994). The mixing and sharing among people up and down the altitudinal gradients the author and his colleagues are studying in Tanzania's North is partly driven by differences in density at different sites on the gradient and a wide range of other factors – interactions of population with other systems.

Economic systems

In the Tanzanian study sites, food security seems only partly about growing the family's annual or semi-annual grain supply. To be able to purchase food and, in extreme circumstances, to command political or social entitlements to food, is critical. Think of the Maasai household growing and selling water melons. In the South Pare Mountains, a low coffee price and long bouts of coffee tree disease have seen many farmers turning to cultivation of root ginger, which has a good price and can be processed easily for added value. Climate may be changing in complex ways, but will this kind of innovation at farm scale keep ahead of the pressures of environmental change? This essential question is not being asked by modelers such as the World Institute for Development Economic Research (WIDER) team that visited Tanzania (Arndt *et al.*, 2012) because their econometric paradigm blinds them to the creativity and innovativeness of the small farmers and herder.

Technological systems

In northern Tanzania a number of technologies have opened up economic opportunities in the face of pressure on traditional or customary livelihoods: transportation infrastructure, pre-paid mobile phones, tissue culture that allows efficient multiplication of new crop varieties, resurgence of organic farming techniques that reduce input prices and damage to ecosystems, availability of small mills for rice and sunflower, and growing use of decentralized energy sources (solar and wind). Local radio stations broadcasting in vernacular dialects bring weather and climate forecasts, and mobile phones are used for warning of floods.

Socio-cultural systems

Socio-cultural systems are important to risk and vulnerability in at least two ways. Social organization is one of these. The spontaneous adaptation to change discussed in this chapter is facilitated by many layers of social organization, some informal such as extended family and clan, and some formal such as civil society organizations. Such social capital has already been discussed above, and is used in a variety of ways. These organizations may be agents of change and spontaneous climate change adaptation.

Culture and language are also vital to understanding risk and vulnerability. "Climate" doesn't translate well into Kiswahili, the lingua franca of Tanzania. The term *tabia nchi* (a neologism created by government to correspond to "climate" as opposed to weather = *hali ya hewa*) connotes the characteristics of place, not just climate. Participants in our study "misunderstand" the elite neologism; however, there is wisdom in the vernacular. Climate change only manifests and affects people, evoking adaptation, at the scale of landscapes and specific rural and urban sites. This is how farmers and herders experience climate change as one of many changes their parents and grandparents have had to live with. These sites of risk and opportunity – specific slopes, internal valleys, spring fed wetlands –

are known by their flora and fauna, their water, grazing, soil characteristics, topography, accessibility, and, above all, histories. Linear groves of coconut palms and mango trees along water courses in some parts of Tanzania tell the story of slave caravans that repeatedly passed that way many years ago. It is difficult to speak with people about "natural hazards" in such a rich linguistic universe without also invoking "resources" and "opportunities," and arriving through the momentum of discussion at subjects including "access" and "power."

Political systems

Risk management at local scale and reduction of vulnerability requires partnerships between local political authorities, civil society, and communities. However, in many countries there is rhetorical and even legislated commitment to decentralization of government while human and financial resources are still retained by the central state. In the Highly Indebted Poor Countries (HIPC) and failed states, it is entirely possible that the government of the day has no motivation to devote resources to disaster risk reduction or to climate change adaptation at the local scale. National level training, workshops, travel overseas and, above all, mega-projects such as sea walls, dams, large scale agri-business schemes involving overseas corporations are much more profitable to the elite, whose cronies receive contracts and who have numerous opportunities for personal enrichment. These governments count on the fact that their rural and urban poor will not protest when disaster strikes. The elite will blame nature (and blame climate change) and sit back while the humanitarian community sorts things out, reestablishing the status quo *ante* (Susman *et al.*, 1983; Mascarenhas and Wisner, 2012).

New agendas

There is a need not just for new agendas, but for new types of agendas that address not just the generic narratives about risk and vulnerability, but the underlying factors that contribute to them. This calls for transformative approaches, rather than for mere reforms based on generalized narratives.

Research agenda

The devil is in the details. Research must dig deep into local knowledge and practice and be open to surprises, especially the way in which people spontaneously adapt to interacting changes (including environmental and climate change). People are not inert or passive victims of hazard events resulting in disaster. They are agents with knowledge and capacity. The question is how to combine that local knowledge with external specialist knowledge better to provide social and personal protection. Although the IPCC's study of climate extremes and disasters involved a number of social scientists, the study is still driven by top down modeling. Social science is the "junior partner." However, if transformation is the goal, knowledge and capacity at the local scale has got to be the starting point (GNDR, 2009, 2011; Wisner *et al.*, 2012c).

Even as the IPCC's SREX was published, a team from WIDER presented a study in Tanzania that uncritically "downscaled" climate models to the sub-national scale (Arndt *et al.*, 2012). This practice is highly questionable to begin with (e.g. Rosentrater, 2010). However, these researchers proceeded to combine questionable cropping and yield data and questionable rural income data with highly questionable climate projections. Their conclusion was that Tanzanian food security and economic development would suffer due to climate change. This macro view, even if backed by solid evidence and reliable models, still leaves the local adaptation of farmers out of the question.

Tanzanians farmers and herders understand their local reality and generally have managed to use this local knowledge to feed themselves. All of Tanzania's over 100 language/cultural groups have sought food security by making use of the country's great physical and cultural diversity. Food security practices have taken many forms: land preparation sensitive to micro-climate and use of a wide variety of agro-climatically suitable staples including rice, millet, sorghum, bananas, cassava and maize. This mosaic of dominant staple food crops was prevalent up to the 1970s when it was mapped by Mascarenhas (1973). Historically, there have been localized food shortages triggered by drought, floods, insect pests and conflict. However, at no time did the whole or even a large part of the country suffer from famine because of agroclimatic diversity (Mascarenhas, 1967). Agroclimatic and ethnic diversity is such in Tanzania that local markets may draw produce from a number of zones. Such choice at local markets provides another element of food security from the household point of view.

Policy agenda

National governments, donors, and international organizations have to treat disaster risk and vulnerability as development issues, not as specialist or technical problems to be "solved" with tech fixes and clever engineering. The large unspoken development question is often, "what policies are actually blocking local initiatives and spontaneous adaptation to change?" In the end, in many cases, it is often possible to do more with less – especially if local governments have the resources necessary to join in partnerships with civil society and communities. This is certainly the case in the villages along the altitude transect in our study in northern Tanzania. Farmers are experimenting with new crops and expanding traditional irrigation works (*ndiva*). In the lowest sites Maasai are supplementing herding with irrigation farming and trade. At the highest sites, clan-based protection of sacred forests (*mbungi*) – a tradition handed down for several centuries – now may become the basis for community access to funding from the UN's REDD programme (Reduction of Emissions from Deforestation and Forest Degradation) (Bwenge *et al.*, 2012).

But will governments take decentralization seriously and pass on international resources to local authorities and communities where this sort of innovation and spontaneous adaptation is occurring? There are 198 governments on Planet Earth. Some 168 of these signed on to the Hyogo Framework for Action (HFA), the detailed work plan created at the World Conference on Disaster Reduction held in Kobe, Japan in 2005. A good deal of new legislation, money spent on training at the national level, new government departments created, and some mega-projects later, little safety has trickled down to localities (GNDR, 2009; 2012). Just two years shy of the 2015 horizon of the HFA, as well as the Millennium Development Goals (MDGs), the post-HFA and post-MDG institutional architecture at the global scale seems set to continue business as usual. Although the Rio+20's briefing papers (UNCSD, 2012: 4) include mention of the transfer of knowledge, technology and expertise to enhance capacity building for disaster risk reduction; the sharing of best practices and lessons learned; and the flow of appropriate support to and between developing countries for enhancing governance for disaster risk reduction and awareness at all levels, the same institutions have so far failed to coax national governments to take localities and people's own knowledge and initiatives seriously. Further they have failed to get agreement on transformative measures in Rio, and failed to hold CO_2 emissions in check. There is new language, to be sure. "Resilience" is the new buzzword, and concepts such as inclusion, empowerment, and social protection are evident throughout the documents (UNISDR, 2012; UNCSD, 2012). Yet one grows weary of the same international bureaucracies repeating the same inspiring words. "Community participation" in disaster reduction has been a recurrent theme in the international discourse since 1994, when UNICEF, IFRC and civil society demanded a community perspective half way through an overwhelmingly engineering-dominated UNISDR.

Action agenda

International non-governmental organizations should lobby for the policies mentioned, and performance of donors, governments, and multi-lateral organizations should be monitored and measured not by the volume of the rhetoric, inputs at the national level and deliverables, but by outputs at the local scale, where people live. This is the minimum agenda for action. However, it will only ameliorate the situation, not deal with the root causes of vulnerability. For that a global campaign for transformation is required. Governments are simply too deeply interpenetrated by corporations and too committed to the economic growth capitalism demands by its very nature (Kovel, 2007; Swartzman, 2009, 2012).

The transformation required must be comprehensive, embracing production, consumption and, above all, governance. Wainwright and Mann (2012) describe what they call Climate Leviathan, a climate change control regime in the near future that would still be dominated by corporate interests although turned over by governments to an international entity. Clearly this would not constitute "transformation" in any sense and likely do little to address climate change. Going even further than Wainwright and Mann, Bigger states (2012: 1):

> The ecological state of emergency is here and the imperative to envision and enact alternatives has been laid bare by Wainwright and Mann. The authors show that Climate Leviathan … [is] unwilling and incapable of restructuring global political economy to avert catastrophe. So the question is what institutions or other means of social organization are best positioned to intervene in the geography of atmospheric carbon and provide hope for transformation of the magnitude necessary, not only to ensure the continued survival of recognizable socio-nature, but to *improve the quality of life of the greatest number of people*? The need for improvements in material standard of living for *vast swaths of the world* beyond climate stabilization alone is not incidental. Successful radical political movements typically have in common at least the *promise of tangible material benefits for the populations that make the radical change* [Emphasis added].

Conclusion

Thinking about the farmers and herders I have been visiting over the past five years and their difficulty in making sense out of the abstract, neo-Swahili terms for climate change and adaptation, they clearly belong to the "vast swath" of humanity that Bigger (2012) mentions. These Tanzanians understand climate change experientially, not abstractly, and they see and speak about it as one change among many. They also report that despite changes in the climate, their lives have improved over the past 10–30 years. Transformation for these rural, risk-adverse people is most likely to be acceptable if (a) it is incremental and (b) it brings material benefits with each increment.

While most climate change adaptation efforts so far have been sectoral, incremental and reactive, transformation should, in principle, be comprehensive, systemic and proactive (O'Brien, 2012). Frankly, that's a very high goal. Since the stakes are so high for people and societies, it seems that transformation is possible, even if seemingly as an incremental process (O'Neill and Handmer, 2012) as long as governments are badgered by their citizens and civil society organizations to bridge sectoral "silos" and all stakeholders work hard proactively across the full range of changes, environments and securities with which this chapter began (Pelling, 2010; Wisner *et al.*, 2012b). In 2010 Bolivia hosted the World People's Conference on Climate Change and the Rights of Mother Earth, and this platform has continued to inform and mobilize, producing damning criticism of the weak results coming from the Durban COP and Rio +20 (World People's Conference, 2012). In a variety of ways the Occupy movement, creation of a Japanese Green Party and German popular pressure to phase out its nuclear power stations and fully support renewable energy sources are also positive signs. Specifically dealing with natural hazards and climate change, the Global Network of Civil Society Organisations for Disaster Reduction (GNDR) now includes seventy countries and has become a major voice for local initiatives based on local knowledge creation and South-South cooperation (GNDR, 2012). There are thus already signs that transformation is possible, but whether it is realized at the scales needed for disaster risk and vulnerability reduction will depend on whether dominant thinking can change.

Note

1 The author has been working with farmers, herders, local officials and other outside specialists in an NSF funded study since 2007, the Local Knowledge and Climate Change Adaptation Project (LKCCAP). This chapter incorporates data obtained in work supported by the National Science Foundation under Grant No. 0921952. Any opinions, findings, and conclusions or recommendations expressed in this material are those of the author and do not necessarily reflect the views of the National Science Foundation.

References

Adger, N. (2006) 'Vulnerability', *Global Environmental Change*, 16: 268–281.

Anderson, M. and Woodrow, P. (1989) *Rising from the Ashes*, Boulder, CO: Lynne Rienner.

Arndt, C., Farmer, W., Strzepek, K. and Thurlow, K. (2012) *Climate Change, Agriculture and Food Security in Tanzania*, Helsinki: UN-WIDER.

Arnstein, S. (1969) 'A ladder of citizen participation', *Journal of the American Planning Association*, 35 (4): 216–224.

Bigger (2012) *Responses to Wainwright and Mann's Climate Leviathan*, July 2012 Red Terror on the Atmosphere, On-line Antipode Symposium. Online. Available HTTP: http://radicalantipode.files.wordpress.com/2012/07/bigger_climate-leviathan.pdf (accessed 6 August 2012).

Bird, K., McKay, A. and Shinyekwa, I. (2010) *Isolation and Poverty: The Relationship between Spatially Differentiated Access to Goods and Services and Poverty*, ODI Working Paper 322/CPRC Working Paper 162. London: ODI. Online. Available HTTP: http://www.odi.org.uk/resources/docs/5516.pdf (accessed 8 July 2012).

Blaikie, P. and Brookfield, H. (1987) *Land Degradation and Society*, London: Routledge and Kegan Paul.

Boulding, K. (1985) *The World as a Total System*, Beverley Hills, CA: Sage.

Büscher, C. and Mascareño, A. (2012) 'Mechanisms of systematic risk production' in D. Heinrichs, K. Krellenberg, B. Hansjurgens and F. Martinez (eds) *Risk Habitat Megacity*, Heidelberg: Springer: 39–63.

Bwenge, C., Smucker, T., Mascarenhas, A., Munishi, P., Wangui, E.,Wisner, B., Weiner, D. and Sinha, G. (2012) 'A politicised cloudscape: disconnected discourses, differentiated livelihoods and climate change adaptation in Tanzania', *Geographical Journal*, under review.

Cannon, T. (2000) 'Vulnerability analysis and disasters', in D. Parker (ed.) *Floods*, London: Routledge: 57–74.

Chambers, R. (1983) [1995 reprint] *Rural Development: Putting the Last First*, New York: Prentice-Hall.

Clark, D. (2005) 'The capability approach: Its development, critiques and recent advances', GPRG-WPS-032, Global Poverty Research Group. Online. Available HTTP: http://www.gprg.org/pubs/workingpapers/pdfs/gprg-wps-032.pdf (accessed 28 March 2013).

Demeritt, D. (2002) 'What is the "Social construction of nature"?' *Progress in Human Geography*, 26 (6): 767–790.

Doocy, S., Rofi A., Moodie, C., Spring, E., Bradley, S., Burnham, G. and Courtland Robinson, C. (2007) 'Tsunami mortality in Aceh Province, Indonesia', *Bulletin of the World Health Organization*, 85 (4): 273–278.

Enarson, E. and Morrow, B. (1998) *The Gendered Terrain of Disaster: Through Women's Eyes*, New York: Praeger.

Eriksen, S., Brown, K. and Kelly, M. (2005) 'The dynamics of vulnerability: locating and coping strategies in Kenya and Tanzania', *The Geographical Journal*, 171 (4): 287–305.

Fordham, M. (2012) 'Gender, sexuality and disaster', in B. Wisner, J.C. Gaillard and I. Kelman (eds) *The Routledge Handbook of Hazards and Disaster Risk Reduction*, London: Routledge: 424–435.

Gaillard, J. (2012) 'Caste, ethnicity, religious affiliation and disaster', in B. Wisner, J.C. Gaillard and I. Kelman (eds) *The Routledge Handbook of Hazards and Disaster Risk Reduction*, London: Routledge: 459–470.

Gaillard, J. and Cadag, J. (2009) 'From marginality to further marginalization: experiences from the victims of the July 2000 Payatas trashslide in the Philippines', *Jamba: Journal of Disaster Risk Studies*, 2 (3) (December): 197–215.

Gaillard, J.-C., Wisner, B., Benouar, D., Cannon, T., Créton-Cazanave, L., Dekens, J., Fordham, M., Gilbert, C., Hewitt, K., Kelman, I., Morin, J., N'Diaye, A., O'Keefe, P., Oliver-Smith, A., Quesada, C., Revet, S., Sudmeier-Rieux, K., Texier, P. and Vallette, C. (2010) 'Alternatives pour une réduction durable des risques de catastrophe', *Human Geography*, 3 (1): 66–88.

GNDR (2009) *Clouds but Little Rain: Views from the Front Line 2009*, London: Global Network for Disaster Reduction. Online. Available HTTP: http://www.preventionweb.net/english/professional/publications/v.php?id=9822 (accessed 7 July 2012).

GNDR (2011) *If We Do Not Join Hands: Views from the Front Line 2011*, London: Global Network for Disaster Reduction. Online. Available HTTP: http://www.globalnetwork-dr.org/images/documents/vfl2011_report/VFL2011_Core_Report_en.pdf (accessed 7 July 2012).

GNDR (2012) Global Network home page. Online. Available HTTP: http://www.globalnetwork-dr.org/ (accessed 1 November 2012).

Goldman, M. (2011) 'Strangers in their own land: Maasai and wildlife conservation in northern Tanzania', *Conservation and Society*, 9 (1): 65–79.

Haghebaert, B., Wisner, B., Schaerer, M. and Arnold, M. (2008) 'ProVention Consortium Toolkit: community-based methods for vulnerability/capacity assessment (January – May 2005; secondphase January 2006–2008)', Geneva: ProVention Consortium. Online. Available HTTP: http://www.proventionconsortium.org/?pageid=39.

Hart, R. (1992) 'Children's participation from tokenism to citizenship, Florence, Italy: UNICEF Innocenti Research Centre'. Online. Available HTTP: http://www.unicef-irc.org/publications/pdf/childrens_participation.pdf (accessed 5 December 2012).

Hartmann, B. (1998) 'Environment, population and security: a new trinity', *Environment and Urbanization*, 10 (2): 116–127.

Heinrichs, D., Aggarwal, R., Barton, J., Bharucha, E., Butsch, C., Fragkias, M., Johnston, P., Kraas, F., Krellenberg, K., Lampis, A., Ling, O.G. and Vogel, J. (2011) 'Adapting cities to climate change: opportunities and constraints', in D. Hoornweg, M. Freire, L. Sugar, C. Anderson, P. Bhada, C. L. Trejos, R. Dave, and M. Lee (eds) *Cities and Climate Change : Responding to an Urgent Agenda*, Washington, DC: World Bank: 193–224.

Igoe, J. (2003) 'Scaling up civil society: donor money, NGOs and the pastoralist land rights movement in Tanzania', *Development and Change*, 34 (5): 863–885.

IPCC (2012) *Managing the Risks of Extreme Events and Disasters to Advance Climate Change Adaptation*, A Special Report of Working Groups I and II of the Intergovernmental Panel on Climate Change, C.B. Field, , V. Barros, T.F. Stocker, D. Qin, D.J. Dokken, K.L. Ebi, M.D. Mastrandrea, K.J. Mach, G.-K. Plattner, S.K. Allen, M. Tignor, and P.M. Midgley (eds), Cambridge, UK: Cambridge University Press.

IPU (2010) *Disaster Risk Reduction: An Instrument for Achieving the Millennium Development Goals*, Geneva: Inter-Parliamentary Union and UNISDR.

Kovel, J. (2007) *The Enemy of Nature: The End of Capitalism or the End of the World?* London: Radical International Publishing.

Mascarenhas, A. (1967) *Famines and Food shortages in Tanzania*, presentation at the East African Social Sciences Conference, Nairobi: University of Nairobi.

Mascarenhas, A. (ed.) (1973) 'Special number: studies in famines and food shortages', *Journal of the Geographical Association of Tanzania* 8 (October), Dar es Salaam: Geography Department, University of Dar es Salaam.

Mascarenhas, A. and Wisner, B. (2012) 'Politics: power and disasters', in B. Wisner, J.C. Gaillard and I. Kelman (eds) *The Routledge Handbook of Hazards and Disaster Risk Reduction*, London: Routledge: 48–60.

Maskrey, A., Lavell, A., Oliver-Smith, A., Mansilla, E., Ramirez, F., Wilches-Chaux, G., Gellert, G., Herzer, H. and Argüello, M. (1997) *Miami Declaration on Disaster Reduction and Sustainable Development,* Hemispheric Congress on Disaster Reduction and Sustainable Development, 1997, Miami: Florida International University, International Hurricane Center and La Red.

Mercer, J. (2012) 'Knowledge and disaster risk reduction', in B. Wisner, J.C. Gaillard and I. Kelman (eds) *The Routledge Handbook of Hazards and Disaster Risk Reduction*, London: Routledge: 97–108.

Naudé, W., Santos-Paulino, A. and McGillivray, M. (eds) (2009) *Vulnerability in Developing Countries*, Tokyo: UNU Press.

NCD (2009) *Effective Emergency Management: Making Improvements for Communities and People with Disabilities*, Washington, DC: National Council on Disability. Online. Available HTTP: http://www.ncd.gov/newsroom/publications/2009/NCD_EmergencyManagement_HTML/EffectiveEmergencyManagement.html#toc (accessed 5 August 2012).

Ngo, E. (2012) 'Elderly people and disaster', in B. Wisner, J.C. Gaillard and I. Kelman (eds) *The Routledge Handbook of Hazards and Disaster Risk Reduction*, London: Routledge: 447–458.

O'Brien, K. (2012) 'Global environmental change II: from adaptation to deliberate transformation', *Progress in Human Geography*, 36 (5): 667–676.

O'Neill, S. and Handmer, J. (2012) 'Responding to bushfire risk: the need for transformative adaptation', *Environmental Research Letters*, 7 (1) 1–7.

Paavola, J. (2004) *Vulnerability to Climate Change in Tanzania: Sources, Substance and Solution*, Maputo: Southern Africa Vulnerability Initiative.

Paavola, J. and Adger, N. (2002) *Justice and Adaptation to Climate Change*, Tyndall Centre Working Paper No. 23 October, Norwich, University of East Anglia.

Pearce, F. (2012) *The Landgrabbers: The New Fight over Who Owns the Earth*, London: Transworld/Eden Project Books.

Pelling, M. (2010) *Adaptation to Climate Change: From Resilience to Transformation*, London: Routledge.

Reuters AlertNet (2009) Crisis Centre, Women. Online. Available HTTP: http://www.trust.org/alertnet/crisis-centre/subtopic/women (accessed 26 February 2013).

Richards, P. (2009) Keynote address, World Conference on Humanitarian Studies, University of Groningen, Netherlands, 8 February.

Rosentrater, L. (2010) 'Representing and using scenarios for responding to climate change', *WIRES Climate Change*, 1: 253–259.

Scott, J. (1998) *Seeing Like a State*, New Haven, CT: Yale University Press.

Sen, A. (1999) *Development as Freedom*, Oxford: Oxford University Press.

Smucker, T. and Wisner, B. (2008) 'Changing household responses to drought in Tharaka, Kenya: persistence, change, and challenge', *Disasters* 32 (2): 190–215.

Susman, P., O'Keefe, P. and Wisner, B. (1983) 'Global disasters: a radical perspective', in K. Hewitt (ed.) *Interpretations of Calamity*, Boston, MA: Allen and Unwin: 263–83.

Swartzman, D. (2009) 'Ecosocialism or ecocatastrophe?', *Capitalism Nature Socialism,* 20 (1): 6–33.

Swartzman, D. (2012) Commentary on Wainwright and Mann, On-line Antipode Symposium. Online. Available HTTP: http://antipodefoundation.org/2012/07/19/symposium-on-geoff-mann-and-joel-wainwrights-climate-leviathan/ (accessed 9 July 2012).

Tiffen, M., Mortimore, M. and Gichuki, F. (1994) *More People, Less Erosion: Environmental Recovery in Kenya*, London: Wiley.

UNCSD (2012) A Guidebook to the Green Economy, available http://www.uncsd2012.org/content/documents/528Green%20Economy%20Guidebook_100912_FINAL.pdf (accessed 26 February 2013).

UNFCCC (2012) *National Adaptation Programmes of Action*, United Nations Framework Convention on Climate Change. Online. Available HTTP: http://unfccc.int/national_reports/napa/items/2719.php (accessed 9 August 2012).

UNISDR (2005) *Hyogo Framework for Action*. Online. Available HTTP: http://www.unisdr.org/we/inform/publications/1037 (accessed 5 July 2012).

UNISDR (2009) *Global Risk Assessment 2009,* Geneva: UNISDR. Online. Available HTTP: http://www.preventionweb.net/english/hyogo/gar/report/index.php?id=9413 (accessed 8 July 2012).

UNISDR (2011) *Global Risk Assessment 2011*, Geneva: UNISDR. Online. Available HTTP: http://www.preventionweb.net/english/hyogo/gar/2011/en/home/index.html (accessed 8 July 2012).

UNISDR (2012) Summary report: post-HFA mayors' and local government consultation, Bonn, Germany, 14 May. Online. Available HTTP: http://www.preventionweb.net/files/27499_summaryreportposthfamayorsandl.pdf (accessed 9 August 2012).

United Republic of Tanzania (2007) *National Adaptation Programme of Action* (NAPA), Dar es Salaam, Vice President's Office Division of Environment.

Wainwright, J. and Mann, G. (2012) 'Climate leviathan', *Antipode*, 45 (1): 1–22.

Wangui, E., Smucker, T., Wisner, B., Lovell, E., Mascarenhas, A., Solomon, M., Weiner, D., Munna, A., Sinha, G., Bwenge, C., Meena, H. and Munishi, P. (2012) 'Spontaneously integrated development, risk management and community-based climate change adaptation in a Mountain-Plains System in Northern Tanzania', *Journal of Alpine Research, 100* (1).

White, G. (1945) *Human Adjustment to Floods*, Chicago: University of Chicago Press.

Wisner, B. (1978) *The human ecology of drought in Eastern Kenya*, Unpublished PhD dissertation, Worcester, MA: Clark University, Graduate School of Geography.

Wisner, B. (1988) *Power and Need in Africa*, London: Earthscan.

Wisner, B. (1993) 'Disaster vulnerability: scale, power, and daily life', *GeoJournal*, 30 (2): 127–140.

Wisner, B. (2003) 'Turning knowledge into timely and appropriate action', in O. Cardona (director) *Information and Indicators Program for Disaster Risk Management,* Manizales, Colombia: National University-IDEA/IADB/ECLAC.

Wisner, B. (2004) 'Assessment of capability and vulnerability', in G. Bankoff, G. Frerks and T. Hilhorst (eds) *Vulnerability: Disasters, Development and People*, London: Earthscan: 183–193.

Wisner, B. (2009) 'Local knowledge and disaster risk reduction', Keynote, Side Meeting on Indigenous Knowledge, Global Platform for Disaster Reduction Geneva, 17 June.

Wisner, B. (2010) 'Climate change and cultural diversity', *International Social Science Journal,* 61: 131–140.

Wisner, B. and Uitto, J. (2009) 'Life on the edge: urban social vulnerability and decentralized, citizen-based disaster risk reduction in four large cities of the Pacific Rim', in H. Brauch et al. (eds) *Facing Global Environmental Change*, Berlin: Springer Verlag: 217–234.

Wisner, B. and Walker, P. (2005) *Beyond Kobe: A Proactive Look at the World Conference on Disaster Reduction, Kobe, Japan, 18–22 January, 2005,* Commissioned by the Swiss Corporation for Development. Online. Available HTTP: http://www.preventionweb.net/english/professional/publications/v.php?id=2609 (accessed 3 December 2012).

Wisner, B., Blaikie, P., Cannon, T. and Davis, I. (2004) *At Risk: Natural Hazards, People's Vulnerability and Disasters*, 2nd edition, London: Routledge.

Wisner, B., Gaillard, J.C. and Kelman, I. (eds) (2012a) *The Routledge Handbook of Hazards and Disaster Risk Reduction*, London: Routledge.

Wisner, B., Gaillard, J.C. and Kelman, I. (eds) (2012b) Framing disaster: theories and stories seeking to understand hazards, vulnerability and risk', in B. Wisner, J.C. Gaillard and I. Kelman (eds) *The Routledge Handbook of Hazards and Disaster Risk Reduction*, London: Routledge: 18–34.

Wisner, B., Gaillard, J.C. and Kelman, I. (2012c) 'Challenging risk: has the left foot stepped forward?' in B. Wisner, J.C. Gaillard and I. Kelman (eds) *The Routledge Handbook of Hazards and Disaster Risk Reduction*, London: Routledge: 789–793

World People's Conference on Climate Change and the Rights of Mother Earth (2012). Online. Available HTTP: http://pwccc.wordpress.com/ (assessed 8 August 2012).

Empowerment

24 Individual and community empowerment for human security

Susanne C. Moser

Introduction

Human security in the face of global environmental change can be considered

> a state that is achieved when and where individuals and communities have the options necessary to end, mitigate or adapt to threats to their human, environmental and social rights; have the capacity and freedom to exercise these options; and actively participate in pursuing them.
>
> (O'Brien *et al.*, 2008: 6)

This notion involves building on people's strengths, aspirations, capacities and capabilities to respond to threats to human, environmental and social rights by creating the political, social, environmental, economic, military and cultural systems that together give people the building blocks of survival, livelihood and dignity (Barnett *et al.*, 2010; Commission on Human Security, 2003). Definitions of human security thus bear significant resemblance to, and sometimes overlap with definitions of "empowerment." Intuitively, one might expect that empowerment is an essential *means* to human security as the *end*, in the same way that development, poverty eradication, greater disaster resilience and adaptive capacity are necessary means to support and achieve human security. In fact, in this book, empowerment is framed as a "break-through condition" for human security.

A vast and diverse multidisciplinary body of literature on empowerment warrants a careful look at its relationship to human security. There are many different ways in which empowerment can be encouraged, supported, and brought about, different forms and degrees in which it manifests in community actions, and different ways in which it is constituted within the community, at the level of individuals, and in relationship to outside forces and networks. This chapter aims to advance a more critical and differentiated view on empowerment as a "break-through condition" for human security through a selective survey of and commentary on some of the pertinent literature on empowerment. It will define and delineate the multiple meanings of this concept, explore common traits of empowerment, and discuss its causes and the ways in which it can be fostered. To achieve this, it is important to place empowerment and achieving human security at the individual and community levels *in relation to* other individuals or communities elsewhere trying to achieve their own goals (security and otherwise), pointing to the challenge of identifying ways to achieve human security through actions oriented toward the common interest.

What is empowerment?

Empowerment as an intensely discussed and debated concept has been around for at least four decades. With principal roots in the liberation pedagogy of Paulo Freire (2008) and studies in gender equality, pertinent work comes from a wide range of fields and disciplines, including organizational change and management studies, psychology, governance and civic studies, philosophy, and in particular economic development and poverty discourses (see Luttrell *et al.*, 2009, for a concise and useful overview). To proponents, empowerment is an essential and subversive process necessary for lasting social transformation of existing dehumanizing, oppressive, or unjust circumstances (e.g., Friedman, 1992). Others lament that empowerment has become one of those "all-purpose terms made plastic by indiscriminate use" (Jouve, 2009: 2), resulting in considerable confusion of meaning, and maybe even rendering the concept meaningless. Yet other critics believe empowerment has become co-opted by neo-liberal agendas (Cornwall and Brock, 2005). Against the backdrop of these perspectives, ongoing scholarly debates focus on whether empowerment is a means or an end in itself, whether it is about structure or agency (or both), what dimensions constitute it and which conditions or interventions are necessary to achieve it, and consequently what would be appropriate ways to measure this enticing, but also elusive concept (Thomas and Velthouse, 1990; Zimmerman, 1995; Kabeer, 1999; Narayan, 2002; Beteta, 2006; Narayan-Parker, 2006; Drydyk, 2008). Rather than whole-heartedly embracing or rejecting the concept, this section aims to lay bare some of the dimensions of empowerment so that its relationship to human security can come into clearer focus.

At the most basic level, empowerment is about attaining the power to make choices. Thus, we must realize first that empowerment implicitly acknowledges and assumes a lack of, or differential in, power or capacity. As such, empowerment is fundamentally about human rights (Jouve, 2009). Empowerment as a relational concept reflects the fact that some individuals or social groups have been deprived and marginalized in some way or another, and can or should be mobilized to obtain that which they heretofore lacked. Only if there is a prior lack of power is empowerment even possible or necessary. Or, as Kabeer (1999: 437) puts it, "People who exercise a great deal of choice in their lives may be very *powerful*, but they are not *empowered* [...] because they were never disempowered in the first place."

The second immediate implication of even just an intuitive examination of empowerment is that it describes a change process, and more specifically, an emancipatory change process: from having less power to having more. This change affects the thoughts, beliefs, sense of self (in relationship to others and one's context), and actions of individuals and communities, but can and often must involve a change in the formal and informal institutions and resources that influence or manipulate people's beliefs and enable or constrain their actions (Conger and Kanungo, 1988; Zimmerman, 1995).

The normative intention behind empowerment, however, goes even further. It calls for more than just a change in the power to make different choices. It also calls for a change in outcomes. Differently put, it would not be enough to assume an individual or community is empowered if they just had greater capacity and freedom to determine their lives. Only if and when that enhanced capacity and freedom is used, and if and when it results in the empowered actors actually coming closer to the ways they value being and to the things they value doing, is empowerment actually achieved (Alkire, 2005; Robeyns, 2005; Drydyk, 2008). This emphasis on capabilities and achievements reflects the intrinsic and instrumental values of empowerment (Alsop *et al.*, 2007).

The final general point to raise here is the meaning of *power* at the core of empowerment. Rowlands (1997) distinguishes four types of power, each with quite different implications for the meaning of empowerment. First, empowerment could give one "*power over*." This kind of power describes the ability to influence and coerce, having the formal or informal authority over something or someone. Empowerment in this case would mean changes in the resources and power to challenge one's constraints. It also assumes power to be a finite or zero-sum resource, where changes in power imply that some gain power, while others inevitably lose it. Empowerment consequently can be expected to entail conflict, struggle, or at least opposition (Drydyk, 2008). Because it assumes existing hierarchical economic and political structures to remain unchanged (i.e., only a shift in who participates in them), this form of empowerment does not entail deep structural changes (Wong, 2003; Luttrell *et al.*, 2009). Second, having the "*power to*" is an enabling power, for example, to organize or to change existing hierarchies. Empowerment here means increased capacities, access to decision-making, or opportunities to make certain changes. This type of power can entail structural changes, but emphasizes procedural changes, such as providing services (such as education) that build strengths and abilities, or developing opportunities to participate in decision-making. The third way to think of power is as "*power with*," (i.e., a power that arises from collective action and mutual support). Empowerment in this case reflects increased solidarity, for example, in challenging norms, assumptions, social roles and so on. This conceptualization, contrary to power over, "stresses the way in which gaining power actually strengthens the power of others rather than diminishing it" (Luttrell *et al.*, 2009: 6), often critical for "creating the conditions for change and [...] reducing the costs for the individual" to step out of traditional oppressive or disempowered positions (Kabeer, 1999: 457). Finally, "*power from within*" pertains to a change in consciousness, attitude, or confidence. Empowerment in this sense might imply an increased awareness, desire, self-esteem and psychological readiness for change. It may also entail changes in perceptions about one's rights, options, and one's potential and abilities (Gutierrez *et al.*, 1995; Luttrell *et al.*, 2009).

One can easily imagine that an individual may have one or several of these forms of power, but not others. For example, it is possible for one and the same person to simultaneously have power over another person (say in the household), but not to have power with other community members to affect a change in the economic situation, nor any power from within to actually feel strong, capable and good about him/herself. Alternatively, another person may not have any power over other individuals, yet with strong self-esteem finds her/himself at a community meeting to work with others to develop the skills and capacities to improve health care for children.

These varying interpretations of power reflect the different emphases that scholars have placed on structure versus agency, capacities vs. outcomes, individual vs. collective change, empowerment's intrinsic vs. instrumental value, and its inner (e.g., psychological) and outer (e.g., institutional or resource) dimensions. Some interpretations of power are clearly complementary, maybe even mutually supportive and necessary; none is obsolete or inherently invalid, though they do not get equal weight in public projects and discussions (e.g., see the analysis of the World Bank's use of power in Wong, 2003).

Human security is intrinsically related to the components of empowerment discussed above. Table 24.1 places side by side some common definitions of human security and empowerment to illustrate the parallels. Both sets of definitions focus on humans and the dichotomous tension between them and their social, natural, political and/or economic environment (Dalby, 2009). Both address individual and collective choices; both in fact go beyond choice, emphasizing achievements; and both speak to the capacity and freedom to exercise choice, and the active participation in that choice. Importantly, however, one notes from this comparison that the definition of human security does not necessarily imply an emancipatory change process, which, as shown above, lies at the heart of empowerment. To put it differently, already powerful and newly empowered actors can achieve security, but those denied power cannot. Only for them is empowerment a break-through condition for human security. And only for them is it a double triumph, as this break-through is often achieved against the will or resistance of those already in positions of power.

Kabeer (1999), in discussing "strategic life choices," distinguishes between first-order and second-order choices. First-order choices are those that are critical for people to live the lives they want (including, presumably, fulfilling one's basic human needs and ascertaining one's safety and security, but also choice of livelihood, whether and whom to marry, and whether to have children). Second-order choices, by contrast, are less consequential ones – important for one's quality of life, but not constituting its defining parameters. Moreover, the power to make these choices (resulting from all meanings of power discussed above) is not a thing one does or does not possess, but a capacity one exercises to varying degrees, depending on the capabilities and assets available, the opportunity structure supporting or hindering engagement, and the individual's or community's perceived sense of and realized agency (i.e., making choices) in turning opportunities into achievements (Alsop *et al.*, 2007; Crocker, 2007). The degree of empowerment thus can be assessed by examining the three aspects of choice:

Table 24.1 Definitions of human security and empowerment

Human security	*Empowerment*
UNDP (1994): Human security is, "first, safety from such chronic threats as hunger, disease and repression. And second, it means protection from sudden and hurtful disruptions in the patterns of daily life – whether in homes, in jobs or in communities. Such threats can exist at all levels of national income and development" (UNDP, 1994: 23).	Drydyk (2008): "People are empowered to the extent that: (a) they exercise enhanced decision-making and influence over strategic life-choices and barriers to agency and well-being freedom; (b) their capacity for such decision-making and influence has also been enhanced; and (c) they are capable of making these gains prevail, given (i) the capabilities they have and assets they control, individually or collectively, and (ii) the opportunity structure in which they act" (Drydyk, 2008: 231).
Narayan (2002): The *Empowerment and Poverty Reduction: A Sourcebook* defines empowerment as "the expansion of freedom of choice and action ... [or:] the expansion of assets and capabilities of poor people to participate in, negotiate with, influence, control, and hold accountable institutions that affect their lives" (Narayan, 2002: 14). O'Brien et al (2008): "A state that is achieved when and where individuals and communities have the options necessary to end, mitigate or adapt to threats to their human, environmental and social rights; have the capacity and freedom to exercise these options; and actively participate in pursuing these options" (O'Brien *et al.*, 2008: 6).	Kabeer (1999): "Empowerment is the expansion of people's ability to make strategic life choices in a context where this ability was previously denied to them. The ability to exercise choice can be thought of in terms of three interrelated dimension: a) Resources – material, human, and social; including actual allocations and future claims and expectations of having the necessary means and choices. b) Agency – the ability to define one's goals and act upon them; including decisions, actions, bargaining, negotiation, manipulation, resistance, reflection, etc. c) Achievements – the particular realized ways of being and doing people value* (adopted from Kabeer, 1999: 437–438). * A. Sen, father of the human capabilities approach underlying much empowerment thinking, combined (a) and (b) under "capabilities" and (c) under "functioning achievements" (Sen, 1985).

- Does an opportunity to make a choice actually exist? (existence of choice, or resources)
- Does a person or group actually use the opportunity to choose? (use of choice, or agency)
- Does the choice result in the desired outcome? (achievement of the desired results due the choices made)

(adapted from Alsop *et al.*, 2007: ix)

A careful assessment of which dimensions of choice are wanting can provide useful insights then – particularly on those first-order choices – into whether or not empowerment is a break-through condition for individual and/or collective human security, and what the possible interventions might be to increase the degree of empowerment and security.

The sustaining forces of dis-empowerment

Speaking rather generically about power and empowerment eventually begs the question: what is at the root of disempowerment? The literature covers a wide and diverse terrain of forms of oppression, exploitation and disrespect – a well-known list ranging from colonialism to capitalism, globalization, imperialism, clientelism and paternalism, from sexism and patriarchy to racism, class and caste systems, state or interpersonal violence, war and related traumatization, other forms of abuse and victimization, and more subtle, but no less impactful legacies of upbringing and family dynamics. While situational triggers for overt violence and deprivation may vary widely (e.g., a change in government, a regional conflict, a natural disaster), sustained forms of oppression and disempowerment typically are supported and further maintained by chronic conditions such as hunger, poverty, land and resource exploitation and degradation, lack (or lack of a particular kind) of education and institutional factors such as landlessness, women's rights, particular market policies, as well as social and cultural norms.

This list of examples of immediate and chronic underlying causes makes clear that the vicious cycle of disempowerment and oppression begets further disempowerment and oppression until and unless it is broken by people who are empowering themselves, maybe assisted by those who are working in solidarity with them. Whatever different theorists' philosophical commitment (e.g., feminist, Marxist, humanist, critical), the common thread among all those who have criticized and analyzed oppressive social systems is their finding that such systems dehumanize people. Moreover, as this book and this chapter in particular argue, oppressive disempowering systems make people less secure. They rob individuals of the most basic human rights and freedoms, and they systematically undermine their ability to acquire the education, skills, and means to fully engage their personal and collective power and influence when dealing with others or with institutions on matters related to their lives and livelihoods, for example, finding the root causes of oppression in exploitative economic development:

> No matter how dynamic, an economic system that has little or no use for better than half of the world's population can and must be radically transformed. Broadly speaking, the objective of an alternative development is to humanize a system that has shut them out, and to accomplish this through forms of everyday resistance and political struggle that insist on the rights of the excluded population as human beings, as citizens, and as persons intent on realizing their loving and creative powers within. Its central objective is their inclusion in a restructured system that does not make them redundant.
>
> (Friedman, 1992: 13)

Individuals stuck in a system that deprives them of their power are Objects being acted upon by visible or invisible oppressors, their circumstances, externally imposed changes, natural or technological powers; they are not Subjects or creators and shapers of their own lives (Freire, 2008). What's more, such disempowered individuals are prisoners of a "circle of certainty," in which they see neither the causes of their disempowerment, nor the possibilities of an escape (Freire, 2008: 39). The space for personal will and individual thought, much less choice and action, is narrowly circumscribed as that prescribed and permitted by the power holders. The dominant consciousness does not allow, or rather, makes it impossible for such individuals to see the possibility (and necessity) of transformation. The truly disempowered thus have become one with what Freire (2008) called a "Culture of Silence" in which there is no consciousness of, and no permission to name or question, the deeply engrained existence of injustice, exploitation and human insecurity.

Importantly, as feminism, liberation psychologies and theologies, political ecology, as well as liberation pedagogy have argued, all oppression has an external and an internal dimension (e.g., Conger and Kanungo, 1988; Thomas and Velthouse, 1990; Prilleltensky and Gonick, 1994; Zimmerman, 1995; Peet and Watts, 1996; Gutierrez, 1998; Bryant, 2001; Forsyth, 2002; Schroeder *et al.*, 2006; Freire, 2008; Watkins and Shulman, 2008). The importance of this finding cannot be overstated. As Prilleltensky and Gonick (1994: 149) argue:

> It is extremely difficult to rise above the collective disempowerment of one's reference community to enjoy these rights [of self-determination, distributive justice, and collaborative and democratic participation], because the individual's ability to claim self-determination, justice, and a voice is tied to the group's capacity to secure these values.

If external forces deprive the powerless of the material access and resources to be active shapers of their lives, the internal disempowerment stems from the almost always present internalization of inferiority and resulting self-rejection, hopelessness and psychological immobility (see Watkins and Shulman, 2008 for a careful discussion of the deep interrelatedness of external and internal oppression for the perpetrator, the victims, and the bystanders). It is from this recognition that we can now turn to the question where emancipatory change may come from.

Empowerment from within and without

From an interdisciplinary perspective, it is somewhat surprising that there is much debate about whether empowerment comes from within or from without. Yet the dichotomous discussion persists and is in fact quite prevalent in a number of pertinent literatures (e.g., development, poverty and community psychology studies) (Gruber and Trickett, 1987; Davis, 2007). Whether writers privilege initiation of change from within or from without, many warn against a simplistic endogenous or exogenous determinism. As early as the 1970s, Freire recognized the need to address both dimensions:

> The oppressed suffer from the duality which has established itself in their innermost being. They discover that without freedom they cannot exist authentically. Yet, although they desire authentic existence, they fear it. They are at one and the same time themselves and the oppressors whose consciousness they have internalized. The conflict lies in the choice between being wholly themselves or being divided; between ejecting the oppressor within or not ejecting them; between human solidarity or alienation; between following prescriptions or having choices; between being spectators or actors; between acting or having the illusion of acting through the actions of the oppressors; between speaking out or being silent, castrated in their power to create and re-create, in their power to transform the world.
>
> (Freire, 2008: 48)

It is from this recognition of internalized oppression that it becomes clear that no attempt at empowerment can succeed if the deeply absorbed and personalized oppression and dehumanization is not also and directly addressed. Individuals have to become Subjects of their own lives; they have to free themselves from being Objects of others determining their lives. Differently put, the principal goal of individual, internal empowerment is to find one's voice in the Culture of Silence (Drydyk, 2008; Freire, 2008; Maton, 2008).

If this form of endogenous empowerment or liberation is to succeed, one must acknowledge the need for, and possibilities of, interrelated economic, political, sociocultural, spiritual and psychological transformation. This holistic perspective, as Watkins and Shulman (2008: 46) remind us, "helps us resist thinking that one could be psychologically liberated or individuated[1] while economically or culturally enslaved." Thus, changes at the collective, contextual, structural level are also needed to ultimately ensure the empowerment of individuals. Rowlands (1997), for example, challenges researchers and practitioners involved in development initiatives to not make "unrealistic claims about the positive changes which can result from their [empowerment] work. [She] also questions how far genuine empowerment is possible for women without major changes in social attitudes."

Yet how does an individual ever create enough space to consider his or her situation and what to do about it in a context that demands unwavering attention to the needs of day-to-day survival, for meeting all but the most basic needs? Alternatively one may ask, how can change in the contextual conditions come about that would allow an individual to even contemplate a better, more secure, more dignified life? This tension between individual agency and broader economic, political and social structures may always persist and the "chicken-or-egg" question forever remain unanswered, were it not for the fact that empowerment is not an all-or-nothing state of affairs. Rather, empowerment is at all times and in all places realized to varying degrees by different people. Thus, even in places where most people are powerless and dehumanized by the political, economic and social forces of oppression, there are community settings and conditions that assist in the empowerment of individuals (Maton, 2008); there are always some individuals that are slightly more empowered, that have already achieved some degree – as Freire would say – of conscientization.[2] There are always some who are somewhat less under the spell of the myth created by the power holders of what the social order ought to be, who have more of a voice already, who have access to some useful resources, and who have begun the hard work of decolonizing their minds and growing a critical consciousness that allows them to see that theirs is not a life conditioned by fate, but by historical circumstance that can be transformed (Freire, 2008). These individuals can become the true allies of those still less conscious and worse off.

Support from those in true solidarity with the disempowered – be it in the form of material, social, psychological, or legal assistance – thus may well be the necessary condition to create an opening from which internal and external empowerment of individuals and of the broader community can commence. In fact, it may be such true, committed solidarity – as opposed to missionary zealotry, temporary humanitarian aid, however benignly-intended cultural invasion, or economic development in the (disguised) name of further entrenchment in the structures and institutions of the power holders – that is the foundational shift required to help bring about empowerment and human security.

True solidarity then is first and foremost a matter of "power with" between those already more and those less empowered. Yet true solidarity also implies that empowerment cannot be done *for* another. While the door can be opened, the opportunity space created, ultimately the disempowered have to make an emancipatory move and take part in their own liberation from the conditions that keep them in less powerful positions. This is the moment where individuals become Subjects in their lives.

It is from this vantage point that the issue of participation has become central to the question of empowerment, development and related discourses, and thus why it is also central to the question of enhancing human security (Prestby *et al.*, 1990; Parfitt, 2004; Cornwall and Brock, 2005; Crocker, 2007). Participation is variously and ambiguously discussed as a means to empowerment and development or as an end in itself. It also has become a contested notion in no small part because of the demands it makes on those already burdened unfairly, and maybe for fear that it opens the door to those not used to wielding any power and who thus may misuse it (Cooke and Kothari, 2001; Williams, 2004; Hickey and Mohan,

2005; Kamete, 2009). This point will be addressed in the next section. Suffice it to say here that while allies can be essential in linking individuals and communities to wider networks and resources (including employment opportunities, financial resources, or information), and while they can provide social and political support in the face of resistance or opposition, true empowerment only occurs when individuals become the active and self-directed co-creators of their lives. Thus, the notion of empowerment advanced here recognizes free will, full personal capacity, and the freedom to participate in the creation of a dignified and secure life. It does not, however, impose a Western ideal of the individual, autonomous (atomized) self. How individuals define themselves as Subjects in relation to others in their varying communities is largely conditioned by personal history and culture, and as such reflects collective notions of the balance between individual autonomy and empowerment. But it also makes space for an individual's capacity to continually renegotiate his or her position with others.

Empowerment for the common good

The definitions of empowerment advanced above suggest that the achieved outcome of empowerment – namely a real and durable improvement in the life circumstances of the empowered – is a critical element and indicator of its presence. Yet it is legitimate to ask about incidences and circumstances in which successful individual or local empowerment may work against the broader common good at the interpersonal, regional or even global scale. Empowered and well-meaning action at the individual or local level does not necessarily ensure human security for others elsewhere and thus can still be counterproductive. For example, community empowerment that leads to local economic growth and greater consumption but also related waste (e.g., water, materials, or emissions of greenhouse gases) ultimately contributes to environmental degradation that undermines human security for all. Or consider an example from a developed-country context: empowerment for the "common good" here may be defined by one community as attaining the resources to build security fences, alarm systems, golf courses and swimming pools, which can easily turn into a "common bad" for another community (i.e., lack of water, pesticide pollution, lack of access to land, etc.).

These examples make apparent the double-edged normative character of empowerment: those favoring and working for empowerment clearly have a normative agenda, namely to distribute power more justly among people. They present having power – choice, capability, and freedom – as a desirable means and outcome. Yet once such power is ascertained, the second normative imperative of empowerment comes into play. If this power is to be wielded not just for personal ends, but for the common good, the use of that power becomes morally and ethically constrained. It demands that one asks: who is "we" and who is "other"? Who does one feel in communal bond with, who is one's community? How far does one's power reach, and thus what responsibility comes with this new-found power? As Jouve (2009) so aptly pointed out,

> Empowerment relies on a general hypothesis that it is possible and necessary to (re-)build local communities gathering individuals who share common interests, same identities. The empowerment strategies implemented, both in developed and developing countries, put the emphasis on this social dimension in which the building and activation of territorial and/or social identities, from which it is possible to generate collective action, are key factors of success, or failure.
>
> (Jouve, 2009: 6)

Clearly, there is always a danger of empowerment becoming hijacked by new forms of self-interest, be they personal or collective, i.e., where empowerment is for one person's or one group's narrow and immediate good at the expense of another's. Often it is neither easily determined that this is happening, nor is it necessarily intentional or conscious on the part of those who have just empowered themselves enough to act in their self-interest. Freire (2008: 45) recognized this risk, stating that, "almost always, during the initial stage of the [empowerment] struggle, the oppressed, instead of striving for liberation, tend themselves to become oppressors, or 'sub-oppressors'." Or, as Prilleltensky and Gonek (1994) noted, "[a]n empowerment agenda that seeks primarily to acquire more power for *my* group is bound to disempower other oppressed groups, thereby undermining the principle of distributive justice" (168–9). In a globalized cultural context that favors Cartesian, consumptive and individualistic beliefs and values, those empowering themselves are easily tempted to adopt the prevalent model of how to be human, powerful and free to do as one pleases from those who have previously subjugated them. Thus, the ideals of Western, modern society – reiterated and spread through the media – can easily become the uncritically adopted aspirations of the newly empowered. Watkins and Shulman (2008), however, warn against the simple exchange of material deprivation for psychic emptiness. Thus, the newly gained autonomy must be tempered with a recognition of our social and ecological interdependence, even if many don't do so. It is, indeed, a difficult dichotomy to hold. But only when we can manage to do so can "community" be understood not as a group fighting or asking for rights to be granted from another, but as a collective linked by mutual obligations (Esteva, 2006, cited in Watkins and Shulman, 2008: 48).

Empowerment of one group or individual *without* diminishing the power, humanity, dignity and integrity of another (or of non-human nature) thus emerges as a crucial normative modification to the very basic need for empowerment as a break-through condition for human security (see also Wong, 2003). Consequently, the interpretation of power as "power over" becomes a less desirable, maybe even counterproductive, form of power and should not be the (sole) goal of empowerment if it is to enhance human security as a common good for all.

This is not to presume idealistically that there are any easy answers or solutions that would allow us to achieve broadly beneficial outcomes. But the recognition of mutual obligation and interdependence (including with the natural environment), and the problematic dual ethical imperative involved in empowerment demands that we critically question any action for the alleged common good from a multi-scalar, systems perspective. It demands that empowerment first and foremost build the critical consciousness and self-affirmation that helps people resist the enticing values of those who exploited them before. It must also foster the requisite interpersonal and communication skills, and create spaces, for dialogue and negotiation of mutually beneficial or acceptable solutions. Given the inevitable blindness to some outcomes of our actions, the constant changes in the social and natural environment, and the distance between the local and other communities elsewhere, empowerment actions must also establish in individuals and groups a commitment to reflexivity and social learning to make adjustments to their choices in the future as they become necessary.

Conclusion: toward reciprocal empowerment and human security

This chapter developed an understanding of empowerment as an emancipatory change process, in which those previously deprived of power attain the opportunities, capabilities and freedom to make strategic life choices that positively improve their lives and livelihoods, and thus their human security. In principle, the powers that are being gained in the process can be power over, power to, power with, and power-from-within. In recognizing that empowerment stands in the ever-present tension between structure and agency, it has become clear that true solidarity between the powerless and the already-more empowered – that is a "power with" – may well be the foundational shift required to enable further empowerment. Moreover, the danger of uncritically using one's power and adopting the very values and action options prescribed by the exploitative culture once individuals or communities obtain a certain degree of empowerment brings to the fore the necessary tempering of power with a recognition of mutual interdependence and obligation. The relational aspects of power and empowerment thus privilege forms and uses of power that do not diminish the humanity and integrity of others.

The struggle for empowerment, as depicted here, is the path from silence to finding or restoring voice, from talk-back to dialogue, from compliance to disruption, from resisting another's vision to dreaming one's own, from deconstruction and critique of the dominant paradigm to imagining, building, and celebrating something new. This work is rarely clean and it is never easy. It may occasionally involve overt struggle in the streets, fields or halls of power, but it always requires the difficult psychological work of freeing oneself from internalized oppression and dehumanization. In this sense, true empowerment is deeply radical (i.e., it has to address the root of the matter to succeed). Ultimately, the most radical aspect of empowerment may be that it cannot be achieved alone or forever. In the context of ever-changing social, economic, and environmental relationships, reciprocal empowerment is not an end that can be achieved once and for all but a continual struggle among individuals and communities. If empowerment can begin from a place of "power with" among people in solidarity with one another, it must lead to the "power to" make structural changes, or else – if undertaken without a critical awareness of interdependence – may only lead to a power swap or a "power over," but not true liberation for all. Thus, the deepest task of empowerment is always to work toward "power-from-within" so as to dismantle the internal and external structures of exploitation that undermine human security.

Notes

1 Individuation refers to the process of psychological differentiation, by which a person becomes a more stable, internally integrated, mature, and self-actualized person (Sharp, 1991).

2 Conscientization, from the Portuguese word *conscientização*, denotes a state of critical consciousness individuals can attain that reflects an awareness and in-depth understanding of the world with all its social and political contradictions (Freire, 2008).

References

Alkire, S. (2005) 'Why the capability approach?' *Journal of Human Development,* 6 (1): 115–133.

Alsop, R., Dudwick N., Bertelsen M.F. and Jones, V. N. (2007) *Empowerment in Practice: Analysis and Implementation. A World Bank Learning Module,* Washington, DC: The International Bank for Reconstruction and Development, The World Bank.

Barnett, J., Matthew, R. and O'Brien, K. (2010) 'Introduction: Global environmental change and human security', Chapter 1, in R. Matthew, J. Barnett, B. McDonald and K.L. O'Brien (eds) *Global Environmental Change and Human Security,* Cambridge, MA: MIT Press.

Beteta, H.C. (2006) 'What is missing in measures of women's empowerment?', *Journal of Human Development,* 7 (2): 221–241.

Bryant, R. (2001) 'Political ecology: A critical agenda for change?' in N. Castree and B. Braun (eds) *Social Nature,* Oxford: Blackwell: 151–169.

Commission on Human Security (2003) *Human Security Now,* New York: Commission on Human Security. Online. Available HTTP: http://ochaonline.un.org/humansecurity/CHS/finalreport/English/chapter1.pdf (accessed 24 September 2012).

Conger, J.A. and Kanungo, R.N. (1988) 'The empowerment process: Integrating theory and practice', *The Academy of Management Review,* 13 (3): 471–482.

Cooke, B., and Kothari, U. (eds) (2001) *Participation: The New Tyranny?* London: Zed Books.

Cornwall, A. and Brock, K. (2005) 'What do buzzwords do for development policy? A critical look at "participation", "empowerment" and "poverty reduction"', *Third World Quarterly,* 26 (7): 1043–1060.

Crocker, D.A. (2007) 'Deliberative participation in local development', *Journal of Human Development,* 8 (3): 431–455.

Dalby, S. (2009) *Security and Environmental Change,* Cambridge: Polity.

Davis, T.W.D. (2007) 'The real world of "community empowerment" in "international development"', Paper presented at the *2007 Annual Meeting of the American Political Science Association* Chicago, IL.

Drydyk, J. (2008) 'Durable empowerment', *Journal of Global Ethics,* 4 (3): 231–245.

Forsyth, T. (2002) *Critical Political Ecology: The Politics of Environmental Science,* London: Routledge.

Freire, P. (2008) *Pedagogy of the Oppressed,* 30th anniversary edn, New York, London: Continuum.

Friedman, J. (1992) *Empowerment. The Politics of Alternative Development,* Cambridge: Blackwell.

Gruber, J. and Trickett, E.J. (1987) 'Can we empower others? The paradox of empowerment in the governing of an alternative public school', *American Journal of Community Psychology,* 15 (3): 353–371.

Gutierrez, G. (1998, 1971) *A Theology of Liberation: History, Politics, and Salvation* (15th anniversary edition with a new introduction by the author), Maryknoll, NY: Orbis Books.

Gutierrez, L.M., DeLois, K.A. and GlenMaye, L. (1995) 'Understanding empowerment practice: Building on practitioner-based knowledge', *Families in Society: The Journal of Contemporary Human Services,* 76 (9): 534–542.

Hickey, S. and Mohan, G. (eds) (2005) *Participation – From Tyranny to Transformation? Exploring New Approaches to Participation in Development,* London: Zed Books.

Jouve, B. (2009) 'Call for papers for an international conference entitled "The Empowerment of Civil Society in Urban Policies: For Whom? For What Projects?"' Lyon, France, University of Lyon.

Kabeer, N. (1999) 'Resources, agency, achievements: Reflections on the measurement of women's empowerment', *Development and Change,* 30: 435–464.

Kamete, A.Y. (2009) '"For enhanced civic participation in local governance": Calling tyranny to account in Harare', *Environment and Urbanization,* 21 (1): 59–75.

Luttrell, C., Quiroz S., Scrutton, C. and Bird, K. (2009) Understanding and operationalising empowerment. *Working Paper #308,* London: Overseas Development Institute.

Maton, K. (2008) 'Empowering community settings: Agents of individual development, community betterment, and positive social change', *American Journal of Community Psychology,* 41 (1): 4–21.

Narayan, D. (2002) *Empowerment and Poverty Reduction: A Sourcebook.* Washington, DC: World Bank.

Narayan-Parker, D. (2006). *Measuring Empowerment: Cross-Disciplinary Perspectives,* Washington, DC: World Bank.

O'Brien, K., Sygna, L., Leichenko, R., Adger, W.N., Barnett, J., Mitchell, T., Schipper, L., Tanner, T., Vogel, C. and Mortreux, C. (2008) 'Disaster risk reduction, climate change adaptation and human security: A commissioned report for the Norwegian Ministry of Foreign Affairs, *GECHS Report,* Oslo, Norway: University of Oslo.

Parfitt, T. (2004) 'The ambiguity of participation: A qualified defence of participatory development', *Third World Quarterly,* 25 (3): 537–555.

Peet, R. and Watts, M. (eds) (1996) *Liberation Ecologies: Environment, Development, Social Movements,* London: Routledge.

Prestby, J. E., Wandersman, A., Florin, P., Rich, R. and Chavis, D. (1990) 'Benefits, costs, incentive management and participation in voluntary organizations: A means to understanding and promoting empowerment', *American Journal of Community Psychology,* 18 (1): 117–149.

Prilleletensky, I. and Gonick L. S. (1994) 'The discourse of oppression in the social sciences: Past, present, and future', in E. J. Trickett, R. J. Watts and D. Birman (eds) *Human Diversity: Perspectives on People in Context,* San Francisco: Jossey-Bass Publishers, 145–177.

Robeyns, I. (2005) 'The capability approach: A theoretical survey', *Journal of Human Development,* 6 (1): 93–114.

Rowlands, J. (1997) *Questioning Empowerment: Working with Women in Honduras,* Rugby, UK: Oxfam GB and Practical Action Publishing.

Schroeder, R.A., Martin, K.S. and Albert, K.E. (2006) 'Political ecology in North America: Discovering the Third World within?', *Geoforum,* 37 (2): 163–168.

Sen, A.K. (1985) *Commodities and Capabilities,* Amsterdam: North-Holland Publishing Co.

Sharp, D. (1991) *C.G. Jung Lexicon: A Primer of Terms and Concepts,* Toronto: Inner City Books.

Thomas, K.W. and Velthouse, B.A. (1990) 'Cognitive elements of empowerment: An "interpretive" model of intrinsic task motivation', *Academy of Management Review,* 15 (4): 666–681.

United Nations Development Program (UNDP) (1994) *Human Development Report 1994,* New York: Oxford University Press.

Watkins, M. and Shulman, H. (2008) *Towards Psychologies of Liberation,* New York: Palgrave Macmillan.

Williams, G. (2004) 'Evaluating participatory development: Tyranny, power and (re)politicisation', *Third World Quarterly,* 25 (3): 557–578.

Wong, K-F. (2003) 'Empowerment as a panacea for poverty – old wine in new bottles? Reflections on the World Bank's conception of power', *Progress in Development Studies,* 3 (4): 307–322.

Zimmerman, M.A. (1995) 'Psychological empowerment: Issues and illustrations', *American Journal of Community Psychology,* 23 (5): 581–599.

25 Community empowerment

Opportunities and challenges for Bolivia's water sector

Paola Ramallo and Elena Ostrovskaya

Introduction

Large parts of the world population live in areas where the freshwater supply is not secure. A recent study found that as much as 80 percent of the world's population is exposed to high levels of threat to water security (Vörösmarty *et al.*, 2010). According to a UNICEF/WHO report (2012: 2) almost 780 million people lack access to safe drinking water and 2.5 billion – 40 percent of the world's population – have no access to improved sanitation. Adding to this grim picture is the threat of climate change and a growing population. Degradation of water bodies and changes in rainfall patterns, which affect the hydrological cycle, increase the vulnerability of regions affected by intense floods and droughts (UNESCO, 2009; Huntjens and Pahl-Wostl, 2010). Failing to ensure access to water and sanitation has immense human costs, both in social and economic terms, and therefore, water resources can drive or limit social and economic development especially in developing countries. The global crisis in water consigns large segments of humanity to lives of poverty, vulnerability and insecurity (UNDP, 2006) and has led to a crisis in water management in terms of delivery of water services in many countries, where competition and potential disagreement between two or more stakeholder groups over the use of the scarce resource is observed (Steelman and Ascher, 1997; UNDP, 2004). Ensuring access to water and sanitation for all people is not simply a question of availability of water resources, technology, and infrastructure, but also of setting priorities, tackling poverty and inequality, addressing societal power imbalances, and, above all, political will (UNESCO, 2009).

In this context, water managers are looking for ways to maintain and improve the state of water resources under a changing environment (Pahl-Wostl, 2007). So far this has been a challenging task since water management systems are characterized by complex governance and need coordination of different components of the resource, interaction of various actors at all levels, consideration of links in the water chain and establishment of complex institutional arrangements (Beierle, 1998; Meadowcroft, 2002; Butterworth *et al.*, 2010). The traditional "command-and-control" approach typically targeted at high predictability and controllability has proven to fail in times of uncertainty and increasing complexity of environmental problems and global change. In the past two decades, new and more integrated and participatory approaches to water management have been developed to overcome these shortcomings, such as Integrated Water Resources Management (IWRM) and adaptive water management (Pahl-Wostl *et al.*, 2007; Merrey, 2008). IWRM has become quite popular among international cooperation organizations and water sector professionals, and has been implemented in many countries around the globe as a main governmental strategy to manage water resources (Watson, 2004; Merrey, 2008). The Global Water Partnership (GWP) (2000: 22) defines IWRM as "a process which promotes the coordinated development and management of water, land and related resources, in order to maximize the resultant economic and social welfare in an equitable manner without compromising the sustainability of vital ecosystems". One of the main characteristics of this approach is its promise to bring real participation and transparency to the processes in the water sector and consider basins as the most suitable geographic unit for implementation (Coelho *et al.*, 2008; Merrey, 2008). It claims that participatory approaches can contribute to better informed policy, more effective water governance, and social, psychological, and political empowerment of poor and vulnerable populations that have traditionally been marginalized sectors of the population (Steelman and Ascher, 1997; Feás *et al.*, 2004; EMPOWERS Partnership, 2007).

However evidence from the ground indicates that increased participation in the water sector can have mixed results, and many considerations should be taken into account to achieve real engagement and avoid conflicts (Michener, 1998; Ramírez, 1999). To be effective, IWRM should consider the specific institutional, cultural, and socioeconomic contexts (Pahl-Wostl *et al.*, 2007; Ingram, 2008), as well as address the interests and concerns of the different stakeholders at the moment of negotiation and decision making because each one of them assigns different values to different ecosystem services and risks (Lebel *et al.*, 2006). However, this is rarely achieved in practice because power is unevenly distributed among the stakeholders involved in (water) management (Ingram, 2008). As power can be defined as the degree of control over material, human, intellectual and financial resources exercised by different sections of society (VeneKlasen and Miller, 2002), the understanding that the control of these resources becomes a source of individual and social power is crucial. In some contexts, powerful political interests attempt to discredit disadvantaged groups, making it impossible for citizens without resources or affiliation to get their voices heard, even if they represent a substantial population. Then the battle over access to resources can become a driver for marginalized groups to challenge existing system and fight for their inclusion and for bringing their needs to the decision making table.

The concept of empowerment is central to the understanding that imbalances in power relations affect people's capacity to make effective choices and benefit from poverty reduction efforts. The World Bank defines empowerment as increasing the capacity of individuals and groups to make choices and transform these choices into desired actions and outcomes (Power, Rights and Poverty, 2004).

This chapter discusses the role that conflicts over water played in the empowerment of indigenous communities in Bolivia. The chapter draws on lessons from a water project in the Ravelo River Basin that provides most of the drinking water to the city of Sucre, Bolivia's historic capital. We argue that the disputes over control and access to water resources have resulted in recent national social reforms that in turn have contributed to reshaping the relations among stakeholders in the water sector through the redistribution of power. Our case shows how the legal inclusion of the indigenous villagers has raised their consciousness, spurred them to action to claim their rights over the water resources, and finally brought about social change in the country.

Social reforms and the empowerment of indigenous people in Bolivia

Bolivia is a South American country with high levels of inequity and poverty. Almost two-thirds of its population lives under the poverty line, about 60 percent have their basic needs unsatisfied, illiteracy and child mortality rates are higher than the world's average, and ethnic, gender and regional divisions are very deep.

The Bolivian environment is now facing serious threats caused by uncontrolled exploitation of natural resources, deforestation due to the agro-industrial expansion, and water pollution from domestic and industrial sources (PROSALUS, 2009). Nevertheless, Bolivia remains a global leader in certified forestry and one of the countries with greatest water resources availability on the planet (PROSALUS, 2009).

In spite of general availability of water resources in Bolivia almost half of the territory (western and southern part) is experiencing water scarcity. There are multiple reasons for this: increased competition between water users, lack of water infrastructure, pollution of water bodies from the mining and oil industries, and climate change. The situation is especially hard for the rural population almost 80 percent of which are involved in agriculture as their main economic activity, and therefore highly dependent on water resources (Van Damme, 2002). The country is also confronted by adverse impacts from climate change with implications for water security including glacial retreat, and more frequent and intense floods (UNDP, 2004). Recently the demand for water in the country from the agricultural, domestic and industrial sectors has risen equally in the different regions giving rise to new sources of conflicts (Van Damme, 2002).

Bolivia is currently facing a transition stage after a period of social and political instability. The distribution of power favors some individuals or groups over others, creating an environment of inequity and discrimination (Boelens, 2002). Almost two-thirds of the country's population (60 percent) is composed of indigenous people of whom the majority live below the poverty line (INE, 2008). Over the years, one of the main conflict drivers for social crisis has been the constant disputes over the rights to use and benefit from natural resources for marginalized sectors of society. During the last 10 years, this once ignored majority has found new forums to express their opinions and concerns, to consolidate their efforts, and to assert their rights (Zibechi, 2005; Fuentes, 2008). The ever growing strength of the indigenous people's social movement in the country has triggered social change (Zibechi, 2005). Deep legal and institutional reforms at national level have resulted in the empowerment of the country's indigenous communities that have heavily impacted the means of their involvement into the decision-making in the water sector. Nowadays, the country is struggling for the democratization of society through the broader involvement of its citizens in the decision-making processes and social inclusion of previously ignored indigenous people.

Briefly reviewing the main events of this process of change in Bolivia, the starting point can be found in the early 2000s with continuous protests and public demonstrations of organized indigenous groups, claiming access to natural resources, especially to water and natural gas (Bustamante, 2002). After the "Water War" in 2000, in which social movements were able to overturn the privatization of the water utilities in the city of Cochabamba and contributed to the election of Evo Morales as the first indigenous president of the country in 2005, Bolivia and its profound social reforms gained international attention. The once marginalized indigenous people became important players in the national decision-making process (Crabtree 2005; Zibechi, 2005).

After these events, the water sector was used as a platform to empower indigenous people through legal and institutional reforms. In this regard, the Ministry of Water (now the Ministry of Water and Environment) was established and water was recognized as an essential element for people's wellbeing and the country's economic development. Later, the concept of integrated water resources management (IWRM) was introduced as a strategy to achieve broader stakeholder and public involvement in planning and decision-making. Specific articles on water management and indigenous rights were included in the National Constitution adopted in 2009. The constitution declares human rights to water, stipulates that all Bolivians are entitled to have access to clean drinking water and sewer systems, and prohibits privatization or granting concessions on them. It confers special rights to the local indigenous communities to use, allocate and manage their natural resources. As a result, indigenous people in the country are officially recognized as legitimate actors and have started openly expressing their willingness to participate in the decision-making.

However, the recent empowerment of local communities in Bolivia has brought not only the opportunities, but also challenges for both the communities themselves and other stakeholders. According to Bartle (2008), social and political change require corresponding change in mindsets of the entire society, and in Bolivia not all actors are prepared to embrace or acknowledge the rights of indigenous people due to the country's long history of unequal relations.

We use a case study on the construction of a dam in the Ravelo River Basin to supply water to Sucre and the conflict around it in order to analyze how the recent remarkable political and legal reforms at the national level have empowered marginalized local communities, driven their inclusion into the negotiation process, and provoked social change leading to the their recognition as the equal stakeholders in the water decision-making at the basin level.

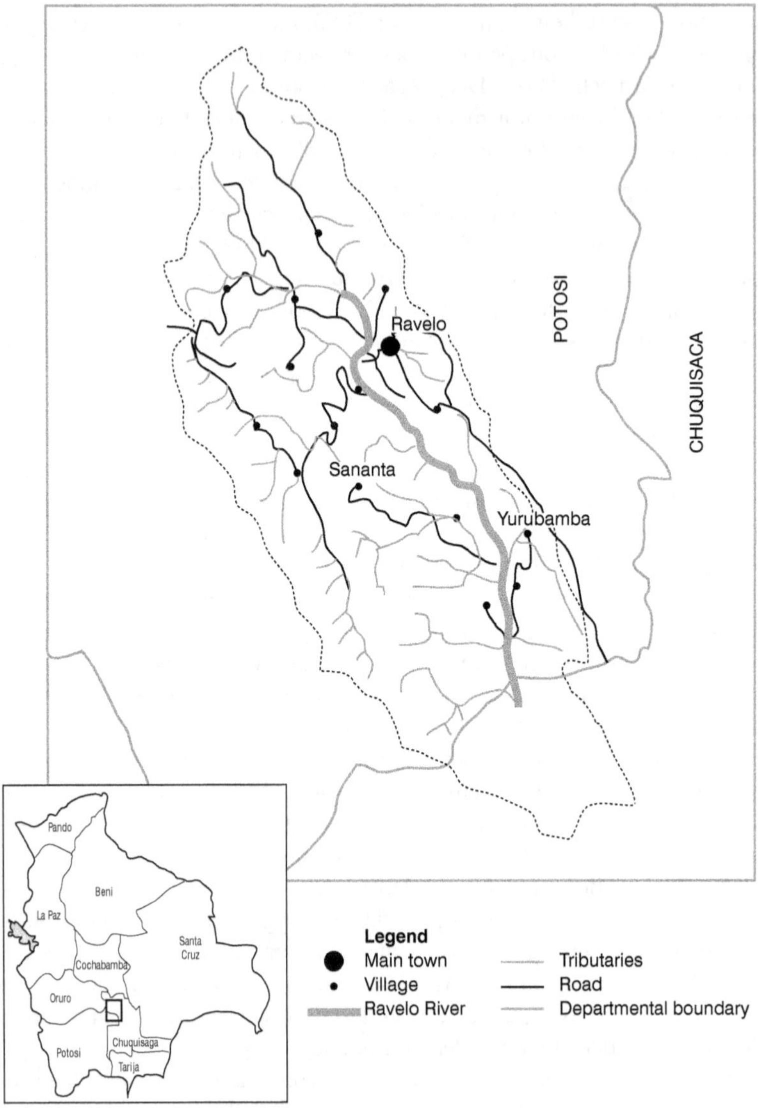

Figure 25.1 Ravelo River Basin, Bolivia. Source: COANDINA (2009)

The Ravelo River Basin case study

The Ravelo River Basin is situated in the south east of Bolivia, and is shared by two neighboring municipalities, Ravelo and Sucre, located in the Departments of Potosí and Chuquisaca respectively (Figure 25.1). The municipality of Ravelo (Figure 25.2), where the villages of Sasanta and Yurubamba are located, has a population of 20,536 habitants of which, 100 percent live in rural areas, 68

Figure 25.2 Summer landscape in the Municipality of Ravelo. Photo by Paola Ramallo, 2009

percent are below the extreme poverty line and 78 to 93 percent of the inhabitants have no access to safe drinking water (INE, 2008).

The municipality of Sucre (Figure 25.3), located in the Department of Chuquisaca, with a population that increased to 230,000 habitants, 90 percent of whom live in urban areas and 27 percent are below the poverty line (HAMS, 2007; INE, 2008). In terms of basic services, 90 percent of the population had access to safe drinking water provided by the local water company ELAPAS (HAMS, 2007).

With the aim to improve water supply in Sucre in the long-term, the city's water company "ELAPAS" started working on a project to construct a reservoir with a capacity of 6.5 million m^3 in the Ravelo River between the villages Sasanta and Yurubamba in 2003 (ELAPAS, 2009). However, the project was not implemented until 2010 due to strong opposition of the villagers. According to meeting records, the initial negotiations in 2003 involved only the representatives of both municipalities and some local leaders, neglecting the opinion of indigenous people who anticipated losing their land and homes because of construction of the reservoir. Later, in 2004 and 2007, the city's water company tried to promote the project through a socialization program that aimed to persuade the villagers to accept the dam construction. This acceptance was required by the national and international donors to receive financial aid. The indigenous people from Sasanta and Yurubamba requested the first socialization meetings to be carried out in the local communities in order to facilitate the participation of most of the villagers and therefore to allow them to reach a decision at a community level. Nevertheless,

Figure 25.3 Panoramic view of the city of Sucre (photo: Paola Ramallo)

the water company failed to comply with this requirement and instead, decided to hold the meetings in Sucre. As a result, the villagers refused to attend those meetings, delaying the project approval once again. The negligence shown by the city's water company over the years and its resistance to acknowledging the local villagers' interests and to recognizing them as active participants in the project resulted in growing mistrust among the villagers and the loss of their intention to engage in further negotiations.

By the end of 2008, the city water company managed to secure funding for the project and tried to start the construction work, but met strong opposition from the rural villagers in Sasanta and Yurubamba. The latter claimed that the decision to implement the project was illicit because it was taken without any consultation with the affected people who would lose their land and properties. Although the project had been endorsed by previous governments, the villagers argued that it would not be implemented without their approval. Relying on the new constitution that stated their rights and acknowledged their role as legitimate stakeholders in the allocation of water in the basin, the villagers fought back.

Later, in 2009, the villagers' position was strengthened even further by the broad social reforms in the country. The Bolivian government committed itself to protecting the rights of indigenous peoples and to support their demands in this and similar cases. A representative from the Ministry of Water and Environment was appointed to support the villagers in the negotiations that followed. Consequently, the water company was forced to find ways to bring the villagers' demands to the negotiating table in order to find a compromise to make the project implementation viable. The villagers mainly demanded compensations for the loss of their fields, which were going to be inundated after construction of the reservoir, and for water withdrawal, which was to be allocated to Sucre. Besides, during the negotiations, it was clear that they also expected to gain some benefits if they decided to approve the dam construction, in terms of compensation such as the implementation of development projects for the municipality of Ravelo like the improvement of their public infrastructure (roads, water supply systems, etc.). They realized that, if the quality of life of the inhabitants of Sucre would be improved with the dam construction, they could also find ways to use the project to improve their lives.

First, communal meetings were held in Yurubamba and Sasanta where people discussed among themselves and decided what compensation would be fair to them. After that, several rounds of negotiation were organized engaging representatives of both the water company and the communities. The representatives from the Ministry of Water and Environment played the crucial role of mediators in these negotiations. They provided technical advice to the villagers helping them to understand the proposals of the water company on one hand, and on the other they guarded the villagers' interests and rights. Finally, the conflict was largely resolved in 2010 after a compromise had been achieved, and the financing for the project implementation was approved. The official version of the Sasanta-Yurubamba Project shows considerable benefits that the villagers of Sasanta and Yurubamba will gain from the project implementation. Among others, the water company, the regional government of Chuquisaca, and the municipality of Sucre agreed to buy their lands at a fair market price in order to compensate for the loss of indigenous terrains due to inundation, to finance the construction of health, road, public and educational infrastructure; and to implement small development projects such as improved agricultural techniques, irrigation, and pisciculture.

Conclusions

To some extent, the water sector in Bolivia has served as one of the driving forces behind the recent drastic social changes. IWRM as a strategy to broad stakeholder and public involvement in planning and decision-making was put in place. As well, legal and institutional frameworks for the sector have drastically changed with newly introduced national policies, legislation, the creation of the Ministry of Water, and the inclusion of indigenous people as the main stakeholders in the decision-making processes. This has led to the empowerment of the once marginalized indigenous communities that now also have the support of governmental institutions. These radical reforms undertaken to empower indigenous people revived their willingness to participate in the decision-making, where they are now recognized as legitimate actors.

In the Ravelo River Basin, the conflict between the water company and the local indigenous communities delayed the implementation of the important water supply project for the country's capital city for seven years. The national political and legal reforms have reshaped relations among the stakeholders through the redistribution of power and have made it possible for the community of Ravelo to stand up and claim their rights. This reactivated the negotiation and resulted in finding a compromise that was accepted by all parties involved.

However, the biggest challenge is still ahead, and that is to create an environment conducive to closing the gaps between promise and practice since changed policies do not necessarily mean they will be effectively and fairly implemented. On one hand, community empowerment requires significant cultural changes representing a complicated and long-term process while also having to cope with the existing social problems in the country such as poverty, illiteracy, and mistrust. On the other hand, the recently approved legislation in the water sector in Bolivia lacks detail and a clear strategic view on the sector's development that poses some challenges for its implementation.

References

Bartle, P. (2008) 'Human factor and community empowerment', *Review of Human Factor Studies*, 14 (1): Special Issue.

Beierle, T.C. (1998) *Public participation in environmental decisions: An evaluation framework using social goals*, Resources for the Future, Discussion Paper 99-06. Online. Available HTTP: http://ageconsearch.umn.edu/bitstream/10497/1/dp990006.pdf (accessed 15 November 2012).

Boelens, R. (2002) *Water law and indigenous rights: research, action and debate*, Wangeningen, the Netherlands: University of Wageningen.

Bustamante, R. (2002) *Legislación del agua en Bolivia. Centro andino para la gestión y uso del agua*, Cochabamba, Bolivia: Centro Agua, UMSS..

Butterworth, J., Warner, J., Moriarty, P., Smits, S. and Batchelor, C. (2010) 'Finding practical approaches to Integrated Water Resources Management', *Water Alternatives*, 3 (1): 68–81.

COANDINA (2009) 'Integrated management of the Ravelo river basin', Potosí: Bolivia.

Coelho, A.C., Fontane D.G., Vlachos E. and Maia R. (2008) 'Delineation of water resources units to promote IWRM and facilitate transboundary water conflicts resolution', paper presented at IV International Symposium on Transboundary Waters Management, Thessaloniki, Greece, October 15–18t, 2008.

Crabtree, J. (2005) *Patterns of protest: politics and social movements in Bolivia*, London: Latin America Bureau.

ELAPAS (2007–2009) 'Executive summary: "Construction of a system for the regulation and increase of the water discharge in the concession of the Ravelo River for the city of Sucre"', Sucre: Bolivia.

EMPOWERS Partnership (2007) *The EMPOWERS approach to water governance: guidelines, methods and tools*. Online. Available HTTP: http://waterwiki.net/images/d/d2/EMPOWERS_Guidelines%2C_Methods_and_Tools.pdf (accessed 15 October 2010).

Feás, J., Giupponi C. and Rosato P. (2004) 'Water management, public participation and decision support systems: the MULINO approach', paper presented at iEMSs International Conference Complexity and Integrated Resources Management, 14–17 June 2004 University of Osnabrück, Germany.

Fuentes, F. (2008) *Bolivia: the struggle for change*, Global Research, 2 December 2008. Online. Available HTTP: http://www.globalresearch.ca/bolivia-the-struggle-for-change/11269 (accessed 10 June 2009).

GWP (2000) *Integrated water resources management*, GWP Technical Committee Background Paper 4, Stockholm.

HAMS (2007) *Plan de desarrollo municipal 2003–2007*, Sucre, Bolivia.

Huntjens, P. and Pahl-Wostl C. (2010) 'Climate change adaptation in European river basins', *Regional Environmental Change*, 10: 263–284.

INE (2008) *Instituto Nacional de Estadística de Bolivia.* Online. Available HTTP: www.ine.gov.bo (accessed 11 September 2008).

Ingram, H. (2008) 'Beyond universal remedies for good water governance: A political and contextual approach', paper presented at the Rosenberg Forum for Water Policy, Zaragoza, Spain, June 25–26.

Lebel, L., Anderies J.M., Campbell, B., Folke, C., Hatfield-Dodds, S., Hughes, T.P. and Wilson, J. (2006) 'Governance and the capacity to manage resilience in regional social-ecological systems', *Ecology and Society*, 11 (1): 19.

Meadowcroft, J. (2002) 'Politics and scale: some implications for environmental governance', *Landscape and Urban Planning*, 61: 169–179.

Merrey, D.J. (2008) 'Is normative integrated water resources management implementable? Charting a practical course with lessons from Southern Africa', *Physics and Chemistry of the Earth*, 33: 899–905.

Michener, V. (1998) 'The participatory approach: contradiction and co-option in Burkina Faso', *World Development*, 26 (12): 2105–2118.

Pahl-Wostl, C. (2007) 'Transitions towards adaptive management of water facing climate and global change', *Water Resources Management*, 21: 49–62.

Pahl-Wostl, C., Sendzimir J., Jeffrey, P., Aerts, J., Berkamp, G. and Cross, K. (2007) 'Managing change toward adaptive water management through social learning', *Ecology and Society*, 12 (2): 30.

Power, Rights, and Poverty: Concepts and Connections (2004) A working meeting sponsored by DFID and the World Bank March 23–24, 2004, Ruth Alsop (ed.) World Bank, Washington, DC.

PROSALUS (2009) *Análsis de la realidad de Bolivia.* Online. Available HTTP: http://www.prosalus.es/gestor/imgsvr/publicaciones/doc/An%C3%A1lisis%20de%20la%20realidad%20Bolivia.pdf (accessed 3 September 2012).

Ramirez, R. (1999) 'Stakeholder analysis and conflict management', in D. Buckles (ed.) *Cultivating Peace: Conflict and Collaboration in Natural Resource Management*, International Development Research Centre, Ottawa, Canada.

Steelman, T. and Ascher W. (1997) 'Public involvement methods in natural resource policy making: Advantages, disadvantages and trade-offs', *Policy Sciences*, 30: 71–90.

UNDP (2004) *Water governance for poverty reduction: key issues and the UNDP response to Millennium Development Goals*, United Nations Development Programme. Online. Available HTTP: http://www.undp.org/water/pdfs/241456_UNDP_Guide_Pages.pdf (accessed 1 November 2008).

UNDP (2006) Human Development Report 2006. Beyond scarcity. Power, poverty and the global water crisis, United Nations Development Programme. Online. Available HTTP: http://hdr.undp.org/en/media/HDR06-complete.pdf (accessed 15 January 2010).

UNESCO (2009) *Outcome of the international experts' meeting on the right to water*, UNESCO, Paris, 7–8 July 2009, SEP Nimes, France,16 p.

UNICEF, WHO (2012) *Progress in drinking-water and sanitation: 2012 update*, WHO/UNICEF Joint Monitoring Programme for Water Supply and Sanitation (JMP), 2012.

Van Damme, P. (2002) *Disponibilidad, uso y calidad de los recursos hídricos en Bolivia*, World Summit on Sustainable Development, Johannesburg.

VeneKlasen, L. and Miller, V. (2002) *A new weave of power, people and politics, the action guide for advocacy and citizen participation*, Oklahoma City, OK: World Neighbors,.

Vörösmarty C.J., McIntyre, P.B., Gessner, M.O., Dudgeon, D., Prusevich, A., Green, P., Glidden, S., Bunn, S.E., Sullivan, C.A., Reidy Liermann, C. and Davies, P.M. (2010) 'Global threats to human water security and river biodiversity', *Nature*, 467: 555–561.

Watson, N. (2004) 'Integrated river basin management: a case for collaboration', *International Journal of River Basin Management*, 2 (4): 243–257.

WWAP (2009) *The United Nations World Water Development Report 3: Water in a Changing World*, World Water Assessment Programme Paris, UNESCO, and London: Earthscan. Online. Available HTTP: http://www.unesco.org/water/wwap/wwdr/wwdr3/pdf/WWDR3_Water_in_a_Changing_World.pdf (accessed 15 January 2010).

Zibechi, R. (2005) *Bolivia: dos visiones opuestas del cambio social.* Online. Available HTTP: http://www.cipamericas.org/es/archives/1231 (accessed 20 February 2009).

26 Human-environmental integration and social power in global environmental change research

David Manuel-Navarrete

Introduction

The last three or four generations are causing and/or enduring unprecedented rates of global environmental change (GEC), a human-induced phenomenon linked to the globalization of capitalist modernity. However, academic and policy arenas tend to formulate and represent it as a geophysical, chemical and bio-ecological (natural) event with a [3]human dimension[2] (see for instance: Committee on Global Change Research, 1999; Munn, 2002). Such representations acknowledge that GEC requires policy and market responses, but these responses are not necessarily related to the socio-political phenomena that traditionally concern social sciences (e.g., power, space, inequality, coloniality, identity).This "human dimension" perspective both deters, and is a consequence of social scientists' persistent disengagement from GEC research (Lever-Tracy, 2008; Shove, 2010). This situation is starting to change as calls, proposals, and efforts to bring society (and social sciences) into the core of GEC analysis are rapidly multiplying. For more than two decades, research on society's role on GEC has been institutionally channeled through the International Human Dimensions Programme on Global Environmental Change (IHDP), which nurtures a loose interdisciplinary community heavily inspired by complexity, systems thinking and disaster risk analysis. A crucial task for the "IHDP community" is to articulate relational/ecological concepts and grammars to bridge (ontologically, epistemologically and methodologically) entrenched human-environment divisions. The involvement of social scientists in this bridging endeavor, and the incorporation of insights from mainstream social theories, will be crucial as well as exceedingly challenging.

This chapter argues that the increasing popularity of systems thinking and relational perspectives within some social disciplines, and the growing centrality of climate change in socio-political discourse, can contribute to reformulate GEC as a human-environmental phenomenon, instead of an environmental one with a "human dimension." However, two main epistemological obstacles stand in the way of this interdisciplinary reformulation. The first is the pervasive assumption that political (and ethical) question(s) ought to be – and indeed can be – left outside of science (e.g., to civil society or politicians). This assumption hinders the involvement of social scientists interested in explaining socio-political (and human-environmental) phenomena from power-laden and culture-laden perspectives. The second obstacle is the assumption that social change is not significantly influenced by the paucity of traditionally sluggish environmental change, an assumption challenged by the very acceleration of human-induced GEC and the realization that, in the Anthropocene (Crutzen, 2002), human agency is as much bio-geological as it is socio-political.

The next sections review recent developments within the IHDP community to transcend formulations of GEC as a phenomenon with two separable dimensions (i.e., biophysical and human), which can be independently studied and, later on, combined into management and policy-making proposals. Transcending dichotomous understandings of human-environmental relations reformulates the origins of GEC, and repositions it in relation to contemporary socio-political phenomena. This is furthering the incorporation of social theory, ethics and values (O'Brien, 2009). In fact, concepts such as vulnerability, adaptation or resilience, which were either borrowed or adapted from ecology and systems thinking, are being problematized, refined and reconsidered in the light of human behavior, social power and global inequalities (Janssen and Ostrom, 2006; O'Brien *et al.*, 2008; Manuel-Navarrete *et al.*, 2011).

In parallel to IHDP efforts, there is a proliferation of social science interventions in climate change debate. This involvement is perhaps harnessed by the growing policy and public relevance of climate change, which turns GEC into a fully-fledged social phenomenon susceptible of sociological research (Urry, 2011). At the same time, the premise that social change unfolds against the background of a stable environment is being rapidly eroded (Chakrabarty, 2009). Interestingly, such interventions are parallel to social scientists' increasing engagement with systems and relational thinking. It is argued that a key issue to explore further is the role of relational paradigms, such as systems and complexity, in further bridging the gap between biophysical and social sciences and thus advancing formulations of GEC as a fully human-environmental phenomenon (Bhaskar *et al.*, 2010). Yet, relational and systems-based paradigms need to be problematized when applied to socially conflicting issues rooted in power structures and related to global inequality, such as GEC. This "political question" can be framed in terms of political reform, including discourse struggles, behavioral change, ecological modernization, or (post-political) neoliberal governance, but it also opens the floor to critical approaches that question forms of domination embedded in epistemologies, power relations, identities, subjectivities, agency and everyday life (Escobar, 2007; Manuel-Navarrete, 2010).

Towards the human-environmental dimension

Human-environment separability allows interpreting GEC as an environmental problem that can be solved from the status quo of current institutional frameworks (Hulme, 2009). Biophysical sciences become crucial to predict the impact of anthropogenic emissions on biogeochemical processes (including tipping points) and recommend safe mitigation levels, while the "human dimension" may focus on the implementation of mitigation and adaptation measures to cope with predicted and actual changes. Overall, mitigation/adaptation provides a basis for the "scientific management" of GEC (Folke *et al.*, 2005). This is useful to construct and legitimize metrics aimed at guiding legislation, markets and management decisions, or serving as inputs to participatory decision-making processes. However, subsuming all aspects of GEC under the rubric of mitigation/adaptation is largely unattractive to social science because it downplays the internal dynamics of social change, which are driven by conflict, contradiction, subjectivity or politics (Swyngedouw, 2010). In fact, the emphasis in outcomes such as technological fixes, risk assessments, policies, management strategies, or incentives takes current socio-political contexts for granted and tends to analyze away social power.

In the "disasters" literature, Hewitt's edited book (1983) was the first to criticize human-environmental separability and the apolitical construction of natural disasters and development. Wescoat *et al.* (1991) observed that climate change response was dominated by "scenario assessments" in which nature was conceived as a separate entity. As in the case of mitigation/adaptation, assuming human-nature separability enabled representations of environmental risks as threats external to society. More generally, isolating and objectifying nature outside society allows abstracting GEC response as an activity foreign to on-going socio-political processes, and consequently external to social and personal transformations. In fact, the framework of risk assessment can be seen as enabling technocratic and policy-based responses to GEC. This is not to negate GEC's role in increasing and worsening emergency situations (i.e., circumstances marked by exceptionality) in which technocratic action might be the most effective means to ensure security in the short term and restore certain forms of social order. Yet, a broader understanding of GEC is also necessary to transcend specific emergencies and question the socio-political and socio-ecological structures and forms of agency that are increasing the recurrence of emergencies through GEC.

The reinterpretation of GEC as an integrated human-environmental phenomenon has been a major concern in disasters literature and also within the IHDP community. Epistemological integration is proposed to transcend reductionist, impact-driven and sectorial approaches that dominate relief programs and agencies. Such proposal is based on disciplinary integration, but also seeks to amalgamate perspectives and paradigms. For instance, Cardona (2004) distinguishes epistemologies of vulnerability and risk along the following three disciplinary categories: *natural sciences*, *applied sciences*, and *social sciences*. Similarly, Füssel (2007) combines disciplinary and other epistemological criteria in his classification of approaches to climate change research: *pure deterministic*, *mechanistic engineering*, *human ecology*, and *political economy*. In an attempt to rise above disciplinary boundaries, Hilhorst (2004) identified *behavioral*, *structural*, and *mutuality/complexity* paradigms to hazard response. Some authors have included integrative categories as part of their classifications. For instance, Kasperson and colleagues (2005) suggested five approaches to vulnerability: *risk-hazard*, *political economy*, *pressure-and-release model*, *resilience, and "integrated."* In a similar vein, McLaughlin and Dietz (2008) identified five perspectives to the study of vulnerability: *biophysical*, *human ecological*, *political economy*, *constructivist*, and *political ecology*. In addition, these authors propose to transcend this theoretical fragmentation through a critical realist epistemology that interprets social change as an evolutionary process. This is a promising path to formulate GEC as an integrated human-environmental phenomenon in which social structures, human agency and the environment are interpreted as co-constituted. As discussed in the following section, McLaughlin and Dietz's (2008) proposal runs parallel to the recent involvement of critical realist philosophers in the formulation of climate change as an interdisciplinary and complex (open-systemic) phenomenon (Bhaskar *et al.*, 2010).

The application of human-environmental integration has been championed within the IHDP by proposals to take "socio-ecological systems" or "coupled human-environmental systems" as a basic unit of analysis (Young *et al.*, 2006). This has led to rethinking and reformulating key systems concepts such as vulnerability, adaptation and resilience (e.g., Anderies *et al.*, 2004; Folke *et al.*, 2005; Gallopín, 2006). However, this notion of a social-biophysical hybrid unit of analysis has failed to appeal to mainstream social sciences, despite the fact that systems and complexity terminologies are increasingly adopted within some social disciplines (e.g., Walby, 2007). The integration of modern societies and the environment is a traditional concern in anthropology (e.g., Ingold, 2005) and in ecological subfields of social disciplines such as environmental sociology or ecological economics. Mainstream social science is also integrating the environment through influential theories that assume human-nature simultaneous co-enactment (such as Latour's (1993) actor-network theory and cognate developments). The question is then: Are we at the verge of a scientific revolution marked by the integration of the environment into social thinking through systems and relational theories? If so, what will be the role of GEC, and the "human dimensions" community, in setting the agenda? What ontologies and epistemologies of human-environmental interaction are more useful for dealing with GEC challenges? For instance, shall we consider Latour's flat ontology with no a priori privileging of the agency or "effectivity" of any particular categories of beings (Clark, 2010)? Shall we instead assume fundamental incommensurability between (interacting) social and ecological systems (in the sense that the former display self-reflexive properties that the latter lack) as in socio-ecological systems analysis (Young *et al.*, 2006)? The following sections build on recent interventions on climate change debates from influential social scientists. It is argued that a crucial challenge lays in the need to problematize systems thinking by fully incorporating social power and subjectivity (Jasanoff, 2010; Manuel-Navarrete and Buzinde, 2010).

Social science recent interventions

The IHDP promotes research on key social aspects of GEC. In particular, it complements biophysical global scenarios with social information, and analyzes the science-policy interface, technological regimes and socio-technical transitions (Biermann, 2007). Generally, the analysis of institutions, governance and policy has been a major concern. Global environmental governance has been recently proposed to overcome excessive emphasis on governments and account for non-state actors, as well as informal sectors and social movements (Bulkeley, 2007; Newell, 2008; Okereke *et al.*, 2009). The IHDP also draws on social-environmental subfields such as environmental sociology or environmental psychology. The ethical question is acquiring growing relevance as development discourse is problematized (Leichenko and O'Brien, 2008), and agency is reformulated beyond stakeholderism (Betsill *et al.*, 2011). Political science and international relations have been central to discussions of global environmental politics and climate change (e.g., Paterson, 2007; Pettenger, 2007). In addition to all these developments within the IHDP, the growing concern within mainstream social sciences on climate change highlights the role of social power and the political question.

Recent social science interventions on climate change can be clustered into three strands according to the role attributed to human-environmental integration and power. The first cluster endorses conventional formulations of GEC as a biophysical event with a "human dimension." The role of political and social analysis is to enhance and refine adaptation policies and the rapid implementation of green technologies. These analyses assume either current socio-political liberal regimes: "we are in the hands of our political leaders" (Giddens, 2009: 228), or innovation-driven fixes that transcend neo-liberalism as in Urry's (2011) "resource capitalism." These authors theorize capitalist solutions to sharp declines in available resources and multiplying environmental shocks. In Urry's case, complex systems theories are embraced to explain change in both social and environmental systems. This is an important step aimed at bridging the human-environmental chasm, which has been pervasive in sociology, economy or political sciences.

A second cluster of interventions argues that GEC challenges the dominant ways in which knowledge is produced and used (Chakrabarty, 2009, Bhaskar *et al.*, 2010). This requires integration across disciplinary boundaries, but also across cultural paradigms. Interdisciplinary cross-fertilization is already frequent within social science and humanities, and boundary crossings between the biophysical and social realms are no longer uncommon. However, the argument of critical realism goes beyond crossing disciplinary boundaries. It postulates that structures, mechanisms, processes, fields and the other intransitive objects of scientific knowledge (epistemology) are fundamentally distinct from, and irreducible to, the actual patterns of events they represent (Bhaskar, 2010). This highlights the artificial nature of drawing boundaries between biophysical and social systems, which in fact contradicts the ontological assertion that the world is constituted of open human-environmental systems. This critical realist argument is consistent with the so-called "ecosystem approach," in which the world is conceived as being constituted of complex (open-systemic) and self-organizing phenomena, stratified in multi-layered holons (Kay and Boyle, 2008), of which GEC would be just one case.

A third set of interventions resituate GEC as an event that forces humanity to rethink the power relations ingrained in capitalist globalization and modernity (Pelling *et al.*, 2012). On the one hand, the role of social sciences is to unveil power asymmetries embedded in climate politics. This unveiling is crucial to avoid taking for granted a consensual, de-politicized environmental governance regime that "combines apocalyptic environmental visions with an hegemonic neoliberal view of social ordering" (Swyngedouw, 2010: 228). On the other hand, social science shall explore the inextricable connections between GEC and social-ecological inequalities, the drive for economic growth, and representations of prosperity (Barry, 2012). This is not only a matter of justice, but of examining the power relations involved in the making of post-capitalist futures, perhaps based on cosmopolitanism or socio-ecological forms of agency (Beck, 2010; Manuel-Navarrete and Buzinde, 2010).

The power of GEC as a human environmental phenomenon

The need to consider power relations, institutions, and overall political and economic structures has been at the center of political economy approaches to vulnerability and adaptation (Wisner *et al.*, 2004; Boyd *et al.*, 2008). In fact, the importance of power is acknowledged in climate change literature addressing economic globalization (Redclift and Sage, 1998; Eakin and Lemos, 2006). Furthermore, political ecology approaches to adaptation have examined the institutional contexts and discourses that influence the adaptations of individuals, communities and specific social groups (Pelling, 2003). As discussed above, McLaughlin and Dietz (2008) have called for a comprehensive theory of vulnerability based on evolutionary perspectives of social change, and capable of theorizing the interrelated dynamics of social structure, agency and the environment. These calls are crucial and more research on GEC focused on power dynamics is needed.

Anthropologists have criticized vulnerability and the very notion of disasters for being articulated within Western discourses that essentialize other regions as unhealthy, poor and disaster-prone (Bankoff, 2001). The question of justice has been framed in climate change literature in terms of "fair adaptation" and "putting the vulnerable first" (Paavola and Adger, 2006). The idea is to insert social justice criteria within climate adaptation regimes and policies. The "vulnerable" are subjected to, and are recipients of, adaptation policies, but are usually unable to influence their design and implementation. Accordingly, the "climate regime" should be modified in order to guarantee equal participation and increase the accountability of the powerful. The notion of fair adaptation is a step forward acknowledging the role of power in GEC. However, it also highlights the need to combine systems analysis of GEC with critical theories of social power. For instance, the category of "the vulnerable" is problematic as it may be judged as inherently disempowering. Likewise, the idea of "incorporating the vulnerable" into "the" process (versus creating a joint process or incorporating Western processes and practices into those of "the vulnerable") puts them in a disadvantageous starting position. Thus, the use of categories such as "the vulnerable" has the risk of slipping into managerialism and what Duffield (2006: 70) describes as "the biopolitical incorporation of a species-life ['the vulnerable'] that, lacking the insurance-based safety-nets and welfare regimes of mass [Western] society, is cast as essentially self-reliant; in other words, 'non-insured.'" Consequently, development, emergency relief and adaptation policies may become "a set of biopolitical compensatory and ameliorative technologies of security that define and act upon non-insured populations to improve resilience by strengthening self-reliance" (Duffield, 2006: 74). As a corollary, advocating for empowerment of "the vulnerable" and keeping a commendable commitment with justice does not necessarily lead to any fundamental revision of human-environmental power relations.

If we assume that GEC is not external to humanity but a human-environmental phenomenon, then we need proactive approaches to change the social conditions under which GEC is produced. This proactive perspective involves redirecting GEC research towards the interrogation of power itself, its origins, essence, and rationalizing mechanisms. This requires social sciences' involvement to inspire and enable the kind of creative political changes that are required to reformulate dysfunctional power relations and align institutions according to socio-ecological co-evolution, instead of the interests of the powerful few. Arguably, the full consideration of power relations in the complex dynamics of human-environmental systems may open up a universe of unexpected and unpredictable social trajectories that can be characterized by sudden flips and transformations. These trajectories result from interactions between autonomous agents, but these agents are in turn influenced by past evolutions and current structures.

This path-dependent structural coupling of agents and human-environmental systems raises some important political questions. For instance: How do human and non-human forms of agency operate at the multiple interrelated domains of complex human-environmental systems? What room is left for free will and emancipatory politics as well as autonomous biological entities and physical forces? How can we avoid the risk of falling into the "everything is connected" thesis and thus failing to discern the distinctive ways in which human-environmental co-constitution takes place? If dealing with GEC involves a full realization of the inextricably human character of the environment and vice-versa, then: What are the consequences for "partial" forms of engagement with the world (e.g., as consumers, voters, managers, workers, scientists)?

Thoughts towards a research agenda on power relations in human-environmental systems

This chapter argues that the formulation of GEC as an environmental phenomenon with a "human dimension" is increasingly challenged both within the IHDP community and through recent interventions from mainstream social science in climate change debates. The main outcome of this challenge is the reformulation of GEC as a truly human-environmental and interdisciplinary phenomenon. Another outcome is a problematization of dominant climate change epistemologies based on the mitigation/adaptation couplet. These dominant epistemologies rely on structuralist policy concepts, such as governance and regimes, and reduce social power to a design problem rather than a matter of domination, resistance, and emancipation. The question is whether improving institutional arrangements and policy outputs (e.g., mitigation measures, vulnerability reduction, bringing together poverty alleviation and adaptation) will suffice to address GEC. Although some authors believe it to be the case (e.g., Giddens, 2009), this chapter shows the potential of richer interpretations of power emanating from, for instance, constructivist, poststructuralist, Marxist, or postcolonial perspectives (Demeritt, 2006; Glover, 2006; Pettenger, 2007; Chakrabarty, 2009). Within the IHDP, these critical perspectives can enrich the "Earth systems governance" framework, and its exploration of the interplay of agency and structures, by inviting GEC scholars to draw more deeply on social theories of power.

The increasing involvement of social sciences is crucial to foster power analysis and move GEC beyond a focus on emergencies and the amelioration of external vulnerabilities. In fact, the eventual recurrence of crisis, disasters, or extreme weather phenomena brings about unique opportunities to rethink society and culture. As a result, GEC response may have more to do with meaning-making than policy or decision-making (Manuel-Navarrete *et al.*, 2004). Complexity and systems thinking may provide a common language to describe biophysical, social and interrelated human-environmental phenomena. Yet, the use of systems thinking in the analysis of human-environmental phenomena remains problematic. Can concepts such as attractors, patterns, feedbacks, trajectories, or thresholds equally describe the melting of icecaps as much as the collapse of financial institutions? Proposals from the IHDP community to study coupled socio-ecological systems, and some interventions from social scientists on climate change, including critical realist philosophers, seem to embrace this possibility. However, the question of human-environmental integration calls for more thinking on the potentially distinctive self-reflexive character of human systems (i.e., the capacity to consciously reflect on, and act upon, systems' evolution) and challenges the simple transposition of systems concept from biophysical to social sciences without reflecting on the specificities of human agency and social power.

References

Anderies, J.M., Janssen, M.A. and Ostrom, E. (2004) 'A framework to analyze the robustness of social-ecological systems from an institutional perspective', *Ecology and Society*, 9 (1): 18.

Bankoff, G. (2001) 'Rendering the world unsafe: "vulnerability" as Western discourse', *Disasters*, 25 (1): 19–35.

Barry, J. (2012) 'Climate change, "the cancer stage of capitalism" and the return of limits to growth: towards a political economy of sustainability', in *Climate Change and the Crisis of Capitalism: A Chance to Reclaim Self, Society and Nature*, New York: Routledge.

Beck, U. (2010) 'Climate for change, or how to create a green modernity?', *Theory, Culture & Society*, 27 (2–3): 254–266.

Betsill, M., Pattberg, P. and Dellas, E. (2011) 'Editorial', *International Environmental Agreements: Politics, Law and Economics*, 11 (1): 1–6.

Bhaskar, R. (2010) 'Contexts of interdisciplinarity. Interdisciplinarity and climate change', in R. Bhaskar, C. Frank, K.G. Hoyer, P. Naess, and J. Parker (eds) *Interdisciplinarity and*

Climate Change. Transforming Knowledge and Practice for our Global Future, London: Routledge, pp. 1–24.

Bhaskar, R., Frank, C., Hoyer, K.G., Naess, P., and Parker, J. (eds) (2010) *Interdisciplinarity and Climate Change: Transforming Knowledge and Practice for our Global Future*. London: Routledge.

Biermann, F. (2007) '"Earth system governance" as a crosscutting theme of global change research', *Global Environmental Change*, 17 (3–4): 326–337.

Boyd, E., Osbahr, H., Ericksen, P.J., Tompkins, E.L., Lemos, M.C. and Miller, F. (2008) 'Resilience and "climatizing" development: examples and policy implications', *Development*, 51: 390–396.

Bulkeley, H. (2007) 'Reconfiguring environmental governance: towards a politics of scales and networks', *Political Geography*, 24 (8): 875–902.

Cardona, O.D. (2004) 'The need for rethinking the concepts of vulnerability and risk from a holistic perspective', in G. Bankoff, G. Frerks and T. Hilhorst (eds), *Mapping Vulnerability: Disasters, Development and People*, London: Earthscan.

Chakrabarty, D. (2009) 'The climate history: Four theses', *Critical Inquiry*, 35: 197–222.

Clark, N. (2010) 'Volatile worlds, vulnerable bodies: Confronting abrupt climate change', *Theory, Culture & Society*, 27 (2–3): 31–53.

Committee on Global Change Research, Board on Sustainable Development, Policy Division, National Research Council (1999) *Global Environmental Change: Research Pathways for the Next Decade*, Washington, DC: National Academies Press.

Crutzen, P.J. (2002) 'The "anthropocene"', *Journal de Physique* 4, 12 (10): 1–5.

Demeritt, D. (2006) 'Science studies, climate change and the prospects for constructivist critique', *Economy and Society*, 35 (3): 453–479.

Duffield, M. (2006) 'Racism, migration and development: The foundations of planetary order', *Progress in Development Studies*, 6 (1): 68–79.

Eakin, H. and Lemos, M.C. (2006) 'Adaptation and the state: Latin America and the challenge of capacity-building under globalization', *Global Environmental Change*, 16 (1): 7–18.

Escobar, A. (2007) 'Worlds and knowledges otherwise', *Cultural Studies*, 21 (2): 179–210.

Folke, C., Hahn, T., Olsson, P. and Norberg, J. (2005) 'Adaptive governance of social-ecological systems', *Annual Review of Environment and Resources*, 30 (1): 441–473.

Füssel, H.-M. (2007) 'Vulnerability: a generally applicable conceptual framework for climate change research', *Global Environmental Change*, 17: 155–167.

Gallopín, G. (2006) 'Linkages between vulnerability, resilience, and adaptive capacity', *Global Environmental Change*, 16 (3): 293–303.

Giddens, A. (2009) *The Politics of Climate Chang*, London: Polity Press.

Glover, L. (2006) *Postmodern Climate Change*, London: Routledge.

Hewitt, K. (1983) *Interpretations of Calamity from the Viewpoint of Human Ecology*, Boston: Allen & Unwin.

Hilhorst, D. (2004) 'Complexity and diversity: Unlocking social domains of disaster response', in G. Bankoff, G. Frerks and T. Hilhorst (eds) *Mapping Vulnerability: Disasters, Development and People*, London: Earthscan, pp. 52–66.

Hulme, M. (2009) *Why we Disagree about Climate Change: Understanding Controversy, Inaction and Opportunity*, Cambridge: Cambridge University Press.

Ingold, T. (2005) 'Epilogue: towards a politics of dwelling', *Conservation and Society*, 3 (2): 501–508.

Janssen, M.A. and Ostrom, E. (2006) 'Resilience, vulnerability, and adaptation: A cross-cutting theme of the International Human Dimensions Programme on Global Environmental Change', *Global Environmental Change*, 16 (3): 237–239.

Jasanoff, S. (2010) 'A new climate for society', *Theory, Culture & Society*, 27 (2–3): 233–253.

Kasperson, J.X., Kasperson, R.E., Turner II, B.L., Hsich, W. and Schiller, A. (2005) 'Vulnerability to global environmental change', in J.X. Kasperson and R.E. Kasperson (eds), *Social Contours of Risk, Vol. II: Risk Analysis Corporations and the Globalization of Risk*, London: Earthscan, pp. 245–285.

Kay, J.J. and Boyle, M. (2008) 'Self-organizing, holarchihc, open systems (SOHOs)', in D. Waltner-Toews, J.J. Kay and N.-M. Lister (eds) *The Ecosystem Approach: Complexity, Uncertainty, and Managing for Sustainability*, New York: Columbia University Press, pp. 51–78.

Latour, B. (1993) *We Have Never Been Modern*, Cambridge, MA: Harvard University Press.

Leichenko, R. and O'Brien, K.L. (2008) *Environmental Change and Globalization: Double Exposures*, Oxford: Oxford University Press.

Lever-Tracy, C. (2008) 'Global warming and sociology', *Current Sociology*, 56 (3): 445–466.

Manuel-Navarrete, D. (2010) 'Power, realism, and the humanist ideal of emancipation in a climate of change', *Wiley Interdisciplinary Reviews: Climate Change*, 1 (6): 781–785.

Manuel-Navarrete, D. and Buzinde, C.N. (2010) 'Socio-ecological agency: from "human exceptionalism" to coping with "exceptional" global environmental change', in M.R. Redclift and G. Woodgate (eds) (second edition) *The International Handbook of Environmental Sociology*, Cheltenham: Edward Elgar, pp. 306–337.

Manuel-Navarrete, D., Kay, J.J. and Dolderman, D. (2004) 'Ecological integrity discourses: Linking ecology with cultural transformation', *Human Ecology Review*, 11 (3): 215–229.

Manuel-Navarrete, D., Pelling, M. and Redclift, M. (2011) 'Critical adaptation to hurricanes in the Mexican Caribbean: development visions, governance structures, and coping strategies', *Global Environmental Change*, 21: 249–258.

McLaughlin, P. and Dietz, T. (2008) 'Structure, agency and environment: toward an integrated perspective on vulnerability', *Global Environmental Change*, 18: 99–111.

Munn, R.E. (ed.) (2002) *Encyclopedia of Global Environmental Change*, New York: John Wiley & Sons.

Newell, P.J. (2008) 'The political economy of global environmental governance', *Review of International Studies*, 34: 507–529.

Okereke, C., Bulkeley, H. and Schroeder, H. (2009) 'Conceptualizing climate governance beyond the international regime', *Global Environmental Politics*, 9 (1): 58–78.

O'Brien, K.L. (2009) 'Do values subjectively define the limits to climate change adaptation?' in W.N. Adger, I. Lorenzoni and K.L. O'Brien (eds), *Adapting to Climate Change. Thresholds, Values, Governance*, New York: Cambridge University Press, pp. 164–180.

O'Brien, K.L., Sygna, L., Leichenko, R..A.W.N., Barnett, J., Mitchell, T., Schipper, L., Tanner, T., Vogel, C. and Mortreux, C. (2008) 'Disaster risk reduction, climate change adaptation and human security', *GECHS Report 3*. Oslo: Global Environmental Change and Human Security (GECHS) Project.

Paavola, J. and Adger. W.N. (2006) 'Fair adaptation to climate change', *Ecological Economics*, 56 (4): 594–609.

Paterson, M. (2007) 'Environmental politics: sustainability and the politics of transformation', *International Political Science Review*, 28 (5): 545–556.

Pelling, M. (2003) *The Vulnerability of Cities: Natural Disasters and Social Resilience*, London: Earthscan.

Pelling, M., Manuel-Navarrete, D. and Redclift, M. (eds) (2012) *Climate Change and the Crisis of Capitalism: A Chance to Reclaim Self, Society and Nature*, New York: Routledge.

Pettenger, M.E. (2007) *The Social Construction of Climate Change: Power, Knowledge, Norms, Discourses*, Aldershot: Ashgate.

Redclift, M. and Sage, C. (1998) 'Global environmental change and global inequality: North/South perspectives', *International Sociology*, 13 (4): 499–516.

Shove, E. (2010) 'Social theory and climate change: questions often, sometimes and not yet asked', *Theory, Culture & Society*, 27 (2–3): 277–288.

Swyngedouw, E. (2010) 'Apocalypse forever?: Post-political populism and the spectre of climate change', *Theory, Culture & Society*, 27 (2–3): 213–232.

Urry, J. (2011) *Climate Change and Society*, Cambridge: Polity Press.

Walby, S. (2007) 'Complexity theory, systems theory, and multiple intersecting social inequalities', *Philosophy of the Social Sciences*, 37 (4): 449–470.

Wescoat Jr., J.L. (1991) 'Managing the Indus river basin in light of climate change. Four conceptual approaches', *Global Environmental Change*, 1 (5): 381–395.

Wisner, B., Blaikie, P., Cannon, T. and Davis, I. (2004) *At Risk: Natural Hazards, People's Vulnerability and Disasters*, London: Routledge.

Young, O.R., Berkhout, F., Gallopin, G.C., Janssen, M.A., Ostrom, E. and van der Leeuw, S. (2006) 'The globalization of socio-ecological systems: an agenda for scientific research', *Global Environmental Change*, 16 (3): 304–316.

27 Human security and personal responsibilities in light of climate change defeatism and complacency

Thomas Heyd

Introduction

The concept of human security, as elaborated by GECHS,[1] concerns 'the freedom to take actions that *promote wellbeing* in response to changing environmental conditions' (emphasis added, Lonergan *et al.*, 2000). More specifically, human security is closely linked to norms, values, rights and entitlements and it draws attention to notions of empowerment, protection and responsibilities. It is about fulfilling people's vital freedoms and the development of capacities to create satisfying lives for all people (O'Brien *et al.*, 2010).

Human security, in other words, entails not only an *end state*, described in terms of an increase in wellbeing and a decrease in exposure to certain threats, but also a *particular process* of achieving this end state, namely, the active participation of individuals and communities in achieving this end state. As such, the definition of human security is explicitly normative in a double sense, since it makes reference to both ends and means. It also explicitly calls upon individuals as well as society at large to *take responsibility* concerning problematic environmental conditions, such as may be expected from climate change.

Indeed, there are good reasons to believe that individuals as well as societies have responsibilities to act on climate change. Paradoxically, knowledge of the increasing certainty about human-induced climate change, and of the severe harm that it may bring to large portions of the world population, may lead individuals and communities to fall into a paralysing *defeatism*. Such defeatism, furthermore, may be accompanied by *complacency*, due to the assumption that, given our societies' institutional, scientific and technical capabilities, we can simply wait to address the problems posed by climate change until they actually become troublesome in our proximate, local, environment.

Here, I begin by explaining a little further the grounds for taking responsibility concerning climate change. After this, I point out that defeatist and complacent attitudes may interfere with acting on our responsibilities, thereby possibly leading to failure in the application of entirely feasible mitigation and adaptation measures. Next, I note that the result may be significantly greater economic costs and human and ecological harms than would otherwise be incurred. I close by pointing toward some of the ways to address these problematic attitudes, and propose that individuals who are aware of this situation should perceive it as a call to help transform the socio-cultural conditions that generate them.

Prudential and ethical responsibilities

A diversity of measures, undertaken by various levels of governments, coordinated nationally and internationally, are doubtlessly necessary in order to create the institutional framework and material infrastructures, and to set the overall course for society, to address climate change satisfactorily through reduction of greenhouse gas emissions, as well as other mitigating strategies, and through adaptation to already inevitable effects (Nihlén Fahlquist, 2009).[2] Though responsibility for action on climate change generally has been discussed in relation to the duties of particular nation-states or blocks of nations (the West or Global North vs. the 'developing' nations of the Global South and the 'emerging' BRIC (Brazil, Russia, India and China) countries), there are good reasons to also consider the responsibilities for action of individual citizens, because of their important role in bringing about change in society (see, e.g., Moser, 2007). Active involvement by concerned citizens is crucial in several respects: as sources of new civil society initiatives, as pressure on governmental institutions to carry through on measures proposed, and as proof of the realisability of the collective transformation of current patterns of production and consumption.

The *Fourth Assessment Report* of the Intergovernmental Panel on Climate Change (IPCC, 2007) strongly supports the view that climate change processes are accelerating, and that future as well as present generations are going to experience very significant increases in severity of disastrous, and possibly catastrophic, phenomena associated with rapid environmental change, which will bring about significant economic costs and harm to human and non-human beings. Alarmingly, measured trends in the rise in global average temperatures and sea levels are seen to be even exceeding the *worst-case* scenarios of this latest IPCC Report.

One analysis of recent developments, published in *PNAS*, states that (2000–2006) trends in the growth in greenhouse gas emissions due to increased economic activity, greater carbon intensity, and lowered efficiency of natural sinks 'characterize a carbon cycle that is generating stronger-than-expected climate forcing sooner than expected' (Canadell *et al.*, 2007:18869, also see Chen *et al.* 2011 on faster than expected range shifts of species due to climate change). Based on such observations, Will Steffen concludes in a 2009 report for the Australian Government that 'risks associated with the upper range of the IPCC projections of climate change for this century need to be considered seriously' (Steffen, 2009: 4). Most recently it has been found that 'Recent temperature measurements show that the Arctic is warming twice as fast as the rest of the planet, with sea ice this summer shrinking to its smallest extent on record' (D'Andrea *et al.*, 2012).

So, given that in the middle to long run everyone, and his or her descendants, can be expected to be at risk from at least some of the harmful effects of climate change, a strong case can already be made for taking personal responsibility for reasons of *prudence*. This by itself means that, reasonably, those individuals who are capable of it should actively support measures directed at mitigation and adaptation, to be implemented both locally and globally across societies and geographical boundaries, and possibly should go even further and develop initiatives to mobilise society about the impending peril (see Garvey, 2008).

There are also very good reasons for supposing that addressing climate change should be considered a matter of *ethical* responsibility. These reasons are diverse, and have been discussed in detail elsewhere (see Gardiner, 2004; Garvey, 2008). They range from taking responsibility for the harmful effects that one causes, to fairness in distribution of burdens; and from assuming responsibility on the basis of a commitment to a general harm prevention principle, to the humanitarian requirements of solidarity with the more vulnerable. I explain these reasons a bit further here.

Of particular relevance in this discussion is the fact that the industrialised countries have produced most of the anthropogenic greenhouse gases now in circulation, and continue to do so at accelerating rates. This means that they bear a special responsibility from the point of view of justice for the harm that these greenhouse gas accumulations in the atmosphere produce (this may be seen as an application of the 'polluter pays' principle). Moreover, insofar as responsibility can be shared across all those who in some way benefit from, or uphold, a system that causes harm, not just governments but anyone who lives in the Global North, and is a beneficiary of the industrial development brought about by historical greenhouse gas emissions, bears a responsibility for the harmful effects of climate change (Young, 2006), and ought to actively support mitigation and adaptation strategies.

One may also consider common principles of fairness, which, when applied to the capacity to act on urgent matters of common concern, translate into the principle that those who are *more capable* of acting should do *more* (Garvey, 2008). With regard to climate change, this approach means that those who are better off should assume greater responsibility to reduce fossil fuel consumption and contribute more, proportionally, to mitigation and adaptation strategies than those who have fewer means.

Notably, insofar as the financial, technological and organisational capital accumulated by industrialised nations largely has been built from, or with the help of, the resources of the rest of the world, and since a small fraction of this capital would likely be sufficient to soften the impact of climate change for poorer populations, there is, moreover, reason for supposing that, in fairness, the richer countries, and especially the richer individuals who live there, have an important responsibility to assist the poorer in the implementation of adaptation strategies. Importantly, some adaptation strategies also may make the poorer more vulnerable if not carried out in full consultation with, or under the direction of, those most in need.

Another approach to the question of ethical responsibility draws on the supposition that those who *can* prevent harm from occurring *ought* to do so, out of consideration for the ultimately common condition of vulnerability that we share as human beings (this may be expressed more fully as a general harm prevention principle, Singer, 1979, or as a principle of respect for persons, see Kant, 1785/2002, or as a principle of solidarity). Everyday examples include acts, such as saving a child from being accidentally run over by a speeding car, or rescuing a person from drowning in a nearby pool, but also helping people from dying due to malnutrition, when doing so does not require sacrificing anything of comparable moral value. This means that those of us who are relatively well-off ought to actively address the new risks to present and future human (and possibly to non-human) beings brought on by climate change.

As such, these various ways of thinking of responsibility with regard to climate change, in terms of justice, harm prevention and solidarity, require commitment with those who are more vulnerable and converge on the idea of taking responsibility for human security, since it concerns the promotion of wellbeing in the context of rapid environmental change. Insofar as the idea of human security not only focuses on the end state to be achieved but also on protecting and advancing options and capacities, 'necessary to end, mitigate or adapt to threats to people's human, environmental and social rights' (Lonergan *et al.*, 2000), it also highlights the importance of the capacity to act according to one's own judgement, which is an expression of autonomy.

The ability to direct one's own life according to one's own choices, in addition to various other rights, is widely considered a key ingredient in the ultimate good for human beings, and, as such, is part of the intrinsic value of our lives. In other words, the idea of human security makes clear that, insofar as climate change is a threat, not only to health, life and livelihood, but also to the ability of people to actively respond to such threats according to their own choices, it constitutes a very significant reason to act on the prudential and ethical responsibilities that arise from this process.

Some may object, however, that, even granting any of the ethical principles suggested, citizens should rather focus on other issues that supposedly are *more urgent* than climate change. Someone may ask, for example, why we should invest in mitigation and adaptation while nearly a billion people around the world are malnourished.[3] Similarly, one may ask why finding safe storage for rapidly growing quantities of health-threatening nuclear wastes, which can be radioactive for more than a million years, should not take precedence. Or, why the question of nuclear weapons proliferation should not come first. Or, why the achievement of peace among nuclear armed countries in volatile areas, in the Middle East, the border regions of Afghanistan-Pakistan, Pakistan-India, or India-China, should not take priority (see Homer-Dixon, 2006, for discussion of a 'watchlist' of global stresses).

The key argument in response to such objections is that climate change likely will make most other urgent problems *even more urgent*. As is well-known, a large portion of the world population is vulnerable to food scarcity, droughts, flooding and other effects associated with climate change, and has very limited resilience with which to recover after disastrous incidents. Historically, it is well known, moreover, that environmental factors (be they cold spells or droughts) are often exploited in conflictual situations (see, e.g., Fagan, 2000).

Consequently, food insecurity, safe storage of nuclear wastes in a politically fragile world, the threat of the use of nuclear weapons, the potential for armed conflicts, and so on, are all issues that potentially can become much more problematic *when accompanied by* climate change. Any further delay in committing to thorough-going mitigation and adaptation will foreseeably make many other upcoming problems so much harder to cope with, thereby making climate change an issue that is *at least as urgent as any other*. So, in response to the objection that other matters are more urgent, one may say that, due to the 'multiplier effects' that it entails, climate change constitutes an especially important issue to immediately address.

Another objection may be that the call to action on climate change may seem unrealistic, given the priorities of today's world (such as resolving the current economic crisis). There are, moreover, various types of obstacles for people to make their action effective in terms of carbon reductions that individuals face at the psychological level (see Gifford, 2008; American Psychology Association, 2009), at the informational level (insofar as people often do not know what to do, see Wolf, 2010), and at the societal-systemic level (insofar as action by individuals is limited by the range of choices available as determined by society at large).

In response to these objections one may note that, though, depending on individual circumstances, it may be difficult to quickly transform agriculture, industry or transportation on a societal or planetary scale, there may be a few things that most people *can* undertake to get the transformation of our societies started. For example, those who are fairly well-off can take steps to eat, heat, transport, work and build with renewable types of energy, and to re-use, recycle, return and compost used materials. For all of these activities there are enough technical solutions and resources available in most places, at least in the industrialised Global North, to get started immediately.

These steps require matching supportive policies from governments, of course. Those with influence in politics, business or the media should be expected to take steps to remind our leaders of their responsibility to change the structural barriers to action, by lobbying governments and uniting in civil society pressure groups. As a matter of fact, there are a variety of real life examples, such as the Transition Town movement, that demonstrate that transformation of the building blocks of society is possible. Consequently, there is little excuse for inaction (even if any adjustment of behaviours requires effort, see Garvey, 2008).

Certainly more detailed arguments for taking seriously our responsibilities (prudential and ethical) can be worked out on the basis of rational choice theory, and from the perspective of diverse approaches to ethics, such as utilitarianism, Kant's ethics, Rawls' theory of justice (see Vanderheiden, 2008) or Habermas' discourse ethics (Habermas, 1991). Similarly, the distribution and source of responsibilities, or the assignment of benefits and costs, may be worth considering in significantly greater detail (e.g., see Gardiner, 2004; Attfield, 2010; Garvey, 2008). In this chapter I will, however, forego further discussion of these matters and assume that it is sufficiently evident that there are good reasons to conclude that climate change constitutes an urgent prudential and ethical issue, such that the need for action, in terms of mitigation and adaptation, is sufficiently well established for well-informed ethical reasoners.

Defeatism and complacency

As noted, climate change has to be addressed through strong policies, laws and incentives issuing from governments and inter-governmental institutions (Nihlén Fahlquist, 2009), but probably little action can be expected from such institutions without active participation by citizens who take personal responsibility for the state of our world. This means, among other things, that it is important to understand how information on climate change is disseminated by natural and social scientists, how it is moulded by the media and appropriated by opinion-makers of various political stripes, and what citizens come to understand as a result of this process (Moser and Dilling, 2007).

Certainly there is no single way in which people apprehend climate news (Moser and Dilling, 2007; Hulme, 2009) but, strikingly, despite widespread awareness of climate change, this knowledge still has not led citizens to exert sufficient pressure on decision-makers to adopt appropriate policies (as we can see from the mostly disappointing series of COP meetings in Copenhagen in 2009, in Cancún in 2010, and in Durban in 2011).[4] One of the most troublesome kinds of reaction to increasingly sobering news about accelerating climate change is a kind of 'all or nothing' attitude, which proclaims that, since we cannot any longer prevent global warming *wholesale*, it may be ignored.[5] This attitude can constitute a powerful impediment to the implementation of mitigation policies, which, as any other insurance scheme, entails costs up front.

The 'all or nothing' attitude seems based on a combination of factual misapprehensions regarding the actual risks associated with further increases in greenhouse gases in the atmosphere (and in the oceans, generating increasing levels of acidification), inappropriate priorities, as well as minimalist habits that recommend inaction despite considerable urgency. Notably, while we are already committed to important climate effects (for example, considerable thawing of on-land glaciers, and increases in droughts and in storm intensities), continued emissions at present rates, however, mean still greater potential commitments, such as the thawing of the Greenland and Antarctica icecaps, with further intensification of linear, as well as hard to fully comprehend non-linear, effects that may be leading to surpassing important tipping points. Therefore, defeatism or 'fatalism' in view of climate change may lead to much worse consequences than already are to be expected.

The complementary attitude to the 'all or nothing' attitude is to 'wait and see' in the expectation that the combined power of science, engineering and financial institutions will be sufficient to cope with national or local disturbances, if and when they occur.[6] A complacent 'wait and see' attitude, however, fails to take into account key aspects concerning capacity to act (also see O'Brien *et al.*, 2006). The climate science consensus clearly shows that, without adequate mitigation and adaptation strategies to limit the extent of climate change and to lower vulnerability and increase resilience, simple coping cannot and should not be relied upon as a satisfactory response to the severe physical, social and political consequences of climate change (IPCC, 2007; Adger *et al.*, 2009; Schipper and Burton, 2009).

The 'wait and see' attitude overlooks that even where societies have considerable scientific capacities, technical know-how and financial power, effective coping with severe environmental events generally requires previous application of these resources through antecedent adaptation strategies. It also overlooks that mitigation may constitute the best kind of adaptation, since mitigation should lead to a significant reduction of the eventual human harm, as well as decrease the financial costs of adaptation, and of the subsequent costs of coping when actual disturbances occur.

A 'wait and see' attitude, furthermore, ignores that there are very large populations in the less industrialised parts of the world that lack most of the means to cope with the consequences of climate change, such as increased floods, droughts, and the expected resulting food shortages, greater spread of diseases, and so on. There are, moreover, considerable populations in the industrialised countries, such as poorer urbanites that are homeless or in temporary housing, who are vulnerable to phenomena such as heat waves, severe winter cold, or flooding. There are multiple cases of significant hardship and even mortality caused by such phenomena in Europe and North America, for example, the 2003 European heat wave, the extraordinarily cold European winter of 2009–2010, the 2005 Katrina hurricane and the flooding of New Orleans, as well as the flooding of Genoa, Italy in 2011). Hence, it is reasonable to suppose that, if those who are better-off and have the capability to act forgo vigorous mitigation and adaptation then the more vulnerable will increasingly will be in harm's way.

Both the defeatist 'all or nothing' and the complacent 'wait and see' attitudes may lead to unnecessary and supplementary difficulties, and bring about important additional costs, because the problems posed by climate change are taken on *too late* in the game (see Stern, 2007). These attitudes may, in fact, play into the hands of those who would rather opt for 'solving' climate change through costly new geo-engineering mega-projects – even if the solutions proposed are untested and may subject world populations to incalculable new risks (Gardiner, 2010).

Hence, allowing for the development of policy under the sway of either defeatist or complacent attitudes may lead to projects that contravene the precautionary principle (widely endorsed by international law), which demands that populations not be subjected to actions that represent *additional* levels of risk (see Gardiner, 2004, also see the 1998 Wingspread Statement on the Precautionary Principle). To this we may add that it is to be expected that species extinction rates will accelerate as climate change proceeds, and that coping activities, even if they involve grand schemes to help in the re-location of some threatened species, can do very little to stem this permanent loss of biological diversity.

Summation and conclusion

Insofar as our ethical and prudential responsibilities are supported by a variety of strong reasons that converge on the idea that we have good grounds to concern ourselves with human security, our personal responsibilities include *making human security possible*. However, both the defeatism of the 'all or nothing' attitude, of doing nothing because action supposedly is futile, and the complacency of the 'wait and see' attitude, of inaction until coping is inevitable, mean that human and other living beings will be exposed to increasing, very serious, harm, a large part of which is still preventable.

Since addressing climate change will require everybody's participation (see, e.g., Weber, 2006), these attitudes in society should lead to two questions: *how can these attitudes be effectively addressed?* and, for individuals who are aware of this situation, *what are their responsibilities under these circumstances?* Though there is no space here to speak to these questions in depth, with regard to the first question we may note recent research that suggests that such attitudes are the product of a combination of individual, social and material factors, which may be viewed as 'barriers' or as 'obstacles' to action (see Leiserowitz, 2006; Lorenzoni *et al.*, 2007; Gifford, 2008; Adger *et al.*, 2009; American Psychology Association, 2009).

Notably, defeatist and complacent attitudes may be the result of misapprehensions regarding the facts, troublesome ways of prioritising aims and objectives, or the behaviour-shaping power of problematic habits, all of which together may be the consequences of the dominance of certain unreflected lifestyle choices and common socio-cultural expectations (see, e.g., Lorenzoni *et al.*, 2007). As such, addressing these attitudes calls for measures such as better modes of communication, greater scientific literacy, and more open debate on policy options.

More specifically, the defeatist and complacent attitudes may perhaps be defused by drawing attention to concrete events with local effects, which, as such, are, or may become, available for everyday experience. One way to counter the 'all or nothing' attitude, for example, may be by clearly showing that significant additional, local, adaptation costs will have to be incurred by various proximate levels of government if mitigation is not taken on, and that, if adaptation is postponed, coping with severe events will be even more costly. Moreover, to address the 'wait and see' attitude it may be useful, for example, to help people recognise that climate change processes are already taking place, including nearby. This may be facilitated through guided tours to nature reserves, where the vertical or latitudinal displacement of species may be observable, or to weather observatories where records of local weather trends are displayed.

Another way to counter these attitudes may consist in creating opportunities for reflection on the fact that vulnerable sectors of our societies, locally and globally, will be directly hurt by climate change. Such facts collide with common humanitarian values that require solidarity with members of the larger community who are at risk, and call for proactive or remedial actions (see, e.g., Heyd, 2007: Chapter 2). Such processes of reflection may help people realise that climate change is an issue of prudential and ethical concern.

With regard to the second question, concerning one's personal responsibilities under these circumstances, we may note that such responsibilities do not, in any case, disappear in the presence of general inaction. Inaction by society should rather be grounds for renewed creative engagement in strategies that will get other citizens involved. As I have argued elsewhere (Heyd, 2011), insofar as inaction is the result of inadequacies in our *cultural frameworks*, addressing it may require a fundamental reassessment of these background conditions, which include the basic beliefs, values, habits, practices and lifestyle choices that control our everyday. Consequently, responsibility to act on climate change translates to responsibility to help *transform* the cultural frameworks that underlie our capacities (or lack of capacities) to take action (see Heyd, 2011). This applies to individuals who understand some of the various barriers and obstacles to action, such as ourselves.

Acknowledgements

I gratefully acknowledge the support of, and valuable discussions at, the Basque Centre for Climate Change (BC3), Bilbao, where I was Research Fellow in May 2011. I also express my appreciation to the School of Environmental Studies, University of Victoria, for providing me with an enriching environment for contextualising my research.

Notes

1 GECHS stands for the Global Environmental Change and Human Security Project of IHDP (International Human Dimensions Programme).
2 Henceforth I will refer to these ways of addressing climate change simply as 'mitigation and adaptation'.
3 According to the FAO, 868 million people were malnourished in 2010–2012 (FAO, 2012).
4 COP stands for Conference of the Parties (COP17) to the United Nations Framework Convention on Climate Change (UNFCCC).
5 This attitude may even be found among otherwise apparently well-informed people. See Baron (2006: 146), who speaks of 'a futile war against global warming', and suggests that it may be more 'cost-effective' to devote financial resources to the promotion of the development of poor countries to prepare them for the now inevitable consequences of climate change. There are reasons to be sceptical, in any case, about the value for the poor of many development and adaptation strategies, which primarily benefit already well-to-do minorities. See, for example, Kates (2009).
6 For example, see Baron (2006: 146), who speculates that 'it may turn out that new technology will make the war against climate change winnable after all'.

References

Adger, W.N., Lorenzoni, I. and O'Brien, K. (eds) (2009) *Adapting to Climate Change: Thresholds, Values and Governance*, Cambridge: Cambridge University Press.

American Psychology Association (2009) *Psychology and Global Climate Change: Addressing a Multi-faceted Phenomenon and Set of Challenges; A Report by the American Psychological Association's Task Force on the Interface Between Psychology and Global Climate Change*. Online. Available HTTP: http://www.apa.org/science/about/publications/climate-change.aspx (accessed 17 August 2011).

Attfield, R. (2010) 'Global warming, equity and future generations', *Human Ecology Review*, 17 (2): 102–105.

Baron, J. (2006) 'Thinking about global warming', *Climatic Change*, 77: 137–150.

Canadell, J.G., Le Quéré, C., Raupach, M.R., Field, C.B., Buitenhuis, E.T., Ciais, P., Conway, T.J., Gillett, N.P., Houghton, R.A., and Marland, G. (2007) 'Contributions to accelerating atmospheric CO_2 growth from economic activity, carbon intensity, and efficiency of natural sinks', *Proceedings of the National Academy of Sciences (PNAS)*, 104 (47): 18866–18870.

Chen, I-C., Hill, J.K., Ohlemüller, R., Roy, D.B., and Thomas, C.D. (2011) 'Rapid range shifts of species associated with high levels of climate warming', *Science* (19 August), 333 (6045): 1024–1026.

D'Andrea, W.J., Vaillencourt, D.A., Balascio, N.L., Werner, A., Roof, S.R., Retelle, M., and Bradley, R.S. (2012) 'Mild Little Ice Age and unprecedented recent warmth in an 1800 year lake sediment record from Svalbard', *Geology*, published online September 18, 2012.

Fagan, B. (2000) *The Little Ice Age: How Climate Made History 1300–1850*, New York: Basic Books.

FAO (2012) *FAO Hunger Portal*, Food and Agriculture Organization. Online. Available HTTP: http://www.fao.org/hunger/en/ (accessed 14 October 2012).

Gardiner, S.M. (2004) 'Ethics and global climate change', *Ethics*, 114: 555–600.

Gardiner, S.M. (2010) 'Is "arming the future" with geoengineering really the lesser evil?', in S.M. Gardiner, S. Caney, D. Jamieson, and H. Shue (eds) *Climate Ethics*, Oxford: Oxford University Press.

Garvey, J. (2008) *The Ethics of Climate Change: Right and Wrong in a Warming World*, London: Continuum International Publishing Group.

Gifford, R. (2008) 'Psychology's essential role in climate change', *Canadian Psychology/Psychologie Canadienne, 49:* 273–280.

Habermas, J. (1991) *Moral Consciousness and Communicative Action*, Cambridge, MA: MIT Press.

Heyd, T. (2007) *Encountering Nature: Toward an Environmental Culture*, Aldershot: Ashgate.

Heyd, T. (2011) 'Ethical and prudential responsibilities, culture and climate change' in V.P. Nanda (ed) *Climate Change and Environmental Ethics*, Piscataway, NJ: Transaction Publishers.

Homer-Dixon, T. (2006) *The Upside of Down: Catastrophe, Creativity, and the Renewal of Civilization*, New York: Knopf/Vintage.

Hulme, M. (2009) *Why We Disagree About Climate Change: Understanding Controversy, Inaction and Opportunity*, Cambridge: Cambridge University Press.

IPCC (2007) *Climate Change 2007: Synthesis Report*, Contribution of Working Groups I, II and III to the Fourth Assessment Report of the Intergovernmental Panel on Climate Change (Core Writing Team, Pachauri, R.K and Reisinger, A. (eds)) Geneva: IPCC.

Kant, I. (1785) *Grundlegung zur Metaphysik der Sitten*, trans. A.W. Wood (2002) *Groundwork for the Metaphysics of Morals*, New Haven, CT: Yale University Press.

Kates, R.W. (2009) 'Cautionary tales: adaptation and the world's poor', in L. Schipper and I. Burton (eds), *The Earthscan Reader in Adaptation to Climate Change*, London: Earthscan.

Leiserowitz, A. (2006) 'Climate change risk perception and policy preferences: the role of affect, imagery, and values', *Climatic Change*, 77: 45–72.

Lonergan, S., with Cocklin, C., Gleditsch, N.P., Langeweg, F., Brklacich, M., Gutierrez-Espeleta, E., Soroos, M., Narain, S. and Matthew, R. (2000) *Global Environmental Change and Human Security: Science Plan*, Bonn: IHDP Report 11, IHDP: The Global Environmental Change and Human Security Project.

Lorenzoni, I., Nicholson-Cole, S. and Whitmarsh, L. (2007) 'Barriers perceived to engaging with climate change among the UK public and their policy implications', *Global Environmental Change*, 17: 445–459.

Moser, S.C. (2007) 'More bad news: the risk of neglecting emotional responses to climate change information', in S.C. Moser and L. Dilling (eds) *Creating a Climate for Change*, Cambridge: Cambridge University Press.

Moser, S.C. and Dilling, L. (2007) *Creating a Climate for Change*, Cambridge: Cambridge: University Press.

Nihlén Fahlquist, J.A. (2009) 'Moral responsibility for environmental problems – individual or institutional?' *Journal of Agricultural and Environmental Ethics, 22 (2):* 109–124.

O'Brien, K, Eriksen, S., Sygna, L. and Næss, L.O. (2006) 'Questioning complacency: climate change impacts, vulnerability, and adaptation in Norway', *Ambio*, 35 (2): 50–56.

O'Brien, K., St. Clair, A.L., and Kristoffersen, B. (2010) *Climate Change, Ethics and Human Security*, Cambridge: Cambridge University Press.

Schipper, L, and Burton, I. (2009) *The Earthscan Reader on Adaptation to Climate Change*, London: Earthscan.

Singer, P. (1979) *Practical Ethics*, Cambridge: Cambridge University Press.

Steffen, W. (2009) *Climate Change 2009: Faster Change and More Serious Risks*, Canberra: Commonwealth of Australia.

Stern, N. H. (2007) *Stern Review on the Economics of Climate Change*, London: Cambridge University Press.

Vanderheiden, S. (2008) *Atmospheric Justice: A Political Theory of Climate Change*, Oxford: Oxford University Press.

Weber, E. (2006) 'Experience-based and description-based perceptions of long-term risk: why global warming does not scare us (yet)', *Climatic Change*, 77: 103–120.

Wolf, J. (2010) 'Ecological citizenship as public engagement with climate change', in L. Whitmarsh, S.J. O'Neill and I. Lorenzoni (eds) *Public Engagement with Climate Change – Behaviour Change and Communication*, London, Earthscan: 120–137.

Young, I. M. (2006) 'Responsibility and global justice: A social connection model', *Philosophy and Social Policy*, 23: 102–30.

28 The social handprint of sustainable citizenship[1]

Bronwyn Hayward

Introduction

This chapter examines the implications of a turn to active citizenship for sustainability, captured in the shift from ecological "footprints" to "handprint" imagery. I critique the way handprint metaphors are increasingly used in environmental education, environmental movements and the green economy, highlighting the limiting assumptions about "individual agency" which are implicit in many of these images of personally responsible citizenship. I then sketch an alternative concept of a "social handprint" as collaborative citizen action across space and time, offering this as more effective and democratic response to global power, social injustice, intergenerational inequity and dangerous environmental change.

Footprints and handprints

In 1996, Wackernagel and Rees published their ground-breaking Ecological Footprint model, a tool that graphically represented the physical space required for meeting the needs of a given population through manufacturing, distribution, consumption and waste (Wackernagel and Rees, 1996: 51–52). The model sparked other novel ways of measuring human consumption and impact on the environment, including carbon or water foot printing footprints (see for example Hoestra and Chapagain, 2007; Wiedman and Minx, 2007; Druckman and Jackson, 2009).

These footprint models have also inspired a variety of more recent handprint metaphors in environmental education, social campaigns and the green economy. In each case citizens are urged to *reduce* their ecological footprint and *increase* their handprint. In this context the use of handprint imagery suggests it is morally imperative and possible for individuals to take effective action to address environmental and social injustice. The handprint symbols carry an implicit if not explicit message, "if environmental degradation or social injustice is the negative effect of my life style … my handprint is what I do about it" (see for example, Edinburgh University Transition, 2010).

To give three examples, first, handprint images are increasingly used in environmental education programs. One of the earliest and most prominent education handprint models was launched at the Fourth International

Environmental Education Conference in 2007 (UNESCO, 2007), and subsequently refined at the Knowledge Conference in Kuala Lumpur in Malaysia in 2008 as a "hands for change" program (CEE, 2008a, 2008b; Gunawardene, 2008). The Centre for Environmental Education (CEE) program aimed to raise awareness about sustainability and assist individuals to assess the efficacy of their actions to address ecological degradation and social injustice. CEE developed an online web tool to measure individuals, ecological handprint, asking for example,

> Do you make conscious efforts to conserve/save water? Do you eat locally grown fresh food? Are you involved in activities that strengthen relationships between different generations in your community? Do you ever share any of your income or resources with other people (not your family or employees) or charitable organisations?
>
> (CEE, 2008b)

Used in this way in environmental education, handprints represent an individual's "action competence", that is the capability and skills of individuals to effect desired change (Jensen and Schnack, 1997; Bishop and Scott, 1998). The CEE "handprint calculator"; allocates more "positive agency points" for sustainable actions taken by individuals than actions taken on city, community, nation or state level (CEE, 2008b).

Second, handprints symbols are increasingly used by *non-governmental organizations* in campaigns to mobilize citizen action. For example the "Your positive handprint campaign" aimed to "commit 1 million people globally to look beyond reducing their carbon footprint, … to take positive action towards helping restore the environment" (Hickman, 2009). Similarly Catholic peace and justice organization CAFOD mobilized members join a protest march with handprint imagery arguing "our climate is in our hands" in London before the Copenhagen Summit in December 2009 (CAFOD, 2009).

Third, handprint imagery is also often used in the green economy. For example the company Carbon Handprint UK argues, "Your carbon footprint is your effect on our planet, your handprint is what you do about it" (Carbon Handprint UK, 2009). The website encourages consumers and business to enlarge their "carbon handprint" by engaging in a range of actions to offset carbon consumption from "off setting Christmas shopping" to adopting an "Eco-code" in the work place (Carbon Handprint, 2009) or investing in pro-environmental technology development to reduce carbon (Viridian, 2006).

Understanding the politics of handprint imagery

The political implications of these handprints images are ambivalent. In the Aristotelian vision, wellbeing is achieved by active participation in public life (Barry, 1999). Seen in this light the emerging handprint models resonate with the powerful political symbolism of citizen agency (the ability to both imagine and effect change in desired directions), which has inspired a long tradition of civil rights and indigenous people's movements. As Barry (1999) notes, this approach differs from the dominant yet passive views of citizen action as voting or ethical consumption.

However, individual handprint images can also reinforce unexamined notions of self-help citizenship, whereby individuals attempt to "fix" environmental symptoms in ways that leave systemic causes of these problems unchallenged. Some models of the handprint (for example, those promoting green consumers or local "civic gardeners") merely cajole citizens to purchase environmentally-friendly products or engage in local, small-scale ecological restoration projects. These efforts may be worthy and important, but largely ineffectual if their effect is to leave wider systemic patterns and processes of unjust investment, consumption and economic or social inequality unquestioned (Seyfang, 2005).

Psychologist Albert Bandura (2007: 14) illustrated the underlying limitations of consumer-citizenship in particular when he argued:

> As long as consumers' daily needs are met, they have little incentive to examine the humaneness of the working conditions, the level of pollution by the production processes, and the costs exacted on the environment to produce, ship, and market the profusion of goods and dispose the wastes. Under these modernized conditions, lifestyle practices are disconnected in time and place from the very ecological systems that provide the basis for them.

Bandura suggested however that citizens can and should be encouraged to reconnect lifestyle practice with time and spatial consequences of their agency, a link disconnected through globalization and export economies that make it difficult for people to consider the indirect effects of their choices and everyday practices on countless distant others (Bandura, 2007).

The strength and limits of individual agency

Individual handprints are therefore potentially powerful if ambivalent images. One implicit suggestion in this imagery is that individual actions to ameliorate environmental degradation and social injustice are possible, effective, and morally desirable. We cannot deny personal responsibility matters. The sociologist Zgymunt Bauman reminds us democracy depends on the courage of individuals to think critically and act morally in the face of pervasive injustice (see Bunting 2003). Yet excessive focus on personal responsibility fails to account for the way the actions and choices of individuals may also be constrained or mediated by their context, including social institutions, norms, habits and the structures of the economy (Seyfang, 2005; Jackson 2008; Giddens, 2009; 50–57). Despite their best efforts and intentions, citizens may find themselves wearing ill-fitting "institutional shoes" that distort the size of their ecological footprint, or "social gloves" that hamper their ability to effect change. Viewed in this light, for citizens to flourish, their capability for taking action also needs to be supported (Nussbaum, 2011).

The image of an individual handprint also risks shifting the focus of our attention towards the responsibilities of individuals and thus blinding us to the responsibilities of other actors (including private companies and public institutions). Individual handprints also risk depoliticizing our vision of agency, diminishing the potential of creative democratic action to "pro-environmental behavior" or "choices" made by citizen-consumers. This impoverished view of citizenship as individual responsibility for private action in the market can be understood in light of many years of neo-liberal policy. Neo-liberalism is used in a variety of ways, but commonly refers to an ideological project which aims to extend free market values (including values of efficiency, competition and choice) to new areas of public life including citizenship (Larner, 2000; Ong, 2004; O'Brien and Hochachka, 2010; Hayward, 2012). Neoliberalism equates the "good" citizen with a citizen-entrepreneur or citizen-consumer, at the expense of visions of social citizenship, communitarianism and the common good (Dobson, 2010b; Wolf, 2011).

With an excessive emphasis on citizenship as personal responsibility, our shared responsibility for creating a common future for human security may be lost from view. An overemphasis on personal responsibility can also divert attention from the complex systemic causes of problems. For example, handprints are often considered to be incriminating evidence at the scene of a crime. This direct assignment of blame is unhelpful, however, if it restricts our sense of responsibility to our immediate local environment or community, or impoverishes our understanding of our shared responsibility for effecting change. Likewise, the temptation to play ecological detective, identifying "whodunit" in environmental crimes, can distract our attention from the more challenging task of unraveling and putting right, the complex chains of responsibility that caused a local injustice. Accountability in decision making is important, but simply centering blame for complex ecological problems on bad managers or greedy bankers when things go wrong risks diverting our focus from the wider structural or systemic nature of injustice or risk rendering local struggles irrelevant (Young, 2006; Hayward, 2008). The centered symbol of a handprint narrows our idea of shared responsibility away from the vision of suffering as a complex chain of human interactions (Barry, 2005; Dobson, 2007: 64–5; Young, 2011). In this respect, work by McNeill and St. Clair (2009) helpfully reframes blame as "response-ability," reminding us that some have greater capability to respond more effectively than others because of their privileged economic and social position in a complex social world.

Individual handprint imagery in sustainability politics also potentially overstates the value of agency as action at the expense of imagination and reflection (Hayward, 2012). While no one denies that we have to do something, John Elliot (2009) recently argued after Arendt (1958) that taking time to pause and create space to engage in dialogue is also a critical way of exercising agency. Making room for listening and thinking seems a prudent strategy if we wish to develop an idea of "determinative morality," or a vision of what we *ought* to do and not to do (Dobson, 2003, 2007). Making space for reflection before launching into action is vital, given the wide reach and strong grasp of some citizens. Aided by global communications and strengthened by financial investment and international infrastructure, the capacity of some to define the problem, identify solutions and leave indelible handprints on the futures and pasts of distant others is both inestimable and undesirable. For example, although Restoration of the Earth calls for citizen action to rescue "great ecosystems," mobilizing a global campaign also threatens to marginalize the values of small indigenous communities (Hayward, 2008).

Given the limits of our understanding, our aim should not always be to *increase* our handprint, but to ensure we have a light touch, taking action in ways that are reversible and humble rather than heroic, ensuring our actions should be taken with the consent of those affected (Freeden, 2009). The experience of indigenous communities also reminds us that sometimes no amount of "action" can "put right" a past suffering, loss or injustice (Walker, 2004). In these situations listening, empathy and compassion also matter. For example, the Australian Sea of Hands campaign for reconciliation aims to highlight shared responsibility for past injustices of war, colonization, domination and genocide (Sea of Hands, 2008). This campaign reminds us that actions do not always speak louder than words. Bearing witness in formal truth, justice and reconciliation tribunals is also important (Dobson, 2010a).

The social handprint: rethinking citizenship as decentred, collaborative agency

The return to Aristotelian concepts of active citizenship in the face of environmental change, symbolized in the emergence of handprint thinking, should be welcomed. However, I argue that our actions should not be measured as individual acts, taken at one place and time, but should be thought of as collective efforts across time and space, taken with others and with consent. Iris Young rightly reminded us that given power is decentered, our political strategies ought to be as well (Young, 2006). Examples of a decentered approach to agency are seen in the Occupy or Indignados movements. These loosely networked citizen actions challenge the traditional green mantra of "think globally, act locally" by suggesting effective action in the face of global injustice should be regional and global responses as well, and taken in ways that are tentative, mindful of the limits of our knowledge and informed by local consent.

Thinking about the potential for citizens to decenter their action and connect empathetically with others (Hayward, 2008, 2012), enriches our view of sustainable citizenship beyond autonomy, reminding us that citizenship can also be expressed in relationships of mutual dependency and human connectedness, perhaps better expressed in images of "holding hands" in our interdependency and vulnerability.

Rethinking the plethora of handprint models emerging in discussions about environmental change provides a chance to rethink our impact as citizens and ways our actions might advance social and ecological justice. Our handprint could be that of a community of stewards or guardians. Taking action at multiple levels helps ensure that our proposed solutions are not rendered irrelevant, isolated gestures, or that they simply displace or obfuscate the problems. Non-governmental organizations, the media and citizens' tribunals have a role to play in this process of decentering action and opening it to wider scrutiny. While endless deliberation can drain a local minority's resources, informed public reasoning and inclusive deliberation help to ensure that our actions are taken in ways that are also sensitive and accountable to local communities and locally defined needs and concerns.

Viewed in this light, we have an obligation to ameliorate the wider harm we have caused (Pogge, 2002: 30–31), collaborating to make a difference "for good" (Dobson, 2003). In this broader and more politicized view of citizenship as shared responsibility, thinking about our social handprint is a way to raise awareness about our responsibility to collaborate, exercising agency in concert with others across space and generations to effect change.

Dobson argues that as citizens, our actions for sustainability should be motivated by moral reasoning and might have to be asymmetrical, that is we should not take action in expectation others will reciprocate, but simply because we think it's the right thing to do, "I will even if you won't" (Dobson, 2010b: 22–23). Dobson's work around new forms of sustainable citizenship is rich and thoughtful; however at the risk of an overstatement of difference, my point of departure is grounded in the context of indigenous communities who have experienced the devastating legacy of actions by former colonial powers. I suggest citizen actions should be informed by the consent of local communities to whom we feel obligation (Pateman and Mills, 2007; O'Brien and Hochachka, 2010; Hayward and O'Brien, 2010). In the New Zealand situation, for example, a treaty and tribunal hearing process has proven to be a very important moral vehicle for on-going reflection about grievances of colonization and for legitimating the actions of amelioration (Hayward and Wheen, 2004; Walker, 2004). Acting without such mechanisms to ensure that any action has the consent of those we hope to assist to avoid simply exacerbating the original injustice.

The power of social handprints

The rise of the Occupy and Indignados movements illustrates the potential for decentered citizen action, yet the struggle of those movements to maintain momentum reminds us that decentering loosely networked citizen actions is never enough. If citizen efforts are not to peter out in the face of hostile defensive reaction or become co-opted by a dominant rhetoric of personal responsibility, we also need to nurture stronger communitarian bonds to sustain effective social action. The alternative social handprints sketched in this chapter are normative, and suggest an image of citizenship as communitarian agency and resistance.

A handprint can be a valuable metaphor for agency, reminding us of that in exercising our citizenship we discover our capacity to imagine and effect change. In exercising agency, however, our aim might not be to always "expand our individual handprints" but to take collaborative action that connects with others across space and time, recognizing our mutual interdependency (Hayward, 2012). In questioning our vision of citizenship as individual responsibility, I am not calling for a loss of courage to act but the vision to enlarge our field of compassion; seeing thoughtful, humble, networked actions as a means to secure the capacity for citizens to flourish on a finite planet.

Note

1 Earlier versions of this chapter were presented at seminars at the University of Oslo, University of Surrey and University College London, and Hayward, B. (2010) 'The social handprint: Understanding decentred citizen agency and UK Uncut', RESOLVE Working Paper Series, 06-10 University of Surrey: Guildford. For an extended discussion see Hayward (2012).

References

Arendt, H. (1958/1998) *The Human Condition*, Chicago, IL: University of Chicago Press.

Bandura, A. (2007) 'Impeding ecological sustainability through selective moral disengagement', *International Journal of Innovation and Sustainable Development*, 2 (1): 8–35.

Barry. J. (1999) *Rethinking Green Politics*, London: Sage.

Barry, J (2005) 'Resistance is fertile: From environmental to sustainability citizenship', in D. Bell and A. Dobson (eds) *Environmental Citizenship: Getting from Here to There?*, MIT Press: 21–48.

Bishop, K. and Scott, W. (1998) 'Deconstructing action competence: developing a case for a more scientifically-attentive environmental education', *Public Understanding of Science*, 7 (3): 225–236.

Bunting, M. (2003) 'Passion and pessimism', *The Guardian*. Online. Available HTTP: http://www.guardian.co.uk/books/2003/apr/05/society (accessed 10 March 2012).

CAFOD (2009) *Our Climate is in Our Hands: Just One World*. Online. Available HTTP: http://www.cafod.org.uk/news/uk-news/greenbelt-2009-04-09(accessed 10 March 2012):

Carbon Handprint UK (2009) *Your Carbon Footprint is Your Effect on Our Planet. Your Handprint is What You do About It*. Online. Available HTTP http://www.carbonhandprint.co.uk/ (accessed 10 March 2012).

CEE (2008a) *Increase Your Hand Print: Decrease Your Footprint*, Gujarat, India: Centre for Environmental Education. Online. Available HTTP: http://www.handsforchange.org/handprinttools.html (accessed 12 December 2011).

CEE (2008b) *Hands for Change: Handprint Calculator*, Gujarat, India: Centre for Environmental Education. Online. Available HTTP: http://www.handsforchange.org/handprintquiz.aspx (accessed 5 January 2012).

Dobson, A. (2003) *Citizenship and the Environment*, Oxford: Oxford University Press.

Dobson, A. (2007) *Green Political Thought*, 4th edition, New York, USA: Routledge.

Dobson, A. (2010a) 'Democracy and nature: speaking and listening' *Political Studies*, 58 (4): 752–768.

Dobson, A. (2010b) 'Environmental citizenship and pro-environmental behaviour: rapid research and evidence review', London: Sustainable Development Research Network.

Druckman, A. and Jackson, T. (2009) 'The carbon footprint of UK households 1990–2004: a socio-economically disaggregated, quasi-multiregional input-output model', *Ecological Economics*, 68 (7): 2066–2077.

Edinburgh University Transition (2010) *Learn About our Impact on the Environment and What We Can Do About It: Handprint Social*. Online. Available HTTP: http://www.transitionedinburghuni.org.uk/projects/meetings-events/handprint-socials/ (accessed 1 November 2010).

Elliot, J. (2009) 'Building educational theory through action research', in S. Noffke and B. Somekh (eds) *The SAGE Handbook of Educational Action Research*, London: Sage.

Freeden, M. (2009) 'Failures of political thinking', *Political Studies*, 57 (1): 141–164.

Gunawardene, N. (2008) *From Footprints to Handprints: Treading Lightly on the Earth*, Online. Available at HTTP: http://movingimages.wordpress.com/2008/03/02/from-footprints-to-handprints-treading-lightly-on-earth/ (accessed 13 September 2009).

Gibney, M., Howard-Hassmann, R., Coicaud, J. and Steiner, N. (2008) *Facing Up to the Past*, Philadelphia, PA: University of Pennsylvania Press.

Giddens, A. (2009) *The Politics of Climate Change*, Cambridge: Polity.

Hayward, B. (2008) 'Let's talk about the weather: decentering democratic debate about climate change', *Hypatia*, 23 (3): 79–98.

Hayward, B. (2012) *Children, Citizenship and Environment: Nurturing a Democratic Imagination in a Changing World*, London: Earthscan/Routledge.

Hayward, B. and O'Brien, K. (2010) 'Social contracts in a changing climate: security of what, and for whom?', in K. O'Brien, A. St. Clair, and B. Kristoffersen (eds) *Climate Change, Ethics and Human Security*, Cambridge: Cambridge University Press: 451–521.

Hayward, J. and Wheen, N. (2004) *The Waitangi Tribunal: Te Roopu Whakamana I te Tiriti o Waitangi*, Wellington: Bridget Williams Books.

Hickman, L. (2009) 'Can Joanna Lumley make us listen when it comes to the environment?' *The Guardian* 29 June 2009. Online. Available HTTP: http://www.guardian.co.uk/environment/blog/2009/jun/24/joanna-lumley-climate-change (accessed 15 November 2012).

Hoestra, A. and Chapagain, A. (2007) 'Water footprint of nations: water use by people as a function of their consumption pattern', *Water Resource Management*, 21: 35–48.

Jackson, J. (2008) 'Where is the well-being dividend? Nature, structure and consumption inequalities', *Local Environment*, 13 (8): 703–723.

Jensen, B. and Schnack, K. (1997) 'The action competence approach in environmental education', *Environmental Education Research*, 3 (2): 163–178.

Larner, W. (2000) Neoliberalism: Policy, ideology, governmentality, *Studies in Political Economy*, 63: 5–25.

McNeill, D. and St. Clair, A.L. (2009) *Global Poverty, Ethics, and Human Rights: The Role of Multilateral Organisations*, New York: Routledge.

McNeill, D. and St. Clair, A. (2011) 'Poverty, human rights, and global justice: the response-ability of multilateral organizations', *Globalizations*, 8 (1): 97– 111.

Nussbaum, M. (2011) *Creating Capabilities: The Human Development Approach*, Cambridge MA: Belknap/Harvard University Press.

O'Brien, K. and Hochachka, G. (2010) 'Integral adaptation to climate change', *Journal of Integral Theory and Practice*, 5 (1): 89–102.

Ong, A. (2004) *Neoliberalism as Exception: Mutations in Citizenship and Sovereignty*, North Carolina: Duke University Press.

Pateman, C. and Mills, C. (2007) *Contract and Domination*, Cambridge: Polity Press.

Pogge, T. (2002) *World Poverty and Human Rights*, Cambridge: Polity Press.

Sea of Hands (2008) *Sea of Hands Event*, European Network for Indigenous Australian Rights. Online. Available HTTP: http://www.eniar.org/events/seaofhands.html (accessed 20 November 2010).

Seyfang, G. (2005) 'Shopping for sustainability: can sustainable consumption promote ecological citizenship?' *Environmental Politics*, 14 (2): 290–306.

UNESCO (2007) *Action Towards Sustainability Launch Fourth International Environmental Education Conference in November 2007*. Online. Available HTTP: http://www.tbilisiplus30.org/index.htm (accessed 13 March 2010).

Viridian (2006) *Viridian Note 00480: The Algae Hummer*, by Bruce Sterling. Online. Available HTTP: http://www.viridiandesign.org/2006/12/viridian-note-00480-algae-hummer.html (accessed 20 November 2010).

Wackernagel, M. and Rees, W. (1996) *Our Ecological Footprint,* Gabriola Island, Canada: New Society Press.

Walker, R. (2004) *Ka Whawhai Tonu Matou: Struggle without End*, Auckland: Penguin Books.

Wiedmann, T. and Minx, J. (2007) 'A definition of "carbon footprint"', in C. Pertsova (ed.) *Ecological Economics; Research Trends*, Hauppauge, NYNova Science Publishers: 1–11.

Wolf, J. (2011) 'Ecological citizenship as public engagement with climate change', in L. Whitmarsh, S. O'Neill and I. Lorenzoni (eds) *Engaging the Public with Climate Change*, London: Earthscan.

Young, I. (2006) *Too Much Blame, Not Enough Responsibility*, Dissent Magazine (Winter). Online. Available HTTP: http://www.dissentmagazine.org/article/katrina-too-much-blame-not-enough-responsibility (accessed 15 November 2012).

Young, I. (2011) *The Responsibility for Justice*, Oxford: Oxford University Press.

29 Disaster risk reduction informing climate change adaptation

Social capital in Agüita de la Perdiz

Paulina Aldunce, Virginia Levín and Alejandro León

Introduction

Climate change is a major issue that human societies are facing in the twenty-first century. The Intergovernmental Panel of Climate Change (IPCC) presents evidence of changes in climate-related hazards and extreme events, as well as for future projections in terms of frequency, magnitude and duration (IPCC, 2007; IPCC, 2012). Even if ambitious mitigation efforts were undertaken, it is expected that societies will face significant and changing risks, and more uncertainty (Reilly *et al.*, 1994; Sterr, 2000). Therefore, adaptation has emerged as critical when responding to climate change and climate-related hazards events (Debels *et al.*, 2009).

Multiple links between climate change adaptation (CCA) and disaster risk reduction (DRR) have been reported in the literature (Thomalla *et al.*, 2006; Aldunce *et al.*, 2008; O'Brien *et al.*, 2008; Adger *et al.*, 2009a; Pelling and Schipper, 2009). One of the most salient of these links is that CCA could be enhanced by DRR research, experience, and practice (Pelling and Schipper, 2009). The local level is an area in which the DRR community has a long history and a large amount of experience (O'Brien *et al.*, 2008). It has also been stressed that the local level is where DRR has been most effective and where the links to adaptation are most apparent (Pelling and Schipper, 2009). Further, it is becoming increasingly apparent that adaptation efforts need to be implemented at the local level. Therefore, there is much to gain from the work of the DRR community at the local level. What has been especially instructive are studies that focus on enabling conditions and barriers for an effective response to climate-related disasters.

This chapter presents a local case study from Agüitas de la Perdiz Sur, a community in Chile that in 2005 was affected by a destructive and extreme rainfall event that had no precedent in 145 years.

Contrary to what was expected for such a severe hazard that was coupled with extremely vulnerable conditions, no deaths and only a few injuries were reported. Therefore, we were curious to explore what enabled this community to overcome the multiple barriers to an effective response. Information gathered in the field reveals that social capital was the decisive element that helped the community to protect their lives and to cope with a disaster of such magnitude. How this happened and how social capital was useful for them to overcome those barriers is discussed in this chapter. We argue that there are important lessons to be learned from this case to inform CCA.

First, the chapter examines the links between DRR and CCA that are reported in the literature. Second, the methodological framework is presented, as well as the case study in terms of geographical and socio-economic characteristics. Third, the discussion focuses on social capital, and aspects related to it, as this is the most relevant enabler for overcoming the barriers for an effective response presented in the case study. Fourth, some reflections are presented.

Linking disaster risk reduction and climate change adaptation

The patterns of some climate-related hazards are acknowledged as changing (IPCC, 2012), and this partially explains the increase in hydro-meteorological disasters over the last two decades (UN/ISDR, 2004; CRED, 2009; Peduzzi *et al.*, 2009), with rising damage costs of extreme hydro-meteorological events (Okuyama and Chang, 2004; Pielke *et al.*, 2008). Nevertheless, changes in hazards should not be considered as the only driver for increasing disastrous events (Gaillard, 2010; Lavell, 2011). There are multiple stressors and drivers that generate vulnerability (Eriksen *et al.*, 2011; Lavell, 2011) and changes in climatic conditions tend to disproportionately affect more severely vulnerable populations and developing countries (Haddad, 2005; Roberts and Parks, 2006), as their vulnerability is determined by physical, social, economic and environmental factors (Wisner *et al.*, 2004; Leichenko and O'Brien, 2008). This changing risk picture and the uncertainty that comes with it complicate the decision making processes. Additionally, in many countries where extreme events occur so frequently, coping capacities are already overwhelmed and this has negative effects on long-term development (Thomalla *et al.*, 2006).

At the start of the century the disaster risk reduction and the climate change adaptation communities worked in isolation (Thomalla *et al.*, 2006), but recently more efforts have aimed to link these two fields (see for example Schipper and Pelling, 2006; Venton and La Trobe, 2008; O'Brien *et al.*, 2008; Pelling and Schipper, 2009). The special report of the IPCC "Managing the risk of extreme events and disasters to advance climate change adaptation" (IPCC 2012) is a good example of such collaboration, where more than one hundred researchers and stakeholders were brought together to assess the current knowledge and experience on responses to changing risks in a changing climate.

There are multiple ways, in both theory and practice, in which CCA and DRR are linked. First, some authors consider that adaptation is a matter of risk management; it is about the future possibilities of the occurrence of a threat and its consequences and impacts (Pelling and Schipper, 2009; Birkmann *et al.*, 2010). Second, disasters open a window of opportunity for change. Individuals and organizations tend to be more open to changing behaviors after a disaster, and this is an opportunity to learn and adapt, and to include climate change considerations

(Pelling and Schipper, 2009). Third, using lessons learned by the DRR community can enhance CCA policy and practice. The DRR community has a long history of responding and adapting to climate-related hazards (O'Brien *et al.*, 2008) and many studies have focused on lessons learned from disasters (see for example McEntire, 2002; Lavell, 2009; Moore *et al.*, 2009; Birkmann *et al.*, 2010). These studies can be useful in examining and informing literature and praxis about enablers and barriers for CCA (Adger *et al.*, 2009b). By using these experiences, the CCA community can draw on some existing mechanisms, institutions and organizations created for DRR, avoiding re-inventing the wheel, wasting time, and creating overlaps (Pelling and Schipper, 2009). Following the recognition that CCA could be enhanced by DRR experiences, we discuss a case study that focuses on examining enablers and barriers for DRR in the next sections.

Background and methodology

The methodology is a case study conducted in the community of Agüita de la Perdiz Sur, located in downtown Concepción, the second largest city in Chile with a population of 848,023 habitants (INE, 2005). On June 26, 2005, an above-normal precipitation event of 162.2 mm in 24 hours took place, representing a record in 142 years. In this area the normal precipitation for the month of June is 218 mm, and in June 2005 it was 511 mm. The event affected the population and infrastructure. Among the 282 houses in the community, almost 100 houses were completely or partially destroyed by landslides (see Figure 29.1) (DMC, 2005; ONEMI, 2005).

The rationale for selecting Agüita de la Perdiz Sur as a case study is based on the following arguments. Climate change will exacerbate the inequity and poverty conditions of the most vulnerable population in developing countries (Barnett, 2003), and as in the case of Agüita de la Perdiz, many of these populations are already experiencing frequent disasters that are overwhelming their coping capacities (Thomalla *et al.*, 2006). Agüita de la Perdiz Sur is an example where people not only live in a risky area, but are also exposed to multiple stressors and barriers for an effective response.

Houses in this area are built on the high sloped Caracol hill (see Figure 29.2), classified as a high landslide risk area (Mardones and Vidal, 2001; Hauser, 2005). Nonetheless, as in many other places in the developing world, the area has been colonized by illegal settlers searching for a place to live (Tribuna del Biobío, 2005). The population of this community is also socially vulnerable because of poverty, inadequate living conditions and low educational levels. With a hazard of this magnitude that developed mostly during the night and affected a population living in a risky area, it is remarkable that only a few injuries and no deaths were reported (ONEMI, 2005). Therefore, this case study is an opportunity to learn about enablers and barriers for DRR.

A multilevel analysis was carried out with emphasis on the local context. The methods consisted of a document review; data base analysis; walks on the site; and semi-structured in-depth interviews with the affected population and government personnel involved in DRR.

Figure 29.1 (a) and (b) Landslides in Agüita de la Perdiz, June 2005. Source: Paulina Aldunce

Figure 29.2 (a) and (b) Houses built on high slopes, Caracol hill, Agüita de la Perdiz. Source: Paulina Aldunce

Social capital as an enabler and barrier in Agüita de la Perdiz

The case of Agüita de la Perdiz Sur illustrates well some enabling conditions for DRR. The research findings show that the most relevant enabler that helped people to respond to the June 26 disaster is undoubtedly social capital. There are multiple definitions of this contested concept (see for example Loury, 1977; Coleman, 1988; Putnam, 1993). We used Bourdieu's (1985: 248) definition of social capital, understood as the aggregate of the actual or potential resources which are linked to a durable network of relationships. In the context of disasters, three key dimensions of social capital are sense of community, place attachment, and citizen participation (for definition of theses concepts see Tartaglia, 2006; Norris *et al.*, 2008). Interviewees expressed arguments that are related to each of these three dimensions.

Sense of community and place attachment

Sense of community and place attachment has a strong presence within Agüita de la Perdiz Sur and we consider them mainly as enablers. However, they have also contributed to create barriers, as discussed below. Most respondents expressed that they feel part of and integrated into the community, and many also explained that they have close relationships with other members of the community as family or friends. Something that emerged strongly is how the physical characteristics of the place, in which this community lives, have shaped and contributed to a sense of belonging. On the one hand, interviewees stressed that they share living in a risky place, and consequently they all share being exposed to disasters and need each other in order to survive. On the other hand, they explained that because the place is surrounded by hills and because the only way to enter is through two main streets, this has created a sense of living in a small village, where everybody knows each other, even though they are in the downtown area of a big city. Having this sense of belonging has resulted in a proactive community that takes care of each other and that has established strong networks within the community. For example, during the June landsides community members organized a shelter in the local school and in some houses to gather affected neighbors. The community helped evacuate older people, and in general was effective in self-responding in a timely manner. Not having to wait for external help was crucial during this disaster as the city was in chaos and the emergency services were overwhelmed.

These results can also be found in the literature. A sense of belonging decreases people's vulnerability and improves their effectiveness in responding to disasters (Paton and Johnston, 2001; Pearce, 2003). This is achieved by increasing access and utilization of social networks, and by facilitating community involvement in disaster prevention, mitigation, and responses (Paton and Johnston, 2001; Norris *et al.*, 2008).

Undoubtedly, sense of community has had positive outcomes in the case study. Nevertheless, it has also had a negative implication as it has constrained the relocation of the community to less risky areas; we consider this implication as a barrier to effective responses. As stated in the literature, poverty conditions force people to use marginal and risky terrains for housing and other needs (Osti, 2004). This is the case for people in the community of Agüita de la Perdiz, as they have illegally occupied terrains that are classified as high risk for landslides. In the 1970s and 1980s, several attempts were made by the government to move people from Agüita de la Perdiz Sur and give them houses outside the city. However, people returned to their homelands, mainly for two reasons. First, they wanted to be closer to downtown Concepción to search for work opportunities and to decrease costs such as transportation. Second, being part of the social networks in the Agüita de la Perdiz community is part of their coping strategies against poverty. The implication of these results concur with what has been stated in the literature; that people living in risky areas frequently return to disaster prone areas (Pelling and Schipper, 2009), as in many cases the strong sense of belonging and levels of poverty result in decisions to live in places at risk (Christoplos *et al.*, 2009).

Citizen participation

Citizen participation is a key element to decrease people's vulnerability and to improve DRR by means of improving problem identification and problem-solving strategies within local communities (Pearce, 2003; Osti, 2004). On-site verification shows that community participation exists in Agüita de la Perdiz Sur, that people in the community self-organize in the face of disasters, and that a number of different members of the community play a meaningful role. Therefore, based on this case study, we consider citizen participation as an enabler for DRR. A critical factor that underpins citizen participation in the study area is the presence of strong leaders within the community. Results of our research concur with what has been described elsewhere (Norris *et al.*, 2008) in the sense that these leaders have a strong place attachment.

Leadership not only emerged as an important enabler for DRR in the case study, but also has helped to overcome one of the most salient barriers presented; the insufficient consideration and inclusion of the community by government agencies in charge of disaster management. Even though a formal institution that supports community participation exists, the Civil Protection National Plan (CPNP) (the CPNP replaced the former "Emergency National Plan"), local people have not been sufficiently included in the managerial system. The CPNP states that "participation should constitute a process by itself, starting at the local level and led by the county, because this is the administrative level that is closest to the local people" (República de Chile, 2002: 10).

Interviewees mentioned the importance of different members of the community taking a lead in the responses during and after the disaster. For example, community leaders organized the emergency brigade that was constituted by local people and ensured that all affected neighbors received adequate and timely aid and support. We find, as elsewhere in the literature, that when civilians are insufficiently included in formal institutions, as in the case of

Agüita de la Perdiz, informal leaders have an even more important role to play (Moser and Ekstrom, 2010). The presence of community leaders has also been reported as one important enabling condition for adaptation to climate change (Moser and Ekstrom, 2010).

Even though the presence of leaders within the case study community has been an important enabler, this should not be used as a justification to not invest greater efforts into improving formal participation by the community. Arguments that underpin this are first, that DRR is best facilitated when government initiatives include working in partnership with civil society at different levels (Pelling and Schipper, 2009). For example, a critical aspect in achieving effective DDR is that community members, participating in formal committees, have the opportunity for their opinions and point of view to be considered (Newport and Jawahar, 2003). Second, in Agüita de la Perdiz Sur the link between community and public agencies has been weak, mainly because public agencies do not effectively engage the community in DRR decision making and actions. CCA literature also points out that in order to scale-up local efforts and successes, it is necessary to have supporting institutions at the local and national level (Pelling and Schipper, 2009).

Third, participation and cooperation during disasters should also be framed as an opportunity to build relationships, and to dispel mistrust and hostility between public organizations and also within the community (Trim, 2004). For the successful involvement of the local community in disaster risk reduction, roles, responsibility and coordination needs to be clearly defined and agreed upon (Newport and Jawahar, 2003). In Chile, the CPNP proposes a way of organizing tasks that includes definition of roles, as well as political, legal, technical and operational responsibilities. Nevertheless, different stakeholders working on DRR in Agüita de la Perdiz Sur have not always followed what the CPNP states, and we observed also that cooperation and coordination between different stakeholders does not always exist. Lack of both coordination and collaboration hamper the planning and the management of disasters.

Local knowledge

Another pattern that strongly emerged from the interviews is the role of local knowledge as an enabler for DRR. Over time people from Agüita de la Perdiz Sur have developed different ways of coping with disasters based on knowledge and understanding of the environment that surrounds them. For example, interviewees explained that the experience gained from having faced several landslides during past winters, has contributed to accumulate knowledge in regards to which places are more susceptible to landslides, and who are the most vulnerable within the community. This knowledge served them well during the extreme rainfall in 2005, as people were able to prioritize the evacuation starting with those most vulnerable or living in the most exposed places within the area. People also stressed that, based on the experience of living in the place, they realized during the first hours of the event that this was outside anything they had experienced before. This allowed them to act in a timely manner by alerting others and organizing the response, avoiding a situation that could easily end in complete chaos, especially as the event happened during the night. These results concur with what has been stated in the disaster management literature, that knowledge gained by experiencing recurring disasters can result in positive outcomes (Paton *et al.*, 2000; Pearce, 2003). In a similar way, local knowledge based on observations of the physical environment has been described as a fundamental condition for adaptation (Olsson and Folke, 2001; Berkes, 2007; Eriksen *et al.*, 2011).

Final remarks

The case Agüita de la Perdiz Sur presented in this chapter offers a concrete example in which enabling conditions for and barriers to disaster risk reduction (DRR) can be extrapolated to climate change adaptation (CCA). It shows that social capital manifested through a sense of community, place attachment and citizen participation play a key role for enabling DRR. Additionally, local knowledge is found to be an important enabler for DRR in the study area. This local knowledge, based on the experience of living in a disaster-prone area and on observing the local physical environment, is a fundamental condition for DRR and is useful in informing CCA.

Even though a sense of community and place attachment are crucial enablers for DRR in this case study, they can also become barriers. Both sense of community and place attachment underpin the decisions made to live in and return to a risky place and thus ignoring the options of relocation offered by government agencies. Another important barrier is that existing governance arrangements have been unsuccessful in sufficiently engaging the affected population.

To improve DRR in this area further efforts are needed. The most urgent one is to foster engagement of citizens in formal institutions. We argue that efforts to tackle this should recognize the special characteristics of this community such as the presence of natural and informal leaders, strong social capital, a sense of community, and local knowledge.

The case study presented here is an example of how a self-organized community that has developed social capital can be successful in decreasing their vulnerability. However, this should never be used as an excuse to delay further efforts towards sustainable adaptation; including aspects of poverty reduction, social justice, and environmental integrity. There is an urgent need to overcome these barriers not only due to the high vulnerability, but also because future vulnerability is expected to increase due to climate change.

Acknowledgement

We would like to thanks the participants from the case study, for their generosity and willingness to participate. This publication received the support of and is a contribution to FONDAP #1511009.

References

Adger, W., Lorenzoni, I. and O'Brien, K. (2009a) *Adapting to Climate Change: Thresholds, Values, Governance,* Cambridge: Cambridge University Press.

Adger, W.N., Dessai, S., Goulden, M., Hulme, M., Lorenzoni, I., Nelson, D.R., Naess, L.O., Wolf, J. and Wreford, A. (2009b) 'Are there social limits to adaptation to climate change?', *Climatic Change,* 93: 335–354.

Aldunce, P., Neri, C. and Debels, A. (2008) 'Aplicación del índice de utilidad de prácticas de adaptación en la evaluación de dos casos de estudio en América Latina', in Aldunce, P., Neri, C. and Szlafsztein, C. (eds) *Hacia la evaluación de prácticas de adaptación ante la variabilidad y el cambio climático,* Belem: NUMA/UFPA.

Barnett, J. (2003) 'Security and climate change', *Global Environmental Change-Human and Policy Dimensions,* 13: 7–17.

Berkes, F. (2007) 'Understanding uncertainty and reducing vulnerability: Lessons from resilience thinking', *Natural Hazards,* 41: 283–295.

Birkmann, J., Buckle, P., Jaeger, J., Pelling, M., Setiadi, N., Garschagen, M., Fernando, N. and Kropp, J. (2010) 'Extreme events and disasters: A window of opportunity for change? Analysis of organizational, institutional and political changes, formal and informal responses after mega-disasters', *Natural Hazards,* 55: 637–655.

Bourdieu, P. (1985) 'The forms of capital', in J. Richardson (ed.) *Handbook of Theory and Research for the Sociology of Education,* New York: Greenwood.

Chistoplos, I., Anderson, S., Arnold, M., Galaz, V., Hedger, M., Klein, R. and Le Goulven, K. (2009) *The Human Dimension of Climate Change: The Importance of Local and Institutional Issues,* Stockholm: Commission on Climate Change and Development.

Coleman, J. (1988) 'Social capital in the creation of human capital', *American Journal of Sociology,* 94: 95–120.

CRED (2009) *Estimated damages (US$ billion) caused by reported natural disasters 1900–2009,* Center for Research on the Epidemiology of Disasters, EM-DAT The International Disaster Database.

Debels, P., Szlafsztein, C., Aldunce, P., Neri, C., Carvajal, Y., Quintero-Angel, M., Celis, A., Bezanilla, A. and Martinez, D. (2009) 'IUPA: a tool for the evaluation of the general usefulness of practices for adaptation to climate change and variability', *Natural Hazards,* 50: 211–233.

DMC Dirección Meteorologica de Chile (2005) 'Boletín climatológico', in G. Torres (ed.) *Volumen XXI N°6 Junio de 2005,* Santiago: Dirección Meteorológica de Chile.

Eriksen, S., Aldunce, P., Bahinipati, C.S., D'Almeida, R., Molefe, J. I., Nhemachena, C., O'Brien, K., Olorrunnfemi, F., Park, J., Sygna, L. and Ulsrud, K. (2011) 'When not every response to climate change is a good one: Identifying principles for sustainable adaptation', *Climate and Development,* 3: 7–20.

Gaillard, J.C. (2010) 'Vulnerability, capacity and resilience: Perspectives for climate and development policy', *Journal of International Development,* 22: 218–232.

Haddad, B. (2005) 'Ranking the adaptive capacity of nations to climate change when socio-political goals are explicit', *Global Environmental Change,* 15: 165–176.

Hauser, A. (2005) *Informe geológico geotécnico, preliminar: sectores Agüita de la Perdiz y Cerro La Pólvora, Concepción, VIII Región.* Santiago, Chile: SERNAGEOMIN.

INE (2005) *Censo 2002: ciudades, pueblos y aldeas,* Santiago de Chile: Instituto Nacional de Estadísticas de Chile.

IPCC (2007) *Climate Change 2007: The Physical Science Basis,* Contribution of Working Group I to the fourth assessment report of the Intergovernmental Panel on Climate Change, in S. Solomon, D. Qin, M. Manning, A. Chen, M. Marquis, K. Averyt, M. Tignor and H. Miller (eds), Cambridge and New York: Cambridge University Press.

IPCC (2012) *Managing the Risks of Extreme Events and Disasters to Advance Climate Change Adaptation,* A special report of Working Groups I and II of the Intergovernmental Panel on Climate Change, Cambridge and New York: Cambridge University Press.

Lavell, A. (2009) *Local Disaster Risk Reduction: Lessons from the Andes,* Lima: Secretaría General de la Comunidad Andina, Proyecto PREDECAN.

Lavell, A. (2011) *Desempacando la adaptación al cambio climático y la gestión del riesgo: buscando las relaciones y diferencia, una crítica y construcción conceptual y epistomológica,* Secreataria General de la FLASCO y La Red para el Estudio Social de la Prevención de Desastres en América Latina.

Leichenko, R. and O'Brien, K. (2008) *Environmental Change and Globalization: Double Exposure,* New York: Oxford University Press.

Loury, G.C. (1977) 'A dynamic theory of racial income differences', in P.A. Wallace and A.M. La Mond (eds) *Women, Minorities and Employment Discrimination*, Lexington, MA: Lexington Books.

Mardones, M. and Vidal, C. (2001) 'La zonificación y evaluación de los riesgos naturales de tipo geomorfológico: Un instrumento para la planificación urbana en la ciudad de Concepción', *EURE,* 27: 97–122.

McEntire, D. (2002) 'Coordinating multi-organisational responses to disasters: Lessons from the March 28, 2000, Fort Worth Tornado', *Disaster Prevention and Management,* 11: 369–379.

Moore, M., Trujillo, H.R., Stearns, B.K., Basurto-Davila, R. and Evans, D.K. (2009) 'Learning from exemplary practices in international disaster management: A fresh avenue to inform US policy?', *Journal of Homeland Security and Emergency Management,* 6: Article 35.

Moser, S.C. and Ekstrom, J.A. (2010) 'A framework to diagnose barriers to climate change adaptation', *Proceedings of the National Academy of Sciences,* 107: 22026–22031.

Newport, J. and Jawahar, G. (2003) 'Community participation and public awareness in disaster mitigation', *Disaster Prevention and Management,* 12: 33–36.

Norris, F.H., Stevens, S.P., Pfefferbaum, B., Wyche, K.F. and Pfefferbaum, R.L. (2008) 'Community resilience as a metaphor, theory, set of capacities, and strategy for disaster readiness', *American Journal of Community Psychology,* 41: 127–150.

O'Brien, K., Sygna, L., Leichenko, R., Adger, N., Barnett, J., Mitchell, T., Schipper, L., Tanner, T., Vogel, C. and Montreux, C. (2008) *Disaster Risk Reduction, Climate Change Adaptation and Human Security,* A Commissioned Report for the Norwegian Ministry of Foreign Affairs, Oslo: University of Oslo.

Okuyama, Y. and Chang, S. (2004) *Modeling Spatial and Economic Impacts of Disasters,* Berlin, Heidelberg, New York: Springer.

Olsson, P. and Folke, C. (2001) 'Local ecological knowledge and institutional dynamics for ecosystem management: A study of Lake Racken Watershed, Sweden', *Ecosystems,* 4: 85–104.

ONEMI (2005) *Informe consolidado. Sistemas frontales sucesivos: 10 mayo–15 julio 2005,* Santiago, Chile: Oficina Nacional de Emergencia, Ministerio del Interior.

Osti, R. (2004) 'Forms of community participation and agencies' role for the implementation of water-induced disaster management: protecting and enhancing the poor', *Disaster Prevention and Management,* 13: 6–12.

Paton, D. and Johnston, D. (2001) 'Disasters and communities: vulnerability, resilience and preparedness', *Disaster Prevention and Management,* 10: 270–277.

Paton, D., Smith, L. and Violanti, J. (2000) 'Disaster response: risk, vulnerability and resilience', *Disaster Prevention and Management,* 9: 173–179.

Pearce, L. (2003) 'Disaster management and community planning, and public participation: How to achieve sustainable hazard mitigation', *Natural Hazards,* 28: 211–228.

Peduzzi, P., Dao, H., Herold, C. and Mouton, F. (2009) 'Assessing global exposure and vulnerability towards natural hazards: The Disaster Risk Index', *Natural Hazards and Earth System Sciences,* 9: 1149–1159.

Pelling, M. and Schipper, L. (2009) 'Climate adaptation as risk management: Limits and lessons from disaster risk reduction', *IHDP Update,* 2: 29–34.

Pielke, R., Gatz, J., Landsea, C., Collind, D., Saunders, M. and Musulin, R. (2008) 'Normalized hurricane damage in the United States: 1900–2005', *Natural Hazards Review,* 9: 29–42.

Putnam, R.D. (1993) *Making Democracy Work: Civic Transition in Modern Italy,* Princeton, NJ: Princeton University Press.

Reilly, J., Hohmann, N. and Kane, S. (1994) 'Climate-change and agricultural trade – who benefits, who loses', *Global Environmental Change-Human and Policy Dimensions,* 4: 24–36.

República de Chile (2002) *Plan nacional de protección civil,* Santiago de Chile: Ministerio del Interior.

Roberts, J. and Parks, B. (2006) *A Climate of Injustice: Global Inequality, North-South Politics, and Climate Policy,* Cambridge, MA: MIT Press.

Schipper, L. and Pelling, M. (2006) 'Disaster risk, climate change and international development: Scope for, and challenges to, integration', *Disasters,* 30: 19–38.

Sterr, H. (2000) 'Implications of climate change on sea level', in J. Lozan, H. Grassl and P. Hupfer (eds) *Climate of the 21st Century: Changes and Risks,* Hamburg: Wissenschaftliche Auswertungen.

Tartaglia, S. (2006) 'A preliminary study for a new model of sense of community', *Journal of Community Psychology,* 34: 25–36.

Thomalla, F., Downing, T., Spanger-Siegfried, E., Han, G.Y. and Rockstrom, J. (2006) 'Reducing hazard vulnerability: towards a common approach between disaster risk reduction and climate adaptation', *Disasters,* 30: 39–48.

Tribuna del Biobío (2005) 'Agüita de la Perdiz: lo que puede hacer la comunidad organizada', Concepción, *Tribuna del Biobío.* Online. Available HTTP: http://www.tribunadelbiobio.cl/one_news.asp?IDNews=543&NewsEditions=19 (accessed 14 September 2006).

Trim, P. (2004) 'An integrative approach to disaster management and planning', *Disaster Prevention and Management,* 13: 218–225.

UN/ISDR (2004) *Living with Risk: A Global Review of Disaster Reduction Initiatives,* Geneva: United Nations International Strategy for Disaster Reduction.

Venton, P. and La Trobe, S. (2008) *Linking Climate Change Adaptation and Disaster Risk Reduction,* Teddington: Tearfund.

Wisner, B., Blaikie, P., Cannon, T. and Davis. I. (2004) *At Risk: Natural Hazards, People's Vulnerability and Disasters* (2nd ed), London and New York: Routledge.

30 Social network actors and novel information for adaptive capacity

Fiona J.Y. Rotberg

Introduction

Social network theory (SNT) and social network analysis are used in this chapter as a theoretical frame to examine how social networks function and if/how community members share information within the networks that then facilitate their coping or adaptive capacity to climate vulnerability. SNT has been used to interpret human relationships through nodes and ties, where a node is the individual actor in the network and ties are the relationships between the actors. Together with social network analysis, the theory has been applied to map human connections, patterns of communications, and their implications (Barnes, 1954; Granovetter, 1973; Rogers, 1986; Wasserman and Faust, 1994; Haythornthwaite, 1996).

Mark Granovetter´s Theory of the Strength of Weak Ties (1973) anchored the notion that most innovative or novel information comes from weak ties or bridging ties that connect two disparate groups. Expanding on Granovetter's theory, Ronald Burt (2004, 2005) established that network brokers are those who create the bridging ties between groups and that brokers have access to information which leads to competitive advantages for groups to develop new or novel ideas for change (2004, 2005). In the context of human adaptation to environmental dimensions of change or climate related impacts (such as extreme and unexpected flooding) these theoretical foundations allow two key questions to be posed: Do brokers inside and outside community networks provide innovative information or knowledge that villagers use to adopt short-term coping, and/or long-term adaptive strategies to climate stresses? Are there other network actors that are important in group innovation? Social network theory provides a frame to discover who provides innovative knowledge and whether communication methods change during different weather events. The field data presented in this chapter addresses these questions while also examining the attributes of different network actors.[1]

Theoretical framework

Social network theorists have shown that within industries, villages, regions, and administrative units, people often specialize their roles in groups or clusters to create advantages for individual or group change (Barnes, 1954; Granovetter, 1973; Rogers, 1986; Burt, 2004; Wasserman and Faust, 1994; Haythornthwaite, 1996; Putnam, 2000; Borgatti, 2003). How knowledge is transferred among network members has been shown in violent gang networks (Grann, 2011), natural resource management structures (Crona and Bodin, 2006), social-ecological systems and adaptive governance (Olsson *et al.*, 2006), water resource management (Adger, 2003; Olsson, *et al.*, 2004); and student social movements (Gladwell, 2010). Olsson and colleagues (2004) provide evidence that key leaders can create opportunities to assist resource users in adaptive resource governance, and that leadership and networks are needed for successful transformations in social-ecological systems. There is also data to support that groups integrate and build knowledge through brokers who create bridges across the groups (Krishna, 2002; Burt, 2005). Fernandez and Gould (1994) exemplified how brokers can mediate flows of information between two disparate actors or groups and thus wield influence. Brokers within social networks have been shown to play a role in interactions between investment bankers, between real estate agents, and in small business transactions where the internal networks were strongest combined with someone who brokered deals outside of the group (Burt, 2005).

Granovetter's Theory of the Strength of Weak Ties focuses on the cohesiveness of the ties one person has to another, and theorizes that most novel information is devised from liaison actors who have weak and bridging ties that connect two networks or groups together. These individuals are 'best placed to diffuse [...] innovation' (Granovetter, 1973: 1367). Granovetter argues that we know little about interactions within small groups and how they aggregate to form larger patterns (1973: 1360), yet these small interactions are important for sharing knowledge and information, as well as for community social cohesion. If one views climate change adaptation as adjustments in group behavior for the purpose of reducing vulnerabilities to climate related impacts, then brokers who provide this novel information that spurs either autonomous or planned actions can serve as agents to enhance community adaptive capacity. Brokers can not only transfer crucial information between groups, but they are able "to see how a belief or practice in one group could create value" in another group (Burt, 2004: 355). These network actors have a competitive advantage in that they are able to see good ideas by being on the network periphery. "What matters is the value produced by the idea, whatever the source." People with connections across structural holes have early access to diverse information "which gives them a competitive advantage in seeing and developing good ideas" (Burt, 2004: 388). Entrepreneurial and innovative opportunities in networks are created by brokers, whether they serve as a liaison, representative, gatekeeper, or coordinator who brings about change (Fernandez and Gould, 1989; Burt, 2000: 12). In understanding more fully who communicates information (and what attributes they possess) in community level social networks, what information is communicated, and how the information leads to change, learning can be garnered about how communities cope and adapt to climate variability.

A social network is defined here as a group of individuals that form a constellation of relationships (Magsino, 2009). Actors within networks include key individuals, brokers, and members. A key individual is a person who has stature within the social network and is viewed as influential (through formal or informal means) by other network members. A broker is a network actor who creates bridges that link one group to another group. The bridge is the method by which the broker shuttles information or knowledge between two groups or networks. Bridges help to facilitate communication and encourage innovation. A formal broker is a person who is a part of the main social network, while an informal broker is positioned outside the main network, but someone who possesses information that can be useful to members in the main network. Members are individuals in the network, in this case, the village members.

Climate change impacts include social, spatial and temporal risks that lead to various coping and adaptive methods and practices, including livelihood diversification, migration, pooling of resources, and use of social networks to share risk (Agrawal and Perrin, 2009: 355, 449). Adaptation is viewed here as a dynamic process that involves multiple actors in society, and as a wide array of strategies that individuals, communities or groups employ when faced with hazards or vulnerabilities (IPCC, 2007). Adaptive capacity is viewed as the ability of groups to adjust to extreme events or stresses, to moderate potential damages (IPCC, 2007), and to manage risks and impacts through the development of new knowledge and effective approaches. Social networks are one mechanism through which communities harness or mobilize their adaptive capacity, and new knowledge development within social networks can lead to innovative tools for coping with climate impacts.

Data from a Bangladeshi village provides useful insights about how a rural social network is composed; who plays which role in the network, what attributes the actors display, and that network actors are potentially important agents in coping and adaptive capacity.

Methods

Bangladesh is one country that will very likely suffer severe climate impacts in the future. Projected impacts will exacerbate the current natural hazards that affect Bangladesh today. The country will witness more frequent and severe tropical cyclones; heavier and more erratic rainfall in some areas and decreased rainfall in others; warmer and more humid weather; and sea level rise (Bangladesh Ministry of Environment and Forests 2008: 13). Several of these impacts will lead to higher river flows, an increase in riverbank erosion and increased sedimentation in riverbeds, affecting the agriculturally dependent population, and further challenging their coping mechanisms and adaptive capacities. The 2005 Bangladesh National Adaptation Programme of Action (NAPA) lays out the Bangladesh governmental plan for the support of local and national future adaptation mechanisms to reduce adverse effects of climate change. One of the planned adaptation strategies suggested in the NAPA is to disseminate adaptation information to vulnerable communities for emergency preparedness measures and awareness raising (Ministry of Environment and Forest Government, 2005: xvi).

This strategy indicates that social networks and their actors could serve as sources of information, as well as dissemination mechanisms to affected groups.

Social network data was collected during 2008 and 2009, during both the wet and dry seasons in the village of Tartapara, Bangladesh. Methods employed include focus group interviews (three groups of 25–30 villagers), in-depth interviews (eight interviews with identified key individuals) and 15 household questionnaires. These tools were used to collect social network data and attribute data characterizing the network actors; to map who acts as a broker within the social network; assess the role of bridges and linkages within the village and other linked networks and organizations; and to identify the key individuals and brokers who villagers turn to both during a climatic event and during non-event periods. Interviews were conducted in Bangla, through an interpreter, and questionnaires were translated from English to Bangla and answered by household members with the assistance of the translator.

Results

Tartapara is located in Jamalpur District, 180 km from Dhaka, on the edge of the Brahmaputra River (locally referred to as the Jamuna River) and is composed of approximately 400 households, and 2,200 people. Roofs are often made of dried grass and leaves, while some homes have corrugated iron roofs. The literacy level of the population is 8 percent. Employed villagers work as day laborers both in Tartapara and nearby villages, and in seasonal agriculture. Those who do possess assets own their houses and/or cattle, hens or goats. Several female villagers described cattle as precious and wealth determining as they served as a risk covering asset that they could sell "at any crisis moment" (Focus group 1, 2008) such as during a flood. Infrastructure around and within Tartapara is extremely limited, and during the rainy months, a boat is often used as transport to and from the village. The Jamuna River creates much lateral erosion and river flooding from June to September, and in Tartapara, villagers reported that the river "had split itself" (Focus group 1, 2008) creating two major rivers around small swathes of land.

The villagers are primarily focused on securing their next meal and planning the next crop, or obtaining a small loan. Annual floods are a fact of life for them, having learned to cope with and respond to annual floods as a natural process (Huq, 2009; Rahman, 2009). However, a widely held view among villagers was that the Jamuna River is "more dangerous" now and that it "can kill": Floods are "unpredictable now" and are occurring more frequently (Focus groups 1 and 2, 2008). For example, villagers spoke about experiencing severe floods two to three times per year (as in 2007). In the past, the villagers expected an annual flood and prepared for it like their ancestors: "We watch the ants!" (Elected Member B interview, 2008). One of the village elders explained that for as long as he can remember, they have relied on the ants to indicate when it is time to move the houses to higher ground. The ants moved their houses up to safety and the villagers followed, just as far as the ants had relocated. Villagers can no longer solely rely on the ants, because the floods come "with no warning" and more often. As one villager said, "we can be flooded at any time of night!" (Elected Member B interview, 2008).

Focus group discussions revealed eight key individuals who were identified by the villagers as the people who they depended on before, during, and after flood periods. Follow up in- depth interviews garnered information about attributes and qualities of the key individuals, and what motivates the linkages within the village network, as exhibited in Table 30.1. Interview findings indicated that the Elected Local Government Chairman (Chairman) and Elected Local Government Member B (described as the Matabbar in Bangla, meaning Social Leader) were the key individuals who served as the formal leader and the formal broker, respectively, within the social network. Elected Local Government Member B (Elected Member B) is described as an "honorable wise leader" by other villagers (Focus groups 1, 2, and 3, 2008; in depth interview Elected Government Member B, 2009) in part perhaps because of his age (68 years) and because he is an elected member of the Union Parishad.[2]

Elected Government Member A (Elected Member A), the only female key individual, served as the main contact for the other female members of the community. Informally, the women gather at the evening market to discuss concerns. If there is an issue that the women feel is not being addressed by the Chairman (for example, they wished to press the Chairman to request governmental authorities to build a dam up river) they voice their concerns through Elected Member A. Elected Member A, in turn, communicated directly to Elected Member B, who then communicated with the Chairman.

All of the interviewees agreed that the formal leadership in the village rests with the Chairman. They also agreed that non-governmental organization (NGO) Worker A serves as an informal broker who creates a bridge for flow of information and employment opportunities. NGO Worker A is informally linked to the network during both dry and wet seasons. During the dry season, he explained, he can reach the 150 villagers in his programme[3] directly, but during

> a natural disaster or flood, I contact the local government leader (Chairman) and social leader (Elected Member B) in the village, as during flood time they make the decision as to what the community needs most ... because of their power, we keep in contact with them.
>
> (In-depth interview, NGO Worker A, 2009)

In-depth interviews with the eight key individuals revealed that honesty (five mentions), acceptance by others (four mentions), and trust (four mentions) were cited most frequently as those leadership qualities that key individuals valued most. When asked how leadership is exhibited in Tartapara, "informal" was used by all of the respondents, both in describing the leaders and the leadership structure. The interviewee responses about the key individual attributes indicate that authority, trust, and accessibility in a crisis appear to be important for brokers to possess for villagers to turn to them when in need, as evidenced in Table 30.1.

Table 30.1 Social network actors and their attributes, Tartapara, Bangladesh

Key individuals	*Gender male ((M)/female (F))*	*Social network role*	*Descriptive attributes*
Elected Local Government Chairman	M	Formal leader	Formal and informal power, authority, respect, mobilizer, trust
Elected Local Government Member A	F	Informal link for women in the village	Formal power and trust
Elected Local Government Member B	M	Formal broker	Authority, wise, respected, honorable
Villager A	M	Actor	Educated, respected, able to assist, informal power
Villager B	M	Actor	Connected, always available, trusted
Villager C	M	Actor	Able to assist, informal trust, connected to NGO
NGO Worker A	M	Informal broker	Provider, accessible in a crisis, problem solver
NGO Worker B	M	Actor	Accessible in a crisis, respected, connected, informal mobilizer

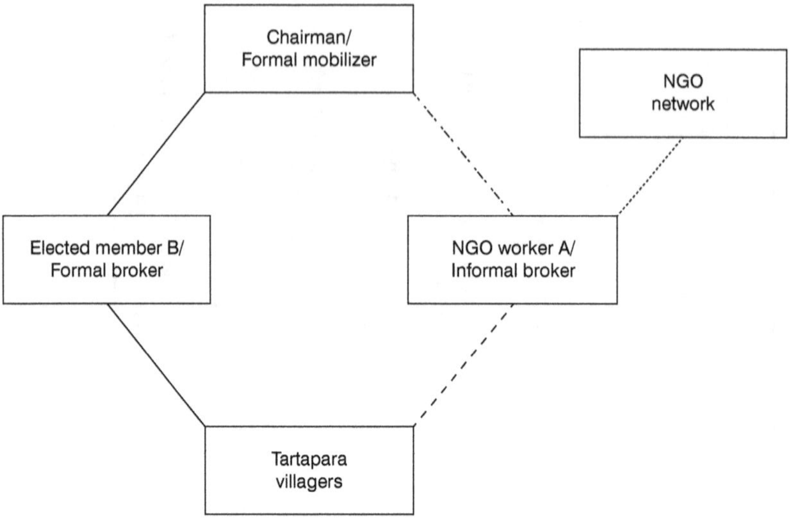

Figure 30.1 Depiction of the social network in Tartapara during wet and dry seasons

Interviews and questionnaires, and consultations with village members, allowed the author to outline the functioning social network of Tartapara during wet and dry seasons, as depicted in Figure 30.1, and shows that the Tartapara villagers are connected to the Chairman (the formal leader), through the Elected Member B (the formal broker). The direct link between the villagers and these two individuals are indicated by a solid line. The villagers are also informally linked to NGO Worker A (informal broker), as shown by a dashed line. In turn, NGO Worker A is connected to the Chairman, as shown by the broken line. Lastly, the entire network is loosely linked to NGO Worker A's own NGO network, as indicated by the dotted line.

Figure 30.2 depicts the Tartapara network during a "hothat ban" scenario, or when a sudden flood impacts the village, and indicates that the brokers have limited roles during this scenario. Figure 30.2 also introduces the presence of "spontaneous young leaders" who are young, strong men in the community, described by the other villagers as those who saved the community from literally falling into the river during a sudden flood.

Villagers described communication and the action steps that the community follows during both seasons. Table 30.2 depicts these steps, and an additional set of steps followed when the floods "come suddenly" (hothat ban), during the wet season. Before 2006, villagers explained that Tartapara witnessed annual, predictable "water rising for three days or so" (Focus Groups 1, 2, and 3, 2008) that forced villagers to relocate and resettle. All respondents reported that they had relocated at least once, either with the entire village, or from another village. Many members talked about having relocated three or more times. In 2007, however, sudden floods began (and continued to occur in 2008 and 2009); occurring in both late June and then again in mid-July. The latter persisted for over two weeks,

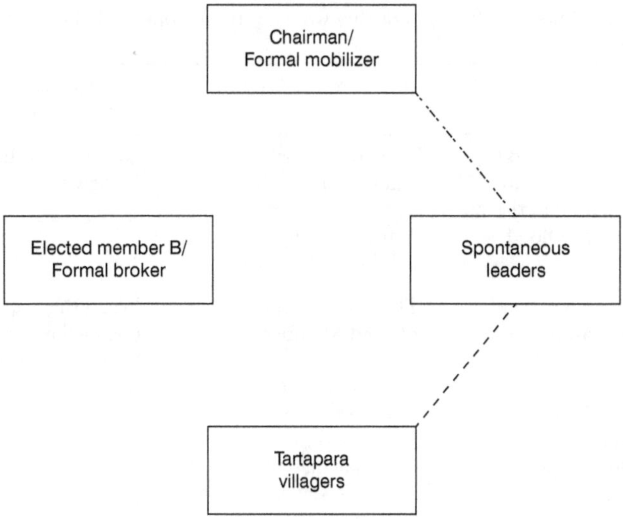

Figure 30.2 Depiction of Tartapara network during "hothat ban" scenario

rather than a few days: villagers then realized they could no longer rely on floods to be predictable. Instead, they had to be ready to act at any time. The data here presents information from three years, making it impossible to know whether reactive adaptive measures will be become more formalized amongst the villages in future years. It can be suggested, however, that sudden floods will become increasingly likely with climate change.

Discussion and conclusion

Data about the personality attributes of the social network actors in Tartapara, as depicted in Table 30.1, indicate that Elected Local Government Member B was described in terms such as "authority, wise, respected, and honorable," while NGO Worker A was described as a "provider, accessible in a crisis, and problem solver." These descriptions indicate that there are no distinctive overlaps between the attributes of these two brokers, providing no conclusive information about which types of people serve best as bridging brokers (Burt, 2005). More data in similar settings needs to be gathered to clarify which attributes effective brokers possess and that are most important for the diffusion of innovative information. Respect, trust and accessibility during a crisis, however, were important attributes for key network actors to possess, not only the brokers. The change in network structure presented in Figure 30.2 indicates that communication methods and network structure do change during unpredictable floods. The social network structure breaks apart and roles change. Both the formal and informal brokers are pushed to the periphery of, or completely outside, the network. Instead, the young spontaneous leaders in this community play a distinctive role during sudden flood events, working to save the community members. The young leaders' relationship ties and attributes

Table 30.2 Social network function during wet and dry seasons, and "hothat ban" (sudden flood) scenario

I. Wet season	*II. Dry season*	*III. Wet season hothat ban scenario*
1 Villagers are warned by megaphones or radio by local agencies that the river is likely to flood.	1 The villagers talk amongst themselves about a concern, for example, their desire to build a dam up river.	1 No warning that the river is likely to flood.
2 Elected Member B and the Chairman mobilize the network: They organize the villagers to move to the river embankment and to shelters for temporary resettlement.	2 The villagers approach Elected Member B, as the social leader, to express their concern. They ask Elected Member B to approach the Chairman with their concerns, depending on the topic.	2 When the water encroaches on the village, the villagers organize themselves to relocate.
3 After the flood recedes, the Chairman manages the resettled community; which includes tending to new needs and providing linkages with outside agencies. For example, the village often finds shelter in a school building close by. The Chairman has direct communication with the school authorities and arranges for school to be suspended and formally appropriated by the villagers for use as a shelter for a period of time.	3 Elected Member B approaches the Chairman, and acts as a broker of information between the villagers and the Chairman. NGO Worker A contacts villagers on a daily basis, providing them with an opportunity for daily labor, as well as longer-term economic development options. He provides them with information from his own network outside the village.	3 The network between Elected Member B and the Chairman has no time mobilize. Spontaneous young leaders work to save the villagers and their possessions. People try first to rescue their family and assets, and then they turn to assist others. After rescuing others, the villagers regroup at a shelter.
	4 The Chairman renders a decision on which action will be taken in the village.	4 After the village resettles, and only after the resettlement, the network between Elected Member B and the Chairman sets itself in motion.

need to be further studied to understand if they too act as brokers (during certain unpredictable climatic events) whose linkages help to diffuse innovative information to the other networks actors in the village. To date, data has not been gathered from these actors about their knowledge and methods of information diffusion to other villagers: We do not yet know what long-term effect the young leaders have on the strength of ties within the network or on long-term adaptation in the community, however they appear to play an important role in community coping during unpredictable natural events. These young leaders should also be drawn upon when devising future development and adaptation policies as they most likely have important knowledge for local innovation and change.

Table 30.2 exemplifies not only steps that the village network actors follow but also illuminates how brokers and the mobilizer share information and knowledge between networks and among network actors. For example, NGO Worker A has the benefit of not only being on the periphery of the network, but he is connected to his professional external network that provides an avenue for innovative ideas. This network actor spoke about the importance of having "loose linkages" to the village network between the village Chairman, the village elder, the villagers, and of not "disturbing the formalities" (in-depth interview, NGO Worker A, 2009) of the village network, indicating that his connections to the formal network were weak, yet important to the daily lives of the villagers. Elected Local Government Member B has the advantage of having strong ties to various smaller groups within the community and thus serves as a broker who approaches the Chairman with ideas/concerns from other community members. Elected Local Government Member B also shuttles and interprets information and decisions from the Chairman back to the villagers.

The theoretical frame presented here (Granovetter, 1973; Burt, 2005) suggests that the combination of Elected Local Government Member B's linkages to the villagers and the Chairman, coupled with the weak linkages from the NGO Worker A, could lead to the creation of new and innovative knowledge that could in turn lead to the development of group adaptive actions. However, from the data collected to date in Tartapara, one cannot conclude that it is these linkages that lead directly to adaptive action. The theory maintains that network brokers have a competitive advantage (Burt, 2004) in that they have access to diverse information which leads to new ideas and strategies: NGO Worker A's weak ties to the network enhances his ability to provide innovative knowledge (Burt, 2005) to the village network actors for the benefit of short-term coping and potentially long-term adaptive capacity. The Chairman and the brokers potentially facilitate action as they use their positions to provide necessary leadership and organizational management during flood times, while they also enable a communication structure where new knowledge for innovative change can be exchanged. They provide the framework for villagers to communicate, mobilize and voice concerns, and they create a link to local organizations and attend to the various and changing needs of villagers.

Acknowledgments

This research was made possible by a grant from the Swedish International Development Cooperation Agency (Sida). Tartapara network data presented here was derived from: Rotberg, F. (2010) 'Social networks and adaptation in rural Bangladesh', *Climate and Development*, 2: 65–72.

Notes

1 Burt offers the theory that a gap in the theoretical discussion surrounds the idea of who becomes a broker, and what general attributes brokers possess that helps them lead others towards change. Burt argues that attribute data need to be collected to understand and to assess whether a broker is more effective if he/she engages in informal or formal relationships between the different groups (2005: 28, 50).

2 There are three levels of local government in Bangladesh: 1) Upazila (Sub-District), 2) Union Parishad (City Council) and 3) Village Council. All of those serving on these levels are directly elected by public vote, for a term of five years. The Elected Local Government Chairman and the two members referenced in Table 30.1 are elected by the people to the local Union Parishad, or City Council. The City Council is composed of one Chairman and nine members.

3 The program was focused on "eco development," where the villagers were provided with day labor opportunities, such as helping to build a government road, or longer-term loan options, providing them with the ability to purchase animals. Villagers were also exposed to training and capacity building programs.

References

Adger, W.N. (2003) 'Social capital, collective action and adaptation to climate change', *Economic Geography*, 79 (4): 387–404.

Agrawal, A. and Perrin N. (2009) 'Climate adaptation, local institutions, and rural livelihoods', in N. Adger, I. Lorenzoni, and K. O'Brien (eds) *Adapting to Climate Change*, Cambridge: Cambridge University Press.

Bangladesh Ministry of Environment and Forests, Government of the People's Republic of Bangladesh (2008) 'Bangladesh climate change strategy and action plan 2008', Dhaka: Ministry of Environment and Forests.

Barnes, J. (1954) 'Class and committees in a Norwegian Island Parish', *Human Relations*, 7: 39–58.

Borgatti, S. P. (2003) 'The key player problem', in R. Breiger, K. Carley and P. Pattison (eds) *Dynamic Social Network Modelling and Analysis: Workshop Summary and Papers*, Washington, DC: The National Academies Press.

Burt, R. (2000) 'The network structure of social capital', in R. Sutton, R. and B. Staw (eds) Research in Organizational Behaviour, 22, Greenwich, CT: JAI Press.

Burt, R. (2004) 'Structural holes and good ideas', *American Journal of Sociology*, 110 (2): 349–399.

Burt, R. (2005) *Brokerage and Closure: An Introduction to Social Capital*, New York: Oxford University Press.

Crona, B. and Bodin, O. (2006) 'What you know is who you know? Communication patterns among resource users as a prerequisite for co management', *Ecology and Society*, 11 (2): 7–40.

Elected Member Government Member B (2009) Personal interview, Tartapara, Bangladesh.

Focus Group 1 (2008) Village of Tartapara, Bangladesh.

Focus Group 2 (2008) Village of Tartapara, Bangladesh.

Focus Group 3 (2008) Village of Tartapara, Bangladesh.

Gladwell, M. (2010) 'Small change', *The New Yorker*, October 4: 42–49.

Grann, D. (2011) A murder foretold, *The New Yorker*, April 4: 2–61.

Fernandez, R. and Gould, R. (1994) 'A dilemma of state power: brokerage and influence in the natural health policy domain', *American Journal of Sociology*, 99 (6): 1455–1491.

Granovetter, M. (1973) 'The strength of weak ties', *American Journal of Sociology*, 78 (6): 1360–1380.

Haythornthwaite, C. (1996) 'Social network analysis: an approach and technique for the study of information exchange', *Library and Information Science Research*, 18: 323–342.

Haythornthwaite, C. (1998) 'A social network study of the growth of community among distance learners', *Information Research*, 4 (1).

Huq, S. (2009) International Institute for Environment and Development, personal communication, Dhaka, Bangladesh, February.

IPCC (2007) 'Summary for policymakers', in M.L. Parry, O.F. Canziani, J.P. Palutikof, P.J. van der Linden and C.E. Hanson (eds) *Climate Change 2007: Impacts, Adaptation and Vulnerability, Contribution of Working Group II to the Fourth Assessment Report of the Intergovernmental Panel on Climate Change*, Cambridge: Cambridge University Press.

Krishna. A. (2002) *Active Social Capital. Tracing the Roots of Development and Democracy*, New York: Columbia University Press.

Magsino, S. (2009) *Applications of Social Network Analysis for Building Community Disaster Resilience*, Washington, DC: National Academies Press.

Ministry of Environment and Forest Government of the People's Republic of Bangladesh (2005) National Adaptation Programme of Action, Final Report, November.

NGO Worker A (2009) Personal interview, Tartapara, Bangladesh.

Olsson, P., Folke, C. and Hahn, T. (2004) 'Social-ecological transformation for ecosystem management: The development of adaptive co-management of a wetland landscape in Southern Sweden', *Ecology and Society*, 9 (4): 2–26.

Olsson, P., Gunderson, L., Carpenter, S., Ryan, P., Lebel, L., Folke, C. and Holling, C.S. (2006) 'Shooting the rapids: Navigating transitions to adaptive governance of socio-ecological systems', *Ecology and Society*, 11 (1): 18–39.

Putnam, R. (2000) *Bowling Alone*, New York: Simon and Schuster.

Rahman, A., Bangladesh Centre for Advanced Studies. Personal Communication, Dhaka, Bangladesh, April 2008 and February 2009.

Rogers, E.M. (1986) *Communication Technology: The New Media in Society*, New York: Free Press.

Smit, B. and Wandel, J. (2006) 'Adaptation, adaptive capacity and vulnerability', *Global Environmental Change*, 16: 282–292.

Wasserman, S. and Faust, K. (1994) *Social Network Analysis*, Cambridge: Cambridge University Press.

Integration of knowledge and action

31 Understanding how to respond to climate change in a context of transformational change

The contribution of sustainable adaptation

Siri Eriksen

Introduction – problematizing adaptation

With the increasing prominence of climate change in the public and policy agendas, there are growing calls for critical thought regarding the type of response measures that are formulated (Barnett and O'Neill, 2010). For example, it has been pointed out that adjustments to reduce potential damages and realize new opportunities that may arise with such changes are not necessarily all benign (Adger *et al.*, 2007; Eriksen and Selboe, 2013). Like any other intervention or action, adaptation aimed at one particular group or purpose can have negative effects on other groups or in the long term (Eriksen and O'Brien, 2007; Barnett and O'Neill, 2010). This concern is the motivation for the term "sustainable adaptation," that is "adaptation that contributes to socially and environmentally sustainable development pathways, including both social justice and environmental integrity" (Eriksen *et al.*, 2011: 8). How can the term sustainable adaptation inform how we respond to climate change?

A key problem with the way that adaptation to climate change has been conceived so far is the tendency to see adaptation as actions or policies directed only at climate change in isolation from societal change processes at large. Adaptation is often assumed to take place more or less automatically as a largely unproblematic process once the right policies and interventions are in place (Eriksen and O'Brien, 2007). There is increasing attention directed at the need to see vulnerability as a result of multiple stressors, including non-climatic social and environmental changes (Eakin, 2006; Reid and Vogel, 2006; Leichenko and O'Brien, 2008; O'Brien *et al.*, 2009). However, so far, we do not really understand how adaptation links with other processes of change in society, nor what it should look like. The failure to frame adaptation and climate responses as part of social change and development processes means that critical thinking regarding development processes and how they take place is not reflected in most understandings and discussions of adaptation.

The failure to link adaptation to an understanding of development and social change manifests itself in two main ways in adaptation efforts: first, some policy responses are formulated to narrowly address climate change as an environmental problem, isolated from societal problems and change at large. Measures, such as those described in national adaptation assessments and programs in wealthy as well as poor countries alike (Agrawal and Perrin, 2009; NOU, 2010) are typically sectoral, top-down (Agrawal and Perrin, 2009; NOU, 2010). They are often focused on avoiding physical or monetary damage, human injuries and death from extreme events, rather than on local needs and strategies (Vincent *et al.*, this volume). They direct attention to questions such as what dimensions should drainable pipes have or what sort of flood early warning systems do we need with projected change in precipitation patterns? Or how many more forest fires may result from future warming? There is increased research focus on community adaptation, local risk management including local actions, and how people live with environmental disasters (Huq and Reid, 2007; Commission on Climate Change and Development, 2009) but there is still a need to link more closely to existing understanding of how people build their livelihoods (Sabates-Wheeler *et al.*, 2008) and what values and aspirations beyond material needs and survival people strive to secure in the face of change (O'Brien, 2009). Adaptation projects that focus on specific manifestations of climate change rather than people's daily actions and felt threats risk leading to measures that are isolated from development needs and efforts.

A second and related challenge emerging from the failure to explicitly link adaptation with development is the risk that a wide range of development actions are uncritically implemented in the name of climate change without being explicit about what aspect of vulnerability they are actually addressing. Based on the assumption that "more development" necessarily reduces vulnerability and is therefore always good, development organizations may be tempted to label current activities as adaptation without a critical examination of whether they address root causes of vulnerability or indeed reinforce vulnerability, how poverty and vulnerability differ, and whose vulnerability they address. This may lead to ineffective or counterproductive measures (Klein *et al.*, 2007). Worse, justifying actions as climate change measures may cloud important political or problematic aspects of these actions. There is a long history of environmental discourses (such as desertification and nature conservation) contributing to a loss of local control over resources to central government institutions, as well as to top-down imposition of rules and regulations that constrain livelihoods and disregard the complexities of socio-environmental processes of change (Benjaminsen *et al.*, 2006). There is a similar danger that climate change discourses lead to mitigation and adaptation policies that are distant from local contexts and concerns and which in effect disempower people.

In order to expand critical thinking regarding adaptation there is an urgent need to build adaptation understanding on development insights. This chapter suggests that the term sustainable adaptation may contribute to addressing this challenge by linking adaptation with sustainable development literatures, in particular linking understandings of adaptation to development pathways, empowerment, distributional issues and strong sustainability (Redclift, 2005; Adams, 2009).

Human security and sustainable adaptation, though not without problems, are terms that may bring in much reflection on adaptation, highlighting people's efforts, options, and capacity to build flourishing lives in the face of multiple stressors, and the role of competing interests and values involved in adaptation processes and decision making, and the need to transform development pathways.

Sustainable adaptation as a transformative change process: a conceptual understanding

The need to think critically about what types of adaptation actions are needed has been reflected in increasing calls over the past decades to target the poor in particular, or creating "pro-poor" adaptation, ostensibly to prioritize measures that focus on challenges met by the poor, compared to the interests of other groups (Eriksen and Næss, 2003; Tanner and Mitchell, 2008). These efforts recognize that adaptation outcomes may differ between groups and that conscious considerations need to be made to ensure that adaptation addresses poverty or at least does not undermine poverty reduction efforts (Yohe *et al.*, 2007; Moser and Satterthwaite, 2008; Vernon, 2008; Jones, 2009). At least implicitly, a normative judgment is made that adaptation should not exacerbate poverty and inequalities.

The term "sustainable adaptation" as used by Eriksen and O'Brien (2007) builds on these insights and emphasizes social equity as a critical part of adaptation. This is important because the tendency of many poor to be highly vulnerable to climate change is often used to justify adaptation measures; however, little thought is then given to whether the proposed measures actually reduce the vulnerability of the poor. Indeed, some adaptation measures may inadvertently increase the vulnerability of the poor. While there have been several efforts to develop tools to guide adaptation (Lim *et al.*, 2005; IUCN, IISD, SEI-US and Intercooperation, 2007; Debels *et al.*, 2009), sustainability is seldom explicitly assessed.

In addition to pointing to the need to prioritize normative outcomes of adaptation, early elaborations of the term sustainable adaptation highlight the importance of processes that lead to these particular outcomes. Eriksen and O'Brien (2007) argue that adaptation cannot be sustainable unless vulnerability-poverty linkages are consciously addressed by:

1 reducing climate risk to well-being of the poor (for example, climate forecast information, disaster management, drought resistant agriculture);
2 strengthening adaptive capacity of the poor (for example, diverse livelihoods, social capital, financial and technological capacity); and
3 targeting the causes of vulnerability of the poor (for example, economic, social and political processes like privatization of resources and loss of rights, power relations and political marginalization, and deteriorating terms of trade).

Hence, the term broadens the focus of climate change adaptation beyond a narrow environmental risk issue to focusing on understanding the social processes creating vulnerability. This is in line with recognition of the need to shift focus from an outcome vulnerability framing, which refers to the likelihood of injury, death, loss, disruption of livelihoods or other harms as the result of stressors or shocks resulting from climate change, to a contextual vulnerability framing. This second interpretation links vulnerability closely to the context in which people experience shocks and stressors related to any type of change, and the processes that create this context (O'Brien *et al.* 2009).

The wide variation in the way that sustainability is understood opens up for misuse of the term, however (Cohen *et al.*, 1998; Brown, 2011). There is the danger that adding the sustainability term to adaptation may make it seem "double good" – not only does it address climate change, it is also sustainable – hiding the need to critically assess the effects and motivation of a particular adaptation effort. Reformist and ecological modernization type approaches to sustainability suggest that improved technologies, practices and economic instruments can combine current economic growth models with social and environmental sustainability (Soltau, 2006). Norwegian economic policy, for example, frames sustainability in terms of maintaining national wealth and incomes for future generations (Johnsen, 2011) while the retreat of Arctic sea ice is seen as an opportunity to expand oil exploitation northwards (Kristoffersen, 2013) perpetuating fossil fuel based economic growth and high emissions.

Hence, in order to avoid "greenwashing" of adaptation (Brown, 2011), *strong* sustainability and political economy approaches need to be promoted, which implies altering current modes of development and economic growth (Redclift, 2005; Eriksen *et al.*, 2011). The climate change problem has brought the relevance of the key message of sustainable development – an environmentalist critique of development that is causing both environmental problems poverty – back on the center stage (Adams, 2009). Both social inequities underlying vulnerability and high energy consumption associated with increasing emissions are generated by and intrinsic to our current development model. Sustainable adaptation involves developing new types of responses that contribute to alternative pathways, rather than reproducing fossil fuel driven economic growth development pathways.

In line with these insights, there is increasing recognition that adaptation cannot be seen merely as a techno-managerial challenge involving incremental adjustment to technologies, regulations, policies and practices in order to sustain current ways of living with change. Instead fundamental shifts in societal systems are required, including "the transformation of energy and agricultural systems, financial systems, governance regimes, development paradigms, power and gender relations, production and consumption patterns, lifestyles, knowledge production systems, or values and world-views" (O'Brien, 2012: 671). Critical here is the need for *deliberate* transformation, that is, to consciously take actions to influence future change towards more sustainable pathways including reducing emissions and vulnerability (contrasting with unintended transformation and outcomes of a process).

Such an understanding of climate change – development relations implies that adaptation can be interpreted as forming part of a wider transformation process.

Adaptation is here seen as a process taking place in the form of interaction between actions, decisions, relations and structures across several scales (in contrast with viewing adaptation as a single policy, project or action). Conceptually, what is termed "responses" in a contextual vulnerability approach (O'Brien *et al.*, 2009) can be unpacked as consisting of an interaction between transformation and development pathways. These in turn affect climatic and societal changes and contextual conditions. In the continuum between deliberate and unintended transformation processes, sustainable adaptation is placed closer to the deliberate transformation end since it aims to achieve particular normative development pathways (in contrast with maladaptive processes, which would be located closer to the unintended transformation end).

Adaptation processes inherent in transformation and the resulting development pathways can lead to both positive and negative outcomes or contextual conditions (such as poverty, welfare, equity, greenhouse gas concentrations, social relations and structures) at particular points in time and space. Even within deliberate forms of change, outcomes will sometimes be negative for some groups or at particular points in time; what is important is the direction and character of the process. Rather than being a particular outcome or isolated action, sustainable adaptation is a *process* characterized by an ability to shape the social and environmental conditions of the future.

The assumption that we can to some extent deliberately choose the kind of world we will end up living in (Robinson and Herbert, 2001) relies on a political adaptation approach, which emphasizes agency in shaping change; for example, current power structures, relations and hegemonic ideas determine decisions and development pathways (Eriksen and Lind, 2009; Cote and Nightingale, 2011; Tanner and Allouche, 2011). Taking as a starting point that transformation is a continuous process shaped by decisions and social structures, the crucial question then is how it can be influenced towards deliberate (rather than unintended) forms, and hence desired development pathways.

Developing a normative vision for change

Recent elaborations of human security and sustainable adaptation concepts provide some important guidance. Deliberate transformation necessarily involves normative judgments about what is a desirable future, and ethical considerations about what outcomes are acceptable for whom.

Importantly, human security elaborates just what a "vision" for society may constitute and specifies that this is much more than meeting materialistic needs or economic goals. Human security can be seen as a condition when and where individuals and communities have the options necessary to end, mitigate, or adapt to risks to their human, environmental, and social rights; have the capacity and freedom to exercise these options; and actively participate in attaining these options (GECHS, 1999).

Well-being, quality of life, and human flourishing entails that people have the freedom to choose and to pursue the outcomes and way of life they want. Drawing from Sen's capabilities approach, O'Brien and Leichenko (2007) hence draw attention to the fact that a sustainable pathway is not just about the opportunities that exist but also the processes that allow freedom of action and decisions. In particular, attention is focused on the importance of individuals' control over own circumstances and feeling socially and culturally secure in one's own life situation, enriching what social equity means beyond normal poverty notions of income or basic needs (Barnett, 2003). O'Brien and Leichenko (2007) propose that the interlinkage between human security and sustainable adaptation revolves around the close connection between equity and connectivity. Differential outcomes and inequity result from processes and relations that link people across space and time. Such outcomes are not just the result of local dynamics or conditions but are linked to global development models, including economic and political processes and structures. Human security highlights the need to shift the climate change discourse, and the way that the climate change issue is addressed, towards approaches that act to empower people, including expanding their options and strengthening their ability to manage a multitude of processes of change. Instead of imposing narrow environmental regulations or specific technical solutions to reduce climate sensitivity, policies must focus on creating an inclusive vision for how society, whether at local or national level, needs to be transformed.

Building on insights regarding the importance for shaping desirable futures of vulnerability context, differential values and competing interests, power in decision making and interconnectivity, recent elaborations of the term sustainable adaptation develop four normative principles to guide the adaptation process (Eriksen *et al.*, 2011).

Widening responses to recognize and address the context for vulnerability is the first of four normative principles identified. A focus on local context in terms of the multiple processes of change that create vulnerability, and people's strategies to manage these processes, is central to understanding what elements societal transformations need to involve. The underlying social, economic, institutional, and cultural conditions that contribute to a wider context for vulnerability need to be examined in order to distinguish direct and indirect consequences of adaptation efforts on different groups over time. Studying the local vulnerability context, for example, may involve considering how trade liberalization is changing the context for agriculture and farming, and the diverse strategies people use in facing both import competition and climate variability and change (Eakin, 2006). It follows that actions to promote sustainable adaptation pathways vary with the local vulnerability context.

The second principle is to acknowledge differing values and interests, and how these are negotiated through social and political relations and processes. Sustainable adaptation pays specific attention to threats posed by climate change or responses to climate change to a wide set of aspirations beyond basic survival and income, such as people's capacity to: earn an income and meet material needs; speak up for oneself and have rights; maintain health and basic education; and maintain a sense of social and cultural affiliation. This principle is closely linked to the human security focus on human flourishing – and the contrasting meanings this may have for different groups – rather than economic growth and other set measures for development. The diversity of values and empowerment aspects highlight that adaptation involves conscious choices, and that these need to be made explicit in order to avoid creating futures through "non-decision-making," that is following business as usual based on existing power structures and development models. In order to contribute to sustainable adaptation, it is important that policy processes recognize competing interests and potential value conflicts, identify how these may influence vulnerability and adaptation outcomes, and make explicit and transparent choices about what values and interests are to be promoted over others. Specific attention should also be paid to the fact that the social process of adaptation can be conflictual. It is often assumed that adaptation to climatic and other changes takes place in peaceful and orderly settings. In fact, adaptation takes place as part of situations of great social flux including conflict and post-conflict settings where fluidity, violence, and uncertainty, rather than order and predictability, dictate people's strategies as well as the politics of adaptation (that is, whose interests count and how competing or conflicting interests are negotiated) (Eriksen and Lind, 2009).

For example, in a study of two dryland areas in eastern and northern Kenya, Eriksen and Lind (2009) found that there were huge local conflicts of interests in terms of adaptation strategies to a situation of frequent drought. Some people benefited from new income opportunities from trade between pastoral and agro-pastoral groups, while others felt threatened by changes in the natural resource access regime and shifts in local power relations that such trade implied. This illustrates the importance of linking democratization and empowerment efforts with those of adaptation in order to reduce the causes of vulnerability in a way that considers the differing and often conflicting adaptation interests. The second principle suggests the need to target power structures and formal and informal decision-making processes more generally and to, more specifically, ensure that groups that are vulnerable to climate variability and change are represented and empowered in such processes.

The third principle highlights that particular knowledge and problem understandings often determine which adaptation interests are prioritized. The principle hence emphasizes the need to integrate local knowledge with other sources of knowledge about climate change into adaptation responses, in particular to counter any discourse tendency to blame environmental degradation on practices of the poor and seeing increased state or private sector control over natural resources as the solution (Benjaminsen *et al.*, 2006). In order to move towards more sustainable pathways, local knowledge and participation, as well as local strategies for living with environmental variability, need to play a central role in formal efforts to build resilience or manage disasters (Wisner *et al.*, 2004; Schipper and Pelling, 2006; Berkes, 2007; van Aalst *et al.*, 2008).

In line with the human security focus on interconnectedness, the fourth principle considers potential feedbacks between local and global processes and the need for responses to recognize the interactions between scales. Adaptation efforts in one place can influence both social justice and environmental integrity over space and time through, for example, economic linkages, and global flows of resources, people, and information (Adger *et al.*, 2007). Local measures need to be interrogated in terms of the potential positive and negative feedbacks, for example, on biodiversity and water access elsewhere as well as greenhouse gas emissions.

Crucially, the four principles are interdependent. Focusing on one principle alone does not effectively lead to more sustainable adaptation pathways unless other principles are also addressed at the same time. The principles guide an approach to understanding and developing adaptation efforts rather than constituting a one-size-fits-all checklist of specific measures, in line with post-development approaches (Ireland and Keegan, this volume). The principles raise a host of new questions and policy challenges. How does one create local empowerment and holistic vision in decision-making processes, incorporating more diverse values and interests than those included in conventional more top-down decision-making processes? Can new policy efforts such as REDD or those initiated by the UN Climate Change Convention (National Adaptation Programmes of Action) or development efforts (PRSPs, MDGs) contribute to local empowerment, or are there problematic aspects that may reinforce disempowerment? Early explorations have provided some illustrations of what sustainable adaptation could mean in a local context (Eriksen and Marin, 2011), though further elaboration through examination of practical cases is needed to understand how the principles can translate into action.

Conclusions

The term sustainable adaptation reflects the urgent need to critically reflect on what types of responses to climate change are desirable. Bringing in sustainable development perspectives helps distinguish such responses. Building on strong sustainability traditions, it emphasizes the need for fundamental societal changes and challenges dominant perspectives that have often seen the question of climate change as a managerial, economic, and technological issue. As pointed out by McNeill and García-Godos (2005: 23), "[s]ustainable development is a political, ethical, social issue as much as it is a technical issue." Addressing structural inequalities, pervasive poverty and disempowerment is central to sustainable adaptation. It frames climate change as intrinsically linked to development pathways.

Sustainable adaptation provides a tool for critically examining how current development efforts and modes may in fact be contributing to vulnerability (and hence should not be relabeled as adaptation) and how climate change responses may be reproducing development pathways currently driving emissions, poverty and inequity. Instead, transforming societal development is required.

Conceptually, sustainable adaptation can usefully be regarded as part of deliberate transformation processes. Determining a vision for what society we want to move towards, in terms of making conscious and explicit choices regarding which interests and values are prioritized in adaptation, is a critical part of this process. Equity includes a wide set of values and aspirations beyond basic needs, as well as the processes through which people seek to achieve these. Hence, the term sustainable adaptation is not just about social equity and environmental integrity *outcomes*, but about altering the adaptation *process*. Strengthening people's ability to choose and achieve their aspirations entails empowering individuals and communities to make decisions about their own adaptation outcomes. This may involve a transformation of decision making, democratizing climate change policy processes to include stakeholders, and reconcile values and competing understandings beyond those embodied by the state and by science, in line with processes previously described by Funtowicz and Ravetz (1993).

Crucially, a political understanding of adaptation is required if sustainable adaptation is to be achieved. Adaptation is not apolitical – and understanding responses and transformations as intrinsically political processes means considering adaptation as a process of continuous negotiation of competing interests between groups as well as inequities in these negotiations. For example, policy attention and investments are often skewed towards commercial farmers and large land owners at the expense of small scale resource users who have a poor negotiating position in labor and trade markets (Eriksen and Silva, 2009). Changing development models requires political and ideological shifts. Furthermore, inequalities, empowerment, and causes of vulnerability can only be addressed by challenging the structures and relations that sustain them.

Achieving such shifts is admittedly a tall order; however, identifying what changes are actually required as well as highlighting the multiple ways in which current development models are *not* leading to sustainable responses represent a first step towards responding effectively to climate change.

References

Adams, W.M. (2009) *Green Development. Environment and Sustainability in a Developing World*, London: Routledge.

Adger, W.N., Agrawala, S., Mirza, M.M.Q., Conde, C., O'Brien, K. Pulhin, J., Pulwarty, R. Smit, B. and Takahashi, K. (2007) 'Assessment of adaptation practices, options, constraints and capacity', in M.L. Parry, O.F. Canziani, J.P. Palutikof, P.J. van der Linden and C.E. Hanson (eds) *Climate Change 2007: Impacts, Adaptation and Vulnerability*. Contribution of Working Group II to the Fourth Assessment Report of the Intergovernmental Panel on Climate Change. Cambridge: Cambridge University Press: 717–743.

Agrawal, A. and Perrin, N. (2009) 'Climate adaptation, local institutions and rural livelihoods'. In W.N. Adger, I. Lorenzoni and K.L. O'Brien (eds) *Adapting to Climate Change. Thesholds, Values, Governance*, Cambridge: Cambridge University Press, pp. 350–367.

Barnett, J. (2003) 'Security and climate change', *Global Environmental Change*, 13: 7–17.

Barnett, J. and O'Neill, S. (2010) 'Maladaptation', *Global Environmental Change*, 20:211–213.

Benjaminsen, T.A., Rohde, R., Sjaastad, E., Wisborg, P. and Lebert, T. (2006) 'Land reform, range ecology, and carrying capacities in Namaqualand, South Africa', *Annals of the Association of American Geographers*, 96 (3): 524–540. doi:10.1111/j.1467-8306.2006.00704.x.

Berkes, F. (2007) 'Understanding uncertainty and reducing vulnerability: Lessons from resilience thinking', *Natural Hazards*, 41: 283–295.

Brown, K. (2011) 'Sustainable adaptation: An oxymoron?' *Climate and Development Journal*, 3 (1): 21–31.

Cohen, S., Demeritt, D., Robinson, J. and Rothman, D. (1998) 'Climate change and sustainable development: towards dialogue', *Global Environment Change*, 8 (4): 341–371.

Commission on Climate Change and Development (2009) *Closing the Gaps: Disaster Risk Reduction and Adaptation to Climate Change in Developing Countries,* Stockholm: Ministry of Foreign Affairs.

Cote, M. and Nightingale, A. (2011) 'Resilience thinking meets social theory: Situating social change in socio-ecological systems (SES) research', *Progress in Human Geography*, 36 (4): 475–489.

Debels, P., Szlafsztein, C., Aldunce, P., Neri, C., Carvajal, Y., Quintero-Angel, M., Celis, A. Bezanilla, A. and Martinez, D. (2009) 'IUPA: A tool for the evaluation of the general usefulness of practices for adaptation to climate change and variability', *Natural Hazards*, 50: 211–233.

Eakin, H. (2006) *Weathering Risk in Rural Mexico. Climatic, Institutional, and Economic Changes*, Tucson: University of Arizona Press.

Eriksen, S. and Lind, J. (2009) 'Adaptation as a political process: Adjusting to drought and conflict in Kenya's drylands', *Environmental Management*, 43 (5): 817–835.

Eriksen, S. and Marin, A. (2011) *Pastoral Pathways: Climate Change Adaptation Lessons from Ethiopia*, Norway: Development Fund.

Eriksen, S. and Næss, L.O. (2003) 'Pro-Poor Climate Adaptation: Norwegian development cooperation and climate change adaptation – an assessment of issues, strategies and potential entry points', Report 2003:02. CICERO.

Eriksen, S. and O'Brien, K. (2007) 'Vulnerability, poverty and the need for sustainable adaptation measures', *Climate Policy* 7: 337-–52.

Eriksen, S. and Selboe, E. (2013) 'Towards sustainable adaptation or paradoxical development pathways? Contrasting understandings and local actions in Øystre Slidre, Norway', in K. O'Brien and E. Selboe (eds) *The Adaptative Challenge of Climate Change*, Cambridge: Cambridge University Press.

Eriksen, S. and Silva, S. (2009). 'The vulnerability context of a savanna area of Mozambique: Household drought coping strategies and responses to economic change', *Environmental Science & Policy*, 12 (1): 33–52.

Eriksen, S., Aldunce, P., Bahinipati, C.S., Martins, R.D'A., Molefe, J.I., Nhemachena, C., O'Brien, K., Olorunfemi, F., Park, J., Sygna, L., Ulstrud, K. (2011) 'When not every response to climate change is a good one: Identifying principles for sustainable adaptation', *Climate and Development*, 3 (1): 7–20.

Funtowicz, S.O. and Ravetz, J.R. (1994) 'The worth of a songbird: ecological economics as a post-normal science', *Ecological Economics*, 10: 197–207.

GECHS (1999) *Global Environmental Change and Human Security (GECHS) Science Plan*, IHDP Report No. 11, Bonn: IHDP.

Huq, S. and Reid, H. (2007) *Community-Based Adaptation. A Vital Approach to the Threat Climate Change Poses to the Poor*, IIED Briefing Papers, London: IIED.

Ireland, P. and Keegan, P.J. (2013) 'Climate change adaptation: Challenging the mainstream', in L. Sygna, K.L. O'Brien and J. Wolf (eds) *A Changing Environment for Human Security: Transformative Approaches to Research, Policy and Action*, London: Earthscan, pp. 224–233.

IUCN, IISD, SEI-US and Intercooperation (2007) Summary of CRISTAL, Community-based Risk Screening Tool – Adaptation & Livelihoods. IISD, Winnipeg. Online. Available HTTP: http://www.iisd.org/pdf/2007/brochure_cristal.pdf (accessed 15 November 2012)

Johnsen, S. (2011) 'Bærekraftig utvikling – en rettesnor for politikken', in N. Bjerkedal, H. Heggenes, R. Malkenes, K. Moum, M. Skancke and S.B. Sæther (eds), *Finansråd i utfordrende tider. Om forvaltning og økonomisk politikk.* Oslo: Norwegian Ministry of Finance, pp. 104–110.

Jones, A. (2009) 'Pro-poor governance of global adaptation funds, a discussion paper', CARE, Germanwatch and Bread for the World. Online. Available HTTP: http://www.careclimatechange.org/files/reports/CARE_AdaptationFunds09.pdf (accessed 15 November 2012).

Klein, R.J.T., Eriksen, S., Næss, L.O., Hammill, A., Robledo, C. and O'Brien, K. (2007) 'Portfolio screening to support the mainstreaming of adaptation to climate change into development', *Climatic Change*, 84 (1): 23–44.

Kristoffersen, B. (2013) 'Opportunistic adaptation: Environmental change, oil and security in the Arctic', in K. O'Brien and E. Selboe (eds) *The Adaptative Challenge of Climate Change*, Cambridge: Cambridge University Press.

Leichenko, R.M. and O'Brien, K.L. (2008) *Environmental Change and Globalization: Double Exposures*, New York: Oxford University Press.

Lim, B., Spanger-Siegfried, E., Burton, I., Malone, E. and Huq, S. (eds) (2005) *Adaptation Policy Frameworks for Climate Change: Developing Strategies, Policies and Measures*, New York: Cambridge University Press.

McNeill, D. and García-Godos, J. (2005) *Doing Interdisciplinary Research on Development and the Environment: Critical Perspectives on SUM's Experience*, Oslo: SUM.

Moser, C. and Satterthwaite, D. (2008) 'Towards pro-poor adaptation to climate change in the urban centres of low- and middle-income countries', *Climate Change and Cities Discussion*, Paper 3 London: IIED.

NOU (2010) *Tilpassing til eit klima i endring Samfunnet si sårbarheit og behov for tilpassing til konsekvensar av klimaendringane*, NOU 2010:10, Oslo, Government of Norway.

O'Brien, K.L. (2009) 'Do values subjectively define the limits to climate change adaptation?' in W.N. Adger, I. Lorenzoni and K.L. O'Brien (eds) *Adapting to Climate Change. Values, Thresholds, Governance*, Cambridge: Cambridge University Press, pp.164–180.

O'Brien, K.L. (2012) 'Global environmental change II: From adaptation to deliberate transformation', *Progress in Human Geography*, 36: 667–676.

O'Brien, K. and Leichenko, R. (2007) 'Human security, vulnerability and sustainable adaptation', Human Development Report 2007/2008. Fighting Climate Change: Human Solidarity in a Divided World. Human Development Report Office Occasional Paper.

O'Brien, K., Quinlan, T. and Ziervogel, G. (2009) 'Vulnerability interventions in the context of multiple stressors: lessons from the Southern Africa Vulnerability Initiative (SAVI)', *Environmental Science and Policy*, 12 (1): 23–32.

Redclift, M. (2005) Sustainable development (1987–2005): An oxymoron comes of age, *Sustainable Development*, 13: 212–227.

Reid, P. and Vogel, C. (2006) 'Living and responding to multiple stressors in South Africa – glimpses from KwaZulu-Natal', *Global Environmental Change*, 16 (2): 195–206.

Robinson, J. and Herbert, D. (2001) 'Integrating climate change and sustainable development', *International Journal of Global Environmental Issues*, 1 (2): 130–149.

Sabates-Wheeler, R., Mitchell, T. and Ellis, F. (2008) 'Avoiding repetition: Time for CBA to engage with the livelihoods literature?' *IDS Bulletin*, 39 (4): 53–59

Schipper, E. L. and Pelling, M. (2006) 'Disaster risk, climate change and international development: Scope and challenges for integration', *Disasters*, 30 (1): 19–38.

Soltau, F. (2006) 'Climate change and sustainable development: Understanding the linkages', *Natural Resources Forum*, 30 (4): 253–255.

St. Clair, A. and Lawson, V. (2013) 'From poverty to prosperity: Addressing growth, equity and ethics in a changing environment', in L. Sygna, K.L. O'Brien and J. Wolf (eds) *A Changing Environment for Human Security: Transformative Approaches to Research, Policy and Action*, London: Earthscan, pp. 203–215.

Tanner, T.M. and Allouche, J. (2011) 'Towards a new political economy of climate change', *IDS Bulletin*, 43 (3): 1–14.

Tanner, T. and Mitchell, T. (2008) 'Introduction: Building the case for pro-poor adaptation'. *IDS Bulletin*, 39 (4): 1–5.

Van Aalst, M.K., Cannon, T. and Burton, I. (2008) 'Community level adaptation to climate change: The potential role of participatory community risk assessment', *Global Environmental Change*, 18 (1): 165–179.

Vernon, T. (2008) 'The economic case for pro-poor adaptation: What do we Know?' *IDS Bulletin*, 39 (4): 32–41.

Vincent, K., Næss, L.O. and Goulden, M. (2013) 'National level policies versus local level realities – can the two be reconciled to promote sustainable adaptation?' in L. Sygna, K.L. O'Brien and J. Wolf (eds) *A Changing Environment for Human Security: Transformative Approaches to Research, Policy and Action*, London: Earthscan, pp. 126–134.

Wisner, B., Blaikie, P., Cannon, T. and Davis, I. (2004) *At Risk: Natural Hazards, People's Vulnerability and Disasters*, London: Routledge.

Yohe, G., Burton, I., Huq, S. and Rosegrant, M.W. (2007) 'Climate change: Pro-poor adaptation, risk management, and mitigation strategies', International Food Policy Research Institute (IFPRI). Online. Available HTTP: http://www.ifpri.org/sites/default/files/publications/beijingbrief_rosegrant.pdf (accessed 15 November 2012).

32 Sustainability and cities

Meeting the grand challenge for the twenty-first century

Peter J. Marcotullio and William D. Solecki

Introduction

Urbanization in the twenty-first century is approaching a tipping point. The scale of increase in urban population and the rate of urbanization are unprecedented, and we are in the midst of a dramatic transition that will impact economic, social and environmental conditions at all scales (UNFPA, 2007; United Nations 2010). All the promises and challenges associated with human development are now increasingly concentrated in cities.

The scope, extent, speed, and timing of contemporary urbanization, combined with the demands for resource efficiency, pollution reduction, social equity and cohesion, and economic growth, create grand challenges for human security. By definition these are interrelated challenges, and they require urgent attention. As we expand current cities and construct new ones, we put into place forces that will constrain future choices. At the same time, there are fundamental questions about the limits to urban size, density, and environmental impact. Moreover, there is no robust theoretical understanding of what urban forms and processes are most economically efficient, socially beneficial, and environmentally benign. Finally, there is no evidence that the planet will be able to maintain an urban population of over 6 billion without severe ecological and/or global system collapse.

As the urbanization process continues, theory building and research into these challenges has become critical. Unfortunately, the trajectory of the knowledge development through the sustainable cities literature is not meeting the challenge. The details of this transition and the effects of increases in human density are not well understood. There remain myths about urbanization that have cast their shadows on policies (Satterthwaite, 2007b; UNFPA, 2007). This, in part, is due to the fragmentation of the sustainable cities literature, with researchers working within distinct frameworks representing different perspectives and understandings of sustainability (Haughton, 1999; McGranahan and Satterthwaite, 2000). A new holistic perspective on urban research is needed, and such a body of knowledge is beginning to emerge in response to policy demands (Bettencourt and West, 2010).

This chapter introduces the wider issues associated with contemporary urban transitions. We first briefly describe the current state and trends in global urbanization and the dynamics underpinning contemporary patterns of urban development. Next we provide a description and analysis of the sustainable cities literature. We discuss the trajectory of the literature and identify how it has splintered. We argue that although a new perspective on urbanization and sustainable development is developing, greater theory building efforts and critical reflection of our current understanding of urbanization processes are necessary before a holistic "urbanization science" will emerge.

State and trends of global urbanization

Traditionally, urbanization has been defined both by the concentration of population in urban areas and the increase in numbers of these centers (Tisdale, 1942). Over time, the average population of large cities around the world has increased. For example, Satterthwaite (2007b) notes that the average size of the world's 100 largest cities increased from 200,000 in 1800 to well over 6 million by 2000. In 1800, there were only two cities in the world with populations of over 1 million, but by 2010 this figure increased to 452 cities (Chandler, 1987; United Nations, 2010). As with the global urban population, much of the growth in both the size and number of large cities occurred after 1950. Up until 1975, there were just three megacities (those with 10 million or more) in the world: New York, Tokyo and Mexico City[1] (United Nations, 2010). Since then, their number has increased markedly such that one of the features of contemporary urbanization is the rise in the number of large cities around the world (Dogan and Kasarda, 1988a, 1988b; Lo and Yeung, 1998; Fuchs *et al.*, 1994; Dogan, 2004).

Global urbanization has been an uneven process; both temporally and spatially. Temporally, three major "waves" of global urbanization can be identified. The first and longest wave began 7,000 years ago, and was defined by slow and sporadic growth. With the emergence of cities, the development of global urbanization waxed and waned for approximately 6,750 years. The second wave occurred over the period between 1750 and 1950, and involved nations in Europe, North America and Japan. These nations experienced an increase in urban population share from 10 to 52 percent, representing an increase in urban population from 15 to 423 million residents (UNFPA, 2007). We are now in the midst of the third wave: 1950–2050. This wave is largely occurring in the developing world and represents an increase in urbanization level from 18 to 66 percent, corresponding to an increase from 302 million to 5.2 billion urban residents.

Spatially, urbanization processes have been highly uneven; during the first wave, urban centers were largely located in Western, Central and South Asia and in Central America. During the second wave, urbanization occurred most intensely in Europe and North America. By the middle of the twentieth century, urbanization shifted to regions in the developing world. By 1970, urbanization in Asia, Latin America and Africa reached 27 percent. As Table 32.1 demonstrates, while the general trend of rapid urbanization was experienced in all developing areas, there were important intra-regional distinctions.[2] Latin America, for example, has always had high urbanization levels due to the particularities of colonial rule. In 1970s, South American countries as a group had already reached

Table 32.1 Distribution of urban population, 1970–2010 (thousands). Source: UN 2009 World Urbanization Prospects, the 2009 Revisions

Region	*1970 Urban population*	*1970 Percent total*	*2010 Urban population*	*2010 Percent total*	*Percent change 1970–2010*
World	1,329,983	100.0	3,486,326	100.0	162.13
More developed regions	651,786	49.0	929,851	26.7	42.66
Less developed regions	678,197	51.0	2,556,475	73.3	276.95
Africa	86,523	6.5	412,990	11.8	377.32
Asia	483,019	36.3	1,757,314	50.4	263.82
Europe	412,386	31.0	533,295	15.3	29.32
Latin America, Caribbean	163,459	12.3	468,757	13.4	186.77
North America	170,691	12.8	288,803	8.3	69.20
Oceania	13,905	1.0	25,167	0.7	80.99

57 percent urbanization. Africa as a whole had reached urbanization levels of approximately 25 percent and in Asia the urban share was 23 percent. It has only been recently that China and India have attended significant urban population shares; by 2010, China reached 50 percent urban population and India 30 percent.

Given the large populations within the developing world, increasing levels of urbanization translates into a dramatic shift of the world's urban population. Between the 1970s and 2010, the less developed countries increased their share of the total global urban population from approximately 51 percent to 73 percent. Asia, which has been experiencing an rapid economic development in the past decades, now has half the world's urban population, followed by Latin America and Africa. Africa's urban population is now larger than that of North America. It is interesting to note that Asia has always been the location of the majority of the world's largest cities. As of 2010, 16 of the world's largest cities were located in Asia, and only six were located in either North America or Europe (Table 32.2). By 2025, when the number of megacities is expected to reach 29, Asia would have gained another five (UN, 2010).

Given the large rural populations in developing countries and given that much of the developed world is already urbanized, the focus of urbanization in the future will continue to center on the developing world. Global urban population is expected to increase by 84 percent by 2050, from 3.4 billion in 2009 to 6.3 billion in 2050, and virtually all of the expected growth will be concentrated in the urban areas of the less developed regions, whose population is projected to increase from 2.5 billion in 2009 to 5.2 billion in 2050 (UN, 2010). It is in these spaces where vulnerabilities to global environmental change may be most significant, as the populations will be largest, most concentrated and have the least resources with which to adapt.

Table 32.2 Largest urban agglomerations, 2010. Source: UN 2009 World Urbanization Prospects, the 2009 Revisions

Country	*Urban agglomeration*	*Population (millions)*
Japan	Tokyo	36.67
India	Delhi	22.16
Brazil	São Paulo	20.26
India	Mumbai (Bombay)	20.04
Mexico	Ciudad de México (Mexico City)	19.46
United States of America	New York-Newark	19.43
China	Shanghai	16.58
India	Kolkata (Calcutta)	15.55
Bangladesh	Dhaka	14.65
Pakistan	Karachi	13.12
Argentina	Buenos Aires	13.07
United States of America	Los Angeles-Long Beach-Santa Ana	12.76
China	Beijing	12.39
Brazil	Rio de Janeiro	11.95
Philippines	Manila	11.63
Japan	Osaka-Kobe	11.34
Egypt	Al-Qahirah (Cairo)	11.00
Nigeria	Lagos	10.58
Russian Federation	Moskva (Moscow)	10.55
Turkey	Istanbul	10.52
France	Paris	10.49
Republic of Korea	Seoul	9.77
China	Chongqing	9.40
Indonesia	Jakarta	9.21
United States of America	Chicago	9.20
China	Shenzhen	9.01
Peru	Lima	8.94
China	Guangzhou, Guangdong	8.88
Democratic Republic of the Congo	Kinshasa	8.75
United Kingdom	London	8.63

Perspectives on sustainable cities

Sustainable development was popularized by the Brundtland Commissions report, *Our Common Future* (1987), which included a chapter on urban issues that provided acknowledgment that cities were important to global development. The chapter primarily addressed urban service deficiencies in fresh water supply, sanitation, housing, waste disposal and transport. Within five years, however, the UN "Earth Summit" in Rio de Janeiro helped to redirect the focus to global issues associated with natural resources, climate change, ozone depletion, trans-boundary pollution, deforestation, soil loss, etc. Urbanization was left out, and poverty and health were termed "cross-sectoral issues."

The literature on sustainable cities grew based upon an attempt to follow the basic definition of the Brundtland Commission. As Campbell (1996: 304) defined it, urban sustainable development is "the long-term ability of a system to reproduce." The central focus was on long-term sustainable development of cities. A variety of books emerged that attempted to bring coherence to the growing literature (Haughton and Hunter, 1994; Leitmann, 1999; Satterthwaite, 1999; Wheeler and Beatley, 2004). Many simply combined a diverse set of writings to point out key issues.

What emerged in the urban literature were two main streams of thought differentiated by their focus on overcoming different sets of tensions. First, there are literatures based upon environmental degradation ("green" agenda issues) versus environmental hazards ("brown" agenda issues) (Hardoy *et al.*, 2001). Second, there are literatures based upon ideological viewpoints about the inherent factors that could create sustainable cities (Haughton and Hunter 1994; Haughton, 1999).

Green and brown agendas

Green agenda

The "green" agenda dominated early studies, as it was compatible with the growing interests in global resource conservation and management and cities were seen as major sources of degradation (Haughton and Hunter, 1994). This interest was underpinned by a long history of urban biophysical studies dating back to before the 1800s, but increasing after the Second World War (Sukopp, 2008). As information on the urban environment generated from individual studies increased, entire book-length treatments emerged on individual cities, such as New York (Kieran, 1972) or London (Fitter, 1945), or on a set of environmental issues associated with the urban development (Detwyler and Marcus, 1972; Gill and Bonnett 1973; Hough 1984; Gilbert, 1989; Platt *et al.*, 1994). These studies examined the "natural" environments of cities focusing on the provision of green space or the ecology of these spaces. Much emphasis was placed on how humans impacted urban biodiversity, hydrology, soils, and atmospheric conditions. The rationale driving these studies was that identifying the processes by which cities impacted the environment would help to inform policies on how to reduce this impact and hence create sustainable development. Few of these studies attempted to integrate the interactions between humans and environmental factors (exceptions are Havlick 1974; Douglas 1981, 1983), but rather saw urban residents as solely driving change. Moreover, environmental hazards associated with urban development were addressed simply as an infrastructure problem.

Brown agenda

The "brown" agenda focused on cities as centers of inequity and injustice (Drakakis-Smith, 1995, 1996, 1997; Hardoy *et al.*, 2001). These studies started by examining the environmental conditions and hazards within poor cities, or within the poorest locations within cities. The poor quality of basic infrastructure and services has been a major theme in the urban literature on development (see for example, Hardoy *et al.*, 1990; Hardoy and Satterthwaite, 1989). Research interest emerged with the growth of squatter settlements starting in the 1960s (Abrams, 1964). These studies pointed out that increasingly larger shares of urban residents of the developing world were living in slum settlements. They emphasized the lack of infrastructure, particularly housing, water supply and sanitation, as the source of health risks. Furthermore, these researchers positioned solutions as embedded in governance, as much as in technical fixes and improved ecological conditions. At the same time, governments were known for associating urbanization with poverty, international development agencies that perceived agricultural development as the first step to development and academic attention on urban policy as preventing nations from developing help to divert attention away from the growing crisis (Satterthwaite, 2007a). As a result, urban poverty has not been addressed and "brown agenda" issues remain an important concern for most cities in the developing world.

Different ideological viewpoints

The literature on sustainable cities also splintered into several different variants based upon ideology. Haughton (1999) and McCarney (1996), for example, have helpfully outlined the different perspectives of the sustainable city. At least four models of sustainable cities emerged within the literature: the free market model, the design model, the self-reliant city model, and the fair share model (Haughton, 1999). Each model has its own set of criteria for evaluating urban sustainable development. Together they form competing visions of the sustainable city and each embodies a different analysis of the problem and alternative attempts to re-order socio-technical relations. The results of each model chart the same city along very different pathways. Moreover, there is little leeway for synthesizing these models into one larger perspective. This is because the models prioritize different fundamental drivers of change, understand feedbacks and internal processes in cities differently, and use different criteria for sustainability. While each provides part of the vision and analysis needed to understand urban sustainable challenges, none alone provide a holistic and integrated (social, environmental and economic) perspective.

Free market perspectives

The externally dependent cities (free-market model) center on the largely non-spatial views of economists seeking to address urban environmental problems through altering market mechanisms. Hence, cities are the sources of economic growth and therefore sustainable development should be based upon their continued and improved economic performance. Those using this model focus on productivity, urban competitiveness and economic growth. Proponents espouse free trade, de-regulation and generally what has been called the "neo-liberal" agenda. This model interprets the sustainable problems of cities, be it poverty or degraded environments, as being largely created by market and regulatory failures (World Bank, 1991; for a critique see Burgess *et al.*, 1997).

Design perspectives

The second model has been termed the "redesigning the city model" and it centers on the physical fabric of the existing city. Researchers examine how changes to the layout and structure of the city can encourage greater resource efficiency, reduced environmental impact, social safety, security and protection, conducive social environments, creativity, prestige, and livability (Frey, 1999). Those examining the city in this light focus on how cities are planned and specifically how this planning has impacted energy flows, usage and conservation (Lyle, 1994). This form of analysis interprets the problems of cities as largely based upon the inappropriate use of materials and lack of (e.g., sprawl, energy inefficient buildings) or misguided (e.g., segregated land use, lack of open space) planning (Calthorpe, 1993; Jenks and Burgess, 2000).

Self-reliant perspectives

The "self-reliant city model" centers on combining a more sensitive approach to nature with a decentralized, community-based, grass-roots politics. This model embraces the deep ecology approach to sustainable development, meaning that all healing activities start with the assumption that living things have an inherent worth aside from their utility to humans. The model emphasizes preservation of natural assets and a philosophical ethics that focuses on reducing human impact to natural systems. Importantly, the model sees the changes needed for urban sustainability as driven by bottom-up forces and that social diversity, like bio-diversity is a "good." Researchers examine how small, autonomous communities reduce ecological footprints, and the use of "appropriate" technologies associated with small-scale economic production. The goal is to create a city that does not need external resources. Urban problems are understood as resulting from inharmonious growth and therefore balance should be achieved of human and natural structures and processes (Walter *et al.*, 1992; Roseland, 1997; Register, 2006).

Fair-share perspectives

The "fair shares model" centers on greater attention to the conditions of trading between a city-region and its hinterland, based on resource consumption and waste-emission trends. This model is best understood as an aggregate approach that draws on many of the features of redesigning the city and the self-reliant city models, such as increase regional autarky, greater urban compaction and improved use of market tools, and adding tools for engaging in more equitable trading relationships with other urban areas. It also emphasizes the importance of domestic and increasingly international trade and investment relationships. Urban problems generate from unequal trade relationships or the dominance of foreign investments (Ravetz, 2000; Marcotullio, 2001).

Reconciling different perspectives

Attempts to reconcile the differences in these streams have made important contributions to the urban sustainable development literature. One major attempt put forth a model that suggested a transition of types of environmental issues with growing wealth, called the urban environmental transition (McGranahan *et al.*, 2001). This model suggests three agendas: "brown," "gray," and "green" (Smith and Ezzati, 2005) and postulates that as cities grow in affluence, urban environmental burdens tend to shift from local, health threatening challenges that have immediate impacts, to global ecosystem threatening, with delayed impacts. Importantly, the model is based upon the notion that the type and scale of the environmental impact change with wealth. Accordingly, as a poor city moves beyond the "brown" agenda, which prioritizes concerns such as inadequate water supply and sanitation, indoor air quality, drainage and solid waste disposal, the environmental impacts of urban activities shift from the household and neighborhood scale to a citywide regional scale. At the metropolitan scale the dominant issues that come into focus include water and air pollution, sometimes called the "gray agenda." As cities continue to grow in wealth, urban environmental burdens shift from metropolitan scale challenges to those most easily observed at the regional and global scale. Global scale challenges are associated with the "green" agenda and include acid rain, water scarcity, greenhouse gas and ozone depleting substance emissions, for example (Marcotullio, 2005).

In this way the impact of cities on the environment is seen along a continuum of increasing wealth with changing but ever present environmental challenges. Moreover, environmental problems are never completely resolved but rather outcomes are displaced across time and space. Historical urban research that examining urban growth and environmental impacts suggests that in the past, urban environmental challenges associated with water supply, sanitation and solid waste management were addressed by simply dispersing the harms to greater scales (Tarr, 1996; Melosi, 2000). Historians have identified changes in type, geographic, and temporal aspects of environmental burdens that are comparable to urban environmental transition theory (Marcotullio, 2005).

Another way of resolving the conflicts between the "brown" and "green" agenda is to include both challenges in a systems approach that allows policy makers to choose the most appropriate solution (McGranahan and Satterthwaite, 2000). When all challenges are considered, priorities and potential alternatives that address both can be examined. First, reconciling these agendas means that different cities or even different neighborhoods within a single city may have different priorities. Second, some agendas can be addressed through carefully crafted policies. For example, air pollution from automobiles that cause health problems which is also related to trans-boundary pollution and global greenhouse gas emissions, can be addressed through transit, traffic and mobility policies. Solid waste problems that create pest infestation and health impacts through oral-fecal material contact can be addressed through sanitation infrastructure, materials storage, recycling and energy policies (McGranahan and Satterthwaite, 2003).

While the urban environmental transition theory is powerful, the focus is not inherently on the urban. Indeed, one could substitute individuals, households, neighborhoods, states, or nations for cities and the theory would generally hold. The reality that the theory represents is change in environmental conditions with increasing wealth, not what is inherently different or unique about cities in this process.

There has been less success in integrating different normative approaches within the sustainable cities literature. Care is required in interpreting attempts to reconcile sustainable city models. Some research that portends universal findings remains biased due to implicit ideological perspectives. Lists of principles for creating sustainable cities, for example, provide a general blueprint to facilitate conforming to one particular agenda (Roseland 1997; Kenworthy, 2006). However, detailed studies of how to manage urban sectors in a sustainable fashion, such as transportation or urban agriculture, typically do not consider cross-sector and cross-scale interactions. Researchers have continued to search for ways to integrate agendas in pursuit of a knowledge base on urban sustainable development (Seto and Satterthwaite, 2010). As Grubler *et al.* (2012) point out, however, there is a lack of understanding of how higher-order organizations of urban systems (i.e., urban functions, urban economy, urban infrastructural setup) shape the scale and intensity of urban activities and hence environmental impact. The problem with much of the current literature is not that it is wrong, but that is remains partial (Holling, 2001; Gunderson and Holling, 2002).

Integrating perspectives on urban sustainable development

What is needed is a more holistic and comprehensive approach; a new science of urbanization. The focus of this science would be on the fundamental laws of the urbanization process including the origin, development, organization, emergent properties and connection to other social and biophysical systems. We see the formation of this science in recent research that attempts to bring together both the human and environmental aspects of the sustainable development agenda. These studies are converging from research fields such as urban ecosystems, coupled human-environment interactions and urban socio-ecological system analysis. Some of the important components of this new approach include a systems approach which attempts to integrate social and biophysical elements, an emphasis on scale and a focus on the higher-order aspects of urbanization. We believe that this literature forms the basis of new scientific understandings of cities and can lead to policies that enhance the promotion of sustainable development both within urban areas and in the distal areas associated with and impacted by urban activities. In this last sub-section we describe these elements and point out types of research that are helping to create this new perspective.

Systems approaches developed during the 1950s and 1960s have been used to study cities in the past and to provide the basis of urban plans (McLoughlin, 1969; Chadwick, 1971). For example, as early as the 1960s, Berry (1964), explained that cities were systems and operated within systems. The contemporary systems approaches are more inclusive of components of the urban system. Urban system thinking now includes economic, social and institutional aspects, as in the past, but now these studies also include biophysical components (Grimm *et al.*, 2000). Cities are perceived as complex systems with important biophysical elements that affect and are affected by the social elements associated with urban life. In this regard, previous urban system models have been under-theorized. These new studies include accounting techniques such as material flows analysis, ecological foot-printing and human appropriation of net primary productivity, and identify the flows and linkages between social and biophysical change. They also allow for a quantification and analysis of the impact of socio-economic activities on the environment as well as what changes in the environment impact human activities. At the same time, they emphasize processes and move away from a focus on places.

Recent studies highlight the lack of consensus on the basic components of urbanization (Seto *et al.*, 2012). Strongly held axiomatic beliefs on the relationship between the population share in urban centers and income generation, or the space for time substitution models associating urban land use and ecosystem service degradation are coming into question. What is desperately needed is an exploration of the most fundamental aspects of urbanization, across space, place, time and cultures. A proactive approach to urbanization and the complex set of socio-ecological systems interactions within and across urban areas demands an understanding of all system components and their interactions (Solecki *et al.*, 2012).

Urban science approaches emphasize the importance of scale. Scale, meaning extent, is perceived as crucial in understanding urban activities, their effects and influence (Sanchez-Rodriguez *et al.*, 2005). The notion of emergent phenomena is central to all complex systems. Emergence is defined by the patterns that arise from a multiplicity of relatively simple interactions (Alberti, 2008). Emergent phenomena may not be visible at a specific scale (i.e., sprawl cannot be understood by examining a single city block). Given the importance of emergence in urban systems, it is necessary to examine urban activities at a variety of scales, for example, the various environmental challenges associated with urbanization at different scales (Satterthwaite, 1997; McGranahan *et al.*, 2001).

Currently, some researchers are examining urbanization at the global level looking for universal patterns. These types of studies were also performed in the past, exemplified by some basic urban laws, such as the rank-size rule (Zipf, 1946, 1947) that states that within a country or region there is a log–linear relationship between the rank of the city and its population size such that the largest city is twice the size of the next largest and three times the size of the third largest and so on. This finding spanned a set of studies that defined national urban systems and subsequent debates over the meaning of different patterns. Exactly why this rule works is still debated (see Fonseca, 1988). Rather than being based upon socially explanatory variables scholars studying this phenomenon in the past used a normative set of underlying assumptions to drive research. Terms such as "parasitic," "over urbanization" (Davis and Golden, 1954; Hoselitz, 1954, 1957), "urban bias" (Lipton, 1977; Todaro, 1981), and "urban primacy" (Jefferson, 1939), were used to describe dysfunctional national urban development patterns because they failed to follow the rank-size rule. Rather than privileging any particular pattern of urban development, today's research critically examines the empirical evidence that presents the urbanization system in-and-of-itself as unique amongst systems (at any scale) and asks how the associated elements and processes influence other systems. Scholars are attempting to identify a theory of urbanization with fundamental and unique components that can withstand scientific scrutiny and produce valuable universal laws and theories (Bettencourt *et al.*, 2007; Bettencourt and West, 2010, 2011).

As the amount of data on cities and urbanization processes at the national level increased, some researchers have re-examined global-scale, urban-related relationships between urban activities and socio-economic variables (Bettencourt *et al.*, 2007). These studies are providing evidence for a theoretical understanding of scaling laws associated with global urbanization. Population size of cities correlates to, for example, rates of innovation, wealth creation, energy consumption, and crime per capita. Some of the relationships are linear, others sublinear, meaning that economies of scale exist (larger cities are more efficient) and other relationships are supralinear, meaning increasing returns with population size (as population increases, the variables increases at a greater rate). This type of research, given its comprehensiveness, can help to identify the relationships between urbanization and other aspects of the Earth system.

Urbanization science focuses on the need to understand how the urbanization process unfolds so that we can develop generalities to larger scales, and with that, understand how this process interacts with local and global environments. For example, while there have been studies of how biodiversity changes in urban to rural gradients and statements on the increasing homogeneity of urban biota, there has not been a study that examines whether and how urbanization has affected, for example, the latitudinal gradient in biodiversity or any of the eco-geographic rules of biodiversity. The answers to this and other questions will lead to a better understanding of how, when, where and at what scale urbanization connects to laws and principles in other sciences.

Other research has examined the differences in socio-ecological systems of cities at the regional scale, examining differences among different groups of cities.

Again, urban studies scholars have long noted the differences between cities, urbanization processes and urban activities of the developed and developing worlds (see for example Weber, 1899). Given the dynamics associated with contemporary urbanization, current studies, using databases over time and across space, emphasize comparative studies of historical transition experiences. While not all urban related experiences are universal, there are seemingly some principles that define differences between those of the developed and the developing worlds. For example, when it comes to urban environmental dynamics, urbanization is related to environmental harms sooner (at lower economic levels), which change faster (over time) and emerge in a more simultaneous manner (Marcotullio *et al.*, 2005). Others have identified different types of environmental dynamics, such as levels of air pollution with different levels of income (Sarzynski, 2012).

Furthermore, there is an increasing interest in urban resilience and vulnerability. Recent events, such as Tropical Storm Sandy that hit the Northeast USA in October, 2012 causing more than $33 billion of damages, demand that a scientific basis for assessing urban climate change adaptation strategies is advanced (Rosenzweig *et al.*, 2011b). Moreover, these types of studies allow for scenario development that facilitates both popular understanding of environmental change and furthers civil society discussion of potential future alternatives.

The goal of urbanization science is to develop relational models (as opposed to end state physical models such as "compact cities") that can help to uncover patterns associated with urbanization including the effect of wider political and commercial processes and the variety (both good and bad) of the potential impacts of new technologies. The outcomes of the analyses provide not only the description of the physical patterns that emerge, but also the patterns of relationships between socio-economic and biophysical variables and processes. Because urbanization unfolds on multiple scales it needs to be addressed at these scales. While there are all too few examples of regional and national urban programs that address urbanization, there is nothing at the global scale. Urbanization science is an effort to fill this gap and therefore support policy development at these larger scales (Rosenzweig *et al.*, 2011a; Grubler *et al.*, 2012). Stakeholders can positively influence urbanization, but only if they have the scientific evidence and understanding of how the system works; how component parts are related, who the winners and losers of change are, and what the drivers are. Given the growing importance of urbanization and the concentration of populations and economies in cities, urbanization is now of global interest to researchers and policies. We believe that the moment has arrived for the development of an urbanization science that will provide the basis for and stimulate local, national and international laws, regulations and agreements.

Conclusion

Urbanization is one of the major trends and challenges of the current century. The increasing strength and level of populations located within urban areas have turned attention to global sustainable development on cities. The literature on sustainable cities has developed rapidly into separate areas of knowledge based upon the different factors influencing urban development and distinct ideologies. This fractured field with partial and biased knowledge is inadequate in addressing the grand challenges associated with current urbanization.

Despite the fracturing in the literature, there are also trends that suggest there is some convergence about how to approach urban development and urbanization. Recent literature is emerging that attempts to overcome the previous imbalances and develop a holistic approach. We call this new approach urbanization science. Specifically, recent studies on urban sustainable development have three important goals. First, models attempt to identify and understand the broad nature of urbanization processes. Given the amount of data and comprehensive nature, they provide an understanding of the contexts in which sustainability issues emerge, are developed and are reinterpreted. Second, the research provides a scientific basis for testing claims for the forms and processes of sustainable urbanization by exploring the socio-ecological effects of different development trajectories. Third, by moving away from a collection of individual city studies, urbanization science identifies different pathways incorporating local, regional and global variables that help to suppress the dominance of inappropriate (unsustainable) agendas and open up opportunities for alternative pathways. While we see much merit in the emerging urbanization science, there is also a need to incorporate an examination of who benefits urbanization patterns and processes. We have noted the valuable work to date on a variety of issues, but there is not enough emphasis, to date, on environmental justice. How to work into these studies a critical reflection of norms and values driving patterns and processes remains an important task for the development of this science.

Notes

1 There has been much written on mega-cities (or those larger than 10 million), which have dominated the popular writings of urbanization. Despite this attention, however, in terms of population, they account for approximately 9.4 percent of the global urban population and only 4.7 percent of the total global

2 Some analysts warn of overgeneralization of regional urbanization levels (see for example, Satterthwaite, 2007b). The point is that in many countries, censuses have been rare and urban population figures are based upon estimates. Moreover, the definition of what is urban varies dramatically between countries. Therefore these figures only provide a glimpse of overall trends.

References

Abrams, C. (1964) *Man's Struggle for Shelter in an Urbanizing World*, Cambridge, MA: MIT Press.

Alberti, M. (2008) *Advances in Urban Ecology: Integrating Humans and Ecological Processes in Urban Ecosystems*, New York: Springer.

Berry, B. J. L. (1964) 'Cities as systems within systems of cities', *Papers of the Regional Science Association*, 13: 147–163.

Bettencourt, L. and West, G. (2010) 'A unified theory of urban living', *Nature*, 467: 912–913.

Bettencourt, L. M. A. and West, G. B. (2011) 'Bigger cities do more with less', *Scientific American*, 305 (3): 52–53.

Bettencourt, L. M. A., Lobo, J., Helbing, D., Kühnert, C. and West, G. B. (2007) 'Growth, innovation, scaling, and the pace of life in cities', *Proceedings of the National Academy of Sciences*, 104 (17): 7301–7306.

Burgess, R., Carmona, M. and Kolstee, T. (eds) (1997) *The Challenge of Sustainable Cities: Neoliberalism and Urban Strategies in Developing Countries*, London: Zed Books.

Calthorpe, P. (1993) *The Next American Metropolis, Ecology, Community, and the American Dream*, New York: Princeton Architectural Press.

Campbell, S. (1996) 'Planning: Green cities, growing cities, just cities? Urban planning and the contradictions of sustainable development', *Journal of the American Planning Association*, 62 (3): 296–312.

Carroll, G. R. (1982) 'National city-size distributions: What do we know after 67 years of research?', *Progress in Human Geography*, 6 (1): 1–43.

Chadwick, G. F. (1971) *A Systems View of Planning*, Oxford: Oxford University Press.

Chandler, T. (1987) *Four Thousand Years of Urban Growth, An Historical Census*, Lewiston, NY: The Edwin Mellen Press.

Davis, K. and Golden, H. H. (1954) 'Urbanization and the development of preindustrial areas', *Economic Development and Cultural Change*, 3: 6–24.

Detwyler, T. R. and Marcus, M. G. (eds) (1972) *Urbanization and Environment*, Belmont: Wadsworth Publishing.

Dogan, M. (2004) 'Four hundred giant cities atop the world', *International Social Science Journal*, 56 (181): 347–360.

Dogan, M. and Kasarda, J. D. (eds) (1988a) *The Metropolis Era, Volume 1, The World of Giant Cities*, Newbury Park, CA: Sage Publications.

Dogan, M. and Kasarda, J. D. (eds) (1988b) *The Metropolis Era, Volume 2, Mega-cities*, Newbury Park, CA: Sage Publications.

Douglas, I. (1981) 'The city as an ecosystem', *Progress in Physical Geography*, 5 (3): 315–367.

Douglas, I. (1983) *The Urban Environment*, London: Edward Arnold.

Drakakis-Smith, D. (1995) 'Third world cities: Sustainable urban development 1', *Urban Studies*, 32 (4–5): 659–677.

Drakakis-Smith, D. (1996) 'Third world cities: Sustainable urban development II – population, labour and poverty', *Urban Studies*, 33 (4–5): 673–701.

Drakakis-Smith, D. (1997) 'Third world cities: Sustainable urban development III – basic needs and human rights', *Urban Studies*, 34 (5–6): 797–823.

Fitter, R. S. R. (1945) *London's Natural History*, London: Collins.

Fonseca, J. W. (1988) *Urban Rank-size Hierarchy*, Fairfax, VA: Institute of Mathematical Geography, George Mason Univeristy.

Frey, H. (1999) *Designing the City, Towards a More Sustainable Urban Form*, London and New York: Spon Press, Taylor and Francis Group.

Fuchs, R., Brennan, E., Chamie, J., Lo, F-C. and Uitto, J. I. (eds) (1994) *Mega-City Growth and the Future*, Tokyo: UNU Press.

Gilbert, O. L. (1989) *The Ecology of Urban Habitats*, New York: Chapman and Hall.

Gill, D. and Bonnett, P. (1973) *Nature in the Urban Landscape, A Study of City Ecosystems*, Baltimore, MD: York Press.

Grimm, N.B., Grove, J. M., Pickett, S. T. A. and Redman, C. L. (2000) 'Integrated approaches to tong-term studies of urban ecological systems', *BioScience*, 50 (7): 571–584.

Grubler, A., Bai, X., Buettner, T., Dhakal, D., Fisk, D., Ichinose, T., Keirstead, J., Sammer, G., Satterthwaite, D., Schulz, N.B., Shah, N., Steinberger, J. and Weisz, H. (2012)

'Urban energy systems', in L. Gomez-Echeverri, T. B. Johansson, N. Nakicenovic, and A. Patwardhan (eds) *Global Energy Assessment: Toward a Sustainable Future*, Laxenburg, Austria: International Institute for Applied Systems Analysis, pp. 1307–1400.

Gunderson, L. H. and Holling, C. S. (eds) (2002) *Panarchy, Understanding Transformations in Human and Natural Systems,* Washington, DC: Island Press.

Hardoy, J. E. and Satterthwaite, D. (eds) (1989) *Squatter Citizen: Life in the Urban Third World,* London: Earthscan Publications.

Hardoy, J. E., Cairncross, S. and Satterthwaite, D. (eds) (1990) *The Poor Die Young: Housing and Health in Thrid World Cities,* London: Earthscan Publications.

Hardoy, J. E., Mitlin, D. and Satterthwaite, D. (2001) *Environmental Problems in an Urbanization World,* London: Earthscan.

Haughton, G. (1999) 'Searching for the sustainable city: Competing philosophical rationales and processes of "Ideological Capture" in Adelaide, South Australia', *Urban Studies*, 36 (11): 1891–1906.

Haughton, G. and Hunter, C. (1994) *Sustainable Cities*, London: Jessica Kingsley Publishers.

Havlick, S. W. (1974) *The Urban Organism, The City's Natural Resources from an Environmental Perspective*, New York: Macmillian.

Holling, C. S. (2001) 'Understanding the complexity of economic, ecological and social systems', *Ecosystems*, 4: 390–405.

Hoselitz, B. F. (1954) 'Generative and parasitic cities', *Economic Development and Cultural Change*, 3: 278–294.

Hoselitz, B. F. (1957) 'Urbanization and economic growth in Asia', *Economic Development and Cultural Change*, 6 (October): 42–54.

Hough, M. (1984) *City Form and Natural Process*, New York: Van Nostrand Riengold Company, Inc.

Jefferson, M. (1939) 'The law of the primate city', *The Geographical Review*, 29: 226–232.

Jenks, M. and Burgess, R. (eds) (2000) *Compact Cities: Sustainable Urban Forms for Developing Countries*, London: SPON Press.

Kenworthy, J. R. (2006) 'The eco-city: ten key transport and planning dimensions for sustainable city development', *Environment and Urbanization*, 18 (1): 67–85.

Kieran, J. F. (1972) *A Natural History of New York City*, Boston, MA: Houghton Mifflin.

Leitmann, J. (1999) *Sustaining Cities, Environmental Planning and Management in Urban Design*, New York: McGraw-Hill.

Lipton, M. (1977) *Why Poor Peole Stay Poor – Urban Bias in World Development*, London: Temple Smith.

Lo, F-C. and Yeung, Y-M. (eds) (1998) *Globalization and the World of Large Cities*, Tokyo: United Nations University.

Lyle, J. T. (1994) *Regenerative Design for Sustainable Development*, New York: John Wiley & Sons.

Marcotullio, P. J. (2001) 'Asian urban sustainability in the era of globalization', *Habitat International*, 25: 577–598.

Marcotullio, P. J. (2005) *Time Space Telescoping and Urban Environmental transitions in the Asia Pacific*, Tokyo: United Nations University Institute of Advanced Studies.

Marcotullio, P. J., Williams, E. W. and Marshall, J. D. (2005) 'Faster, sooner, and more simultaneously: How recent transportation CO_2 emission trends in developing countries differ from historic trends in the United States of America', *Journal of Environment and Development*, 14 (1): 125–148.

McCarney, P. (1995) 'Four approaches to the environment of cities', in R. Stren and J. K. Bell (eds) *Perspectives on the City*, Toronto: Center for Urban & Community Studies, University of Toronto: 229–278.

McGranahan, G. and Satterthwaite, D. (2000) 'Environmental health or ecological sustainability? Reconciling the brown and green agendas in urban development', in C. Pugh (ed.) *Sustainable Cities in Developing Countries*, London: Earthscan, pp. 73–90.

McGranahan, G. and Satterthwaite, D. (2003) 'Urban centers: An assessment of sustainability', *Annual Review of Environment and Resources*, 28: 243–274.

McGranahan, G., Jacobi, P., Songsore, J., Surjadi, C. and Kjellen, M. (2001) *The Citizens at Risk, From Urban Sanitation to Sustainable Cities*, London: Earthscan.

McLoughlin, J. B. (1969) *Urban and Regional Planning: A Systems Approach*, London: Faber and Faber.

Melosi, M. V. (2000) *The Sanitary City: Urban Infrastructure in America from Colonial Times to the Present*, Baltimore, MD: Johns Hopkins Press.

Platt, R., Rowntree, R. A. and Muick, P. C. (eds) (1994) *The Ecological City, Preserving and Restoring Urban Biodiversity*, Amherst, MA: University of Massachusetts Press.

Ravetz, J. (2000) *City Region 2020, Integrated Planning for a Sustainable Environment*, London: Earthscan.

Register, R. (2006) *Ecocities, Rebuilding Cities in Balance with Nature*, Gabriola Island, BC: New Society Publishers.

Roseland, M. (1997) 'Dimensions of the eco-city', *Cities*, 14 (4): 197–202.

Rosenzweig, C., Solecki, W.D., Hammer, S. and Mehrotra, S. (eds) (2011a) *Climate Change and Cities, First Assessment Report of the Urban Climate Change Research Network*, Cambridge: Cambridge University Press.

Rosenzweig, C., Solecki, W.D., Hammer, S.A. and Mehrotra, S. (eds) (2011b) *Climate Change and Cities, First Assessment Report of the Urban Climate Change Research Network*, Cambridge: Cambridge University Press.

Sanchez-Rodriguez, R., Seto, K. C., Simon, D., Solecki, W. D., Kraas, F. and Laumann, G. (2005) *Science Plan*, Urbanization and Global Environmental Change, Bonn: IHDP.

Sarzynski, A. (2012) 'Bigger is not always better: A comparative analysis of cities and their air pollution impact', *Urban Studies*, 49: 3115–3132.

Satterthwaite, D. (1997) 'Sustainable cities or cities that contribute to sustainable development', *Urban Studies*, 34 (10): 1667–1691.

Satterthwaite, D. (ed.) (1999) *The Earthscan Reader in Sustainable Cities*, London, Earthscan Publications, Ltd.

Satterthwaite, D. (2007a) 'In pursuit of a healthy urban environment in low- and middle-income nations', in P. J. Marcotullio and G. McGranahan (eds) *Scaling the Urban Environmental Challenge, From Local to Global and Back*, London: Earthscan Publishers.

Satterthwaite, D. (2007b*) The Transition to a Predominantly Urban World and its Underpinnings*, London: IIED, Human Settlements Discussion Paper Series, Urban Change-4.

Seto, K. C. and Satterthwaite, D. (2010) 'Interactions between urbanization and global environmental change', *Current Opinion in Environmental Sustainability*, 2 (3): 127–128.

Seto, K. C., Reenberg, A., Boone, C. G., Fragkias, M., Haase, D., Langanke, T., Marcotullio, P., Munroe, D. K., Olah, B. and Simon, D. (2012) 'Urban land teleconnections and sustainability', *Proceeding of the National Academy of Sciences of the United States of America*, 109 (18).

Smith, K. and Ezzati, M. (2005) 'How environmental health risks change with development: The epidemiologic and environmental risk transitions revisited', *Annual Review of Environment and Resources*, 30: 291–333.

Solecki, W., Seto, K. C. and Marcotullio, P. J. (2012) 'It's time for an urbanization science', *Environment and Urbanization*, 55 (1): 12–17.

Sukopp, H. (2008) 'On the early history of urban ecology in Europe', in J. M. Marzluff, E. Shulenberger, W. Endlicher, M. Alberti, G. Bradley, C. Ryan, U. Simon and C.

Zumbrunnen (eds) *Urban Ecology, An International Perspective on Interaction Between Humans and Nature*, New York: Springer: 79–97.

Tarr, J. A. (1996) *The Search for the Ultimate Sink, Urban Pollution in Historical Perspective*, Akron: University of Akron Press.

Tisdale, H. (1942) 'The process of urbanization', *Social Forces*, 20 (3): 311–316.

Todaro, M. (1981) *City Bias and Rural Neglect: The Dilemma of Urban Development*, New York: Population Council.

UN-Habitat (2009) *Global Report on Human Settlements 2009, Planning Sustainable Cities*, London: Earthscan.

United Nations (2010) *World Urbanization Prospects: The 2009 Revisions*, New York: Department of Economic and Social Affairs.

Walter, B., Arkin, L. and Crenshaw, R. (eds) (1992) *Sustainable Cities, Concepts and Strategies for Eco-City Development*, Los Angeles: Eco-Home Media.

Weber, A. F. (1899) *The Growth of Cities in the Nineteenth Century*, Ithaca, NY: Cornell University Press.

Wheeler, S. M. and Beatley, T. (eds) (2004) *The Sustainable Urban Development Reader*, London and New York: Routledge.

World Bank (1991) *World Development Report*, New York: World Bank and Oxford University Press.

World Commission on Environment and Development (1987) *Our Common Future*, Oxford: Oxford Press.

Zipf, G. K. (1946) 'The P_1P_2/D hypothesis and intercity movement of persons', *American Sociological Review*, 11 (6): 677–686.

Zipf, G. K. (1947) 'The hypothesis of the minimum equation as a unifying social principle', *American Sociological Review*, 12 (6): 627–650.

33 A tetra-transition away from fossil fuels

Svein Jarle Horn

Introduction

In a historical perspective we live in a very special time. This is the fossil era where humanity is consuming Earth's vast resources of oil, gas and coal. The consumption and depletion of this energy treasure only happens once. Cheap and easily available fossil fuels made the agricultural green revolution possible, which again led to an enormous growth in the world population from 1 billion in 1800 to 7 billion today. The population growth and unsustainable development paths have lead to an explosion in the consumption of resources and energy. But we live on a finite planet; unlimited economic growth is not possible and resource limitations will sooner or later halt the growth.

Our modern civilization is founded on a worldview that emerged with the onset of the Industrial Revolution, when the world was still relatively empty of humans and their infrastructure (Costanza, 2008). Natural resources were abundant and people had limited access to infrastructure and consumer goods. This set of circumstances has been called an "empty" world (Costanza *et al.*, 1997). It was in this empty world that our current worldview of what is desirable and what is possible was shaped. Unending economic growth was envisaged as a real possibility, and was indeed initially possible because we found fossil fuels that could provide the abundant energy which was necessary for this growth. We have been extremely successful the last two centuries, but we have changed our world dramatically along the economic growth path. Our world is now a "full" world, relatively full of humans and their built infrastructure. But our current institutions and technologies are extensions of a long-term trend of adaptation to an empty world. Thus, our modern culture's worldviews, institutions and technologies are failing to meet our needs in a changing world. Global crises such as climate change, peak oil and loss of ecosystem services can be traced back to one problem: we have failed to adapt our current socio-ecological system from an empty world to a full world (Beddoe *et al.*, 2009).

Because we now live in a full world with resource and sink restraints "business as usual" is not an option. Our main problem is arguably our addiction to fossil fuels as a source of energy. The consumption of fossil fuels has to be dramatically reduced within the next two decades, both because of climate change and because of the peaking of world oil production. To be able to respond efficiently to these challenges we must substitute the current fragmented pluralistic views of the world with an integral view. To better understand our stress situation, and to be able to respond as effectively as possible, we need a grand perspective. In this chapter a bigger perspective on the transition away from fossil fuels is outlined by 1) showing how peak oil and climate change are interconnected, and 2) by applying Ken Wilber's integral framework to analyze this transition.

Climate change and peak oil

Carbon dioxide levels are now substantially higher than at any time in the last 800,000 years (Petit *et al.*, 1999; Lüthi *et al.*, 2008). Since the beginning of the industrial revolution anthropogenic production of CO_2, mainly caused by the consumption of fossil fuels, has elevated atmospheric CO_2 levels from a concentration of approximately 280 ppm in pre-industrial times to 394 ppm today (Tans, 2012). If current rates of emissions continue, CO_2 concentrations are projected to reach a range between 535 to 983 ppm by the end of the twenty-first century (IPCC, 2007). It is a broad consensus in the scientific community that this anthropogenic production of CO_2 is affecting the global climate (IPCC, 2007), and the concern is that climate tipping points may be reached where the climate system goes into qualitatively different modes of operation (Lenton *et al.*, 2008).

Climate change, caused by burning of fossil fuels, is only one half of the story. The other half concerns the depletion of these finite energy resources. Extraction of finite non-renewable resources such as e.g. minerals (van der Veen, 2006), phosphorus (Cordell *et al.*, 2009) and fossil fuels typically follow a pattern of growth, peak production and decline. The fact that fossil fuels are finite resources and that their extraction someday will peak was already pointed out in 1949 by the American geophysicist Marion King Hubbert (Hubbert, 1949).

Because of its convenient liquid form and high energy density, oil is much more important for our society and economy than gas and coal. Most of our infrastructure and industry is completely dependent on cheap oil and the transportation sector is run close to 100 percent on liquid fuels. This, together with the fact that world oil production will peak before gas and coal production, is the reason that the main concern related to fossil fuel supply are put on oil.

Oil needs to be discovered before it can be produced. Most oil producing countries have followed a pattern where the peak in discovery occurred 30–40 years before the peak in production. The world as a whole peaked in discovery of oil in 1965, indicating that we may be at or close to the peak in production. We now actually consume about four barrels of oil for every one we discover (Hopkins, 2008). Since mid-2004 world liquid fuels production has been relatively constant around 85 million barrels per day (mb/d) (IEA, 2011; Höök *et al.*, 2009), indicating that new production has only been able to offset the decline in existing production. Oil prices (Brent crude) did rise from $12 a barrel in 1988 to $147 a barrel in July 2008, pointing to an inability of supply to keep up with demand. Following this spike in price was an economic recession and a collapse of the oil price, but prices have again spiked and have for most of 2011 and 2012 been fluctuating between $100 and $120 a barrel. Thus several factors indicate that we may have reached the peak in world oil production. We should also remember that IEA estimates a total oil production of 99 mb/d in 2035, requiring 47 mb/d capacity additions (twice current OPCE Middle East production) just to counter decline in production at existing fields (IEA, 2011).

In 2005 Hirsch and his colleagues published a report commissioned by the US Department of Energy which looked at mitigation strategies for peak oil (Hirsch *et al.*, 2005). In the report peak oil is seen as inevitable and as an enormous problem for our civilization, and also concludes that we need at least a decade to prepare our economy for this transition, preferably 20 years:

> The peaking of world oil production presents the U.S. and the world with an unprecedented risk management problem. As peaking is approached, liquid fuel prices and price volatility will increase dramatically, and, without timely mitigation, the economic, social, and political costs will be unprecedented. Viable mitigation options exist on both the supply and demand sides, but to have substantial impact, they must be initiated more than a decade in advance of peaking.
>
> (Hirsch *et al.*, 2005: 4)

Obviously peak oil will slow down the emissions from liquid fuel use. However, it should be stated that these involuntary limits on fuel emissions are not in themselves anywhere near sufficient to address the climate problem. It has been calculated, though, that some of the IPCC scenarios are unrealistic due to fossil fuel limitations (e.g. the A2 scenario; Kharecha and Hansen, 2008).

Both climate change and peak oil must be given importance in decision-making processes. Long-term planning and investment in infrastructure have to consider that we in the future will have less oil available. It is important that we leave oil before oil leaves us. Thus a transition away from fossil fuels is urgently needed. This will be a transition from a fossil-based economy to a bioeconomy based on biomass and renewable energy.

The integral approach

"Integral" in this context specifically refers to the integral theory developed by the contemporary American philosopher Ken Wilber. Over the last 35 years he has published over two dozen books and gradually developed his integral framework, which may be found in its mature form in his latest books (e.g. Wilber, 2006, 2007). Integral theory weaves together insights from all major human disciplines of knowledge, including the natural sciences, the arts and humanities. In this chapter, two elements of the integral model, *quadrants* and *levels*, are used to analyze the transition away from fossil fuels.

"Quadrants" refer to the observation that realities can be divided into an inside and an outside perspective, and a singular and plural perspective. These perspectives outline the four quadrants. We as human beings have subjective experiences and

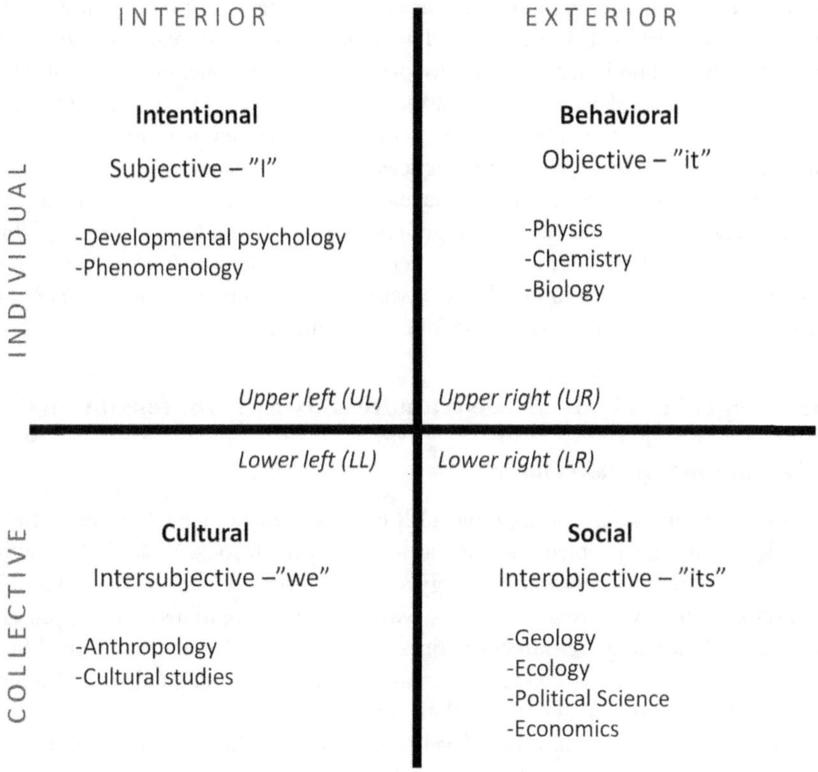

Figure 33.1 The four quadrants and some examples of scientific disciplines

intentionality, our interiors, as well as observable behaviors and physiological components, our exteriors. However, individuals are never just alone, but are members of collectives. The interiors of collectives are known as intersubjective cultural realities. The exteriors of collectives, the interobjective domain, are known as ecological and social systems. These four domains may be represented by the four pronouns I, we, it and its, corresponding to the upper left (UL), lower left (LL), upper right (UR) and lower right (LR) quadrants, respectively (see Figure 33.1). The different sciences and other human endeavors of knowledge are typically focused on one of the quadrants. The traditional hard sciences only study exterior realties. Thus, integral theory provides a framework that shows how different approaches and methodologies relate and fit together. This framework acknowledges methodological pluralism and is very useful for understanding and facilitating interdisciplinary research. It may be used to holistically analyze problems and to prescribe solutions.

Integral theory is a developmental theory, and the term "levels" refers to developmental levels. Evolution and development are seen in all four quadrants. From the time they are born all human beings show outer biological and behavioral development in the UR quadrant, corresponding to different levels of psychological development in the UL quadrant. In a historical perspective human societies have developed from hunter-gatherer, to horticultural, to agricultural, to industrial, to information societies (LR quadrant). This outer collective development typically has a corresponding inner cultural development: from archaic, magic, traditional, modern to post-modern culture (Graves, 2005). Development in the right-hand quadrants is characterized by increasing complexity; development in the left-hand quadrants is characterized by increasing levels of consciousness.

Integral theory stresses that we have, both individually and collectively, an interior reality, a factor that is often ignored by mainstream approaches to global problems. If we do not acknowledge interiority we exclude half of reality. Thus, the integral model helps us check the bases when we approach a phenomenon; it organizes knowledge and helps us identify our blind spots.

An integral analysis of the transition away from fossil fuels

LR-quadrant: system change

Peak oil and climate change are typical LR quadrant phenomena. Our use of fossil fuels is increasingly problematic and causes stress on our socio-ecological systems, and a transition to a renewable energy system is urgently needed. To make this transition smooth we have to act before we experience limitations in oil supply. We need abundant energy resources to create a new renewable energy system. Thus, we both have to adapt to change *and* create change in the LR quadrant. Changes in the LR quadrant will affect the other three quadrants.

Modern civilization has a high level of complexity, which carries a substantial cost in energy and resources. In a future with less energy available we must reduce the complexity of our institutions to increase their resilience. History has shown that favoring simpler institutions increased societies' chances of surviving during times of decreased energy surplus (Tainter, 1988). It should be noted that modern food supply is highly dependent on the use of fossil fuels. It has been estimated that each American needs 2000 l/year in oil equivalents to supply their food, which accounts for about 19 percent of the total energy use in the USA (Pimentel *et al.*, 2008). Thus energy security and food security are closely interlinked.

In the LR quadrant we have to adapt to climate change and limitations in oil supply. We need new institutions that understand and acknowledge that we now live in a "full" world. We have to create new policies, a non growth-based economic system and a renewable energy system. Most paths from potentially sustainable resources to human needs pass through either biomass or renewable electricity, with biomass being the sole foreseeable sustainable source of fuels, chemicals, materials and food (Lynd *et al.*, 2009). Thus sustainable biomass production and biomass processing will be key technologies to master.

The transition away from fossil fuels is a giant challenge for humanity. However, this is not only a question of new infrastructure and technology in the LR quadrant. To do this transition more effectively we need change in all quadrants: system change, cultural change, psychological change and behavioral change; a tetra-change (see Figure 33.2).

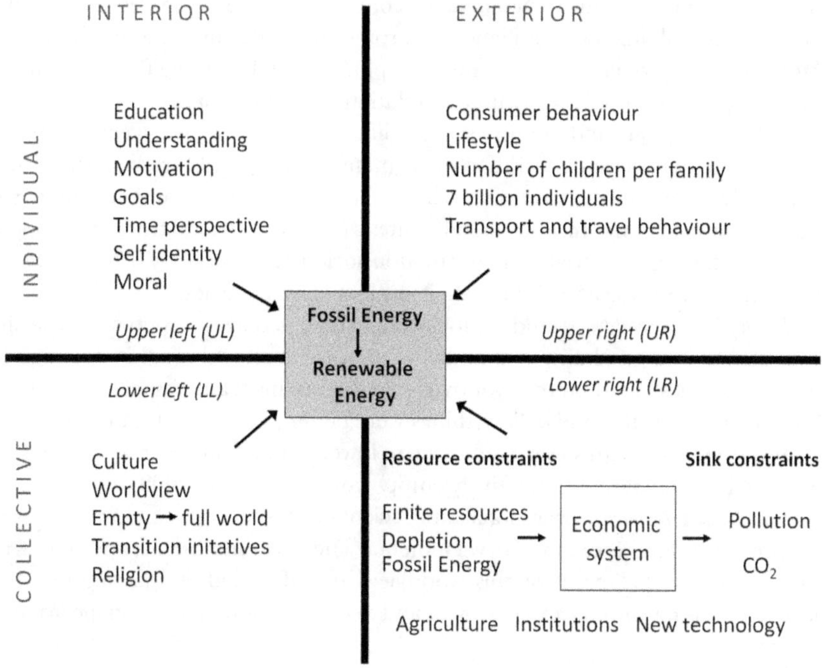

Figure 33.2 Four quadrant analysis of factors important for the transition away from fossil fuels

LL-quadrant: cultural change

In the LL quadrant we find cultural phenomena such as beliefs, stories, ideologies, cultural norms, shared meanings and worldviews. These are unseen phenomena that we take for granted but they inform our actions and decision-making. In the developed world we have a consumer culture where increased consumption and unending economic growth is a goal. Mainstream media and advertising tell us that it is possible and indeed desirable to live a happy life. Obviously, this is not a sustainable worldview for a population of 7 billion people living on a finite planet.

Peak oil and climate change are exterior symptoms of an interior reality. Our civilization's dependence on fuels is the result of choices we collectively have made – choices that are shaped by our understanding of the world, our worldviews. These worldviews are cultural stories that define who we are and they strongly impact on our behaviors. In the modern world we have a story of "progress" that tells us that we as a civilization are advancing further and faster all the time. As long as we hold to this cultural story we believe that "business as usual" is a real possibility.

The challenges of the changes in the LR quadrant will sooner or later lead to changes in the LL quadrant. However, cultural change will happen much faster if we acknowledge our interiority and actively engage in the change process. An interesting example of this is the so called "transition initiative" which originated in the UK. The initiative encompasses community-scale responses to peak oil, focusing on building local resilience by promoting both inner and outer change (Hopkins, 2008). Such local bottom-up responses may be more efficient than the top-down approaches typically used in relation to climate change.

Different regions and countries have different cultures, and even within the same population there exists different segments of culture. Our Western culture is typically composed of traditional-religious, modern-scientific and post-modern-pluralistic worldviews. In a historical context these worldviews trace their origin back to agrarian, industrial and information societies, respectively (LR quadrant). These three worldviews each have their distinct characteristics:

The traditional worldview tends to focus on its own people or group and typically splits reality into good and evil, and is often associated with mythic or religious beliefs. Societies based on this worldview are hierarchical and respect authorities. They are conservative and believe things were better and more sustainable before, but miss the fact that this was mainly due to lower population densities and lack of advanced technology, i.e. a relatively empty world (Costanza, 2008).

The modern worldview is rational and scientific and focuses on progress and technological and economic development. The free individual and the free marketplace are hallmarks of this worldview. It is this kind of thinking that has been the basis for the enormous growth and development we have seen the last two centuries.

The postmodern worldview holds multiple perspectives and understands that context and interpretation are important factors for how we see the world. Postmodernism is anti-hierarchical and emphasizes justice and ethical behavior. Complex issues like climate change are usually well understood, but proposed solutions to problems are not always realistic because they are based on the assumption that soon everybody will share their worldview. Moreover, an emphasis on consensus-building may lead to endless discussions and little action.

Change of worldview takes time (Kegan, 1994) and it is not realistic to believe that everybody will share the postmodern worldview anytime soon. Instead we have to understand and accept that individuals and cultures inhabit different worldviews. Promotion of change thus requires flexibility in communication because change is motivated differently at different worldviews. It is important to know whom one is talking to, what they believe and how they construct meaning.

UL-quadrant: psychological change

Our individual subjectivity encompasses phenomena such as awareness, spirituality, feelings, values, motivation, morals, intellectual capacity, knowledge and understanding. Developmental psychology tells us that complexity of thinking and the ability to take many perspectives changes across a person's lifetime. People tend to become more inclusive and better able to take multiple perspectives as they mature.

Our world needs responsible people, and particularly leaders, that can take a grand perspective on our interconnected global problems. This also includes the ability to hold longer time frames than typical political election cycles; we need politicians that simultaneously work within the limited political scales of the governance system and yet can also envision and set long-time goals over time spans of 50–100 years. Education and upbringing are extremely important to facilitate the growth of people to their maximum potential. Understanding, motivation and morals are decisive for how people act in the world.

Information about our global problems affects people differently because each human being has a unique perspective. It is also important to acknowledge that people develop and change perspectives, beliefs, values and cognition during their lifetime. Individual psychological change must first involve recognition of the problems associated with fossil fuel addiction. This will change attitudes and motivate people to act differently in the world. Problems and challenges might actually promote the development of new mindsets better able to prescribe solutions.

Many people will lack an understanding of the situation or show indifference. However, other factors such as religion, new laws or economic incentives may also motivate change. It is important to keep in mind that the same behavioral change might be motivated by different causes.

UR-quadrant: behavioral change

Ultimately global problems like population growth, loss of ecosystem services, climate change and resource depletion are caused by human behavior: the number of children per family, what and how much we consume, travel habits, which politicians we vote for – all human behavior. The question then is, of course, which factors are affecting human behavior? The answer from integral philosophy is "all of the three other quadrants!" Ecosystem and infrastructure (LR), culture and worldviews (LL), psychological development and education (UL) – all are important factors affecting human behavior. Factors such as health, nutrition, skills and habits in the UR quadrant itself also affect behavior. We see that the behavioral change needed for the transition way from fossil fuels is more effectively facilitated if it is viewed as a *tetra-transition*.

Human behavior has created our global problems, and we have to respond and adapt through behavior. We need to transform our behavior by adopting a lifestyle based on sustainable products and renewable energy. We as individuals have to "walk the talk" and reduce our ecological footprint.

Conclusions

Climate change and peak oil are caused by the global economy's dependence on fossil fuels. They are two sides of the same phenomenon and the transition away from fossil fuel use will be more efficient if this is acknowledged. Indeed, this is the challenge of this century. We are living beyond our means at a time when the supplies of many critical resources will soon be insufficient to enable our current way of living. Is it even possible to produce enough food for the world population in a world marked by climate change and energy limitations?

The analysis carried out in this chapter only serves as an introduction to the use of integral theory, and more comprehensive integral analyses should be carried out in the future (see e.g. Esbjörn-Hargens, 2009). However, the analysis shows how such a holistic framework may help us identify gaps in action and planning, and also gives leads on how different disciplines could work together to solve complex problems. Especially the integration of interiority (from individual interiors such as feelings, values and motivation to collective interiors such as worldviews) and knowledge disciplines like social sciences, psychology and cultural studies, is a valuable contribution of integral theory. The framework also presents a "view from above" where it is clearly seen that no single discipline can see the whole picture of a phenomenon.

Thus, this chapter suggests that the needed transition away from fossil fuels is best facilitated by using a holistic framework for analysis, planning and action. This transition is a complex issue and a transdisciplinary approach is needed to grasp the whole picture. By using Ken Wilber's Integral Model it was shown that the transition can be viewed as a tetra-transition, where individual, collective, subjective and objective factors must be considered and changed.

References

Beddoe, R., Costanza, R., Farley, J., Garza, E., Kent, J., Kubiszewski, I., Martinez, L., McCowen, T., Murphy, K., Myers, N., Ogden, Z., Stapleton, K. and Woodward, J. (2009) 'Overcoming systemic roadblocks to sustainability: the evolutionary redesign of worldviews, institutions, and technologies', *Proceedings of the National Academy of Sciences,* 106 (8): 2483–2489.

Cordell, D., Drangert, J-O. and White, S. (2009) 'The story of phosphorus: Global food security and food for thought', *Global Environmental Change,* 19 (2): 292–305.

Costanza, R. (2008) 'Stewardship for a "full" world', *Current History,* 107 (705): 30–35.

Costanza, R., d'Arge, R., de Groot, R., Faber, S., Grasso, M., Hannon, B., Limburg, K., Naeem, S., O'Neill, R., Paruelo, J., Raskin, R., Sutton, P. and van den Belt, M. (1997) 'The value of the world's ecosystems services and natural capital', *Nature,* 387: 253–260.

Esbjörn-Hargens, S. and Zimmerman, M.E. (2009) *Integral Ecology: Uniting Multiple Perspectives on the Natural World,* Boston, MA: Random House/Integral Books.

Graves, C.W. (2005) *The Never Ending Quest: Dr. Clare W. Graves Explores Human Nature: A Treatise on An Emergent Cyclical Conception of Adult Behavioral Systems and their Development,* C. Cowan, and N. Todorovic (eds), Santa Barbara, USA: ECLET Publishing.

Hirsch, R.L., Bezdek, R.H. and Wendling, R.M. (2005) *Peaking of World Oil Production: Impacts, Mitigation and Risk Management,* United States Department of Energy (DOE), National Energy Technology Laboratory (NETL).

Höök, M., Hirsch, R. and Aleklett, K. (2009) 'Giant oil field decline rates and their influence on world oil production', *Energy Policy,* 37 (6): 2262–2272.

Hopkins, R. (2008) *The Transition Handbook: From Oil Dependency to Local Resilience,* United Kingdom: Green Books Ltd.

Hubbert, M.K. (1949) 'Energy from fossil fuels', *Science,* 109 (2823): 103–109.

International Energy Agency, IEA (2011) World Energy Outlook 2011.

IPCC (2007) *Climate Change 2007: The Physical Science Basis. Contribution of Working Group I to the Fourth Assessment Report of the Intergovernmental Panel on Climate Change,* S. Solomon, D.

Qin, M. Manning, Z. Chen, M. Marquis, K.B. Averyt, M. Tignor and H.L. Miller (eds), Cambridge and New York: Cambridge University Press.

Kegan, R. (1994) *In Over Our Heads. The Mental Demands of Modern Life*, Cambridge, MA: Harvard University Press.

Kharecha P.A. and Hansen, J.E. (2008) 'Implications of "peak oil" for atmospheric CO_2 and climate', *Global Biogeochemical Cycles,* 22 (3): GB3012.

Lenton, T.M., Held, H., Kriegler, E., Hall, J.W., Lucht, W., Rahmstorf, S. and Schellnhuber, H.J. (2008) 'Tipping elements in the Earth's climate system', *Proceedings of the National Academy of Sciences,* 105 (6): 1786–1793.

Lüthi, D., Le Floch, M., Bereiter, B., Blunier, T., Barnola, J.-M., Siegenthaler, U., Raynaud, D., Jouzel, J., Fischer, H., Kawamura, K. and Stocker, T.F. (2008) 'High-resolution carbon dioxide concentration record 650,000–800,000 years before present', *Nature,* 453: 379–382.

Lynd, R.L., Larson, E., Greene, N., Laser, M., Sheehan, J., Dale, B.E., McLaughlin, S. and Wang, M. (2009) 'The role of biomass in America's energy future: framing the analysis', *Biofuels, Bioproducts & Biorefining,* 3: 113–123.

Petit, J.R., Jouzel, J., Raynaud, D., Barkov, N.I., Barnola, J.M., Basile, I., Bender, M., Chappellaz, J., Davis, M., Delaygue, G., Delmotte, M., Kotlyakov, V.M., Legrand, M., Lipenkov, V.Y., Lorius, C., Pepin, L., Ritz, C., Saltzman, E. and Stievenard, M. (1999) 'Climate and atmospheric history of the past 420,000 years from the Vostok ice core, Antarctica', *Nature,* 399: 429–436.

Pimentel, D., Williamson, S., Alexander, C.E., Gonzelez-Pagan, O., Kontak, C. and Mulkey, S.E. (2008) 'Reducing energy inputs in the US food system', *Human Ecology, 36 (4): 459–471.*

Tainter, J.A. (1988) *The Collapse of Complex Societies*, Cambridge: Cambridge University Press.

Tans, P. (2012) *Trends in Atmospheric Carbon Dioxide*, National Oceanic and Atmospheric Administration, Earth System Research Laboratory, U.S. Department of Commerce, NOAA/ESRL. Online. Available HTTP: www.esrl.noaa.gov/gmd/ccgg/trends (accessed 17 October 2012).

Van der Veen, C.J. (2006) 'Reevaluating Hubbert's prediction of U.S. peak oil', *Transactions American Geophysical Union*, 87 (20): 199–202.

Wilber, K. (2006) *Integral Spirituality: A Startling New Role for Religion in the Modern and Post-Modern World*, Boston, MA: Shambhala.

Wilber, K. (2007) *The Integral Vision: A Very Short Introduction to the Revolutionary Integral Approach to Life, God, the Universe, and Everything*, Boston: Shambhala.

34 Uncovering the essence of the climate change adaptation agenda

The policy sciences as a problem-oriented approach

Carolina E. Adler and Amanda H. Lynch

Introduction

Over the last few years, calls for research into climate change adaptation have seen a general shift from defining vulnerability to likely impacts of future climate change (Kelly and Adger, 2000) towards broader and more problem-oriented approaches that focus on present conditions and experiences (Schröter *et al.*, 2005; O'Brien *et al.*, 2007; Roman *et al.*, 2011). This broader approach includes climate change as a process operating within a biophysical and socio-economic system and recognizes the intricate and far-reaching interactions that occur in the coupled human-biophysical system and beyond (O'Brien *et al.*, 2007; Patt, 2009). Adaptation in this broader context is also better placed to serve the common interest, given the emphasis that is placed on the importance of context and congruent valued outcomes for those concerned (Lynch and Brunner, 2007; Lynch *et al.*, 2008). For the purposes of this discussion, the common interest is broadly understood as interests widely shared by and beneficial to the members of a community. It must be constructed in each community on the basis of their interests that are both appropriate (consistent with basic community values) and valid, i.e. consistent with the evidence available (Roman *et al.*, 2011).

An important consideration when differentiating between common and special interests, rests on the values that underpin these interests and how these form part of the context in which a given problem unfolds. Values are increasingly being recognized as influential and critical in advancing our understanding of how action on climate change unfolds, or could potentially unfold (e.g. see Morss, 2005; de Chazal *et al.*, 2008; O'Brien, 2009; O'Brien and Wolf, 2010), notwithstanding the many challenges in problem framing and definition when distinct worldviews and values are implicated (Roman *et al.*, 2011; Adler *et al.*, 2012). Also to be excluded are pre-conceived "solutions" that masquerade as problems, i.e. when "people see a condition as problematic only because it is the mirror image of a solution they already believe in" (Bardach, 1981: 164). As Bardach (1981) points out, the definition of the problem should be as general and tactful as possible by removing causal and prescriptive qualifiers of the problem and instead focusing on identifying the feelings of discontent, effectively describing the perspectives and myths entrenched in them.

In this chapter we present insights on the common interest, based on conclusions and lessons learned from earlier versions of work conducted in Alpine Shire, Victoria, Australia (see Roman, 2010; Roman *et al.*, 2010; Roman *et al.*, 2011). The insights discussed here center on key methodological and conceptual considerations for how we used problem-oriented approaches to move forward on adaptation in the context of irreducible uncertainties as well as cognizant of values and interests at stake. We begin this chapter by first providing an overview of the main features of the policy sciences as a problem-oriented approach that clarifies and focuses on relevant and evidence-based factors in a given context. This is then followed by a case study application, and finally some of the main conclusions and reflections from this experience.

The policy sciences as a problem-oriented approach

Problem-oriented approaches are said to reflect philosophical underpinnings in pragmatism, a discourse characterized as an extension of critical theory and referring specifically to the usefulness, workability, and practicality of ideas, policies, and proposals as criteria of their own merit, even though it is not necessarily committed to any one philosophy or reality (de Leon, 1994; Creswell, 2007; de Leon and Vogenbeck, 2007). Instead of focusing meticulously on methods alone, the important aspect of research in pragmatism is the problem being studied and the questions asked about this problem (Creswell, 2007). A fundamental point of difference that differentiates pragmatism from other philosophical stances is the ethical dimension that is embraced (Brunner, 2006). In pragmatism, "change" is treated as an inevitable condition of life, so pragmatist approaches seek for ways in which change can be directed for compatible individual and social benefits, both in the present and for the future (Clark *et al.*, 2000; Brunner, 2006). Its distinguishable notions of community, dialogue, and democracy have significant implications for and application "in a world characterized by uncertainty, change, and instability" (Hytten, 1994: 1). It represents a philosophical stance that is transdisciplinary, whereby ethical considerations define the purpose for present and future actions "as if people matter" (Max-Neef, 2005: 8).

In policy, problems are seen as human constructions of how a situation is perceived versus a preferred outcome (Lasswell, 1971). However, given how uniquely perspectives are shaped by values, a problem can therefore yield many definitions among the individuals concerned thus influencing whether a problem should (or should not) be addressed (Morss, 2005). The policy sciences approach offers a means to analyze and evaluate these complex contexts wherein people derive meaning and perspective and the outcomes that result when interacting with others (Lasswell and McDougal, 1992; Clark *et al.*, 2000). This is not to imply that addressing complexity necessarily translates to a complete analysis, given the inherent human limitations that inhibit complete examination of all details in any given situation. Instead, the policy sciences offer procedures that bring relevant content to the focus of attention through selectivity and comprehensiveness (Lasswell and McDougal, 1992).

The framework is based on three principles (Lasswell, 1971). First is contextuality, recognizing that decisions are derived and made within a larger social process that is also largely values-based. Mapping the social and decision-making processes (the context) provide the detail and focus required to inform on perceived discrepancies between the status quo and goals aspired to (the problem). Second is problem orientation, an analytical task used to procedurally identify the essence of a problem. In problem orientation, the emphasis is on synthesizing, rather than sequencing intelligence centered on a problem (Auer, 2007). This requires a multi-method and interdisciplinary approach, one that derives knowledge from a number of intellectual foundations. Third is diversity in the methods employed to collect and analyze the empirical evidence that clarifies a problem. Freedom through insight, rather than predictions based on generalizations, is how the role of "science" is characterized in policy sciences, highlighting the need for integrated approaches for problem orientation (Lasswell and Kaplan, 1950; Brunner, 1991, 2001; ABS, 2009). It is a comprehensive yet sufficiently precise approach that directs attention to all the important features of the total context (Lasswell and McDougal, 1992).

The policy sciences approach has been widely applied across many disciplines, often only making implicit references to the framework itself even though elements of the framework are embedded and recognizable (see examples cited in Roman *et al.*, 2011). These include a number of studies that have explored problems of environmental focus, including climate change. The significance of the approach through its applications is that it places emphasis on the goals and values of all participants with valid interests, rather than their risks, vulnerabilities, or resilience even though these factors form part of the study, providing a flexible and powerful tool for analysis.

The policy sciences framework also contributes to a better understanding of the different motivations and even myths that operate in a local community, clarifying the fine distinction that should be made when describing special versus common interests. From a policy sciences perspective, myth is akin to a stance or worldview (Lasswell, 1971), and is described by its doctrine – core beliefs, formula – preferred ways of achieving doctrinal outcomes, and miranda – symbols that manifest core beliefs (see for example Clark *et al.*, 2008).

A case study of the tourism sector at Alpine Shire, Victoria, Australia

Framing the problem

In Australian mountain regions, much of the research conducted on vulnerability to climate change has largely focused on the biophysical impacts that climate change is likely to impose, for example the issue of diminishing winter snow in the Australian alpine environment (Worboys and Pickering, 2002; Hennessy *et al.*, 2003;

Pickering *et al.*, 2004). Adaptation to climate change has also gained attention as a significant strategic issue for regions that rely on tourism as a principal economic activity (Becken and Hay, 2007; Dwyer *et al.*, 2009), however given the prevailing focus on biophysical impacts, the consideration of viable climate adaptation options for these mountain regions has been somewhat narrow and devoid of commuunity values in context (Roman *et al.*, 2010). Previous work carried out in Alpine Shire identified the importance of tourism as a vital activity for economic development and viability (Tryhorn, 2008). Expanding from those initial findings, the aim of the research reported through this case study is to further explore the concept of contextual vulnerability and implications for adaptation with a view to include values and perspectives, using the policy sciences approach.

Alpine Shire, located 270 km northeast of Melbourne (see Figure 34.1), covers 4397 km^2, and with a resident population of 12,988 (DPCD, 2007). Recreational skiing and the formation of local skiing clubs in the 1880s were the first activities to promote tourism (Webb and Adams, 1998), followed by sporting, cultural and musical events initiated by the mining community when gold was discovered in the mid- to late 1800s. Over the years, much of the tourism development in the region has been attributed to improvements in infrastructure such as rail and sealed roads, allowing visitors access from centers such as Melbourne (Webb and Adams, 1998).

Nowadays, tourism in Alpine Shire is characterized by a diverse range of products, experiences and services that include food and wine, festivals and events, sport, sightseeing, nature and adventure-based activities, touring and four-wheel driving, and historical attractions (ARTB, 2009). Various local actors coalesce to facilitate tourism in the region, including government at various scales – federal, state and local, tourism operators, service providers, regional tourism organizations (RTOs) and committees, chambers of commerce, and neighboring alpine resorts and national parks. Tourism in the Shire is spread across all four seasons of the year but peaks during the summer months. Whilst winter is not identified as a significant tourist season in the valleys, there are four areas within the Shire's geographical boundaries that are reliant on winter snow-based tourism. These areas include the two alpine resorts of Mount Hotham and Falls Creek, which are managed externally from the Alpine Shire Council by the Alpine Resorts Coordinating Council (ARCC) under state-level jurisdiction. The third is a smaller snow-play recreational area within Mount Buffalo National Park, managed by Parks Victoria, and the fourth and only alpine area entirely within the Shire's jurisdiction is Dinner Plain Village located just outside Mount Hotham.

Methodology – data collection and analysis

We based our analysis on primary and secondary sources of information in order to construct our description of the social and decision-making processes for tourism in Alpine Shire. Whilst the case study is centered on tourism in Alpine Shire, it is not necessarily bounded by its jurisdiction given that there are valid interests that are non-resident (e.g. stakeholders in neighboring alpine resorts and national parks) and invalid ones within the Shire itself (e.g. individuals promoting

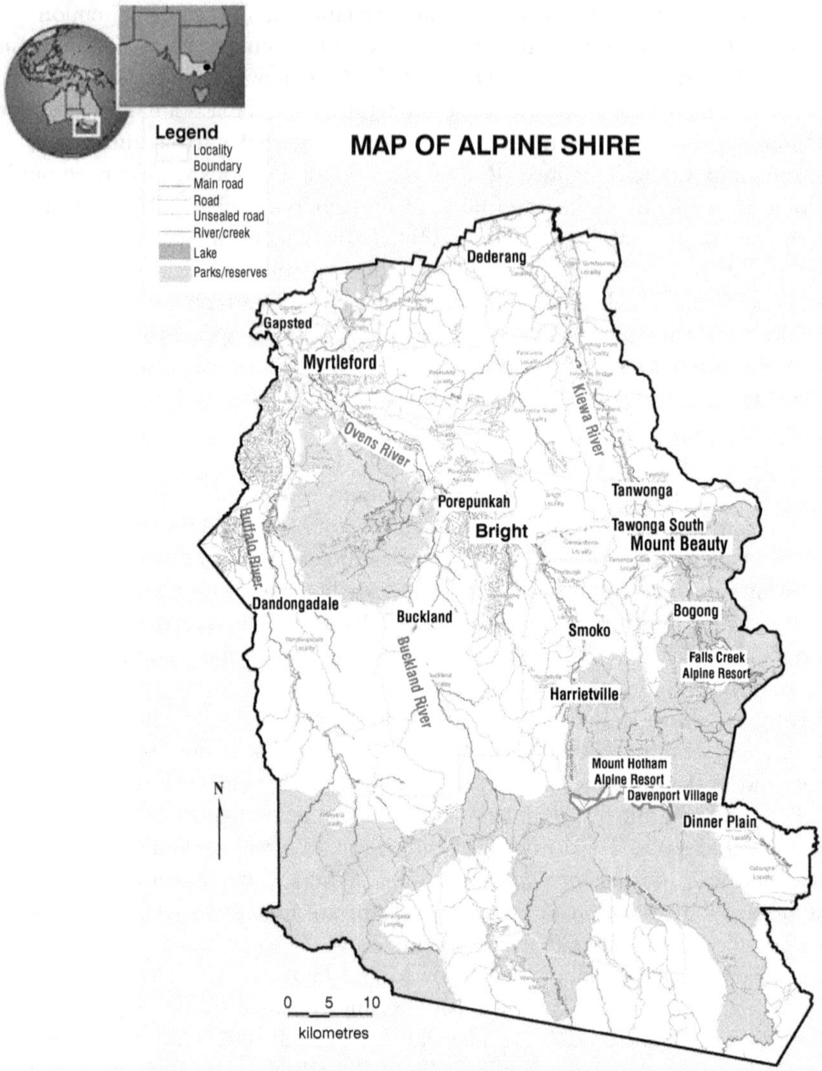

Figure 34.1 Location of Alpine Shire, Victoria, Australia. Source: Adapted from Roman *et al.* (2010), with permission from the Victorian Electoral Commission, 2004

a special interest that is incompatible with the common interest of the tourism sector). For these reasons, input was sought from a diverse set of actors such as tourism operators and service providers operating in the Shire, Alpine Shire government leaders, Council officers, chambers of commerce, national park rangers, and alpine resort operators. Following transcription of these interviews, identification of themes and issues of concern were checked and validated through peer-triangulation amongst the authors as well as an informal focus-group session held in August 2008, with a few of the case study respondents.

Results

Through "a vision of beauty and contentment," the residents of Alpine Shire have documented their vision for their community's future to 2030 (Alpine Shire Council, 2005). This vision points to the preferred future, and in many ways, these would also influence and underpin the type of destination that would be offered to tourists visiting the region. This formed the basic premise of our initial consultation with relevant tourism actors at Alpine Shire. The aim was to try and identify their main issues of concern particularly with respect to potential and actual barriers to those stated aspirations. Of interest to us was to also ascertain the extent to which climate change and/or climatic factors figured within these concerns, and if so, how.

A number of issues of concern emerged and were identified (see Table 34.1). Through these consultations, we identified the initial framework of widely recognized and on-going issues of concern in this region, and therefore provided confidence that further in-depth analysis would bear insightful, relevant and fruitful information as well as remaining open to ongoing refinement and correction.

Table 34.1 Summary of issues of concern (sourced from Roman, *et al.*, 2010)

Issue of concern	*Description of the issue*
Representation and leadership	State-level approach to strategic tourism development difficult to engage with locally, consequently affecting branding and marketing efforts as the local scale, functions of the Alpine Region Tourism Board (ARBT)
Stakeholder relationships	Socio-political barriers evident amongst major town centers in the Shire, as well as conflicting interests for tourism development with external stakeholders such as local alpine resorts
Data and statistics	Lack of consistency in tourist-related data gathering and reporting
Seasonality and weather variability	Season-specific tourism offer, drought conditions resulting in water restrictions, observed extremes and shifts in weather patterns and their variability (e.g. bushfire conditions, onset of frosts in autumn)
Natural disaster management	Post-disaster recovery and access to support for re-building, as well as the media's role in the post-disaster recovery process
Business capacity	Skills shortage, employability and lack of adequate business skills
Infrastructure and transport	Closure of Mt Buffalo Chalet, road signage and tourist attraction signage, touring and public transport options
Exposure to external factors of macro-scale	Fluctuations in economic indicators such as the value of the Australian dollar, interest rates, inflation, GDP growth and fuel prices

Whilst there was some consistency in the way respondents were able to identify and describe the circumstances under which these issues manifest in the Shire, the importance placed on some of these concerns tended to differ depending on geographic or spatial perspective and the actor's own base values and role in the industry. Table 34.2 summarizes some of these distinct perspectives, with exemplar quotations to clarify on the perspective presented. The table also makes reference and links to the concept of myths as used in policy sciences, to illustrate the usefulness of these categories in distinguishing between special and common interests.

These distinct clusters of values and perspectives were broadly grouped as those belonging to Council members and operators in the Ovens and Kiewa valleys. Interestingly, comparable observations were also reported by Wearing (1981), almost 30 years prior, when the Shires of Bright and Mt. Beauty operated independently. During surveys conducted as part of that earlier study, Wearing (1981: 3) reported that "... the Mt Beauty and Tawonga residents were less inclined to criticize council planning than were the Ovens Valley people." It was noted that these differences in and range of needs and opinions must be taken into account, despite the fact that no further elaboration was presented as to possible underlying causes. It is clear from this analysis that values and perspectives that underscore a point of view must not be assumed and should therefore be appraised with evidence and procedure, as well as validated for the context they represent.

Yet, the question remains as to how to legitimize those that address the common interest, if that interest can be constructed. As discussed earlier, special interests that are inconsistent with the common interest, as well preconceived solutions that masquerade as problems, are excluded. Comparable to findings in Mattson *et al.* (2006) and Clark *et al.* (2008), these results suggest an adaptation problem for the Shire that is embedded in unclear strategic directions, uncertain information, and diverse and strongly felt perspectives among actors. Finding common ground was certainly difficult. However, an approximation to the common interest centered on active engagement to raise the business capacity of all tourism operators in the Shire, allowing Council to better facilitate relations with state authorities in order to mobilize resources and make the sector more resilient in the face of climatic and non-climatic shocks (Roman, 2010).

It is clear from the competing formula espoused by actors in Shire's tourism sector that a problem can be defined by the mismatch between desired outcomes and current policies and practices. But just as important, the context that makes up this mismatch, the problem, can also insinuate on pathways to adaptation. An analysis of these actors *in situ* enabled an understanding of behaviors, relationships and developments that contribute to the clarification of the unique context within which the sector operates in Alpine Shire.

Discussion and conclusions

In this chapter, we sought to present insights that center on key methodological and conceptual considerations for how we may use problem-oriented approaches to move forward on adaptation in the context of irreducible uncertainties. We

Table 34.2 Contradicting problem definitions and perspectives prevalent in the Alpine Shire tourism sector, and components associated with their myths (sourced from Roman, *et al.*, 2011)

	Myth		
Perspective	*Doctrine*	*Formulas*	*Miranda*
Council managers: "… the business operators don't see that they have to be a partner in this ..."	Initiative and acumen is required from operators for a collective common good – Council is under-resourced.	Facilitate business skills training by the Economic Development Officer and appointment of new ART Board.	*New ART Board* – new direction and guidance for tourism – run by business people for business people.
Ovens Valley Operator: "… after spending in fact millions of dollars resourcing the tourism decision of the shire, they don't have any numbers as to how many tourists come here …"	Increase accountability, transparency, and "say" regarding use of their resource to serve their needs.	Critical of the current decision process for tourism.	*Special rates* – increasing but not delivering on individuals' expectations on how it should be spent.
Kiewa Valley Operator: "… I believe we have a very good relationship with the Alpine Shire and we get very good value for dollar from the council … Alpine Regional Tourism (ART) I believe is the best thing since sliced bread, for what they can do, what they can offer the whole region	Core values based on rectitude – a duty to the community. Self-sufficiency and cooperative attributes.	Good appraisal of current decision-process, seeking to create synergies within the region (i.e. neighboring resorts).	*Good value* received from special rates paid for Council services.

began by first discussing challenges in problem framing and definition when distinct worldviews and values are implicated in a given context, followed by an overview of the main features of the policy sciences approach as a means to structure and navigate through this complexity. This was then followed by a case study of the tourism sector in Alpine Shire, Victoria, Australia, to illustrate this application and harness lessons learned from that experience.

The policy sciences framework was applied as a problem-oriented and integrative approach to explore underlying problems in context and characterize potential barriers to adaptation. As depicted from the case study, experiences with climate as extreme weather-related events were considered to impact upon this community's ability to develop its tourism industry, however, these experiences were not isolated from a much broader context where non-climatic factors also mattered. The complexity and interaction seen in the types of issues raised provided us with an appreciation for a much broader contextual vulnerability present and the various problem definitions often implicitly articulated by the respondents. However, given the number of problems for adaptability in the tourism sector, the question about how to legitimize those that address the common interest still remains.

An analysis of the myths associated with those issues of concern, and compared against how well they serve the preferred outcome i.e. reducing vulnerability of what is valued, allowed us to distil those concerns that serve a special interest versus that which serves a common interest. Based on our identification of an approximation to the common interest in the Alpine Shire tourism sector, further work involved the presentation of these issues of concern to study participants for correction and refinement and for the development of a comprehensive set of creative policy alternatives that might be explored to resolve the problem (Roman, 2010).

From a methodology perspective, application of the policy sciences has served as a robust framework for structuring the inquiry, as well as understanding and analyzing the empirical data. Admittedly, our experience in defining the problem in the first instance is that it can appear deceptively simple and we concur with Morss (2005) that although it is a crucial key activity defining the problem is also quite analytically challenging and time intensive. Nevertheless, our experience in applying the policy sciences approach has revealed many complex and interacting issues for the Shire's tourism sector to the challenges of global change. These issues represent more immediate concerns for addressing contextual vulnerability and adaptability, and direct attention to the types of policies and measures that would streamline concrete action towards adaptation and sustainability. Provided these policies are treated as hypotheses to be tested (Lynch, 2009), i.e. appraised through action, it would allow the means towards technical, social and institutional learning for sustainable adaptation.

References

ABS (2009) *Cat. No. 3218.0 Regional Population Frowth, Australia 2007–08*, Table 2 – Estimated Resident Population, Local Government Areas, Victoria. Australian Bureau of Statistics, Commonwealth of Australia.

Adler, C., McEvoy, D., Chhetri, P. and Kruk, E. (2012) 'The role of tourism in a changing climate for conservation and development. A problem-oriented study in the Kailash Sacred Landscape, Nepal', *Policy Sciences* published online 16 October, 2012. Online. Available HTTP: http://link.springer.com/article/10.1007%2Fs11077-012-9168-4# (accessed 5 December 2012).

Alpine Shire Council (2005) *Alpine Shire: 2030 Community Vision*, Adopted 7 June 2005, Bright: Alpine Shire Council.

ARTB (2009) *Alpine Region Tourism Strategic Plan 2009–2011*, Adopted February 2009, Alpine Shire, Bright: Alpine Region Tourism Board.

Auer, M.R. (2007) 'The policy sciences in critical perspective', in J. Rabin, W. B. Hildreth, and G. Miller (eds) *Handbook of Public Administration*, New York: CRC Press.

Bardach, E. (1981) 'Problems of problem definition in policy analysis', *Research in Public Policy & Analysis and Management*, 1: 161–171.

Becken, S. and Hay, J.E. (2007) *Tourism and Climate Change: Risks and Opportunities*, Bristol: Channel View Publications.

Brunner, R.D. (1991) 'Global climate change: defining the policy problem', *Policy Sciences*, 24: 291–311.

Brunner, R.D. (2001) 'Science and the climate change regime', *Policy Sciences*, 34: 1–33.

Brunner, R.D. (2006) 'A paradigm for practice', *Policy Sciences*, 39: 135–167.

Clark, T.W., Willard, A.R. and Cromley, C.M. (2000) *Foundations of Natural Resources Policy and Management*, New Haven, CT: Yale University Press.

Clark, D.A., Lee, D.S., Freeman, M.M.R. and Clark, S.G. (2008) 'Polar bear conservation in Canada: Denning the policy problems', *Arctic*, 61: 347–360.

Creswell, J.W. (2007) *Qualitative Inquiry and Research Design: Choosing Among Five Approaches*, second ed, Beverly Hills, CA: Sage Publications Inc.

de Chazal, J., Quétier, F., Lavorel, S. and van Doorn, A. (2008) 'Including multiple differing stakeholder values into vulnerability assessments of socio-ecological systems', *Global Environmental Change*, 18: 508–520.

de Leon, P. (1994) 'Reinventing the policy sciences: Three steps back to the future', *Policy Sciences*, 27: 77–95.

de Leon, P. and Vogenbeck, D.M. (2007) 'Back to square one: the history and promise of the policy sciences', in J. Rabin, W. Hildreth, and G. Miller (eds) *Handbook of Public Administration*, New York: CRC Press.

DPCD (2007) *Local Government Victoria*: Alpine Shire: Department of Planning and Community Development, State Government of Victoria.

Dwyer, L., Edwards, D., Mistilis, N., Roman, C. and Scott, N. (2009) 'Destination and enterprise management for a tourism future', *Tourism Management*, 30: 63–74.

Hennessy, K., Wheatton, P., Smith, I., Bathols, J., Hutchinson, M. and Sharples, J. (2003) *The Impact of Climate Change on Snow Conditions in Mainland Australia*, Australia: CSIRO Atmospheric Research & CERES.

Hytten, K. (1994) *Pragmatism, Postmodernism and Education*, Annual Meeting of the American Educational Studies Association, 13 November, Chapel Hill, NC.

Kelly, P.M. and Adger, W.N. (2000) 'Theory and practice in assessing vulnerability to climatic change and facilitating adaptation', *Climatic Change*, 47: 325-352.

Lasswell, H. (1971) *A Pre-View of Policy Sciences*, New York: Elsevier.

Lasswell, H. and Kaplan, A. (1950) *Power and Society*, New Haven, CT: Yale University Press.

Lasswell, H. and McDougal, M. (1992) *Jurisprudence for a Free Society: Studies in Law, Science, and Policy*, The Hague: Martinus Nijhoff Publishers.

Lynch, A.H. (2009) 'Adaptive governance: how and why does government policy change?' *ECOS Magazine*, 146: 31.

Lynch, A. and Brunner, R. (2007) 'Context and climate change: an integrated assessment for Barrow, Alaska', *Climatic Change*, 82: 93–111.

Lynch, A.H., Tryhorn, L. and Abramson, R. (2008) 'Working at the boundary: facilitating interdisciplinarity in climate change adaptation research', *Bulletin of the American Meteorological Society*, 89: 169–179.

Mattson, D., Byrd, K., Rutherford, M., Brown, S. and Clark, T. (2006) 'Finding common ground in large carnivore conservation: mapping contending perspectives', *Environmental Science and Policy*, 9: 392–405.

Max-Neef, M.A. (2005) 'Foundations of transdisciplinarity', *Ecological Economics*, 53: 5–16.

Morss, R.E. (2005) 'Problem definition in atmospheric science public policy: the example of observing system design for weather prediction', *Bulletin of the American Meteorological Society*, 86: 181–191.

O'Brien, K. (2009) 'Do values subjectively define the limits to climate change adaptation?', in W.N. Adger, I. Lorenzoni, and K. O'Brien (eds) *Adapting to Climate Change: Thresholds, Values, Governance*, Cambridge: Cambridge University Press.

O'Brien, K.L. and Wolf, J. (2010) 'A values-based approach to vulnerability and adaptation to climate change', *Wiley Interdisciplinary Reviews: Climate Change*, 1: 232–242.

O'Brien, K., Eriksen, S., Nygaard, L.P. and Schjolden, A. (2007) 'Why different interpretations of vulnerability matter in climate change discourses', *Climate Policy*, 7: 73–88.

Patt, A.G. (2009) 'Learning to crawl: how to use seasonal climate forecasts to build adaptive capacity' in W.N. Adger, I. Lorenzoni, and K. O'Brien (eds) *Adapting to Climate Change: Thresholds, Values, Governance*, Cambridge: Cambridge University Press.

Pickering, C., Good, R. and Green, K. (2004) *Potential Effects of Global Warming on the Biota of the Australian Alps*, Report to the Australian Greenhouse Office, Canberra.

Roman, C.E. (2010) *On Being Adaptable: Transformative Lessons on Adaptation Through Problem-orientation. A Case of the Tourism Sector in Alpine Shire, Victoria, Australia*, Melbourne: School of Geography and Environmental Science, Monash University.

Roman, C.E., Lynch, A.H. and Dominey-Howes, D. (2010) 'Uncovering the essence of the climate change adaptation problem – a case study of the tourism sector at Alpine Shire, Victoria, Australia', *Tourism and Hospitality Planning & Development*, 7: 237–252.

Roman, C.E., Lynch, A.H., and Dominey-Howes, D. (2011) ,What is the goal? Framing the climate change adaptation question through a problem-oriented approach', *Weather Climate and Society*, 3: 16–30.

Schröter, D., Polsky, C. and Patt, A.G. (2005) 'Assessing vulnerabilities to the effects of global change: an eight step approach', *Mitigation and Adaptation Strategies for Global Change*, 10: 573–595.

Tryhorn, L. (2008) *An integrated assessment of flooding vulnerabilities in the Alpine Shire of Victoria, Australia*, unpublished PhD thesis, School of Geography and Environmental Science, Monash University, Melbourne.

Wearing, R.J. (1981) 'Tourism: blessing for bright? The impact of tourism on a small community and its planning processes', in Australian Institute of Urban Studies, La Trobe University (eds), Canberra: Paragon Printers.

Webb, D. and Adams, B. (1998) *The Mount Buffalo Story, 1898–1998*, Melbourne: The Miegunyah & Melbourne University Press.

Worboys, G.L. and Pickering, C.M. (2002) *Managing the Kosciouszko Alpine Area Conservation Milestones and Future Challenges*, CRC for Sustainable Tourism Pty Ltd, Mountain Tourism Research Report Series: No. 3. Gold Coast.

35 Changing places

Migration and adaptation to climate change

Helen Adams and W. Neil Adger

Introduction

Demographic change is a central part of human-environment interactions; relocation and population mobility are some of the key strategies that make environmental change tolerable. While almost never fully ascribable to environmental change, migration is one of many adaptation strategies to deal with changes in resource productivity and risk to life and livelihood. Therefore, migration can be conceptualized as an adaptation strategy to insecurity caused by environmental and other stressors and an important process in the sustainability of resource use in both sending and receiving areas.

Much of the discussion of migration and environmental change has focused on international migration and the impact of people moving across borders as a threat or challenge to state and regional security (White, 2011). While large-scale displacements of populations are possible in the incoming century triggered by climate and other environmental stresses (Gemenne, 2011), an emphasis on this type of migration hides the significance of individual movement and relocation decisions that are affected by environmental change everywhere.

When adapting to environmental stress means changing location, there is a significant set of psychosocial, cultural and economic challenges to individuals and communities. This makes the decision to migrate a complex process and "rarely the adaptation of first resort" (McLeman, 2009: 297) as changing location in itself involves large financial and psychological costs. In this way, migration and mobility can both contribute to and decrease human security in the context of environmental change.

Environmental change matters because it affects the things people really care about: their health, their homes and neighborhoods and subsequently their sense of place and belonging, through impacts on the local environment to which they attach significance. Sense of place is an important asset that promotes resilience and human security (Adger *et al.*, 2012; Fresque-Baxter and Armitage, 2012; Smith *et al.*, 2012) and the environment is important in its creation (Adams and Adger, 2013).

Hence, in this chapter we examine migration as a social phenomenon that is central to the needs, rights and values of human security. We suggest that environmental change affects the individual's migration decision through the productivity of ecosystem services on which not only he or she depends for income, but also to which he or she forms attachment. Changes in ecosystem services affect the place utility gained from a location and whether the decision is made to stay or relocate. This is a new framework for analysis that emphasizes agency and makes explicit the interaction of individuals with environmental risks within migration decision-making (Adams and Adger, 2013).

Migration is multi-faceted and the delineation of what constitutes migration involves judgments of permanence, distance and duration. Most migration occurs over short distances and on "temporary" cycles ranging from daily to many years. Remittances and social networks play key roles in initiating migration and defining the destination of migrants. The causes of migration are usually various and act upon the migrant both directly and indirectly making the drivers of migration difficult to define.

The next section briefly summarizes the interaction of environmental change and demographic change over the coming decades. The chapter continues by examining how changes in the environment affect migration decision-making through benefits gained from the environment at the level of the individual. Migration as a form of adaptation is then discussed. Finally, policy discourses on adaptation are examined. The chapter concludes by calling for the reframing of migration as an important component of any policy response to environmental change.

Climate change impacts, people and settlements

The world will undergo unprecedented change over the incoming two or three generations in both economic and environmental dimensions. While climate and other environmental change will be ever more significant, demographic changes associated with population growth and stabilization will form the backdrop to how these environmental risks play out in specific countries and regions.

Communities facing a changing climate will have to deal with slow changing variables; with changing risks associated with extreme events; and with unforeseen and possibly rapid onset regime shifts in climate and resource availability. There are three major likely climate change drivers that will directly affect settlements and hence migration flows: changes in sea level; flood and other extreme weather-related risks; and land degradation. Land degradation will be driven through processes such as desertification, glacial retreat and snowpack loss.

Table 35.1 summarizes major projected impacts of climate change in a qualitative way from the Intergovernmental Panel on Climate Change (IPCC) reports in 2007 and suggests where and how these may translate into potential movement of people or changes in the location of economic activities. Many impacts of climate change will first be experienced through weather-related hazards, but longer term processes such as coastal erosion and impacts on biodiversity also directly affect ecosystem services available to people (McLeman and Hunter, 2010). There is growing evidence suggesting that systems may

Table 35.1 Projected fast onset, slow onset and long-term hazards from projected climate change and potential associations with future population displacement and migration. Source: Adapted from McLeman and Hunter (2010) with trends in impacts summarized from IPCC (2007)

Projected changes (from IPCC, 2007)	*Regions affected*	*Associated mobility and displacement implications*
Fast onset hazards		
Increased extent of areas affected by drought	Widespread across semi-arid areas	Loss of productivity resulting in seasonal displacement; specific impacts on pastoralist economies
Decreasing crop productivity due to extreme heat and drought	Dry tropics and seasonally dry regions	Potential for amplification of rural-urban migration flows
More intense precipitation events	Widespread across regions and climates	Risk of flood displacements to settlements in flood plains
Slow onset hazards		
Reduced water availability in regions dependent on mountain snowmelt	Himalayan region including China; western North America; western South America	Implications for agricultural viability and labor demand in agriculture; impacts on urban water use from water scarcity
Increased risk of coastal erosion from sea level change and coastal storminess	All coastal regions, with some already more sensitive than others	Potential for inundation and involuntary relocation for small coastal settlements; large coastal cities likely to be protected and act as population growth poles
Longer term processes		
Decreased snow and ice cover	Arctic	Expansion of shipping and economic activities potentially stimulating population movement
Increases in average river runoff and water availability	High latitudes and wet tropical areas	Risk of flood displacement in settlements on flood plains
Decreases in average river runoff and water availability	Mid latitude and dry tropics	Water availability limiting economic activities and labor demand; new areas prone to such impacts
Ecosystem disturbance and plant and animal species at risk from localized or global extinction	Global	Limit on economic activity for economies dependent on narrow range of resources with demographic implications

be more sensitive to climate change than previously perceived (Füssel, 2009; Smith *et al.*, 2009). Observed and projected sea level rises, for example, may exceed those reported by the IPCC, exacerbating the vulnerability of coastal communities.

The timescales of many of the major climatic risks coincide with other major demographic trends – continued urbanization driven in part by rural depopulation, migration from country to city, and increasing population density at the coast. The impact of climate change on these trends is not clear, but it will be a critical issue (Black *et al.*, 2011). By 2050 there are likely to be an additional 2.6 billion people living in urban areas as a result of both migration and growth (UNDESA, 2012). The majority of this increase is expected in Asian and African cities (Montgomery, 2008).

The big move: making decisions

Will climate change impacts amplify or stall trends in urbanization around the world? The answer will depend on both structural factors and on individual decisions. In general, individuals in resource-dependent, rural areas make a decision to relocate on the basis of expected benefits of moving, relative to staying in the same place, along with factors of security, identity and perceived well-being. In resource dependent societies, much of satisfaction with place (place utility, or the net composite of utilities that are derived from the individual's integration at some position in space (Wolpert, 1965: 162)) is created by the quality of the environment in which people live. Importantly, the environment contributes to both instrumental and affective aspects of place utility; individuals form attachment to nonmaterial aspects of the natural environment, as much as they rely on the environment for income (Adams and Adger, 2013).

Environmental change, such as changing climate, through its negative impacts on ecosystem functioning will alter the utility that a location offers its population. It is changes in these ecosystem services, to follow the terminology of the Millennium Ecosystem Assessment, that mediate between environmental change and the decision to migrate through mechanisms of:

- Productivity – provisioning services from marketed and non-marketed goods in sending and receiving areas;
- Risk – role of regulating services such as flood and micro-climate control in sending and receiving areas;
- Well-being – cultural services and other elements to which individuals and communities form attachment.

Although environmental risks cannot generally be traced through to individual migration decisions, climate change will alter the relative attractiveness of sending rural areas and receiving urban areas. Climate change is likely to reduce productivity within agriculture and have detrimental impacts on other ecosystems in many rural areas, particularly in the semi-arid tropics (Easterling, 2007).

However, increased flood risk or declining water resources may equally diminish the relative pull of urban areas (see Rosenzweig *et al.*, 2011).

Existing studies show a complex relationship between environmental degradation, ecosystem services and rates of outmigration depending on the location, the form of environmental change and the social drivers of migration. Whether migration increases or decreases depends on the duration and distance of the migration and social structures. For example, with respect to desertification, a reduction in resources during drought reduces the possibility of long-distance international migration (Jónsson, 2010) but has been shown to also increase the possibility of short-term migration (Henry *et al.*, 2004). Poor soil quality has been associated with out-migration (McLeman and Ploeger, 2012) but the effects may be gendered (Gray, 2010; Gray and Mueller, 2011). Furthermore, this migration may be taking place in societies that are already highly mobile. In pastoral economies, movement of people cannot necessarily be described as migration but an adaptation to climate variability and harsh climate conditions (Black, 2001; Mortimore and Adams, 2001; Henry *et al.*, 2004; McLemen, 2009).

Interactions between demographic change and environmental change are not unidirectional. Migration can impact negatively on the environment of the sending area through the investment of remittances in unsustainable environmental practices. Migrants can also cause environmental degradation at the receiving area; migration to frontier agricultural areas drives forest loss and other environmental degradation in many parts of the world. In the case of Vietnam, significant rural-rural migration to the highlands as part of the coffee boom in the past two decades has undermined sustainability and made these areas sites for environmental degradation (Winkels, 2008).

Figure 35.1 demonstrates the relationship between environmental factors as both a trigger of migration decisions and consequences of movements themselves for both sending and receiving areas. Extreme events lead to large-scale but generally short distance and temporary migration with a strong environmental weighting in the decision to move. Slow onset events such drought, which gradually affect the natural resource base in an area, tend to lead to migration decisions based on maintaining a livelihood and can involve longer distance more permanent moves of one member of the family (to remit) or the entire household. The weighting of the environment in the decision to migrate varies between these two extremes. When people move, they enhance the linkages between sending and receiving areas, such that remittances and people moving back and forth often benefit both areas (Figure 35.1). Characteristics of the individual and their decision-making process further influence migration. A fall in place utility because of environmental change and the likelihood of migration as a result of that change depends on exposure to hazard, contributors to identity and place attachment and levels of social capital and networks (Smith *et al.*, 2012; Adams and Adger, 2013).

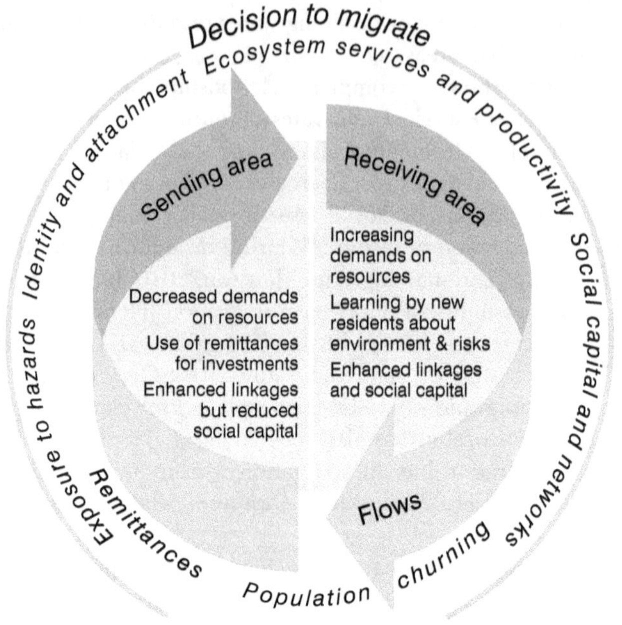

Figure 35.1 Environmental and social implications of migration in sending and receiving areas related to individual migration decision-making.

Migration as an adaptation

Migration in relation to adaptation can be conceptualized in one of two ways. It can be seen as the action of last resort, that which takes place when all other types of adaptation have been implemented and failed. In this framing the end goal of adaptation is to recover livelihoods after an external shock. An alternative framing suggests migration as one of many coping mechanisms used to diversify incomes to reduce risks associated with uncertainty. In reality, the interaction of migration and environmental change produces both migration of the last resort and migration as an adaptation. These two types of migration are discussed further below.

Human-induced climate change is a special case of environmental risk in that it is an imposed harm. Hence, some migrations associated with direct displacement as a result of changing sea level or radically altered productivity of land can only be conceived of as forced and migration a necessity. The government of the Maldives is contemplating such migration. Almost 80 percent of the 1,200 islands of the Maldives are no more than one meter above sea level and, without the large sea defenses that protect Male, the capital, the islands may be uninhabitable if sea levels rise by more than one meter. Hence, the former Maldives President Mohamed Nasheed talked of a "survival deal" on global climate change and the potential for the 400,000 residents to move en masse to a "new home" in Asia or Australia on land purchased through a sovereign wealth fund (New York Times, 2009).

However, more common is for migration to act as a coping mechanism to a changing climate and greater climate variability. There are two obvious means by which migration serves to help a household adapt along a continuum from pre-emptive to responsive adaptation. First, with greater levels of *pre-emption*, migration of a family member on a daily, seasonal or yearly cycle is a rational deployment of human capital across locations and economic sectors to increase income and minimize risk to livelihood sources (de Sherbinin *et al.*, 2008). Such migration patterns have diverse impacts in sending and receiving areas. In sending areas, remittance incomes are often vital and there is evidence across the developing world that such flows tend to be invested in education and other forms of capital (Ellis, 2000). Likewise, the ability of remittances to allow investment (for example in technology or education) in the sending area further increases the resilience of households.

At a more responsive level, a family can elect to send a family member (or additional family members) to seek work elsewhere in order to prevent the entire household from migrating from an area where environmental variability, change or degradation is reducing the productivity of other income sources. Adger *et al.* (2002), using data from rural coastal Vietnam, showed how remittance income was used to spread risk, though increased overall income inequality in sending areas. The authors concluded that for those areas migration was an important element in maintaining social resilience. The second area is migration in *response* to climate change impacts. Here, the important element is the ability of migration to provide access to new livelihoods or a new location in which to carry out a livelihood. In the context of displacement as a result of extreme events, migration provides access to the support of social networks and reduces mortality.

Policy and discourse on adaptation and migration

Dominant discourses on migration in the context of environmental change tend to depict migration as both a negative outcome of climate change impacts and a negative form of demographic change in itself (White, 2011). Promoting migration as an effective form of adaptation requires significant changes in how it is conceived by governments and the public everywhere. The evidence on the benefits of migration for sending areas (via remittances); receiving areas (through a supply of labor) and the migrants (through higher wages) is persuasive. Pritchett (2006) estimates that increasing the rate of South to North international migration by 3 percent over present rates could benefit developing country economies by amounts greater than overseas development assistance, trade and debt relief combined. Putnam (2007) further suggests that international migration increases diversity that has positive spin-offs in terms of creativity and human development.

The Foresight report on Migration and Global Environmental Change (Foresight, 2011) goes some way to promoting the benefits of migration as an adaptation and the need to prepare for a more mobile world under environmental change. If migration is accepted as an established adaptive response to all kinds of changes in the social, cultural and natural environment, then to enable effective adaptation governments will need to create policies to allow and support migration and population mobility (Tacoli, 2009). Suggestions for how this can be undertaken range from ensuring access to non-farm activities in small and intermediate urban centers (Tacoli, 2009) to a global protocol on the protection and resettlement of climate change refugees (Biermann and Boas, 2008).

If governments are to make policies which encourage and enable migration as an adaptation to climate variability then there will be increasing demand for legal definitions of environmental or climate migrants (Dun and Gemenne, 2010; Martin, 2010) since if environmental migrants cannot be defined then they cannot be counted. Various definitions of environmental migrant have been attempted based on the type of environmental change and the length and distance of migration (Barnett and Webber, 2009; and Warner, 2010). However, any definition of an environmental migrant will be unsatisfactory since the decision to migrate will always be the result of a combination of drivers. A clear definition of "environmental migrants" is, however, only a marginal issue for governments. Their primary concerns with respect to managing the impact of environmental change on society focus on the impacts of migration on sustainable cities, land use planning and the economic viability of agriculture and primary industries.

A further challenge for policy of migration and its human security implications is the level of uncertainty around migration flows, particularly international flows, and the multicausality in individual decisions (Barnett and Webber, 2009). This is compounded by the uncertainty surrounding climate impacts and how interactions between the environment and migration may play out in the future with increased climate change. The standard reaction to uncertainty in scenario studies has been to project climate change and postulate how this may influence migration in the future (Feng *et al.*, 2010; Barbieri *et al.*, 2010). Common criticisms are that estimates of future migrants are based on exposure to risk and not on an analysis of how such risks might interact with the migration decision. The effectiveness of adaptation in ameliorating the impacts and protecting populations and settlements is rarely incorporated. Cultural factors, such as the perceived benefits of "home," are not easily captured in assessments or standard welfare criteria and are often discounted.

Conclusion

This chapter has outlined some of what is known about migration response to environmental change in the context of human security. We have stressed that migration as an adaptation strategy should be examined in the context of both multi-dimensional adaptation to change and of broader demographic changes and trends. We propose that environmental change is transmitted as changes in ecosystem services that directly affect the place utility of individuals through the availability of provisioning, regulating and cultural services.

Adaptation to change aims to enhance, or at a minimum maintain, human security in the face of multiple risks and relocation is just one of these possible actions. Given the social and cultural costs of migration, the evidence concurs with McLeman's (2009) notion that migration is rarely the "first resort" adaptation in the face of specific environmental stressors, yet is an important dimension in the long-term human security of regions, households and individuals. In the same manner, migration as a response to climate change will only alter the macro-scale patterns of population movement that have existed and accelerated throughout human history.

The focus on migration in many climate change policy debates is therefore somewhat misplaced. The discourse on migration focuses primarily, if not exclusively in some circles, on the perceived negative consequences of international migration as a problem to be managed. States seek to protect their borders and manage their labor markets rather than view migration as a humanitarian or human security issue (Biermann and Boas, 2008). In many ways, this skewing of the debate hinders serious consideration of migration, including international migration, as a critical adaptation strategy in the face of climate change. It also underemphasizes the economic dimensions for both sending and receiving countries and areas and for the migrants themselves. With these observations in mind, the debate on migration needs to take a positive turn to realize the benefits of population movement on social resilience and human security.

References

Adams, H. and Adger, W. N. (2013) 'The contribution of ecosystem services to place utility as a determinant of migration decision-making' *Environmental Research Letters* (8 015006).

Adger, W. N., Kelly, P. M., Winkels, A., Huy, L. Q. and Locke, C. (2002) 'Migration, remittances, livelihood trajectories and social resilience', *Ambio*, 31: 358–366.

Adger W. N., Barnett J., Brown, K., Marshall, N. and O'Brien, K. (2012) 'Cultural dimensions of climate change impacts and adaptation', *Nature Climate Change*, 3: 112-117.

Barbieri, A., Domingues, E., Queiroz, B., Ruiz, R., Rigotti, J., Carvalho, J. and Resende, M. (2010) 'Climate change and population migration in Brazil's Northeast: scenarios for 2025–2050', *Population & Environment*, 31 (5), 344–370.

Barnett, J. and Webber, M. (2009) *Accommodating migration to promote adaptation to climate change*, Stockholm: Swedish Commission on Climate Change and Development.

Biermann, F. and Boas, I. (2008) 'Protecting climate refugees: The case for a global Protocol', *Environment*, 50: 8–16.

Black, R. (2001) *Environmental refugees: Myth or reality?* Working Paper No. 34, Geneva: United Nations High Commission for Refugees.

Black, R., Kniveton, D. and Schmidt-Verkerk, K. (2011) 'Migration and climate change: towards an integrated assessment of sensitivity', *Environment and Planning A* 43, 431–450.

De Sherbinin, A., VanWey, L. K., McSweeney, K., Aggarwal, R., Barbieri, A., Henry, S., Hunter, L. M., Twine, W. and Walker, R. (2008) 'Rural household demographics, livelihoods and the environment', *Global Environmental Change*, 18: 38–53.

Dun, O. and Gemenne, F. (2010) 'Defining environmental migration', *Forced Migration Review*, 31: 10–11.

Easterling, W. E. (2007) 'Climate change and the adequacy of food and timber in the 21st century', *Proceedings of the National Academy of Sciences*, 104: 19679–19679.

Ellis, F. (2000) *Rural livelihoods and diversity in developing countries*, Oxford: Oxford University Press.

Feng, S., Krueger, A. B. and Oppenheimer, M. (2010) 'Linkages among climate change, crop yields and Mexico-US cross-border migration', *Proceedings of the National Academy of Sciences*, 107 (32): 14257–14262.

Foresight: Migration and Global Environmental Change (2011) *Final Project Report*, London: The Government Office for Science.

Fresque-Baxter, J. A. and Armitage, D. (2012) 'Place identity and climate change adaptation: a synthesis and framework for understanding', *Wiley Interdisciplinary Reviews: Climate Change*, 3(3): 251–266.

Füssel, H.-M. (2009) 'An updated assessment of the risks from climate change based on research published since the IPCC Fourth Assessment Report', *Climatic Change*, 97: 469–482.

Gemenne, F. (2011) 'Climate-induced population displacements in a 4C+ world', *Philosophical Transactions of the Royal Society A: Mathematical, Physical and Engineering Sciences*, 369: 182–195.

Gray, C. L. (2010) 'Gender, natural capital, and migration in the southern Ecuadorian Andes', *Environment and Planning A*, 42: 678–696.

Gray, C. L. and Mueller, V. (2011) 'Drought and population mobility in rural Ethiopia', *World Development*, 40(1): 134–145.

Henry, S., Schoumaker, B. and Beauchemin, C. (2004) 'The impact of rainfall on the first out-migration: A multi-level event-history analysis in Burkina Faso', *Population and Environment*, 25: 423–460.

IPCC (2007) 'Summary for policymakers', in M.L. Parry, O.F. Canziani, J.P. Palutikof, P.J. van der Linden and C.E. Hanson (eds) *Climate Change 2007: Impacts, Adaptation and Vulnerability. Contribution of Working Group II to the Fourth Assessment Report of the Intergovernmental Panel on Climate Change*, Cambridge, UK: Cambridge University Press.

Jonssón, G. (2010) *The environmental factor in migration dynamics – a review of African case studies*, Working Paper 21, University of Oxford, Oxford: International Migration Institute.

Martin, S. (2010) 'Climate change, migration and governance', *Global Governance*, 16: 397–414.

McLeman, R. (2009) 'Climate change and adaptive human migration: lessons from rural North America', in W.N. Adger, I. Lorenzoni and K. O'Brien (eds) *Adapting to Climate Change: Thresholds, Values, Governance*, Cambridge: Cambridge University Press.

McLeman, R. A. and Hunter, L. M. (2010) 'Migration in the context of vulnerability and adaptation to climate change: insights from analogues', *Wiley Interdisciplinary Reviews: Climate Change*, 1: 450–461.

McLeman, R. and Ploeger, S. (2012) 'Soil and its influence on rural drought migration: insights from Depression-era Southwestern Saskatchewan, Canada' *Population & Environment*, 33: 304–332.

Mendelsohn, R., Basist, A., Kurukulasuriya, P. and Dinar, A. (2007) 'Climate and rural income', *Climatic Change*, 81: 101–118.

Montgomery, M. R. (2008) 'The urban transformation of the developing world', *Science*, 319 (5864): 761–764.

Mortimore, M. J. and Adams, W. M. (2001) 'Farmer adaptation, change and "crisis" in the Sahel', *Global Environmental Change*, 11: 49–57.

New York Times (2009) 'Wanted: A new home for my country', *New York Times*, 8 May 2009. Online. Available HTTP: www.nytimes.com/2009/05/10/magazine/10MALDIVES-t.html (accessed 4 April 2013).

Pritchett, L. (2006) *Let their people come: Breaking the deadlock in international labour mobility*, Washington, DC: Brookings Institution Press.

Putnam, R. D. (2007) 'E luribus unum: Diversity and community in the twenty-first century. The 2006 Johan Skytte Prize Lecture', *Scandinavian Political Studies*, 30: 137–174.

Rosenzweig, C., Solecki, W. D., Hammer, S. A. and Mehrotra, S. (2011) *Climate change and cities: First assessment report of the urban climate change research network*, New York: Cambridge University Press.

Smith, J. B., Schneider, S. H., Oppenheimer, M., Yohe, G. W., Hare, W., Mastrandrea, M. D., Patwardhan, A., Burton, I., Corfee-Morlot, J., Magadza, C. H. D., Füssel, H. M., Pittock, A. B., Rahman, A., Suarez, A. and van Ypersele, J. P. (2009) 'Assessing dangerous climate change through an update of the Intergovernmental Panel on Climate Change (IPCC) reasons for concern', *Proceedings of the National Academy of Sciences*, 106: 4133–4137.

Smith, J. W., Anderson, D. H. and Moore, R. L. (2012) 'Social capital, place meanings, and perceived resilience to climate change', *Rural Sociology*, 77 (3): 380–407.

Tacoli, C. (2009) 'Crisis or adaptation? Migration and climate change in a context of high mobility', *Environment and Urbanization*, 21: 513–525.

United Nations, Department of Economic and Social Affairs Population Division (2012) *World urbanization prospects, the 2011 revision: highlights,* New York.

Warner, K. (2010) 'Governance of migration and environment', *Global Environmental Change*, 20: 402–413.

White, G. (2011) *Climate change and migration security and borders in a warming world*, Oxford: Oxford University Press.

Winkels, A. (2008) 'Rural in-migration and global trade: managing the risks of coffee farming in the Central Highlands of Vietnam', *Mountain Research and Development*, 28: 32–40.

Wolpert, J. (1965) 'Behavioural aspects of the decision to migrate', *Papers of the Regional Science Association*, 15: 159–169.

36 Clumsy solutions to environmental change

Lessons from cultural theory

Michael Thompson

Introduction

"Knowledge is always shaped by the organizational setting in which it is produced" is the cry of those who have taken the social constructionist crock (e.g., Jasanoff and Wynne, 1998). "Of course," reply cultural theorists, "in which case we should press on to enquire into how many organizational arrangements there are, and how it is that each of them constitutes the production line for a particular kind of knowledge." Cultural theorists then argue that there are just four ways of organizing, each of which is, at the same time, a way of disorganizing the other three.[1] Each way of organizing, the theory holds, shapes the knowledge it produces in such a way as to strengthen itself and weaken its rivals (Figure 36.1).[2]

At first glance this may appear to be a daunting diagram, but much of what it contains is fairly orthodox. Indeed, two of its ways of organizing – individualism and hierarchy – have long been familiar to institutional theorists; they are usually referred to as *markets* and *hierarchies* (Williamson, 1975; Lindblom, 1977), with their somewhat tangled roots reaching back to the dualistic distinctions of the masters: Maine's (1861) *contract* and *status*, Tönnies' (1887) *Gesellschaft* and *Gemeinschaft* and Durkheim's (1893) *organic solidarity* and *mechanical solidarity*.[3] The theory's novelty therefore lies in its addition of the other two ways of organizing – egalitarianism and fatalism – and in its making explicit the different social constructions (or *myths of nature*) that sustain and justify these four fundamental arrangements for the promotion of social transactions. For good measure, it also explains the sort of dynamic and contentious processes that generate this fourfold typology; the ways of organizing and their associated ways of knowing can thus be seen as recurrent regularities within an endless flux, like eddies in a fast-flowing stream.

Hierarchies are founded on formal status distinctions and inculcate deference. In consequence, their constituent social beings (to use Durkheim's terminology) tend to place their trust in expert, certified knowledge and to subscribe to a view of nature that is forgiving of many interventions but is vulnerable to those excessive perturbations that may carry the totality beyond its stabilizing limits. Hence the icon (in Figure 36.1) of a ball nestled in a valley between two peaks. Hierarchical actors also tend to a view of human nature as originated in sin but redeemable by firm, long-lasting and nurturing institutions (as in the headmasterly "Give me the

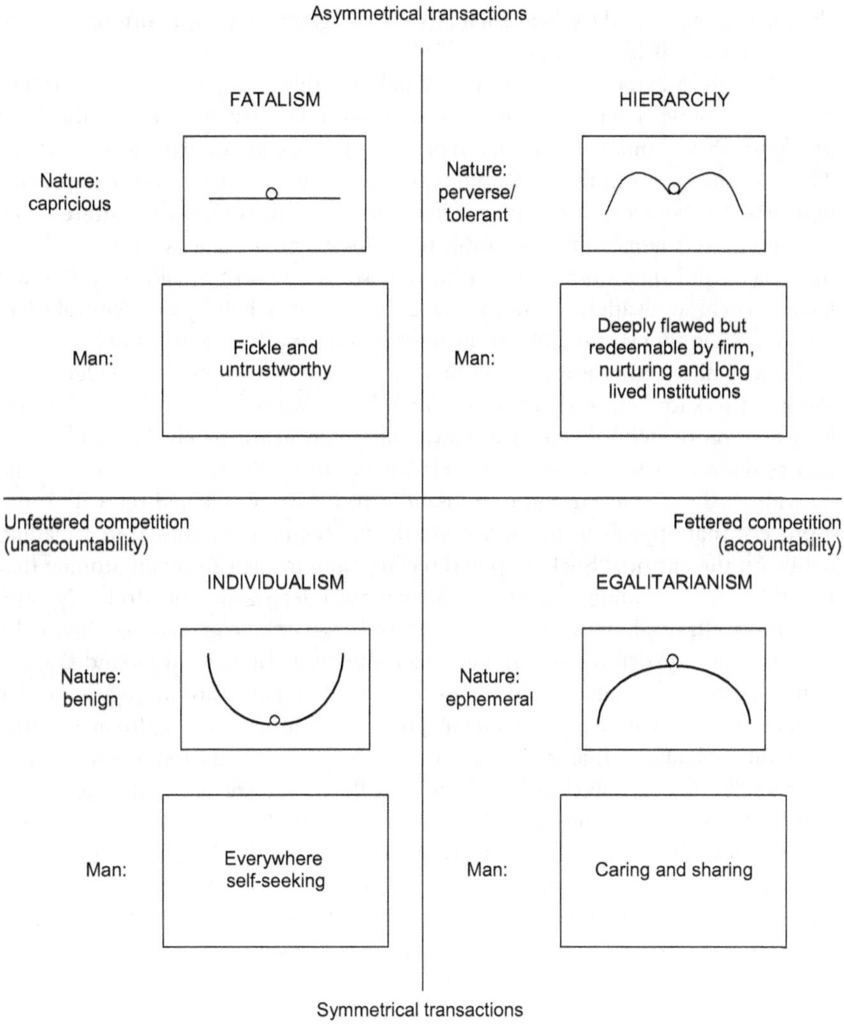

Figure 36.1 The four forms of social solidarity and their associated premises (or myths of nature). Source: Thompson (2003)

boy and I will give you the man"). Steering a careful path between the egalitarian actors and the individualist actors, hierarchical actors are committed to wise guidance and far-sighted stewardship.

Everything, therefore, hinges on their discovering just where the limits are and then ensuring (by statutory regulation, in the case of the nation state, and by more homely, face-to-face sanctions in the case of, say, the Himalayan village) that everyone stays on the right side of them. Whenever we hear talk of "safe limits," "assimilative capacities," "tolerable risks," "carrying capacities," "dangerous climate change" – and we hear such talk in both global and local forums – we are in the presence of hierarchy.

Individualist actors, however, do not talk like this. They see nature as robust enough to bounce back from anything we throw at it – hence the ball in the basin (in Figure 36.1) – and they consider man as inherently competitive and self-seeking. They put their trust in those institutional arrangements – the sort of ego-focused networks that characterize markets – that harness man's self-seeking nature to the benefit of all. Adam Smith's "invisible hand," these actors feel, is all the guidance they need, and they note that its track-record in terms of wealth creation and technological innovation has not yet been brought to a halt by any natural limit. "If something is unsustainable," they reassure one another, "it will stop!"

Individualist actors are not too fussy about the form their knowledge takes. What matters to them is that it works (and that it works for them before it works for their competitors). What does cause them concern is when the hierarchical and egalitarian actors use their knowledge to justify placing restrictions on the individualistic urge to engage in trial-and-error. "Curiosity killed the cat" is not a proverb that appeals to them, nor are they particularly interested in "treading lightly on the earth." "Suck it up and see" is more to their taste: an attitude that results in a never-ending, exuberant and anarchic interrogation of Mother Nature.

Others – the egalitarian actors – see nature as everywhere precarious – hence the ball on the up-turned basin – and consider man as essentially caring-and-sharing, until corrupted by coercive and unjust institutions. Egalitarians are aghast at this business-as-usual complacency, and will point out that the individualist actor's true predicament is akin to that of the man who, having fallen off the top of a skyscraper, remarks "So far, so good" as he passes each floor. Nor are the egalitarian actors prepared to stop off at the hierarchical halfway-house. Nature, they are convinced, allows us no safe limits (see Figure 36.1), and they are deeply distrustful of the sort of "bench science" that is so relied on by the hierarchical actors. Holism – the insistence that "you can never change just one thing" (which, of course, is what those bench scientists claim to do) – is the bedrock of egalitarian science.

Finally, fatalism is the odd one out, in the sense that, unlike with the other three, its rules are made by those who are not themselves the upholders of this way of organizing. Fatalism is a sort of passive margin that is made up of those who find themselves squeezed out from the three "active" ways: unable (like the "undeserving poor") to conform to the expectations of the hierarchy, unable to come up with the entry fee for the market, and unable to muster the communal fervor and engagement that keeps the upholders of egalitarianism equal and united. Nature, they know, operates without rhyme or reason, and man is inherently fickle and untrustworthy. With no way of ever getting in sync with nature (i.e., push the ball this way *or* that and the feedback is everywhere the same (see Figure 36.1)) – or of building trust with others, the fatalist world (unlike the other three) is one in which learning is impossible. "Don't just do something, stand there" is therefore the sensible advice.

Because of the dualistic focus on markets and hierarchies, the other two ways of organizing – egalitarianism and fatalism – have tended to be ignored by social science in general and by policy science in particular. This has resulted in a regrettable loss of explanatory power, as is evident in the case, back in the 1990s, of the enormous Brent Spar oil storage structure, the deep ocean disposal of which was proposed by the market actor – Shell – and approved by the hierarchical actor – the British government's regulatory agency.

Had there been only markets and hierarchies the Brent Spar would now be an artificial reef way out in the Atlantic, but of course it is not. Another actor – Greenpeace – from a third way of organizing (egalitarianism) forced its way in, at the eleventh hour, by audaciously landing a helicopter on the structure as it was being towed towards what the market and hierarchical actors had agreed was to be its final resting place. Such was the media attention that Shell quickly abandoned its disposal plans (motorists, especially in Germany, having stopped buying its petrol) and the British government was left with egg all over its face (John Major, the prime minister at the time, called Shell's senior management "wimps"). Shell then entered into lengthy negotiations with Greenpeace, and the Brent Spar was eventually cut up into huge cylindrical sections to help form a ferry terminal in Norway. Those British citizens who managed to remain ignorant of the entire affair (and they were many) or who found themselves totally convinced by whichever of the protagonists they happened to have last seen arguing their case on television, were evidently bound into none of the "active" ways of organizing and constituted the fourth and rather "inactive" way – fatalism – assuring one another either that ignorance is bliss or that "Nothing we could do would make any difference anyway."

So cultural theory, by doubling up from two ways to four, and by filling in their contentious and non-linear dynamics, is able to make sense – predictive sense – of something that must always remain beyond the explanatory grasp of those who have equipped themselves with the conventional (i.e., dualistic) social and policy science wisdom. That is the practical argument in favor of cultural theory.

The next step along the road to clumsy solutions is to point out that these four ways of organizing, in varying strengths and patterns of pair-wise alliance, are clearly discernible almost anywhere you care to look: in debates over water engineering in South Asia (Gyawali, 2001; Thompson and Gyawali, 2007; Gyawali and Dixit, 2010), in international fora where delegates struggle to do something about climate change (Thompson *et al.*, 1998; Douglas *et al.*, 2003; Verweij, 2006, 2011); in the different ways international regimes cope with trans-boundary risks such as water pollution (Verweij, 2000) and municipalities go about the business of transport planning (Hendriks, 1994); in the various ways households set about making ends meet (Dake and Thompson, 1999); in the different diagnoses of the pensions crisis in countries with ageing populations (Ney, 2009) and in the different panaceas that are variously championed and rejected by theorists of public administration (Hood, 1998), to mention but a few.

In all these examples we find that each way of organizing is generating a storyline that inevitably contradicts those that are generated by the other ways. Each, in rejecting the others, then seeks to impose its *elegant* solution: the solution that is nicely matched to the problem as it is defined in its storyline. Yet, since each storyline distils certain elements of experience and wisdom that are missed by the others, and since each provides a clear expression of the way in which a significant portion of the populace feels we should live with one another and with nature, it is important that they all be taken into some sort of account in the policy process: that they are all able to make themselves heard, and that each is then responded to by the others. That, in essence, is the case for *clumsiness* (Verweij and Thompson, 2006).

This cultural theory framing, we should note, takes us straight to some normative advice – "elegance out, clumsiness in" – without our having to invoke the notion of *sustainable development.* Sustainable development, we can now see, really only makes sense in relation to the hierarchical myth of nature (Figure 36.1). Behavior that lies within the stable trough between the two humps is sustainable; behavior that takes us beyond those humps and onto the downward slopes is unsustainable (and this is consistent with the concept of *assimilative capacity*, which was the basis for the British government's giving of the all clear to the Brent Spar's ocean disposal). But, with the individualist's myth – the ball in the basin – *all* development is sustainable (which is what Shell's scientists argued) whilst, with the egalitarian myth – the ball on the upturned basin – *no* development is sustainable (which is what Greenpeace's scientists were insisting when they argued that, in relation to marine ecosystems, there is no scientific validity to the concept of assimilative capacity).

To rely on the notion of sustainable development, and to require that the policy debate be conducted in terms of it, is simply to impose one social construction – one myth of nature, one storyline – and discard the others. That, of course, is the epitome of elegance and the suppression of clumsiness: the exact opposite of what is needed. A moratorium on the word "sustainability" is therefore the first essential if we are to stand any chance of not making things worse than they already are.

Clumsy approaches to human security

Should this moratorium also be extended to the concept of *human security*? The answer is a qualified "No," in the sense that all humans, be they binding themselves into the hierarchical, individualist, egalitarian or fatalist ways of organizing, are concerned about security, but in very different ways.

For the hierarchical actor, the biggest fear is loss of control. That is why the British government's agency carefully re-examined its involvement in the Brent Spar fiasco and concluded that its initial decision – that the disposal be permitted – had been "flawed," in that it had failed to take account of the "thin end of the wedge" implications: that the precedent would open the way for a host of similar disposals and that they, added together, *would* exceed the ocean's assimilative capacity. In carefully shifting its position in this way, this hierarchical actor was able to regain control. This is the voice that talks of "global stewardship," that readily invokes the *fallacy of composition* (that what is rational for the parts – belt-tightening during a recession, say – may be disastrous for the whole) and that insists that global human security problems (such as climate change) demand global and expertly planned solutions.

For individualistic actors, there is not really anything out there in the environment that could trigger a loss of security. But unwarranted interference with the proper functioning of the market *does* cause them concern. And of course it is the hierarchical actors (with all their statutory regulation) and the egalitarian actors (with their precautionary principles and their "tabooing" of nature) who are all the time striving to do exactly that! So this is the voice that calls for deregulation, for the freedom to innovate and take risks, and for the internalization of environmental costs so as to "get the prices right."

For egalitarian actors, the loss of security is very much inherent in the way we are currently treating the environment, and that, in turn, is very much a consequence of the grossly inequitable ways in which we are treating one another. We (by which they mean those who are binding themselves into the non-egalitarian ways) are drinking in the Last Chance Saloon, sawing off the branch we are all sitting on ... operating Beyond The Limits. With devastating insecurity almost upon us, radical change now, before it is too late, is our only hope. So this is the voice that defines the opposite of development as hospitality, that scorns the idea of "trickle down" and seeks instead to target "the poorest of the poor," that argues for zero growth, and that calls urgently for major shifts in our behavior so as to bring our profligate consumption down within the limits that have been set by Mother Nature.

For fatalist actors (or perhaps we should say non-actors) insecurity is just something you have to live with. Sometimes it hits you, sometimes it doesn't, but, either way, there is nothing you can do about it. So fatalist actors don't really have a voice; if they had they wouldn't be fatalistic. Nevertheless, since time and money spent on something about which nothing can be done is time and money wasted, there is some wisdom here that should not be ignored.[4]

A clumsy solution, it must be stressed, is not a compromise; still less a consensus. Rather, it emerges from a messy, noisy and argumentative process: a *constructive engagement* between the three "active" ways of organizing: hierarchy, individualism and egalitarianism. Fatalism too is involved, in its characteristically "passive" way, becoming ever more extensive as one or more of the active ways are excluded, and shrinking when things shift towards clumsiness. A reduction in fatalism is not the only benefit of clumsy solutions. In a clumsy solution (and this is the counter-intuitive bit) each voice ends up getting more of what it wants, and less of what it doesn't want, than it would have got if it had somehow managed to achieve "hegemony" and go it alone. Let me finish, therefore, by giving a quick example: the Bhattedanda Milkway in Nepal.

A case for clumsy development

Development aid has always been over-elegant – some combination of just hierarchies and markets. But a few years ago a Nepali activist group – the egalitarian voice – managed to get through to Chris Patten who, being at that time a European Commissioner, was able to persuade the European Union to devote €75,000 to a goods-only ropeway, rather than to the road-building programme it normally spends our money on. This egalitarian voice is clearly evident in the project being labeled a "*conservation* ropeway," its primary stated aim being to reduce the pressure on the forest (see Gyawali *et al.*, 2004). But, of course, the project had to find its way through all the hierarchical procedures by which development aid is delivered (baseline studies, screening tools and so on) and it also had to mesh with the market aspirations of the very poor villagers it was designed to help. So all three "active" voices were there, and with little choice, it would seem, but to respond to one another.

Just south of Kathmandu a small, unsurfaced road climbs up to the rim of the Kathmandu Valley and then goes off down the valley on the other side. If you stand on the col and look across half-left you will see, just a mile away, an extensive *danda* – a high, rounded ridge – and this one, Bhattedanda, has several villages on it and rather a lot of cows and water buffaloes which, between them, produce a fair amount of milk. The trouble is that, between the col and the *danda*, there is a 2,000-foot chasm. In consequence, by the time the farmers have carried their milk down and then up and got it onto the milk-lorry to Kathmandu, it is sour. So they have had no option but to laboriously boil it down into a condensed form, *khuwa*, consuming vast amounts of fuel wood in the process. On top of that, *khuwa* fetches a much lower price in Kathmandu than fresh milk.

The secondhand Austrian ropeway, in spanning this chasm, gets the milk to the lorry in just 20 minutes and then all the way to Kathmandu while it is still fresh. The villagers get a better price for their milk. They are relieved of the twin drudgeries of boiling it down into *khuwa* and then carrying it all the way down and up to the road, and the pressure on the forest is massively reduced.

Nor is that anything like the end of it. The ropeway, initially, was powered by a diesel engine but the villagers, after the project had been handed over to them, got onto the internet and worked out how to convert it to electricity. Since the electricity is from hydropower, they have effected the transition from fossil fuel to renewable energy *and* reduced the ropeway's running costs to one quarter of what they were paying for the diesel. The only fossil fuel now is the milk-lorry, and even here there has been a substantial saving because, where in the past it came up empty, it now carries goods destined for the *danda*: sacks of rice-chaff, for instance, as a supplementary feed to increase the fat content (and the market price) of the milk, sheets of corrugated iron to replace the thatched roofs, and bags of white cement for the frescos in the new *gompa* (Tibetan Buddhist temple).

In a somewhat informal survey we asked a number of the villagers how they would rate their quality of life now, if they took 100 as the score for how it was before the advent of the ropeway. One hundred and forty was the average; and that from farmers who, as we all know, are famously pessimistic.

Well, however you choose to measure it, this is *economic growth* – development as defined in the conventional "aid paradigm" – achieved, moreover, by means of a carefully-planned intervention that would never have happened "autonomously." So those who speak with the hierarchical voice have certainly got a lot of what they want. Individualist actors, too, have come out well: incomes increased substantially, along with a host of innovations that have shifted them away from a precarious subsistence agriculture and onto what is, increasingly, a lucrative form of market gardening. Kathmandu, like all South Asian cities, is growing fast, and ropeways such as this one to Bhattedanda can massively expand accessible food production, even with just the existing road system. And those who speak with the egalitarian voice, though they may not have got zero-growth, have secured the conservation of the forest, along with a convincing demonstration of how to effect the transition from fossil fuels to renewables. Moreover, it is the "poorest of the poor" who have benefited most; even more so, now that the ropeway has been extended to the next *danda*: a *danda* so remote and so impoverished that hitherto could not even get its *khuwa* to market. Indeed, it used to be said of it that all it was good for producing was Maoist insurgents.

Notes

1 Cultural theory was pioneered by the celebrated British anthropologist Dame Mary Douglas, and started life as just a heuristic device: grid-group analysis (Douglas, 1978). A wide range of applications were subsequently assembled in "Essays in the sociology of perception" (Douglas, 1982) with this evident success of the heuristic device then justifying its promotion to the rank of theory (Thompson *et al.*, 1990).

2 Properly speaking, the typology is fivefold, in that there is a somewhat "socially detached" solidarity – characterized by the hermit – within which it is possible to contemplate each of the other four sets of premises as "stills" within a cyclical sequence of transitions (see Thompson *et al.*, 1990; Thompson, 2008).

3 For more on these historical roots, and how they can be untangled, see Thompson (2008). For a fuller treatment of cultural theory, and of how it relates to other theories in social science, also see Thompson, Ellis and Wildavsky (1990) and Thompson *et al.* (2006).

4 For instance, James Lovelock (2009) argues that we are already committed to a 4–5°C rise in global temperature, and pinpoints the shortcomings of the IPCC's models which tell us that, with suitably energetic, expensive and globally-coordinated measures, we can limit those changes to just 2°C.

References

Dake, K. and Thompson, M. (1999) 'Making ends meet, in the household and on the planet', *Geo Journal*, 47 (3): 417–424.

Douglas, M. (1978) 'Cultural bias', London: Royal Anthropological Institute, Occasional Paper no. 35. Reprinted in Douglas (1982) *In the Active Voice*, London: Routledge and Kegan Paul.

Douglas, M. (ed.) (1982) *Essays In The Sociology Of Perception*, London: Routledge and Kegan Paul.

Douglas, M., Thompson, M. and Verweij, M. (2003) 'Is time running out? The case of global warming', *Daedalus*, Spring: 98–107.

Durkheim, E. (1993) *De la Division du Travail Sociale*, Paris: Alcan.

Gyawali, D. (2001) *Water in Nepal*, Kathmandu: Himal.

Gyawali, D. and Dixit, A. (2010) 'The construction and destruction of scarcity in development: Water and power experiences in Nepal', in L. Mehta (ed.) *The Limits To Scarcity*, London: Earthscan.

Gyawali, D., Dixit, A. and Upadhya, M. (2004) *Ropeways in Nepal*, Lalitpur, Nepal: Nepal Water Conservation Foundation with Kathmandu Electric Vehicles Alliance.

Hendriks, F. (1994) 'Cars and culture in Munich and Birmingham: The case for cultural pluralism', in D.J. Coyle and R.J. Ellis (eds) *Politics, Policy and Culture*, Boulder, CO: Westview.

Hood, C. (1998) *The Art of the State: Culture, Rhetoric and Public Management*, Oxford: Clarendon Press.

Jasanoff, S. and Wynne, B. (1998) 'Science and decision-making', in S. Rayner and E.L. Malone (eds) *Human Choice and Climate Change, Vol 1: The Societal Framework*, Columbus, OH: Battelle Press.

Lindblom, C.E. (1977) *Politics and Markets: The World's Political-Economic Systems*, New York: Basic Books.

Lovelock, J. (2009) *The Vanishing Face of Gaia*, London: Allen Lane.

Maine, H.S. (1861) *Ancient Law*, London: John Murray.

Ney, S. (2009) *Resolving Messy Policy Problems: Handling Conflict in Environmental, Transport, Health and Ageing Policy*, London: Earthscan.

Thompson, M. (2003) 'Cultural theory, climate change and clumsiness', *Economic and Political Weekly*, XXXVIII (48): 5107–5112. Subsequently republished in: Baviskar, A. (ed.) (2008) *Contested Grounds: Essays on Nature, Culture and Power*. New Delhi: Oxford University Press.

Thompson, M. (2008) *Organising and Disorganising*, Axminster: Triarchy Press.

Thompson, M. and Gyawali, D. (2007) 'Uncertainty revisited' (new introduction), in M. Thompson, M. Warburton and T. Hatley (eds) *Uncertainty on a Himalayan Scale*, Lalitpur, Nepal: Himal Books.

Thompson, M., Ellis, R.J. and Wildavsky, A. (1990) *Cultural Theory*, Boulder, CO: Westview.

Thompson, M., Rayner, S. and Ney, S. (1998) 'Risk and governance part 2: Policy in a complex and plurally perceived world', *Government and Opposition*, 33 (3): 139–166.

Thompson, M., Verweij. M. and Ellis, R.E. (2006) 'Why and how culture matters', in R.E. Goodin and C. Tilly (eds) *The Oxford Handbook of Contextual Political Analysis*, Oxford: Oxford University Press.

Tönnies, F. (1887) *Gemeinschaft und Gesellschaft*, Darmstadt: Wisenschaftlich; translated by C.P. Loomis (1957) *Community and Society*, East Lansing M1: Michigan State University Press.

Verweij, M. (2000) *Transboundary Environmental Problems and Cultural Theory: The Protection of the Rhine and the Great Lakes*, London: Palgrave.

Verweij, M. (2006) 'Is the Kyoto protocol merely irrelevant, or positively harmful, for the efforts to curb climate change?' in M. Verweij and M. Thompson (eds) *Clumsy Solutions for a Complex World*, Basingstoke: Palgrave.

Verweij, M. (2011) *Clumsy Solutions For A Wicked World*, Basingstoke: Palgrave.

Verweij, M. and Thompson, M. (2006) *Clumsy Solutions for a Complex World*, Basingstoke: Palgrave.

Williamson, O. (1975) *Markets and Hierarchies, Analysis and Anti-Trust Implications*, New York: Free Press.

Part III

Transformations: past and future

37 'Clumsy solutions' and 'elegant failures'

Lessons on climate change adaptation from the settlement of the North Atlantic islands

Andrew J. Dugmore, Thomas H. McGovern, Richard Streeter, Christian Koch Madsen, Konrad Smiarowski and Christian Keller

Introduction

The ways that we currently approach and solve problems have profound implications for the future, and the current search for elegant solutions to climate change, biodiversity loss, financial crises, and other social-ecological problems may in fact lead to counterintuitive outcomes. Alternatively, we may also embrace "clumsy solutions" which are those that include many different viewpoints when making a decision or establishing policy (Shapiro, 1988). These may be viewed as "fudges" that muddle along through a series of compromises and the uneasy co-existence of incompatible (or inconsistent) goals (Verweij *et al.*, 2006). Clumsy approaches are the antithesis of focused or "elegant" solutions that are based on a single vision of both the problem and the best way to tackle it. Despite their limitations, clumsy approaches can provide enduring solutions for complex problems (Verweij *et al.*, 2006). Perhaps of most importance, however, is the knowledge of how to avoid "elegant failure," because although the enhancement of human security is a key goal, an overriding need is to avoid making profound mistakes that threaten long-term security.

In this chapter we illustrate the utility of long-term perspectives when assessing human security through an analysis of the contrasting fates of the Norse Settlement in the North Atlantic islands (see Figure 37.1). The Faroe Islands and Iceland are examples of enduring success where the resilience of settlement over century time scales laid the foundations for prosperous modern societies with well-developed human security. In contrast, Norse settlement in Greenland came to an end after some 450 years. This has been seen as an iconic example of failure, of maladaptation by an inflexible northern European farming community and collapse driven by climate change (Diamond, 2005). Here we argue for a quite different interpretation and relevance for human security. We recognize the successful Arctic adaptations of the Greenland Norse through the tenth to fourteenth centuries and in order to assess why an adaptive and successful community could fail we explore the ideas of "clumsy solutions" to complex problems of human security and "elegant failures."

The examples we explore have the notable advantage that they are "completed experiments" that we can track over century scales to evaluate different outcomes at different times. Reflecting on current global economic problems and the impact they are having on human security, very different conclusions could be drawn about the wisdom of late twentieth-century decisions if analysis stopped in 2006 (while economies were booming) rather than continuing to 2012 (a year of deepening economic problems following the worst global economic crisis since the Great Depression of the 1930s).

Long-term perspectives

The past can offer illuminating case studies of human activities and the consequences of particular choices through the investigation of coupled natural and human systems, human ecodynamic modeling and the study of long-term human impacts (McGovern, 2008). Indeed, a sustainable future may be unobtainable without a well-understood past (Crumley, 1994). Misleading short observational series of data can fatally undermine landscape and resource management strategies essential for human security. For example, unsuccessful managers of the North Atlantic cod fisheries began their data sets with the early twentieth century as a pristine baseline, despite hundreds of years of prior human impact, with disastrous results both for cod and people (Hamilton 2003, 2007; Hamilton *et al.*, 2004; Perdikaris and McGovern, 2008a, b).

Key advantages of long-term perspectives include the multi-generational data that show when people's choices can establish resilience and promote enduring human security. We can assess whether some measure of sustainability can be achieved in terms of social practice or the natural environment, and we can explore the circumstances under which changes occur. Some choices may create inherent instabilities that may not become apparent for decades or indeed centuries: examples of this include the "over-optimistic pioneer fringe" (where initial settlement choices may be viable in the short term but have no long-term future) and other types of non-sustainable practices (such as excessive grazing, over hunting and practices leading to reduced soil fertility). If we can identify practices that do establish sustainable and resilient dynamic equilibria over longer timescales, it may be possible to evaluate reasons for instability or change, and identify triggers for change that may be due to both human factors (such as governance or economic practices) and natural variation (such as climate and landscape changes).

The past does, however, have to be studied with care, as there are pitfalls that need to be avoided in order to gain the maximum advantage from the long-term perspectives and rich data that are available. Notable issues relate to the changing nature of society and data resolution. There are, for example, fundamental differences between medieval and modern communities in terms of knowledge,

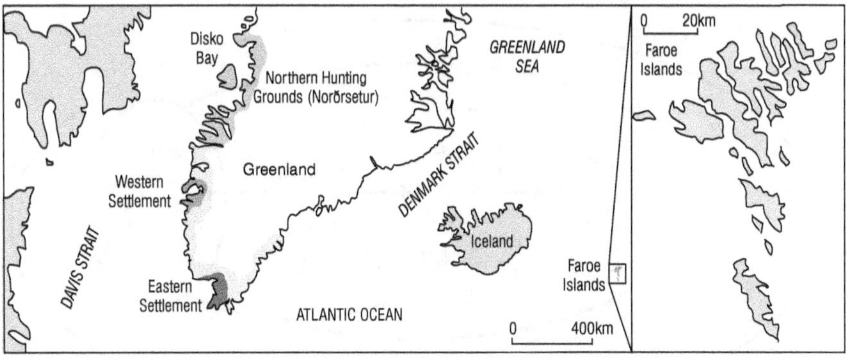

Figure 37.1 The Faroe Islands, Iceland and Greenland were settled during the Viking Age[1]

beliefs and technology, social organization and economic activities, socially-induced goals and measures of success. Indeed these sorts of contrasts are such that many choices in the past cannot be directly related to those of the present or future, but understanding the processes that develop vulnerability or resilience may have great relevance and utility.

Additional concerns relate to the scale of environmental change; indeed one point of view is that likely future global changes are such that the last two millennia lack any examples of comparable magnitude indeed. Temperature changes within the next century are likely to be without precedent in terms of past civilizations, however, the past does contain many examples of acute social stress and demographic shock that may be most instructive: for example, when plague repeatedly killed 10–40 percent of entire societies (as in the Black Death and the Second Pandemic (Herlihy, 1997)), when economic activities underwent fundamental change (as in the development of bulk commodity trading by merchants (Brotton, 2002), when world views changed (as in the adoption of Christianity in northern Europe (Vésteinsson, 2000a; Carver, 2003) and when migrations led to new culture contacts (as in the period of Viking migration (Fitzhugh and Ward, 2000; Graham-Campbell, 2001; Konstam, 2005) and the spread of the Thule culture (Gulløv, 2004: 283–343)). In the past, the scale of some environmental and social changes at local and regional scales is likely to match scales of future change – but with the crucial difference that future change will be *global* in its effects. So although the spatial scale future *environmental* change may be without precedent, the regional and local expression of future *social* change may not be.

We can usefully consider general themes such as the nature of the decision-making process: whether there is much discussion with many voices heard or little discussion and few voices heard; and whether decision making is local or distant. As shown in Figure 37.2, the location and breadth of the decision-making process will affect the practical "schemata" or the Traditional Ecological Knowledge (TEK) that forms the tertiary interface between the deep reservoirs that provide cultural legitimacy for action and the store of social memory that informs and

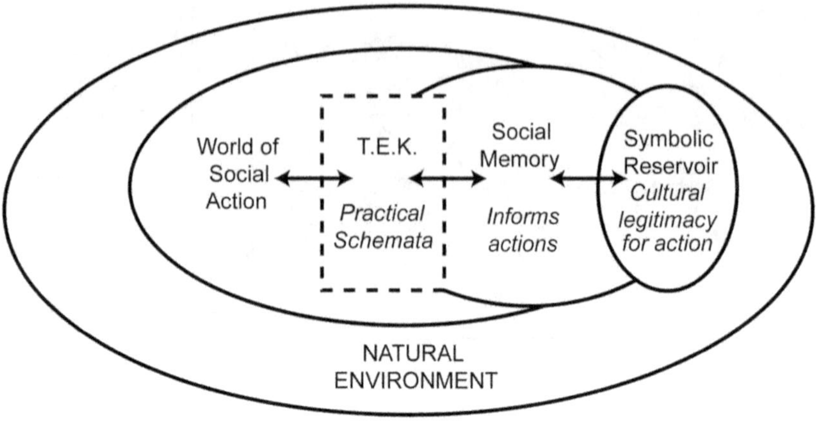

Figure 37.2 Cultural schemata: filtering experience, legitimizing action. Source: after McIntosh and others (2000)

provides context for social action (Figure 37.2, McIntosh *et al.*, 2000). The development of TEK encompasses an understanding of likely seasonal variability in resources, ranges of expected weather, its consequences on the decadal scale and alternate strategies for hard times and good. TEK will in turn shape decisions about specialization versus generalization, focus versus breadth, the development of resilience and sustainability. While the breadth or restriction of TEK doesn't map directly onto "clumsy" and "elegant" strategies, they are likely to be closely related; "elegant" solutions are likely to go hand-in-hand with a restricted development of TEK because they are less inclusive of diverse perspectives, have focused approach and single view of the both the problem under consideration and the solution to that problem.

We also need to beware of misinterpretations that may arise from a lack of data, limited data resolution, or stereotyping. Modern studies can generate very rich data, but as we extend time horizons into the past, data richness generally reduces; a simple lack of data on, for example, social organization, economic activity or health may lead to mistaken conclusions as to the importance of climate change on human security. We may have no doubt that large-scale changes have occurred, such as economic recession (late 1200s A.D.), demographic collapse (the 'Black Death') or environmental disasters (the coldest decades of the "Little Ice Age"); if we focus at the resolution of individual households the effects of the large-scale changes are often difficult, if not impossible to identify. The challenge is therefore to have case studies that cover sufficiently long time scales to allow long term change, adaptation and sustainability to be considered, and yet which have a sufficient richness and granularity of interdisciplinary data to enable complex ideas to be explored. This can be done in the North Atlantic for the period of the Norse settlement; there are environmental archives of unparallel quality from ice cores (Mayewski and White, 2002) to palaeoecology and zooarchaeology (McGovern *et al.*, 2007); there are exceptional dating controls that include high resolution tephrochronology

(Thorarinsson, 1981) and radiocarbon dating used in Bayesian analyses (e.g. Church *et al.*, 2007), a superb historical record (Jones, 1986; Karlsson, 2000; Vésteinsson, 2000a) and increasingly sophisticated landscape-scale archaeological studies (e.g. McGovern *et al.*, 2007). So it is to this arena we will turn our attention.

The case of the North Atlantic: beginnings

The Norse settled across the North Atlantic from their Scandinavian homelands from around 800 A.D., progressing from the British Isles and Ireland across to the Faroes (*c.* 825 A.D.), Iceland (from *c.* 870 A.D), and on to Greenland (*c.* 985 A.D.) – with short-lived visits to North America around 1000 A.D. (Fitzhugh and Ward, 2000; Ingstad and Ingstad, 2001). They took with them a remarkably flexible and durable settlement package of ideas, tools and domesticated animals that was adapted to local conditions and is clearly recognizable as "Viking settlement."

After gaining knowledge of a new island, settlement could proceed rapidly with the creation of a flexible subsistence system based on an introduced domesticated livestock of horses, cattle, sheep, goats, pigs and dogs, closely integrated with a system for harvesting wild resources such as additional fodder, wood, birds, eggs, fish from both fresh water and the sea, shellfish, and marine mammals (McGovern *et al.*, 2001). In favored areas cultivation would take place ranging from cereal production in the eastern North Atlantic to barley and small scale ("walled garden") cabbage and root crop production in Iceland, problematic barley but some flax in Greenland (Videnskab.dk 2012). Greenland (unlike Iceland and the Faroe islands) has indigenous populations of terrestrial grazing mammals including reindeer (caribou) that were hunted by the Norse.

Traveling westwards from Norway is to encounter essentially "more polar" environments, and so, as the Norse established settlements across the North Atlantic, they crossed a series environmental thresholds – such as the limits to forest growth, cereal cultivation, the distribution of seasonally migrating seals and sea ice. Each new island required different adaptations of the "settlement package," saw the development of different approaches to gathering wild resources, and offered different opportunities for trade and exchange (Keller, 2010). The consequences were varying opportunities and related travel costs depending on the local environment, the type of resource gathering, and, ultimately, differing impacts on the environment, consequences of environmental and social change and different outcomes in terms of community survival. This allows us to explore key themes of human security in times of global change.

Understanding differing outcomes

The Faroe Islands

In a pan-Atlantic context, the Faroe Islands are perhaps most instructive in terms of what has *not* happened. Norse colonization of the Faroe Islands did not result in a great overall change in the land cover; the islands had few trees before settlement, grazing was maintained, and there has been no large-scale increase in erosion (Mairs, 2007). Although some destruction of lowland soils and peat has occurred, this is relatively limited and because there were so few trees present before settlement, "deforestation" has had little practical consequence (Lawson *et al.*, 2005). Upland areas have been destabilized with the introduction and long-term activities of grazing mammals, but the overall impact of these changes has been limited (Mairs, 2007). The early Faroese diet included a significant proportion of birds and fish (Church *et al.*, 2005), and whale meat gathered in the *grind*, communal hunts of pods of pilot whales, *Globicephala melas* (Bloch, 1996). In common with Iceland and Greenland early settlement included the introduction of pigs as livestock. These animals were closely managed, but were later extirpated, presumably when the costs of rearing exceeded the benefits gained from their consumption. Outfield grazing was also closely controlled and subject to regulation (for example, the *Sauðabrævið*, 'Sheep letter' rules of 1298 A.D.); prior to this there was a shieling system to manage summer-time grazing (Mahler, 2007). Key trade items included both fish and wool.

The restricted range of favorable farm sites combined with the lack of abandoned settlements and the presence of Viking and early-medieval deposits within modern villages suggested that there has been long-term settlement continuity in the Faroe Islands and a number of factors have helped to promote this; most importantly the Norse subsistence system that effectively combined farming, the utilization of wild resources, exchange and trade into regional provisioning networks. Farms provided cereals, meat, dairy and textiles, and the wild resources exploited by the community provided much else; birds for meat, eggs and feathers, fish for subsistence and trade, and marine mammals, in particular the pilot whale, for food and a basis for a community cohesion (Dugmore *et al.*, 2007a). Subsistence was governed by rules that aided the management of resources, and these resources were generally found in close proximity. Cereal production is viable in the Faroe Islands and has been maintained despite the significant labor required.

While significant, the impacts of climate change did not cross key thresholds within the Faroe Islands and trade links proved to be robust (Mairs, 2007). Faroese settlement was nucleated into clusters surrounded by home fields and bordered by upland grazing, bird cliffs, and sealing beaches managed communally. This may have indirectly promoted community-based "clumsy" approaches. In Faroese communities contact between people was forced on a daily basis because of the settlement pattern. This contrasts with the islands to the west where medieval settlement was characterized by dispersed homesteads, which dramatically reduced daily interactions. The breadth and utility of the TEK developed by the Norse in the Faroe Islands is also likely to have benefited from the diverse origins of the original settlers who would have come from both the wooded coasts of Norway and the open landscapes of the Northern British Isles. Crucially the environments of the Faroe Islands while offering distinctive and special challenges (such as no indigenous land mammals or significant pre-existing settlement infrastructure) do have fundamental similarities to the homelands of the Nordic settlers and certainly contained long-term residents whose local TEK had been developed over century timescales.

Iceland

The long-term outcomes of Norse settlement in Iceland have proved to be both similar to those of the Faroe Islands, and different from them. Despite ambiguous references and some controversial archaeological finds, definite evidence of a pre-Viking age non-Nordic occupation has proved elusive, and most workers place first settlement in the later ninth century (Vésteinsson, 2000b).[2] This ninth-century settlement was both substantial and widespread, apparently involved a mixed Nordic/Celtic population whose TEK had been accumulated in the Northern and Western British Isles and Norway (Vésteinsson, 2012). Settlement of inland and upland districts occurred at the same time as coastal occupation, and there is evidence for planned settlement without village concentrations but with the dispersal of small sites and seasonal shielings to occupy and exploit distant resources from first settlement times onwards (Vésteinsson, 2000b). The Icelandic farms were scattered and mostly land-locked, meaning that there was a different and probably greater mobility cost in the regional organization of subsistence in Iceland compared to the Faroes. In Iceland there was an emphasis on travel by horse and on foot, rather than boat, but there is also convincing evidence for the early development of long-distance intra-island trade and exchange (for example, sea birds eggs and bones from processed sea fish, and marine mammal bones found at inland sites (McGovern *et al.*, 2006)). This allowed for enhanced specialization in the scattered farms and a more complex associated exchange and trade network. Resource management was enshrined in early laws (*Grágás* (Dennis *et al.*, 1980; Simpson *et al.*, 2001)) and in later times laws acted as a constraint on specialization and economic development; as for example, in the requirement for those who fished to also belong to a farm. Limited cereal cultivation occurred in Iceland from settlement through to the sixteenth century, but then the practice died out for reasons that are unknown (Einarsson, 1963; Simpson *et al.*, 2002; Sveinbjarnardóttir *et al.*, 2007). The increased export of stockfish from the thirteenth century probably triggered a comparable import of cereals as the English stockfish trade from 1410 and the Hanseatic trade from 1470 focused on an exchange of cereals for stockfish (KLNM, 1981). Cultivation became focused on providing winter fodder for animals, to enable them to utilize the extensive grazing of inland areas only available in summer months. Iron production flourished in tenth to fourteenth-century Iceland (Espelund, 2007), but it declined as essential woodland resources were degraded through non-sustainable charcoal production (Church *et al.*, 2007).

Unprecedented climatic variability from the fourteenth century onwards triggered increases in soil erosion despite the presence of otherwise effective stock management because cultural systems were unable to anticipate an unknowable future and effectively match the timing of grazing and plant growth (Simpson *et al.*, 2001). These centuries did see a continuation of sustainable practices begun with the initial settlement, such as the effective conservation and management of waterfowl around Lake Mývatn (McGovern *et al.*, 2006). In addition there is evidence of effective responses to environmental change; woodland conservation that probably developed in the fourteenth century ensured that charcoal production for the maintenance of tools essential for farming was successfully continued through to modern times (Dugmore *et al.*, 2006, 2007b). Wool production and stockfish trade were established early and they provided the pillars of the later Icelandic economy (Karlsson, 2000).

Human security and long-term settlement continuity in Iceland owed much to the resilient self-reliant subsistence based on flexible strategies of farming and wildlife exploitation and dispersed but interconnected settlement pattern that connected inland and coastal resources. This proved to be most effective in dealing with periodic setbacks, which included the most profound and persistent problems of unfavorable weather and periodic outbreaks of disease as well as infrequent natural hazards, such as volcanic eruptions, earthquakes and floods. This subsistence flexibility is, however, likely to have been the key reason for the sluggish economic development which characterized Iceland until the end of the nineteenth century. Modifying or abandoning this ability to deal with setbacks in favor of economic development did not seem like an acceptable strategy until modern times.

Why did the flexible system of subsistence persist as long as it did? The oligarchic Commonwealth period (from settlement to 1262 A.D.) may, at least initially, have been characterized by "clumsiness" in so much as the *"goðar"* or chieftains all had a voice in decision making in the sense that each one took his own decision – there was no common instance of decision making. With concentration of power (and the loss of independence) many Icelandic voices of the early settlement period were probably silenced – or the creation of higher tiers of authority blocked out those voices. Despite these changes, in the challenging centuries that followed the loss of independence it is likely that the Icelandic elites as much as the colonial rulers in Norway and Denmark, played a key role in shaping subsistence strategies. This system was successful, resilient and endured, but significant costs are likely to have been paid in terms of limited prosperity, lives not lived and environmental degradation.

Greenland

Norse settlement in Greenland lasted for about 450 years. As in the islands to the west, the regionally-based subsistence networks centered on farms and the communal utilization of wild resources was still the focus of subsistence, but the long-term successes of the Norse in Faroes and Iceland were not to be repeated. The Norse population of Greenland was probably similar in number to that of the Faroe Islands and perhaps an order of magnitude less than Iceland. In contrast to the Faroe Islands, settlement patterns were characterized by scattered farms. Despite geographical isolation Norse Greenland was also characterized by regional subsistence networks and communal efforts to utilize marine mammals for both trade and subsistence. A major focus of the economy was the hunting of walrus in the northern hunting grounds (*Norðrsetur)* which involved journeys of many hundred of kilometers through perilous waters and possible over wintering in hunting camps (see Figure 37.1). This, allied to the time and distance costs involved in harvesting the food resources such as inland caribou and the coastal spring seal migration in a big, sparsely-populated area, led to very large distance costs, with associated labor demands.

Walrus ivory was a key export needed to secure vital imports such as iron, which was not produced within the colonies (Buchwald, 2001; Espelund, 2007). Indeed the trading opportunities presented by large walrus colonies in Greenland, combined with sources of fur and hide not present in Iceland plus the chance to establish new settlements were probably key motivations driving the tenth-century colonization of Greenland. This occurred despite the challenging climate, isolation, extensive winter sea ice and hazardous sea passages to Scandinavian homelands. The walrus tusks collected in the northern hunting grounds were brought back to the permanent settlements of south-west Greenland still embedded in skull bone. The frequency of small chips of tusk ivory accidentally flaked off during tusk extraction drops with time and the numbers of bone fragments increase (McGovern *et al.*, 1996). The implication is that over time and with experience the Norse improved both quality control in ivory production and the volume of ivory collected. Ivory production in Greenland was enhanced (implying a single minded focus) while in Iceland a very different economy based on the trade of fish and wool developed.

In the early centuries of settlement the *c.* 40:60 ratios of marine to terrestrial food in the Norse Greenlanders' diets were not dissimilar from those of their relatives in Iceland and the Faroe Islands, but isotopic records from human remains shows that, through time the Greenlanders went on to develop a remarkably high consumption of marine resources – with up to 80 percent of the diet coming from the sea (Arneborg *et al.*, 1999, 2012). Furthermore, as the known marine archaeofaunal record is dominated by seal bones and fish bones are very rare, two further conclusions may be drawn; either fishing was undertaken and the remains were disposed of in such ways as not to become part of the archaeological records investigated to date, or the marine emphasis developed a specialized, if not single minded focus on seals (McGovern, 1980; Dugmore *et al.*, 2007a). If we follow the latter line of argument then at this western limit of Norse Atlantic settlement the local adaptations became very focused on marine mammals for both subsistence (seals) and trade (walrus).

The initial emphasis of the marine diet on seals rather than fish is likely to have been both pragmatic and shaped by traditions practiced to the east; when the Norse arrived in Greenland they would have encountered for the first time the mass spring migrations millions of harp (*Pagophilus groenlandicus*) and hooded seals (*Cystophora cristata*). These presented a remarkable opportunity and one that the Norse immediately exploited with a communal hunt that may have drawn on Faroese experience of pilot whale exploitation. Effective hunting requires the participation of many members of the community, and to secure cultural legitimacy (Figure 37.2) the benefits of the hunt are spread through the whole community. The Faroese whale hunt, which has provided a significant part of the islanders' staple diet for centuries, may well have provided a social as well as an economic model for communal seal hunts in Greenland (Bloch, 1996; Guttesen, 1996).

The intensification of seal hunting coincided with climatic changes that had a significant impact on both terrestrial food production and the ability of the Norse to maintain their livestock. Globally, the thirteenth century was repeatedly affected by volcanically-driven episodes of cooling (Gao *et al.*, 2008), the most significant of which was a short-lived but profound episode driven by a very large low latitude volcanic eruption in 1258–1259 A.D. (Oppenheimer, 2003). This was followed in the fourteenth century by episodes of both unusually variable and particularly cold weather (Barlow *et al.*, 1997; Mann *et al.*, 2009). These changes would have made pastoralism particularly challenging because they would have massively limited both fodder production and the availability of grazing. In the worse years snow cover in the best farming areas of south-west Greenland probably lasted from September through to June. These difficult years could well explain the shift to a greater emphasis on the spring seal hunt which could have been used to plug a crucial subsistence gap. Success could then have reinforced and consolidated the provisioning shift that is apparent in both the human isotope records and midden archaeofaunal data (Arneborg *et al.*, 2012).

A further consequence of the difficult years of the thirteenth and fourteenth centuries is likely to have been the consolidation of power within the elite; a phenomenon we can observe in contemporaneous Icelandic society. On small farms there would have been a greater risk that the long, cold and snowy winters would have reduced herds and flocks below minimal biological replacement levels. The magnate farms with their large holdings of both livestock and fodder are likely to have played a key role in maintaining the long-term viability of the farming system so consolidating their power and leading to a greater stratification of society. Greater stratification is likely to be associated with greater conformity and an eroding of "clumsy institutions" where many voices are heard.

In the short term, the adaptations and intensifications of the thirteenth and fourteenth centuries seem to have initially served the Norse Greenlanders well. They survived periods of very difficult climate and maintained sustainable subsistence practices of land management and the exploitation of both caribou and potentially vulnerable populations on non-migratory common seals (*Phoca vitulina*). Subsistence was physically separated from the potential environmental impacts of external trade-generating activities because they involved distant walrus colonies. The distance of the walrus hunt also imposed limits on the generation of goods for external trade because of travel time and the hazards of the journey. This situation contrasted with Iceland where there were no effective natural checks on trade-generating activity; Icelandic walrus was wiped out and Icelandic iron and wool production both led to extensive environmental degradation, although the simple loss of natural capital seems to have done little to have actually constrained these latter activities.

If an assessment of Norse Greenland is stopped in the fourteenth century, it could be considered a great success; having survived very demanding times through developing a focus on core activities and also having embarked on an ambitious programme of church construction. Within a century, however, a very different picture emerges with the demise of the whole colony. In the fifteenth century an unexpected series of events occurred and the hither-to successful focus of the Greenlanders became an "elegant failure"; in the aftermath of the Black Death economic change swept across Europe (Herlihy, 1997) and a new proto-world system began to extend into the North Atlantic; climates deteriorated again, but this time with changing circulation patterns and storminess that developed into the Little Ice Age (Meeker and Maywski, 2002). With storms and sea ice disrupting regional networks of communal subsistence activities and trade, declining human populations in Greenland may have reached a critical threshold in terms of social coherence, just as the Norse were having to manage a new set of relations with the incoming Thule peoples (Gulløv, 2008; Dugmore *et al.*, 2012).

Lessons learned

In the North Atlantic we can evaluate contrasting adaptations of the Norse settlement. In the Faroe Islands with their closely placed resources and comparatively benign climates, the chosen balance of terrestrial and marine subsistence allowed great stability. They were favored by their TEK, close proximity to key resources and the continued viability of cereal cultivation throughout the climate changes of the Little Ice Age (albeit with significant labor demands). Communal approaches and a strongly nucleated settlement may have helped "clumsy solutions" to be found.

In Iceland, the early development of bulk trade in wool and stockfish also underpinned settlement continuity and the maintenance of the settlement despite more limited options (such as the initial lack and later absence of cereal cultivation). "Clumsy" approaches may have facilitated the development of resilient, self-reliant strategies that were effective but inhibited economic development. Scattered settlement, the development of hierarchy and the imposition of foreign rule were probably associated with a loss of "clumsiness" in decision making; settlement endured but it was related to extensive land degradation.

In Greenland, the chosen approaches to both trade and the climate challenges in the thirteenth and fourteenth centuries were elegant, a greater utilization of marine resources (seal) for subsistence and a more refined production of ivory for export. Greenlandic adaptations were characterized by *intensification*, whereas Icelandic adaptations involved *fundamental changes* in economic practice. Greenlandic adaptations proved unable to withstand the conjunctures of the mid-fifteenth century and settlement failed. Less focused, "clumsier" solutions may have proved of greater utility. The Greenlanders' elegant "failure" was an inability to anticipate an unknowable future, to have a narrowly focused TEK base, and to be too specialized, too small and too isolated to be able to capitalize on, and compete in, the new proto-world system extending into the North Atlantic in the early fifteenth century.

Inevitabable outcomes?

Was the end of Norse Greenlanders inevitable? Could they have continued their multi-century occupation of the island and maintained their human security? We can clearly identify changes that were not made and consider how circumstances might have been different. Changes could have been made to boost the subsistence base; a broader TEK could have plugged subsistence gaps through winter provisioning, but this would have required the adoption of an additional suite of specialized harpoon-based hunting techniques capable to taking ringed seals through breathing holes in sea ice. A mixed *Métis* population of Norse-Inuit could have provided a route by which ideas might have diffused between the Norse Greenlanders and the incoming Thule Inuit, but there is no evidence of such a group, and we do not understand why this is so.

Through the decades and centuries before the final mid-fifteenth-century ending there is no evidence of economic changes in Greenland: the key products of the early settlement period (ivory hides and fur) are still the same at the end; there is no evidence of attempts to innovate and switch to dried fish or wool production. Fishery production might have been feasible if there was a means of exporting goods to Europe, but commercial wool production might have been impossible given the impact of the frequent bad years that caused high sheep mortality. Data from modern Greenlandic commercial sheep farms that work the land of the old Norse settlements, show the episodic impact of cold and snowy winters on large herds (Madsen, 2008). These not-infrequent surges in mortality require the constant rebuilding of flocks. It is possible that the Norse stocking levels were nicely judged to be those that could be sustained through good years and bad, but these were at a level insufficient to generate surplus for export.

Structural evidence of the mismatch between indoor (overwinter) stalling capacity and the capacity of field pens/livestock enclosures indicates two discrete themes in Norse livestock management within Greenland (Madsen, 2008); the first being the maintenance of a fodder-fed population overwintered indoors (a strategy which could have secured comparatively limited numbers of breeding and milking livestock). In addition there seem to have been free-roaming livestock that may (or may not) have survived the winter outdoors; the total area of the animal enclosures/pens in the Eastern Settlement region from Sermilik fjord to Unartoq Fjord indicates that there was a maximum number of around 15,000 sheep and goats, far above the stalling capacity found at the farmsteads in the same region (Madsen, 2008). This is a strong indication that significant numbers of livestock were being herded extensively, because this estimate of 15,000 is well below the 48,000 ewes and variable numbers of lambs and wethers that were extensively herded in the same area in the 1960s. Norse stocking levels also seem to be significantly below the stable, intensively herded livestock population of some 18,000–21,000 prevailing since the 1980s (Madsen, 2008, 2012).

Following the decline of the Norwegian economy in the aftermath of the Black Death and the rise of the Hanseatic League, motivations for trading voyages into the Atlantic shifted; the Hansa would have had little interest in Greenland's traditional exports as their focus was on stockfish. Changed economic strategies in Greenland might have maintained contact with Europe, by providing goods that would have attracted merchants and maintained flows of both commodities and people.

In medieval Greenland the Norse faced multiple exposures; to climate change, to the economic transformations sweeping Europe, the development of new proto-world systems, plus the challenges of culture contacts brought about by the migration of the Thule Peoples, phenomena which were unrelated to each other. In the face of these pressures it is likely that settlement collapse occurred because of the resilience lost from the subsistence system; specialization and an apparently "elegant" solution to fourteenth-century climate changes might have quite suddenly lost their utility with devastating consequences for human security.

Now and in the near future we face challenges not dissimilar to those encountered in the medieval North Atlantic; how best to respond to climate threats to food production, how to manage the challenges of new economies, the balance to be struck between generating wealth and promoting security, how best to manage cross-cultural contacts and whether to allow the growth of impoverished groups within society and the concentration of power within a small elite. Inclusive approaches to tackling the challenges of living with global environmental change and enhancing human security are not likely to be elegant. They are more likely to be clumsy. While there is no clear blueprint for successful adaptations, there are lessons to be learned from past failures. Focusing on single points of view and simple solutions to complex problems may well lead to seemingly effective short-term solutions but these strategies can also produce unexpected negative outcomes. The intensification of parts of an established system may be a very problematic way to cope with changes over multi-decadal time scales and crises may arise swiftly and threaten the existence of entire communities.

Acknowledgments

We would like to acknowledge support from the UK Leverhulme Trust through an International Fellowship to AJD, grants F/00152/F and F/00152/Q and US National Science Foundation through Grants OPP 0352596; BCS 0001026; OPP 02900001 and OPP 0732327.

Notes

1 The Faroe Islands are hilly and no place is far from the sea. Settlement is nucleated. Further west, medieval settlement was highly dispersed. In Iceland there are opportunities for extensive inland grazing. Greenland is affected by seasonal sea ice and is the most marginal area for pastoralism in the North Atlantic islands. Norse Greenland benefited from annual seal migrations, but the settled areas lay far from the northern hunting grounds for walrus whose ivory was a key export.

2 The analysis of modern mitochondrial DNA indicates a very substantial Celtic maternal contribution to the ancestral Faroese population (Jorgensen *et al.*, 2004; Als *et al.*, 2006) – a factor held in common with Iceland and explained by a substantial settlement contribution from the British Isles. Recently, a suite of AMS C14 dates on charred barley has confirmed pre-Viking Age, seventh-century A.D. settlement in the Faroes, which raises the possibility that some of these Celtic elements originated from a pre-Norse Faroese population (Church *et al.*, 2005) rather than a Celtic subset of the Norse Landnám.

References

Als, T.D., Jorgensen, T.H., Børglum, A.D., Petersen P.A., Mors O. and Wang A.G. (2006) 'Highly discrepant proportions of female and male Scandinavian and British Isles ancestry within the isolated population of the Faroe Islands', *European Journal of Human Genetics*, 14, 497–504.

Arge, S.V., Church, M.J. and Brewington, S.D. (2009) 'Pigs in the Faroe Islands: an ancient facet of the islands' paleoeconomy', *Journal of the North Atlantic,* 2: 19–32.

Arneborg, J., Heinemeier, J., Lynnerup, N., Nielsen, H.L., Rud, N. and Sveinbjörnsdóttir, A.E. (1999) 'Change of diet of the Greenland Vikings determined from stable carbon isotope analysis and 14C dating of their bones', *Radiocarbon,* 41 (2): 157–168.

Arneborg, J., Lynnerup, N., Heinemeier, J., Møhl, J., Rud, N. and Sveinbjörnsdóttir Á. E. (2012) 'Norse Greenland Dietary Economy ca. AD 980-ca. AD 1450: Introduction', *Journal of the North Atlantic*, 3(3): 1–39.

Barlow, L.K., Sadler, J.P., Ogilvie, A.E.J., Buckland, P.C., Amorosi, T., Ingimundarson, J.H., Skidmore, P., Dugmore, A.J. and McGovern, T.H. (1997) 'Interdisciplinary investigations of the end of the Norse Western Settlement in Greenland', *The Holocene*, 7: 489–499.

Berger, J-F., Nuninger, L. and van der Leeuw, S. (2007) 'Modeling the role of resilience in socioenvironmental co-evolution: the Middle Rhone Valley between 1000 BC and AD 1000', in T.A. Kohler and S.E. van der Leeuw (eds) *The Model Based Archaeology of Socionatural Systems,* Santa Fe: School for Advanced Research.

Bloch, D. (1996) 'Whaling in the Faroe Islands, 1584–1994, an overview', *Studia Atlantic,* 1: 49–61.

Brotton, J. (2002) *The Renaissance Bazaar: From the Silk Road to Michelangelo,* Oxford: Oxford University Press.

Buchwald, V.F. (2001) 'Ancient iron and slags in Greenland', *Meddeleleser om Gronland*, 26 Copenhagen.

Carver, M. (ed.) (2003) *The Cross goes North: Processes of Conversion in Northern Europe, AD 300–1300*, York: University of York Medieval Press.

Church, M.J., Arge, S.V., Brewington, S., McGovern, T.H., Woollett, J., Perdikaris, S., Lawson, I., Cook, G.T., Amundsen, C., Harrison, R., Krivogorskaya, K. and Dunbar E. (2005) 'Puffins, pigs, cod, and barley: palaeoeconomy at Undir Junkarinsfløtti, Sandoy, Faroe Islands', *Environmental Archaeology,* 10 (2): 179–197.

Church, M.J., Dugmore, A.J., Mairs, K-A., Millard, A., Cook, G.T., Sveinbjarnardóttir, G., Ascough, P.A., Newton, A.J. and Roucoux, K. (2007) 'Charcoal production during the Norse and early medieval periods in Eyjafjallahreppur, Southern Iceland', *Radiocarbon,* 49 (2): 659–672.

Crumley, C.R. (ed). (1994) *Historical Ecology,* School of American Research, Santa Fe: University of Washington Press.

Dennis, A., Foote, P. and Perkins, R. (translators) (1980) *Laws of Early Iceland, Gragas: The Codex Regis of Gragas with Material from Other Manuscripts,* Winnipeg: University of Manitoba.

Diamond, J. (2005) *Collapse: How Societies Choose to Fail or Survive,* London: Allen. Lane.

Dugmore, A.J., Church, M.J., Mairs, K-A., McGovern, T.H., Newton, A.J. and Sveinbjarnardóttir, G. (2006) 'An over-optimistic pioneer fringe? Environmental perspectives on medieval settlement abandonment in Þórsmörk, South Iceland', in J. Arneborg, B. Grønnow and H. C. Gulløv (eds) *The Dynamics of Northern Societies,* PNM, Publications from the National Museum, Studies in Archaeology and History, 10, Copenhagen.

Dugmore, A.J., Keller, C. and McGovern, T.H. (2007a) 'The Norse Greenland settlement: reflections on climate change, trade and the contrasting fates of human settlements in the Atlantic islands', *Arctic Anthropology,* 44 (1): 12–37.

Dugmore, A.J., Church, M.J., Mairs, K.A., McGovern, T.H., Perdikaris, S. and Vesteinsson, O. (2007b) 'Abandoned farms, volcanic impacts and woodland management: revisting Þjórsárdalur, the "Pompeii of Iceland"', *Arctic Anthropology,* 44 (1): 1–11.

Dugmore, A.J., Gísladóttir, G., Simpson, I.A. and Newton A.J. (2009) 'Conceptual models of 1,200 years of Icelandic soil erosion reconstructed using tephrochronology', *Journal of the North Atlantic,* 2, 1–18.

Dugmore, A.J., McGovern, T.H., Vésteinsson, O., Arneborg, J., Streeter, R. and Keller, C. (2012) 'Cultural adaptation, compounding vulnerabilities and conjunctures in Norse Greenland', *Proceedings of the National Academy of Sciences,* 109 (10): 3658–3663.

Einarsson, Þ., (1963) 'Pollen analytical studies on the vegetation and climate history of Iceland in late and post-glacial times', in Á. Löve and D. Löve (eds) *North Atlantic Biota and their History*, Oxford: Pergamon Press.

Espelund, A. (2007) 'Ancient ironmaking in Iceland Greenland and Newfoundland', *Archaeologica Islandica,* 6: 48–74.

Fitzhugh, W.W. and Ward, E. I. (eds) (2000) *Vikings: The North Atlantic Sag,* Washington, DC: Smithsonian Institution Press.

Gao, C., Robock, A. and Ammann, C. (2008) 'Volcanic forcing of climate over the past 1500 years: an improved ice core based index for climate models', *Journal of Geophysical Research,* 113, D23111.

Graham-Campbell, J. (2001) *The Viking World,* London: Frances Lincoln.

Gulløv, H.C. (ed.) (2004) *Grønlands Forhistorie,* Copenhagen: Gyldendal.

Gulløv, H.C. (2008) 'The nature of contact between native Greenlanders and Norse', *Journal of the North Atlantic,* 1: 16–24.

Guttesen, R. (1996) 'Hvalbøur- largest old farming village', in R. Guttesen (ed.) *The Faeroe Islands Topographic Atlas,* Copenhagen: Kort and Matrikelstyrelsen.

Herlihy, D. (1997) *The Black Death and the Transformation of the West,* Cambridge, MA: *Harvard University Press.*

Hamilton, L.C. (2003) 'Fisheries dependent communities: propositions about ecological and social change', in G. Duhaime and N. Bernard (eds) *Arctic Economic Development and Self-Government,* Quebec City: GETIC, Université Laval.

Hamilton, L.C. (2007) 'Climate, fishery and society interactions: observations from the North Atlantic', *Deep-Sea Research* II.

Hamilton, L.C., Haedrich, R.L. and Duncan, C.M. (2004) 'Above and below the water: Social/ecological transformation in northwest Newfoundland', *Population and Environment,* 25 (3):195–215.

Ingstad, H. and Ingstad, A. S. (2001) *The Viking Discovery of America: The Excavation of a Norse Settlement in L'Anse aux Meadows, Newfoundland,* New York: Checkmark Books.

Jones, G. (1986) *The Norse Atlantic Saga: Being the Norse Voyages of Discovery and Settlement to Iceland, Greenland, and North America,* Oxford: Oxford University Press.

Jorgensen, T.H., Buttenschön, H.N., Wang, A.G., Als, T.D., Børglum, A.D. and Ewald, H. (2004) 'The origin of the isolated population of the Faroe Islands investigated using Y chromosomal markers', *Human Genetics,* 115 (1): 19–28.

Karlsson, G. (2000) *Iceland's 1100 Years: History of a Marginal Society,* Reykjavik: Mál og Menning.

Keller, C. (2010) 'Furs, fish, and ivory: Medieval Norsemen at the Arctic Fringe', *Journal of the North Atlantic,* 3 (1): 1–23.

KLNM (1981) *Kulturhistorisk Leksikon for Nordisk Middelalder 1956–1978* , Copenhagen: Rosenkilde og Bagger.

Konstam, A. (2005) *Historical Atlas of the Viking World.* San Diego: Mercury Books.

Lawson, I.T., Church, M.J., McGovern, T.H., Arge, S.V., Woollet, J., Edwards, K.J., Gathorne-Hardy, F.J., Dugmore, A.J., Cook, G., Mairs, K.-A., Thomson, A.M. and Sveinbjarnardóttir, G. (2005) 'Historical ecology on Sandoy, Faroe Islands: palaeoenvironmental and archaeological perspectives', *Human Ecology,* 33 (5): 651–684.

Leichenko, R. and O'Brien, K. (2008) *Environmental Change and Globalization: Double Exposures,* Oxford: Oxford University Press.

Madsen, C.K. (2008) 'Får, geder og fold i det norrøne Vatnahverfi', *Tidsskriftet Grønland,* 1: 4–14.

Madsen, C.K. (2012) *Pastures Found – Farming in Greenland (re)Rntroduced,* Report from Second Workshop at the Danish National Museum, *Northern Worlds* 2011.

Mahler, D. (2007) 'Sæteren ved Argisbrekka. Økonomiske forandringer på Færøerne i Vikingetid og Tidlig Middelalder', *Annales Societatis Scientiarum Færoensis, Supplemententum* 47, Tórshavn: Faroe University Press.

Mann, M.E., Zhang, Z., Rutherford, S., Bradley, R.S., Hughes, M.K., Shindell, D., Ammann, C., Faluvegi, G. and Ni, F. (2009) 'Global signatures and dynamical origins of the Little Ice Age and Medieval Climate Anomaly', *Science,* 326: 1256.

Mayewski, P.A. and White, F. (2002) *The Ice Chronicles,* Hanover and London: University of New England Press.

Mairs, K-A. (2007) *Islands and human impact: Under what circumstances do people put unsustainable demands on island environments? Evidence from the North Atlantic,* PhD Thesis, University of Edinburgh.

McGovern, T.H. (1980) 'Cows, harp seals, and churchbells: adaptation and extinction in Norse Greenland', *Human Ecology,* 8 (3): 245–275.

McGovern, T.H. (2008) 'Going global: let past serve the future?' *American Anthropologist,* 110 (1): 76–79.

McGovern, T.H., Amorosi, T., Perdikaris, S. and Woollett, J.W. (1996) 'Zooarchaeology of Sandnes V51: Economic change at a chieftain's farm in West Greenland', *Arctic Anthropology,* 33(2): 94–122.

McGovern, T. H., Perdikaris S. and Tinsley, C. (2001) 'Economy of landnám: the evidence of zooarchaeology', in A. Wawn and T. Sigurdardottir (eds) *Approaches to Vinland,* Reykjavik: Sigurdur Nordal Institute Studies .

McGovern, T.H., Perdikaris, S., Einarsson, A. and Sidell J. (2006) 'Coastal connections, local fishing, and sustainable egg harvesting: patterns of Viking Age inland wild resource use in Myvatn district, Northern Iceland', *Environmental Archaeology,* 11 (2): 187–205.

McGovern, T.H., Vésteinsson, O., Fridriksson, A., Church, M.J., Lawson, I., Simpson, I.A., Einarsson, A., Dugmore, A.J., Cook, G.T., Perdikaris, S., Edwards, K.J., Thomson, A., Adderly, W.P., Newton, A.J., Lucas, G., Edvardsson, R., Aldred, O. and Dunbar, E. (2007) 'Landscapes of settlement in Northern Iceland: historical ecology of human impact and climate fluctuation on the millennial scale', *American Anthropologist,* 109 (1): 27–51.

McIntosh, R.J., Tainter, J.A. and McIntosh, S.K. (eds) (2000) *The Way the Wind Blows: Climate History, and Human Action,* New York: Columbia University Press.

Meeker, L.D. and Mayewski, P.A. (2002) 'A 1400-year high-resolution record of atmospheric circulation over the North Atlantic and Asia', *The Holocene,* 12: 257–266.

Oppenheimer, C. (2003) 'Ice core and paleoclimatic evidence for the timing and nature of the great mid-13th century volcanic eruption', *International Journal of Climatology*, 23: 417–426.

Perdikaris, S. and McGovern, T.H. (2008a), 'Codfish and kings, seals and subsistence: Norse marine resource use in the North Atlantic', in R. Torben and J. Erlandson (eds) *Human Impacts on Marine Environments*, Berkeley: UCLA Press Historical Ecology Series.

Perdikaris, S. and McGovern, T.H. (2008b) 'Viking Age economics and the origins of commercial cod fisheries in the North Atlantic', in L. Sickling and D. Abreu-Ferreira (eds) *The North Atlantic Fisheries in the Middle Ages and Early Modern Period: Interdisciplinary Approaches in History, Archaeology, and Biology*, Leiden: Brill Publishers.

Shapiro, M. (1988) 'Introduction: judicial selection and the design of clumsy institutions', *Southern California Law Review*, 61 (1988): 1555–1563.

Simpson, I.A., Dugmore, A.J., Thomson, A. and Vésteinsson, O. (2001) 'Crossing the thresholds: human ecology and historical patterns of landscape degradation'. *Catena*, 42: 175–192.

Simpson, I.A., Adderley, W.P., Gumundsson, G., Hallsdottir, M., Sigurgeirsson, M.A. and Snaesdottir, M. (2002) 'Soil limitations to agrarian land production in premodern Iceland', *Human Ecology*, 30: 423–443.

Sveinbjarnardottir, G., Erlendsson, E., Vickers, K., McGovern, T.H., Milek, K.B., Edwards, K.J., Simpson, I.A. and Cook, G. (2007) 'The palaeoecology of a high status Icelandic farm', *Environmental Archaeology*, 12: 187–206.

Thorarinsson, S. (1981) 'The application of tephrochronology in Iceland', in S. Self, and R.S.J. Sparks (eds) *Tephra Studies*, Dordrecht: Reidel.

Verweij, M., Douglas, M., Ellis, R.J., Engel, C., Hendriks, F., Lohmann, S., Ney, S., Rayner, S. and Thompson, M. (2006) 'The case for clumsiness', in M. Verweij and M. Thompson (eds) *Clumsy Solutions for a Complex World: Governance, Politics and Plural Perceptions*, Basingstoke: Palgrave Macmillan.

Vésteinsson, O. (2000a) *The Christianization of Iceland: Priests, Power, and Social Change 1000–1300*, New York: Oxford University Press.

Vésteinsson, O. (2000b) 'The archaeology of *landnám*. Early settlement in Iceland', in W.W. Fitzhugh and E. Ward (eds) *Vikings. The North Atlantic Saga*, Washington, DC: Smithsonian Institute Press.

Vesteinsson, O. (2011) 'The settlement of Iceland: current archaeological evidence', paper presented at *A Molecular View of Colonization: The Norse Settlement of the North Atlantic Conference*, Carlsberg Akademi, Copenhagen 28–29 April 2011.

Videnskab.dk (2012) Vikingerne dyrkede korn på Grønland, *Videnskab dk*, January 26, 2012. Online. Available HTTP: http://videnskab.dk/kultur-samfund/vikingerne-dyrkede-korn-pa-gronland (accessed 1 August 2012).

38 The future is now

Human security as a choice

Johanna Wolf, Karen O'Brien and Linda Sygna

> Let us take inspiration from the solutions we see around us and from the transformations that we know are possible. Let us prove to future generations that we had the vision to see where we needed to go, and the wisdom to get there.
>
> (UN Secretary Ban-Ki Moon Remarks to UNFCCC COP18, Doha, December 4, 2012)

Introduction

As the biophysical environment of the twenty-first century is changing all around us, so too is the social environment in flux. We live in a period where multiple transformative changes are already underway; while global environmental change is taking place at unprecedented rates, the axis of global geopolitical powers is shifting, global economies are in crisis, and in the midst of it all a new generation is emerging with new ideas about the future. In light of the social and environmental transformations that current greenhouse gas emissions trajectories imply, the ultimate question may not be whether or not transformations are indeed coming, but rather whether and to what extent humanity will mobilize its collective agency to actively participate in, rather than passively endure, transformations (Raskin *et al.*, 2002; Anderson and Bows, 2011; WGBU, 2011; PriceWaterhouseCoopers, 2012). As the above quote by Ban-Ki Moon reminds us, transformations to alternative futures call for both vision and wisdom. Within this context, human security is without a doubt more important than ever.

Human security in a changing world

Human security is both a choice and an imperative. The choice is about what kind of future is desirable, which values are important, and who eventually decides. The imperative emerges from the scope and intensity of the planetary challenges that are melting the boundaries between more and less vulnerable people and places. It is about decisions that are made here and now, rather than somewhere else and in the future. Just as future sustainability can be considered a mirage if it is not connected to the deep needs of the present (Foster, 2008), so is human security in a changing environment spurious if it is not connected to decisions and actions that are taken today. If current trajectories continue, the effects of environmental change will be profound everywhere, and they will change everything. In order to work towards human security in this context, we need to accept that humans are responsible for the changes taking place, challenge assumptions and paradigms, and explicitly engage with the experiential and subjective aspects of environmental change. Equally important, we need to take actions that create the conditions for human security, recognizing that the latter is part of a process, rather than a simple outcome.

This book has discussed the reality of human security in a changing environment, demonstrating that we ourselves are now responsible for many of the outcomes, and for creating deliberate transformations that promote individual and collective human security. The breakthrough conditions that we describe are important in realizing transformative change, and may be considered a missing component of a more holistic, integrated approach. A number of chapters in this book demonstrate the importance of engaging with multiple perspectives, including connections across scales, as this is a prerequisite for understanding what truly is at stake for people and communities around the world – from their point of view – and to begin to envision what a desirable future looks like. Questioning dominant framings and paradigms is a necessity in order to move from reworking existing approaches towards transforming them. Explicitly recognizing the role of individual and collective agency through empowerment can not only help avoid repeating past mistakes, including technological and policy responses that enhance existing power gradients, but also refocus the efforts that must now transpire into humanity's collective hands. Without inspiring collaboration and harnessing the power of the collective there is arguably little chance of accomplishing what is required to mitigate global environmental change. Finally, integrating knowledge and action is a paramount challenge that requires different approaches to knowledge generation, as well as different types of actions.

Transforming knowledge

Past and indeed much of current understandings of environmental change have fallen far short of delivering the insights needed to trigger the transformative changes necessary to halt and reverse current trajectories of greenhouse gas emissions, biodiversity loss, and other negative indicators of sustainability. This shortcoming is as much a symptom of the approaches taken to global change research, as it is evidence that policy making is often not based on or informed by current scientific understandings of human-environment systems. The approaches taken to illuminate our understanding of environmental change have been dominated by piecemeal attempts to use objective measures to estimate changes or impacts, e.g. to assess vulnerability in a specific context through case studies, to describe some of the objective impacts on people's lives and livelihoods, or to suggest avenues to ameliorate these impacts through policies often targeting the *a priori* vulnerable through technical interventions. This piecemeal approach has left significant blind spots that are rarely if ever considered in the objective calculus of human-environment interactions. One blind spot concerns the experiential, subjective dimensions of changes as they are experienced by people, rather than as measured and assessed by researchers. They include the meaning people attach to places, livelihoods, and identity, the goals they pursue in making their lives work, and therefore how changes they perceive in the environment affect their own perception of their lives and security (Wolf *et al.*, in press). These subjective dimensions cannot be explained by models, impact studies or vulnerability assessments because they are underpinned by values, worldviews and the goals to which people aspire (O'Brien and Wolf, 2010). One result of these blind spots is that much of the existing literature on environmental change and human security targets a reality that exists in and indeed is created from the perspective of science and scientists (Hackmann and St. Clair, 2012). Of course, resulting recommendations for policy then can only address this singular "flat" version of what may be better described as a multidimensional reality with many distinct and seemingly irreconcilable perspectives. Therefore, new approaches to understanding environmental change and human security must consider not only such multiple subjective realities but their underpinning values, worldviews and goals.

Academic disciplines have provided the foundations for our present understanding of environmental change and human security, with each discipline highlighting its distinct viewpoints. Yet, one of the key barriers to transforming this disciplinary approach is the lack of overt, active, and direct support of transdisciplinary science in academic environments. Future research that will help transform existing approaches will necessarily be inter- or transdisciplinary in order to fill the blind spots left by what is largely still disciplinary or multidisciplinary science (O'Brien *et al.*, 2013). According to the findings demonstrated in this book, inter- and transdisciplinary research has the greatest capacity to explain hitherto ignored and misunderstood aspects of the relationship between humans and the environment. Such research will be needed at a large scale in order to move beyond a reworking of current approaches and toward a transformation to collective human security.

Transforming policy

The future cannot be extrapolated from the past, and policies need to become much more flexible, innovative, and visionary if they are to facilitate effective, equitable and legitimate responses to environmental change. Current policy institutions largely represent an inheritance from the past that has less relevance in the complexity of today's world. As a result, transforming the ways in which knowledge and policy interact and inform each other is a must. Recent conceptualizations of this interaction already go beyond the "speaking truth to power" model of decisionist and technocratic approaches to science-policy interactions and explicitly recognize that co-production of knowledge and policy already takes place (Millstone, 2005; Hulme, 2009). Such a transformation necessarily involves a conscious harnessing of inclusive, participatory mechanisms that can lead to knowledge-policy co-production by providing a voice to those affected by both environmental change and the policies developed to address them. To this effect, the scientific community has to embrace methods and approaches that enable participation and inclusion in research.

A critical thread emerging from this volume is that the dominant structures and institutions currently shaping human lives are constraining the collective human capabilities in responding to environmental change. Where such structures are fundamentally changed through social reform, collective action, or creative innovation, evidence suggests that a redistribution of power can lead to the empowerment of previously excluded groups. But even then both changes in mindsets and changes in institutional structures are needed for transformation. Transformation therefore hinges on both elements that together facilitate a transition from a mere reworking of current ways, toward new approaches.

Much research suggests that a specific element in the transformation of policy on environmental change is flexibility – a trait that is clearly lacking in current environmental decision making. The rigidity of existing decision-making processes and outcomes acts as a barrier to developing responsive policy systems. Recent thinking on social ecological resilience has pointed to a need for inherently adaptable institutions, rules and regulations that can accommodate and respond effectively and equitably to sudden shifts in social and ecological systems (Olsson *et al.*, 2006).

Existing environmental policy making – especially but not exclusively at the global level – is surprisingly devoid of any vision of the future. Given the scope of the challenge, this is a major deficiency. Long-term vision may certainly be encumbered by the sometimes limited human ability to think forward, but the absence of creative visions in environmental decisions can no longer be justified given the grand challenges we face. Here too, the role of knowledge co-production cannot be underestimated; scientific knowledge that focuses on objective, measurable and quantifiable elements of environmental change has little to say about visions for the future, for it is mostly extrapolating expectations from the past (Miller, 2007). Transforming policy then also means engaging the collective human imagination to envision what a desirable future might look like, and this requires collaborative, transdisciplinary efforts that draw on the social sciences, fine arts and the humanities.

Transforming action

Contrary to current calls for radical behavior and value change, we suggest that it cannot be the objective of research or policy to attempt to alter values, nor to manufacture a new or improved set of values in order to meet environmental or social goals, as important as they might be. Attempts to manipulate beliefs in particular phenomena, deities, or ideas have in the not too distant past had disastrous effects, as evident in some of the legacies of colonialism that still influence the lives of many contemporaries. Therefore, any undertaking in research, policy, or practice, in which the goal is improve lives and livelihoods in the face of environmental change must not only be based on and informed critically by the voices of those whose lives are affected, but ought to be shaped by them. Such engagement, for example, through citizen science, participatory research, and co-production of knowledge and policy, will be a crucial necessity of any future attempts to understand environmental change and human security. This is not to underplay the importance of science and research, as this has helped us to understand the complex challenges that we are now facing, but rather to call for a different type of science for responding to these challenges – a transdisciplinary science that integrates many more perspectives, including those from the social sciences (O'Brien *et al.*, 2010).

One of the many hopeful avenues to work effectively, equitably and legitimately towards a sustainable future lies in global citizenship. Current lives are inextricably connected in a web of interactions and interdependencies around the globe. At the heart of global citizenship lies a realization that due to many aspects of the contemporary world, lives are interconnected through systems and networks that span the earth and through the collective impacts of decisions made by individuals and groups in nearly all locations (Dobson, 2003). In order to shape a global citizenry that is aware of these interconnections and interdependencies, new approaches to primary, secondary and post-secondary education is needed. As Sterling notes, these changes in education do not call for "doing more of the same, but better," but rather for re-thinking systems by "seeing things differently" (Sterling, 2001: 28).

It is simultaneously tragic and symptomatic that the involvement and the power of youth have been all but ignored in global environmental change research and policy. While there is significant theorizing on intergenerational equity, painfully little of it has informed policy and policy-led practice to date. Emerging, however in spite of or perhaps precisely because of this, is a youth movement around the globe that, if enabled, may well be a key factor in transforming how global environmental change and human security are approached. With significant, even ultimate, stakes in the outcomes of past and current decisions, youth not only deserve but represent an absolutely crucial voice in research and policy. Drawing on social media, citizen advocacy, and a sense of connectedness, a new generation is taking ownership over exactly those issues from which they have been systematically excluded. Unified by the significant and specific challenges they face, including unprecedented and dangerous environmental change, weakening democracies, growing social inequality, and a global economy marked by unprecedented youth unemployment and unsustainable resource extraction (Hayward, 2012), youth are attempting to claim their role in shaping a future worth living. The time is ripe to invite and embrace their participation in research, policy and practice that will have a profound effect on their lives.

Perhaps most fundamentally, these elements point to a need for a new type of leadership. Such leadership is more likely emergent than existing now, and includes both decision-making powers as well as collective, networked social agency. A new type of leadership may therefore include non-traditional actors who collectively bring about social movements as well as leaders in decision-making positions. This new leadership distinguishes itself because it openly acknowledges and takes on the responsibility to change the systems and structures that inhibit diverse participation and encumber agency of those affected by environmental change. It is adaptive leadership, in that it calls for helping people to move through critical times while recognizing that old habits, identities, loyalties and beliefs may no longer function in a changing environment (Heifetz *et al.*, 2009). Most importantly, however, new leadership conveys with heart both vision and hope for the future, making explicit rather than ignoring the rising emotional stakes.

Facing a collective challenge

We live in a time where we risk falling into an inevitability trap, believing that environmental change can no longer be addressed, or that human insecurity is an inevitable "fact of life." There is, however, much evidence of an emerging capacity to deliberately transform that which has traditionally been thought of as the responsibility of others. Fostering collective capacities to take responsibility for the future is perhaps the best news for humans and the planet. In the story we have presented in the book, humans are the protagonists of change, able to create human security as a choice, and able to thrive in a changing world.

References

Anderson, K. and Bows, A. (2011) 'Beyond "dangerous" climate change: emission scenarios for a new world', *Philosophical Transactions of the Royal Society A: Mathematical, Physical and Engineering Sciences*, 369 (1934): 20–44.

Dobson, A. (2003) *Citizenship and the Environment*, Oxford: Oxford University Press.

Foster, J. (2008) *The Sustainability Mirage – Illusion and Reality in the Coming War on Climate Change*, London: Earthscan.

Hackmann, H. and St. Clair, A. (2012) *Transformative Cornerstones of Social Science Research for Global Change*, ISSC Report, Paris: International Social Science Council. Online. Available HTTP: http://www.worldsocialscience.org/pdf/ISSC_Transformative_Cornerstones_Report.pdf (accessed 1 December 2012).

Hayward, B. (2012) *Children, Citizenship and Environment: Nurturing a Democratic Imagination in a Changing World*, London: Earthscan.

Heifetz, R.A., Grashow, A. and Linsky, M. (2009) *The Practice of Adaptive Leadership: Tools and Tactics for Changing Your Organization and the World*, Cambridge, MA: Harvard Business Press.

Hulme, M. (2009) *Why We Disagree About Climate Change: Understanding Controversy, Inaction and Opportunity*, Cambridge: Cambridge University Press.

Miller, R. (2007) 'Futures literacy: A hybrid strategic scenario method', *Futures*, 39: 241–362.

Millstone, E. (2005) 'Analysing the role of science in public policy making', in P. van Zwanenberg and E. Millstone (eds) *BSE: Risk, Science and Governance*, Oxford: Oxford University Press: 11–38.

O'Brien, K. and Wolf, J. (2010) 'A values-based approach to vulnerability and adaptation to climate change', *Wiley Interdisciplinary Reviews Climate Change*, 1 (2): 232–242.

O'Brien, K.L., St Clair, A.L. and Kristoffersen, B. (2010) 'Towards a new science on climate change', in K. O'Brien, A.L. St. Clair and B. Kristoffersen (eds) *Climate Change, Ethics, and Human Security*, Cambridge: Cambridge University Press: 215–227.

O'Brien, K., Reams, J., Caspari, A., Dugmore, A., Faghihimani, M., Fazey, I., Hackmann, H., Manuel-Navarrete, D., Marks, J., Miller, R., Raivio, K., Romero-Lankao, P., Virji, H., Vogel, C. and Winiwarter, V. (2013) 'You say you want a revolution? Transforming education and capacity building in response to global change', *Environmental Science & Policy*, 29, 48–59.

Olsson, P., Gunderson, L.H., Carpenter, S.R., Ryan, P., Lebel, L., Folke, C. and Holling, C.S. (2006) 'Shooting the rapids: Navigating transitions to adaptive governance of social-ecological systems', *Ecology and Society*, 11 (1): 18.

PriceWaterhouseCoopers (2012) 'Low carbon economy index 2012 – too late for two degrees?', PWC UK, London. Online. Available HTTP: http://pdf.pwc.co.uk/low-carbon-economy-index-nov12.pdf (accessed 10 November 2012).

Raskin, P., Banuri, T., Gallopin, G., Gutman, P., Hammond, A., Kates, R. and Swart, R. (2002) *Great Transition: The Promise and Lure of the Times Ahead*, Scenario Group, Stockholm Environment Institute, Stockholm: Tellus Institute.

Sterling, S. (2001) *Sustainable Education: Re-visioning Learning and Change*, Schumacher Briefings 6, Bristol: Green Books.

WGBU (2011) *World in Transition: A Social Contract for Sustainability, Summary for Policy-Makers*, Berlin: German Advisory Council on Global Change.

Wolf, J., Allice, I. and Bell, T. (in press) 'Values, climate change, and implications for adaptation: Evidence from two communities in Labrador, Canada', *Global Environmental Change*, 23 (2): 548–562.

Index

Bold page numbers indicate figures, *italic* numbers indicate tables